METHODS IN MOLECULAR BIOLOGY™

Series Editor
John M. Walker
School of Life Sciences
University of Hertfordshire
Hatfield, Hertfordshire, AL10 9AB, UK

For further volumes:
http://www.springer.com/series/7651

Genotoxicity Assessment

Methods and Protocols

Edited by

Alok Dhawan

Institute of Life Sciences, School of Science and Technology, Ahmedabad University, Ahmedabad, Gujarat, India;
Nanomaterial Toxicology Group, Council of Scientific and Industrial Research (CSIR), Indian Institute of Toxicology Research, Lucknow, UP, India

Mahima Bajpayee

Institute of Life Sciences, School of Science and Technology, Ahmedabad University, Ahmedabad, Gujarat, India;
The Science Hub, Lucknow, UP, India

Editors
Alok Dhawan
Institute of Life Sciences
School of Science and Technology
Ahmedabad University
Ahmedabad, Gujarat, India

Nanomaterial Toxicology Group
Council of Scientific and Industrial Research (CSIR)
Indian Institute of Toxicology Research
Lucknow, UP, India

Mahima Bajpayee
Institute of Life Sciences
School of Science and Technology
Ahmedabad University
Ahmedabad, Gujarat, India

The Science Hub, Lucknow, UP, India

ISSN 1064-3745   ISSN 1940-6029 (electronic)
ISBN 978-1-62703-528-6   ISBN 978-1-62703-529-3 (eBook)
DOI 10.1007/978-1-62703-529-3
Springer New York Heidelberg Dordrecht London

Library of Congress Control Number: 2013941725

© Springer Science+Business Media New York 2013
This work is subject to copyright. All rights are reserved by the Publisher, whether the whole or part of the material is concerned, specifically the rights of translation, reprinting, reuse of illustrations, recitation, broadcasting, reproduction on microfilms or in any other physical way, and transmission or information storage and retrieval, electronic adaptation, computer software, or by similar or dissimilar methodology now known or hereafter developed. Exempted from this legal reservation are brief excerpts in connection with reviews or scholarly analysis or material supplied specifically for the purpose of being entered and executed on a computer system, for exclusive use by the purchaser of the work. Duplication of this publication or parts thereof is permitted only under the provisions of the Copyright Law of the Publisher's location, in its current version, and permission for use must always be obtained from Springer. Permissions for use may be obtained through RightsLink at the Copyright Clearance Center. Violations are liable to prosecution under the respective Copyright Law.
The use of general descriptive names, registered names, trademarks, service marks, etc. in this publication does not imply, even in the absence of a specific statement, that such names are exempt from the relevant protective laws and regulations and therefore free for general use.
While the advice and information in this book are believed to be true and accurate at the date of publication, neither the authors nor the editors nor the publisher can accept any legal responsibility for any errors or omissions that may be made. The publisher makes no warranty, express or implied, with respect to the material contained herein.

Printed on acid-free paper

Humana Press is a brand of Springer
Springer is part of Springer Science+Business Media (www.springer.com)

Foreword

"...human beings contain hundreds of thousands of genes and although mutation in any particular one is exceedingly rare, the vast numbers of genes within an individual ensures that most of us carry one or two mutant genes and that we are all, in effect, mutants." Testimony of Dr. Gary Flamm, National Institutes of Environmental Health, USA, at a US Senate Hearing on "Chemicals and the Future of Man", April, 1971.

In the late 1960s a small group of geneticists, including Charlotte Auerbach, Alexander Hollaender, Fritz Sobels, and Takashi Sugimura, representing countries around the globe, began voicing their fears that mutagenic agents might pose serious, and potentially global, threats. As a result, they argued, untold millions of people throughout the world could face genetic assault from what scientists had begun calling "environmental mutagens."

The field of Genetic Toxicology has evolved from these early days into a complex integration of basic and applied research. Focus has been on genetic tools to preemptively disease-causing effects of various chemicals before they are put on the market and/or released into the environment. These genetic toxicology tests determine whether and how substances cause damage or mutations in DNA and genes, which could lead to disease, most notably cancer.

There are three components to the FDA-required genetic toxicology test battery. The tests described so elegantly in *Genotoxicity Assessment: Methods and Protocols* reflect the key, three-tier array of tools for high-level risk assessment of chemical agents. These range from tools to reflect DNA mutation such as can be accomplished with the elegantly simple, yet powerful, Ames test, to in vitro tests looking at chromosomal aberrations and effects, and then on to complex in vivo approaches involving mammals that can determine a broad spectrum of gene alterations that lead to mutations and/or cancer.

I applaud the editors of this important volume and each of the authors for their thorough and excellent presentation of the state-of-the-art tools of genetic toxicology, and how these tools can be used to serve human kind, and our environment, in a remarkable and exquisite manner.

Holland, MI, USA *James M. Gentile*
Tucson, AZ, USA

Preface

The industrial revolution has seen thousands of chemicals being released into the environment. Due to the lack of regulatory framework, several chemicals for which toxicity data is not sufficient still exist in the environment. This has been compounded by the indiscriminate mining, mineral processing, use of fossil fuels and pesticides as well as development of new chemical entities. Hence, there has been a paradigm shift in the way safety assessment of chemicals is conducted. The post human genome era has seen a lot of emphasis being put on development and validation of test systems to assess the interaction of chemicals with macromolecules, especially DNA. A lot of potential drugs have had to be discontinued as they were found to cause DNA damage on prolonged usage. This has resulted in modifications in the existing test systems, which evaluate the risk and predict health effects upon low-level, long-term exposures.

More recently, efforts are being made to minimize the use of animals for safety assessment of chemicals and move towards computational tools and in vitro high-throughput screening in target cells. Computational models and emerging tests systems such as functional genomics and transcriptomics will provide a large database, which will facilitate data mining and prioritize chemical entities for safety assessment. The extensive testing of chemicals in vitro in human cells and cell lines will provide early mechanistic information to the perturbations in toxicity pathways in humans, and extrapolation of the in vitro results in humans will then help us in the understanding of chemical safety and precautions to be taken in case of human exposure.

Genetic toxicology is the study of the toxicity of agents which interact with the hereditary material (DNA) resulting in alterations of the nucleic acids or its components, leading to inactivation and/or modifications in its structure and/or function. The discipline was recognized when geneticists and researchers concerned with the genetic impact of man-made chemicals, formed the Environmental Mutagen Society under the aegis of Alexander Hollaender in 1969. Since then hundreds of chemicals and compounds have been tested for their genotoxic potential using tests that allow an understanding of the interaction of these agents with the DNA and their outcome. Genetic toxicology was earlier used for hazard identification and is now used for integrated risk assessment. It is an intrinsic component of safety testing for the approval of drugs and animal health products and also a regulatory tool used by the U.S. Environmental Protection Agency for environmental exposures.

Alterations in DNA that lead to DNA damage or mutations include covalent binding (adduct formation leading to base pair substitution mutations), intercalations (between the base pairs of DNA leading to insertion/deletions mutations), cross-linking (in intra- or inter-strands producing DNA or chromosomal breaks), or breakage (DNA-strand breaks, unrepaired DNA damage causing chromosomal aberrations). Indirect genotoxicity is caused by alterations in spindle fiber or kinetochore proteins (leading to incorrect chromosomal segregation) and incorporation of DNA analogues (leading to strand breaks). Adequate tests need to be conducted to find the response of a chemical to the wide category of damage. Since no one test system can provide full information about the damage a chemical can cause, work-groups of the European Union (EU), Organization for Economic Co-operation and Development (OECD), and International Conference on Harmonization of Technical Requirements for Registration of Pharmaceuticals for Human Use (ICH) have defined standard battery of tests, including both in vitro as well as in vivo tests which need to be conducted for a compound. Tiered test strategy includes a test for mutations in bacteria (Ames test), an in vitro test in mammalian cells (Mouse lymphoma or micronucleus assays) and a follow-up of a positive result in vivo (bone marrow chromosomal aberrations, germ cell). International workgroups on genetic toxicology (IWGT), COMNET for the Comet assay, and HuMN for micronucleus assay in human cells are playing a pivotal role in conducting inter-laboratory validations of test systems for their acceptance by the regulatory agencies.

This book incorporates several comprehensive genetic toxicology protocols which will serve as a highly useful and ready resource for research students and scientists working in regulatory toxicology as well as biomedical, and pharmaceutical sciences. The authors have actively contributed to peer-reviewed scientific literature in the area of genetic toxicology.

"Protocols in Genotoxicity Assessment" is divided into five sections, each with a comprehensive discussion of the tests included. The first section describes mutation assays conducted in vitro in bacteria (reverse mutation assay) and mammalian cells (recent guidelines from IWGT workshop on mouse lymphoma assay), as well as in vivo mutation assays in transgenic animals and the newer test-*Pig-a* gene mutation assay. The second section includes cytogenetic techniques with detailed chapters on chromosomal aberrations and micronucleus assays conducted in vitro and in vivo in somatic and germ cells. The third section details tests for assessing primary DNA damage, the Comet assay technique, unscheduled DNA synthesis and ^{32}P-post labelling. Alternate to animal models forms the fourth section including bioassays in plants (Tradescantia micronucleus assay) and Drosophila (wing spot test and the Comet assay) used in genotoxicity assessment of chemicals. The recently updated ICH guidelines form the fifth and final section of the book.

This book epitomizes the long-term scientific association that the editors have enjoyed with the authors and is a culmination of their collaborative effort.

Ahmedabad, Gujarat, India   *Alok Dhawan*
Lucknow, UP, India   *Mahima Bajpayee*

Acknowledgements

The editors wish to acknowledge the hard work put in by the authors as well as reviewers to bring out the book in time. We thank Prof. John Walker, series editor, for his critical comments and valuable suggestions.

This endeavor would not have been possible without the funding to the editors from CSIR, New Delhi, under its network projects (NWP34, NWP35) and OLP 009. The funding from UK India Education and Research Initiative (UKIERI) standard award to Institute of Life Sciences, Ahmedabad University, Ahmedabad, India (IND/CONT/E/11-12/217), is gratefully acknowledged. Funding from the European Union Seventh Framework Programme (FP7/2007-2013) under grant agreement no. 263147 (NanoValid—Development of reference methods for hazard identification, risk assessment, and LCA of engineered nanomaterials) is also acknowledged.

Contents

Preface .... *vii*
Acknowledgements .... *ix*
Contributors .... *xiii*

PART I GENE MUTATION ASSAYS

 1 Bacterial Mutation Assays 3
    Errol Zeiger
 2 In Vitro Mouse Lymphoma (L5178Y $Tk^{+/-}$-3.7.2C) Forward
    Mutation Assay 27
    Melissa R. Schisler, Martha M. Moore, and B. Bhaskar Gollapudi
 3 Erythrocyte-Based *Pig-a* Gene Mutation Assay 51
    Jeffrey C. Bemis, Nikki E. Hall, and Stephen D. Dertinger
 4 Detection of In Vivo Mutation in the *Hprt* and *Pig-a* Genes
    of Rat Lymphocytes 79
    *Vasily N. Dobrovolsky, Joseph G. Shaddock, Roberta A. Mittelstaedt,
    Daishiro Miura, and Robert H. Heflich*
 5 In Vivo *cII*, *gpt*, and Spi⁻ Gene Mutation Assays
    in Transgenic Mice and Rats 97
    *Mugimane G. Manjanatha, Xuefei Cao, Sharon D. Shelton,
    Roberta A. Mittelstaedt, and Robert H. Heflich*

PART II ASSAYS FOR CHROMOSOMAL ABNORMALITIES

 6 In Vitro Cytogenetic Assays: Chromosomal Aberrations
    and Micronucleus Tests 123
    *Pasquale Mosesso, Serena Cinelli, Adyapalam T. Natarajan,
    and Fabrizio Palitti*
 7 Analysis of Chromosome Aberrations in Somatic and Germ Cells
    of the Mouse 147
    Francesca Pacchierotti and Valentina Stocchi
 8 Chromosomal Aberration Test in Human Lymphocytes 165
    Christian Johannes and Guenter Obe
 9 In Vivo Micronucleus Assay in Mouse Bone Marrow and Peripheral Blood ... 179
    Sawako Kasamoto, Shoji Masumori, and Makoto Hayashi
10 Micronucleus Assay in Human Cells: Lymphocytes and Buccal Cells 191
    Claudia Bolognesi and Michael Fenech

11 Flow Cytometric Determination of Micronucleus Frequency............ 209
   Azeddine Elhajouji and Magdalena Lukamowicz-Rajska

12 Fluorescence In Situ Hybridization (FISH) Technique
   for the Micronucleus Test................................ 237
   Ilse Decordier and Micheline Kirsch-Volders

13 Comparative Genomic Hybridization (CGH) in Genotoxicology 245
   Adolf Baumgartner

14 The In Vitro Micronucleus Assay and Kinetochore Staining:
   Methodology and Criteria for the Accurate Assessment
   of Genotoxicity and Cytotoxicity........................... 269
   Bella B. Manshian, Neenu Singh, and Shareen H. Doak

PART III TESTS FOR PRIMARY DNA DAMAGE

15 γ-H2AX Detection in Somatic and Germ Cells of Mice............... 293
   Eugenia Cordelli and Lorena Paris

16 Multicolor Laser Scanning Confocal Immunofluorescence
   Microscopy of DNA Damage Response Biomarkers................ 311
   *Julian Laubenthal, Michal R. Gdula, Alok Dhawan,
   and Diana Anderson*

17 The Comet Assay: Assessment of In Vitro and In Vivo DNA Damage...... 325
   Mahima Bajpayee, Ashutosh Kumar, and Alok Dhawan

18 The Comet Assay in Human Biomonitoring...................... 347
   Diana Anderson, Alok Dhawan, and Julian Laubenthal

19 The Comet Assay in Marine Animals........................... 363
   Giada Frenzilli and Brett P. Lyons

20 Unscheduled DNA Synthesis (UDS) Test with Mammalian Liver Cells In Vivo 373
   Fabrice Nesslany

21 ^{32}P-Postlabeling Analysis of DNA Adducts 389
   Heinz H. Schmeiser, Marie Stiborova, and Volker M. Arlt

PART IV ASSAYS IN PLANTS AND ALTERNATE ANIMAL MODELS

22 Micronucleus Assay with Tetrad Cells of *Tradescantia*................ 405
   *Miroslav Mišík, Clemens Pichler, Bernhard Rainer,
   Armen Nersesyan, and Siegfried Knasmueller*

23 The Wing-Spot and the Comet Tests as Useful Assays Detecting
   Genotoxicity in Drosophila................................ 417
   Ricard Marcos and Erico R. Carmona

PART V GUIDELINES

24 Genotoxicity Guidelines Recommended by International Conference
   of Harmonization (ICH) 431
   Gireesh H. Kamath and K.S. Rao

Index.. 459

Contributors

DIANA ANDERSON • *Biomedical Sciences Division, School of Life Sciences, University of Bradford, Bradford, UK*

VOLKER M. ARLT • *Analytical and Environmental Sciences Division, MRC–HPA Centre for Environment & Health, King's College London, London, UK*

MAHIMA BAJPAYEE • *Institute of Life Sciences, School of Science and Technology, Ahmedabad University, Ahmedabad, Gujarat, India; The Science Hub, Lucknow, UP, India*

ADOLF BAUMGARTNER • *Biomedical Sciences, University of Bradford, Bradford, UK; Department of Paediatric Cardiology, Cytometry Group, University of Leipzig, Heart Centre, Leipzig, Germany*

JEFFREY C. BEMIS • *Litron Laboratories, Rochester, NY, USA*

CLAUDIA BOLOGNESI • *Environmental Carcinogenesis Unit, IRCCS Azienda Ospedaliera, Universitaria San Martino- IST Istituto Nazionale Ricerca sul Cancro, Genova, Italy*

XUEFEI CAO • *Division of Genetic and Molecular Toxicology, National Center for Toxicological Research, U.S. Food and Drug Administration, Jefferson, AR, USA*

ERICO R. CARMONA • *Escuela de Ciencias Ambientales, Núcleo de Investigación en Estudios Ambientales, Universidad Católica de Temuco, Temuco, Chile*

SERENA CINELLI • *Research Toxicology Centre, Pomezia (Roma), Italy*

EUGENIA CORDELLI • *Unit of Radiation Biology and Human Health, ENEA, Rome, Italy*

ILSE DECORDIER • *Laboratorium voor Cellulaire Genetica, Vrije Universiteit Brussel, Brussel, Belgium*

STEPHEN D. DERTINGER • *Litron Laboratories, Rochester, NY, USA*

ALOK DHAWAN • *Institute of Life Sciences, School of Science & Technology, Ahmedabad University, Ahmedabad, Gujarat, India; Nanomaterial Toxicology Group, Council of Scientific and Industrial Research (CSIR), Indian Institute of Toxicology Research, Lucknow, UP, India*

SHAREEN H. DOAK • *Institute of Life Science, College of Medicine, Swansea University, Singleton Park, Wales, UK*

VASILY N. DOBROVOLSKY • *Division of Genetic and Molecular Toxicology, National Center for Toxicological Research, U.S. Food and Drug Administration, Jefferson, AR, USA*

AZEDDINE ELHAJOUJI • *Genetic Toxicology and Safety Pharmacology, Preclinical Safety, Novartis Institutes for Biomedical Research, Basel, Switzerland*

MICHAEL FENECH • *CSIRO Preventative Health Flagship, Adelaide ,BC, Australia*

GIADA FRENZILLI • *Department of Clinical and Experimental Medicine, University of Pisa, Pisa, Italy*

MICHAL R. GDULA • *Centre of Skin Sciences, School of Life Sciences, University of Bradford, Bradford, UK; Department of Dermatology, School of Medicine, University of Boston, Boston, USA*

B. BHASKAR GOLLAPUDI • *The Dow Chemical Company, Toxicology & Environmental Research & Consulting, Midland, MI, USA*
NIKKI E. HALL • *Litron Laboratories, Rochester, NY, USA*
MAKOTO HAYASHI • *Public Interest Incorporated Foundation, Biosafety Research Center, Shioshinden, Iwata, Japan*
ROBERT H. HEFLICH • *Division of Genetic and Molecular Toxicology, National Center for Toxicological Research, U.S. Food and Drug Administration, Jefferson, AR, USA*
CHRISTIAN JOHANNES • *Faculty of Biology—Genetics, Center for Medical Biotechnology, University of Duisburg-Essen, Essen, Germany*
GIREESH H. KAMATH • *Syngene International Limited, Bangalore, India*
SAWAKO KASAMOTO • *Public Interest Incorporated Foundation, Biosafety Research Center, Shioshinden, Iwata, Japan*
MICHELINE KIRSCH-VOLDERS • *Laboratorium voor Cellulaire Genetica, Vrije Universiteit Brussel, Brussels, Belgium*
SIEGFRIED KNASMUELLER • *Institute of Cancer Research, Internal Medicine I, Medical University of Vienna, Vienna, Austria*
ASHUTOSH KUMAR • *Institute of Life Sciences, School of Science & Technology, Ahmedabad University, Ahmedabad, Gujarat, India*
JULIAN LAUBENTHAL • *Medical Sciences Division, School of Life Sciences, University of Bradford, Bradford, UK*
MAGDALENA LUKAMOWICZ-RAJSKA • *Institute of Surgical Pathology, Zurich University Hospital, Zurich, Switzerland*
BRETT P. LYONS • *Weymouth Laboratory, CEFAS, Weymouth, Dorset, UK*
MUGIMANE G. MANJANATHA • *Division of Genetic and Molecular Toxicology, National Center for Toxicological Research, U.S. Food and Drug Administration, Jefferson, AR, USA*
BELLA B. MANSHIAN • *Institute of Life Science, College of Medicine, Swansea University, Swansea, Wales, UK*
RICARD MARCOS • *Grup de Mutagènesi, Departament de Genética i de Microbiologia, Facultat de Biociències, Universitat Autònoma de Barcelona, Barcelona, Spain*
SHOJI MASUMORI • *Public Interest Incorporated Foundation, Biosafety Research Center, Shioshinden, Iwata, Japan*
MIROSLAV MIŠÍK • *Institute of Cancer Research, Internal Medicine I, Medical University of Vienna, Vienna, Austria*
ROBERTA A. MITTELSTAEDT • *Division of Genetic and Molecular Toxicology, National Center for Toxicological Research, U.S. Food and Drug Administration, Jefferson, AR, USA*
DAISHIRO MIURA • *Strategic Planning Department, Teijin Pharma Limited, Tokyo, Japan*
MARTHA M. MOORE • *National Center for Toxicological Research, Food and Drug Administration, Jefferson, AR, USA*
PASQUALE MOSESSO • *Dipartimento di Scienze Ecologiche e Biologiche, Università degli Studi della Tuscia, Largo dell'Università, Viterbo, Italy*
ADYAPALAM T. NATARAJAN • *Dipartimento di Scienze Ecologiche e Biologiche, Università degli Studi della Tuscia, Largo dell'Università, Viterbo, Italy*
ARMEN NERSESYAN • *Institute of Cancer Research, Internal Medicine I, Medical University of Vienna, Vienna, Austria*

FABRICE NESSLANY • *Genetic Toxicology Department, Institut Pasteur de Lille, Lille, France*

GUENTER OBE • *Faculty of Biology—Genetics, Center for Medical Biotechnology, University of Duisburg-Essen, Essen, Germany*

FRANCESCA PACCHIEROTTI • *Unit of Radiation Biology and Human Health, ENEA, Rome, Italy*

FABRIZIO PALITTI • *Dipartimento di Scienze Ecologiche e Biologiche, Università degli Studi della Tuscia, Largo dell'Università, Viterbo, Italy*

LORENA PARIS • *Unit of Radiation Biology and Human Health, ENEA, Rome, Italy; Department of Ecology and Biology, University of Tuscia, Viterbo, Italy*

CLEMENS PICHLER • *Institute of Cancer Research, Internal Medicine I, Medical University of Vienna, Vienna, Austria*

BERNHARD RAINER • *Institute of Cancer Research, Internal Medicine I, Medical University of Vienna, Vienna, Austria*

K.S. RAO • *Syngene International Limited, Bangalore, India*

MELISSA R. SCHISLER • *The Dow Chemical Company, Toxicology & Environmental Research & Consulting, Midland, MI, USA*

HEINZ H. SCHMEISER • *Research Group Genetic Alterations in Carcinogenesis, German Cancer Research Center (DKFZ), Heidelberg, Germany*

JOSEPH G. SHADDOCK • *Division of Genetic and Molecular Toxicology, National Center for Toxicological Research, U.S. Food and Drug Administration, Jefferson, AR, USA*

SHARON D. SHELTON • *Division of Genetic and Molecular Toxicology, National Center for Toxicological Research, U.S. Food and Drug Administration, Jefferson, AR, USA*

NEENU SINGH • *Institute of Life Science, College of Medicine, Swansea University, Singleton Park, Swansea, Wales, UK*

MARIE STIBOROVA • *Department of Biochemistry, Faculty of Science, Charles University, Prague, Czech Republic*

VALENTINA STOCCHI • *Unit of Radiation Biology and Human Health, ENEA, Rome, Italy*

ERROL ZEIGER • *Errol Zeiger Consulting, Chapel Hill, NC, USA*

Part I

Gene Mutation Assays

Chapter 1

Bacterial Mutation Assays

Errol Zeiger

Abstract

Bacterial mutagenicity tests, specifically the Salmonella and *E. coli* reverse mutation (Ames) test, are widely used and are usually required before a chemical, drug, pesticide, or food additive can be registered for use. The tests are also widely used for environmental monitoring to detect mutagens in air or water. Their use is based on the showing that a positive result in the test was highly predictive for carcinogenesis. This chapter describes the Salmonella and *E. coli* tests, presents protocols for their use, and addresses data interpretation and reporting.

Key words Salmonella, *E. coli*, Ames test, Mutagenicity testing, Reverse mutation, Carcinogen identification

1 Introduction

Testing for mutation is usually the first step in the selection of chemicals for development, and a required step for the regulatory approval and marketing of a chemical, whether it is an industrial chemical, pesticide, food additive, or a drug. The most widely performed mutation test, and often the only test, is in bacteria, most often in *Salmonella typhimurium* and also in *Escherichia coli*. The Salmonella microsome mutagenicity test, often referred to as the Ames test, was developed by Prof. Bruce Ames in the early 1970s [1, 2]. These, and other laboratory's, initial evaluations of the test correctly identified approximately 90 % of the carcinogenic chemicals (Table 1), and a similar proportion of the noncarcinogens [3–5]. These high predictive values led to the widespread use of the test for screening chemicals as potential mutagens and carcinogens, and to its inclusion in regulatory toxicology requirements.

Subsequent studies that used a wider range of chemical classes and more noncarcinogens reported lower predictive values, i.e., approximately 50 % of the tested carcinogens were not mutagenic in the Ames test, and approximately 80 % of the test positives were carcinogenic [6–11]; Table 1). Even though these values were

Table 1
Prediction of carcinogenicity using Salmonella mutant tester strains*

Chems.	Sensitivity	Specificity	+Predictivity	−Predictivity	Concordance	References	Year
224	0.88	0.75	0.90	0.71	0.84	[3]	1975
139	0.93	0.74	0.89	0.82	0.87	[4]	1976
53	0.68	0.88	0.93	0.54	0.74	[5]	1977
120	0.92	0.93	0.93	0.92	0.93	[6]	1978
60	0.76	0.59	0.69	0.67	0.68	[7]	1985
73	0.45	0.86	0.83	0.51	0.62	[8]	1987
114	0.48	0.91	0.89	0.55	0.66	[9]	1990
363	0.54	0.79	0.77	0.57	0.65	[10]	1998
717	0.59	0.74	0.87	0.37	0.62	[11]	2005

Chems. the number of chemicals evaluated, *Sensitivity* the proportion of carcinogens positive in the test, *Specificity* the proportion of noncarcinogens negative in the test, *+Predictivity* the proportion of positives in the test that are carcinogens, *−Predictivity* the proportion of negatives in the test that are not carcinogens, *Concordance* the overall proportion of correct predictions, *Year* year of publication
*The compilations from 1985 through 2005 contain many of the same test results produced by the US National Toxicology Program

lower than those originally published, they were still sufficiently high to continue the use of the test. Similar studies were not performed with the *E. coli* test.

Bacterial systems are most often the first genetic toxicity tests to be used because they are the least expensive. Also, as had been shown in a number of studies [8–11] they are the least likely to falsely identify a noncarcinogen as a carcinogen. One additional advantage to the use of bacterial tests is the time duration it takes to perform. The tester strain(s) is inoculated into growth media the evening prior to the test, the test itself can be performed in the morning, and the results are available 48 h later.

A number of publications over the years have provided detailed information on the performance of these bacterial mutagenicity tests with respect to their reproducibility and ability to correctly identify carcinogens and noncarcinogens [8, 10, 12, 13]. This chapter describes the Salmonella and *E. coli* tests, the most commonly used plate and pre-incubation procedures, and strategies for their most efficient use. It also provides information to aid in the interpretation of the test performance and test results, and describes minimum requirements for test reports. Other variations of the bacterial mutagenicity tests, e.g., vapor, liquid, and fluctuation, have been used [14–16] but are not addressed here. There are some other useful and informative publications which may be of interest to the reader or the persons planning to perform bacterial mutagenicity testing [17–33].

Table 2
Salmonella and *Escherichia coli* tester strains most commonly used for routine mutagen testing

Tester strain	Mutation	Δuvr	rfa	Plasmid	Specificity
Salmonella					
TA97	*his*D6610	+	+	pKM101	FS; BPS
TA98	*his*D3052	+	+	pKM101	FS
TA100	*his*G46	+	+	pKM101	BPS; FS
TA102	*his*ΔG428	−	+	pKM101; pAQ1 (*his*G428)	BPS; FS
TA1535	*his*G46	+	+	None	BPS
TA1537	*his*C3076	+	+	None	FS
TA1538	*his*D3052	+	+	None	FS
E. coli					
WP2, *uvrA*	*trp*E	+	−	None	BPS
WP2 *uvrA*, pKM101	*trp*E	+	−	pKM101	BPS

Δ*uvr* uvrB deletion (Salmonella strains), *uvrA* deletion (*E. coli* strains); *rfa* deep rough cell wall mutation
FS frameshift mutations, BPS base-pair substitution mutations. The first mentioned is the predominant mutation type

The Salmonella strains were originally developed to study the genetics of histidine biosynthesis [34, 35]. The strains used for testing contain base-pair substitution (BPS) or frameshift (FS) mutations in genes of the histidine (*his*) operon. The *E. coli* strains have a BPS mutation in the tryptophan (*trp*) operon (Table 2). As a consequence the cells are unable to synthesize the amino acids histidine or tryptophan, and cannot grow and form colonies on minimal medium in the absence of the added amino acid. These particular strains used were selected for the testing set because they are readily revertable by chemical mutagens and they represent different molecular targets, i.e., G:C base pairs, A:T base pairs, and addition and deletion frameshifts. After treatment with a mutagen, only the cells that undergo a subsequent mutation that restores their ability to synthesize histidine or tryptophan will be able to grow and form colonies on the minimal agar plates. The number of these revertant colonies formed after treatment provides a measure of the mutagenic potency of the test substance in this test system (Fig. 1).

In addition to the *his* or the *trp* mutations, the strains were engineered to increase their sensitivity to mutagenic chemicals (Table 2). The *uvr* deletion removes a relatively error-free DNA repair system so that the cell has to rely on the more error-prone systems. The *rfa*

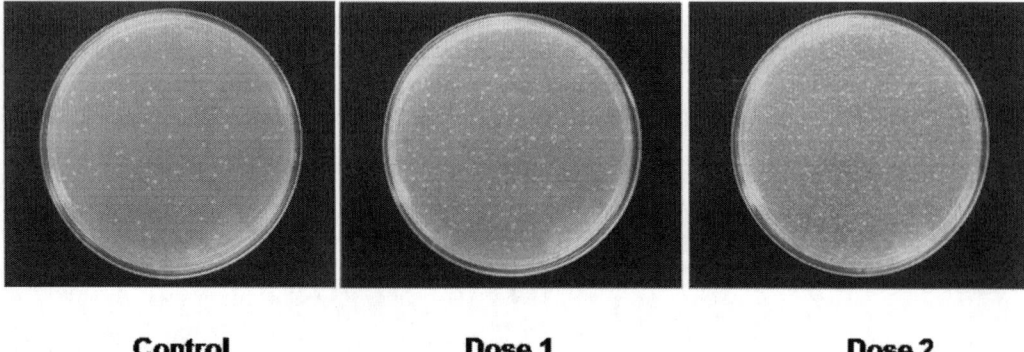

Fig. 1 Bacterial mutagenicity assay dose–response. Control is no added test chemical. Dose 1 and Dose 2 are increasing concentrations of the test chemical

mutation causes the cells to produce a defective cell wall, which makes them more permeable to large molecules. These additional mutations are present in all of the tester strains. Some of the strains also contain the pKM101 plasmid which contains an error-prone recombinational DNA repair pathway, thereby further increasing their sensitivity to mutagens.

Many chemicals are biologically inactive in their native forms and need to be metabolized, usually in the mammalian liver, to an active form. It is this active form that, in many cases, is the causative agent for mutation, cancer, and other effects. Because the bacteria used in the test are not capable of performing this metabolism, Prof. Ames and his colleagues added a rat liver homogenate (called S9, because it is prepared from a $9,000 \times g$ supernatant liver fraction) and enzymatic cofactors, to the test [2]. The S9 is prepared from rats that were pretreated with a polychlorinated biphenyl mixture or phenobarbital-plus-β-naphthoflavone to induce the cytochrome P450 enzymes needed for the metabolic activation [36]. The S9 has also been prepared from mice, hamsters, and other animals, but the rat S9 is the most widely used. Human S9 has also been used [2, 37–39] but there is currently no evidence that it is more effective in identifying mutagens or predicting carcinogenicity than induced rodent S9, and may be less effective for many classes of chemicals.

2 Materials (*See* Note 1)

2.1 Equipment

1. 37 °C Shaker for growing the cell cultures.
2. 37 °C Incubator for incubating the Petri dishes for the test.
3. Variable temperature (37–40 °C) water bath or dry temperature block.
4. Autoclave for sterilizing some of the reagents before use and the bacterial cultures after use.

5. Micropipettes.
6. Automated or manual bacterial colony counter.

2.2 Supplies

1. Sterile microliter disposable pipettes.
2. 100 cm² Radiation-sterilized Petri dishes.
3. Sterile cryogenic tubes for maintaining stock cultures of the tester strains.
4. 13 × 100 mL Sterile test tubes.
5. 0.45 µm Sterilizing filters.

2.3 Bacterial Tester Strains

See Table 2 for strains most commonly used [14–16, 40–49] (see **Note 2**).

1. Salmonella.
2. E. coli.

2.4 Growth Media

1. *Nutrient broth.* 1,000 mL of distilled water, 25 g of nutrient broth (commercially available). Dispense 50 mL of broth in 125 mL flasks, or 5 mL in 100 × 16 mm test tubes. Autoclave the solution for 20 min. Store it in the dark at room temperature.

2. *Nutrient agar plates.* 1,000 mL of distilled water, 15 g of agar, 25 g of nutrient broth. Add the agar to the water in a 3 L flask and heat with stirring until dissolved. Add the nutrient broth powder and stir until dissolved. Autoclave the solution for 20 min (see **Note 3**).

3. *Vogel–Bonner (VB) minimal salts.* 50× concentration stock of VB salts. 650 mL of warm (about 50 °C) distilled water, 10 g of magnesium sulfate ($MgSO_4 \cdot H_2O$), 100 g of citric acid monohydrate, 500 g of potassium phosphate, dibasic, anhydrous (K_2HPO_4), 175 g of sodium ammonium phosphate ($Na_2NH_2PO_4 \cdot 4 H_2O$), 124 mg of biotin (see **Note 4**).

4. *Minimal agar plates.* 900 mL of distilled water, 15 g of agar, 20 mL of 50× VB salts, 50 mL of 10 % (v/v) glucose solution (see **Note 5**). Add the agar to the water in a 3 L flask. Autoclave for 30 min. Let cool to about 65 °C. Add 20 mL of sterile VB salts, mix thoroughly, then add the 50 mL of sterile 10 % (v/v) glucose solution, and mix thoroughly. Dispense the minimal agar, approximately 20 or 25 mL per plate, in 100 × 15 mm Petri dishes. Keep the plates on a level surface and allow the agar to solidify. After the agar solidifies, the plates can be stored at 4 °C for several weeks in sealed plastic bags. Before use, the plates should be allowed to come to room temperature outside the bags and examined for excess moisture. If visible moisture is present on the agar surface, incubate the plates overnight at 37 °C before use (see **Note 6**).

5. *Supplemented minimal agar plates* (*see* **Note 7**). Add the following sterile solution(s), as required, to 1 L of the minimal agar medium; mix well before dispensing.

    (a) Biotin/histidine plates: 8 mL of 0.01 % biotin solution and 0.5 % histidine solution.

    (b) Biotin/histidine/ampicillin plates: Same as biotin/histidine plates and add 3 mL of 8 mg/mL ampicillin solution to a final concentration of 24 µg ampicillin/mL.

    (c) Biotin/histidine/tetracycline plates: Same as biotin/histidine plates, add 0.25 mL of 8 mg/mL tetracycline solution to give a final concentration of 2 µg tetracycline/mL.

6. *Top agar* (*see* **Note 8**). 900 mL distilled water, 6 g agar, 6 g NaCl. Bring to 1 L with distilled water. Heat in autoclave to melt the agar. Add 100 mL of 0.5 mM histidine/tryptophan/biotin solution. Dispense 200-mL aliquots in 500-mL screw-cap bottles. Autoclave the solution for 30 min and store it at room temperature in the dark. When ready to use, melt the top agar in a microwave oven or in boiling water.

2.5 Solutions

1. S9 Mix: (*see* **Note 9**)

    (a) For 10 % S9 add 1.0 mL S9 to 9.0 mL of S9 mix.

    (b) For 30 % S9 add 3.0 mL S9 to 7.0 mL of S9 mix.

2. Sodium phosphate buffer: 0.01 mM, pH 7.4 (*see* **Note 10**): 120 mL of 0.1 M sodium phosphate, monobasic (13.8 g/L $NaH_2PO_4 \cdot H_2O$), 880 mL of 0.1 M sodium phosphate, dibasic (14.2 g/L $Na_2HPO_4 \cdot H_2O$, *see* **Note 11**).

3. Glucose: 10 % glucose solution (*see* **Note 12**). 700 mL of distilled water, 100 g of glucose. After dissolving make up the volume to 1 L.

4. Histidine/tryptophan/biotin solution: 0.5 mM L-tryptophan, 0.5 mM L-histidine HCl, 0.5 mM D-biotin. Add 124 mg of D-biotin, 96 mg of L-histidine HCl, and 102 mg of L-tryptophan, one at a time until dissolved, to 1 L of boiling distilled water. Sterilize the solution by filtration through a 0.45 µm filter or by autoclaving for 20 min. Store at 4 °C in a glass bottle (*see* **Note 13**).

5. Crystal violet: Add 100 mg crystal violet to 100 mL distilled water; store at 4 °C in the dark (*see* **Note 14**).

6. Ampicillin: Add 8 mg ampicillin to 100 mL distilled water; filter sterilize (*see* **Note 15**). Store at 4 °C.

7. Tetracycline: Add 8 mg tetracycline to 100 mL of 0.02 N HCl; filter sterilize the solution (*see* **Note 16**). Store at 4 °C in the dark (light sensitive).

8. S9 Cofactors: 900 mL distilled water, 1.6 g of D-glucose-6-phosphate, 3.5 g of nicotinamide adenine dinucleotide

phosphate (NADP), 1.8 g of magnesium chloride (MgCl$_2$), 2.7 g of potassium chloride (KCl), 12.8 g of Na phosphate, dibasic (Na$_2$HPO$_4$·H$_2$O), 2.8 g of Na phosphate, monobasic (NaH$_2$PO$_4$·H$_2$O) (*see* **Note 17**).

9. Cryopreservant: 10 % sterile glycerol or DMSO v/v in medium.
10. Control chemicals: Solvent controls and their range of responses are listed in Table 3 (*see* **Note 18**). Positive controls typically used in the assay are listed in Table 4 (*see* **Note 19**).

3 Methods (*See* Note 20)

3.1 Preparing and Growing the Bacteria

1. For long-term preservation, the tester strains should be kept frozen at −70 to −80 °C.
2. Upon receipt of the strains, up to five frozen cultures should be prepared from a single colony isolate that has been purified and checked for its genotypic characteristics (*his, rfa, uvr*B-*bio; trp, uvr*A) and for the presence of plasmids pKM101 and pAQ1, when appropriate.
3. These cultures should be considered the frozen permanent strains and should be used only for the preparation of new frozen working cultures or master agar plates.

3.2 Receipt and Growth of the Tester Strains

1. If the new strain is received on a small sterile filter disk embedded in nutrient agar, wipe the disk across the surface of a nutrient agar plate and then transfer the disk to 5 mL of nutrient broth.
2. If the strain arrives as a lyophilized culture, aseptically add 1 mL of nutrient broth, and then transfer the rehydrated culture to 4 mL of nutrient broth.
3. Keep a drop of the rehydrated culture and streak it across the surface of a nutrient agar plate for individual colonies.
4. Incubate the broth culture and the agar plate at 37 °C overnight. The broth culture serves as a backup, in case no growth is observed on the agar plate.
5. After overnight incubation at 37 °C, inspect the agar plates for growth.
6. Pick a healthy looking colony and re-streak it on nutrient agar for individual colonies. This purification step should be repeated at least once, and is recommended to ensure that a pure culture will be used for the preparation of the frozen permanent cultures. Nutrient agar plates or histidine- or tryptophan-supplemented minimal agar plates may be used for this purification; however the use of the supplemented minimal agar reduces the risk of contamination. Although the cells will grow overnight at 37 °C on the nutrient agar plates, good growth on the minimal agar plates may take up to 2 days.

Table 3
Tester strain solvent control ranges

Tester strain	Solvent control range[a]
Salmonella	
TA97	75–200[b]
TA98	20–50
TA100	75–200
TA102	100–400[b]
TA1535	5–20
TA1537	5–20
TA1538	5–20
E. coli	
WP2 *uvrA*	5–20
WP2 *uvrA* pKM101	100–200

[a]Based on results from the author's laboratory and other sources. These are for reference only, and should not be used as the standard against which individual laboratories are evaluated. Each laboratory must develop its own acceptable solvent control range

[b]These strains tend to have a higher range of control values in the presence of S9

Table 4
Standard positive control mutagens for use with and without S9

Tester strain	No S9	+S9
Salmonella		
TA97	9-Aminoacridine	2-Aminoanthracene
TA98	4-Nitro-*o*-phenylenediamine	
TA100	Sodium azide	
TA102	Mitomycin C	
TA1535	Sodium azide	
TA1537	9-Aminoacridine	
TA1538	4-Nitro-*o*-phenylenediamine	
E. coli		
WP2 uvrA	*N*-Ethyl-*N'*-nitro-*N*-nitrosoguanidine (ENNG)	2-Aminoanthracene
WP2 uvrA/ pKM101		

7. The supplemental amino acids and biotin can be applied directly to the surface of the minimal agar plates by delivering the appropriate solutions, which are then evenly distributed over the agar surface by a sterile glass spreader. Alternatively, these solutions can be incorporated in the agar when the plates

are prepared. If both the Salmonella and *E. coli* strains will be used, it is recommended that the histidine, biotin, and tryptophan be added to the same minimal agar plates so that two different sets of plates will not be needed. The added histidine and biotin will not affect the growth of the *E. coli* strains, and the tryptophan will not affect the growth of the Salmonella strains. Additional sets of minimal agar plates containing ampicillin (for the pKM101 plasmid strains), tetracycline (for the pAQ1 plasmid strain), or ampicillin and tetracycline (for strains containing both pKM101 and pAQ1) can be prepared to test for the presence of these plasmids.

8. Pick up to five single colonies from the second purification plate and transfer to a preassigned location on a minimal agar plate supplemented with the appropriate nutrients/antibiotics. These will serve as the master plates for preparing the permanent frozen cultures.

9. Incubate the master plates for 1 (nutrient agar) or 2 (minimal agar) days at 37 °C.

10. When good growth is observed, inoculate 5 mL of nutrient broth with a small inoculum from each of the single isolated colonies on the master plate. After overnight incubation at 37 °C confirm the genotypes (strain check) of the tester strains.

3.3 Preparation of Working Cultures

1. Upon completion of the strain check, select the colony from the master plate that has given the best overall results in terms of phenotypic characteristics, including the best overall spontaneous mutation induction.

2. Transfer a small inoculum into 4.5 mL of nutrient broth and incubate overnight at 37 °C with shaking, till the cells reach a density of $1-2 \times 10^9$ colony-forming units (CFU)/mL (O.D.$_{540}$ between 0.1 and 0.2).

3. Add 0.5 mL of sterile glycerol or DMSO to the culture for a final concentration of 10 %, v/v, as a cryopreservative, mix thoroughly, and dispense 1 mL aliquots in sterile cryogenic tubes.

4. Quick-freeze on dry ice and store at −80 °C. These are the working cultures, and enough tubes of each strain should be prepared to allow for approximately one year of testing.

3.4 Preparation of Working Stock Agar Plates

1. If there is a problem maintaining frozen working cultures, working stock agar plates can be prepared by streaking supplemented minimal agar plates with the tester strains after a strain check.

2. These plates should be stored in the refrigerator wrapped in parafilm to prevent dehydration of the agar.

3. A small inoculum from an area of confluent growth, instead of a single colony, should be used to inoculate the nutrient broth.

4. The plates can be stored up to 2 months, except for the plate containing strain TA102 which should be stored for only 2 weeks because the number of spontaneous revertant colonies increases over time. Therefore, it is recommended that frozen working cultures, rather than working agar plates, be prepared for strain TA102.

3.5 Overnight Cultures for Use in the Tests

1. For each experiment the tester strain cultures are grown overnight in nutrient broth to a density of $1-2 \times 10^9$ CFU/mL. The overnight cultures should not be saved for future day's experiments because the viable cell count will decrease with time, even when kept refrigerated, although the O.D. may not change because of the presence of dead or dying cells. The volume of the cultures will depend on the size of the experiment but is usually between 10 and 50 mL. The culture flask should be at least 3–5 times the volume of the culture medium to ensure adequate aeration and growth of the cells.

2. Individual culture flasks of nutrient broth are inoculated with each strain.

3. When frozen stock cultures are used, allow the culture to thaw at room temperature. Aliquots between 0.1 and 0.5 mL are then transferred into nutrient broth with volumes between 10 and 50 mL, respectively, which gives an initial cell density between 10^6 and 10^7 CFU/mL (1:100 dilution).

4. The remainder of the working culture should be discarded, because frequent thawing and refreezing of the frozen cultures reduce their viability.

5. An alternate procedure for inoculation of the overnight cultures is to use a sterile inoculation loop to scrape a small amount of the frozen culture from the surface of the frozen culture. This procedure should be done quickly to prevent thawing of the culture. Overnight cultures may also be inoculated with a small inoculum taken from a working stock agar plate.

6. The freshly inoculated cultures are placed on a shaker in the dark at room temperature for 4 h without shaking, and then gently shaken at 100 rpm for 11–14 h at 37 °C.

7. The next morning the cultures are removed from the incubator and kept at room temperature away from direct fluorescent light by wrapping them in foil. This procedure is not needed if the laboratory is equipped with yellow or red overhead lights. If not immediately used, the cultures should be kept on ice until they are ready to be used to prevent loss of viability of the strains. The cultures should then be brought to room temperature before they are added to the 43–48 °C top agar.

3.6 Genetic Analysis of the Tester Strains

1. The tester strains should be analyzed for their genetic integrity and spontaneous mutation rate when frozen cultures are prepared.
2. An analysis of the genetic integrity of every strain used should also be performed at the time of every experiment, and in parallel with the experiment, to ensure that each strain used contains the *uvr* and cell wall (*rfa*) mutations, and plasmid, where appropriate.
3. The strain check is performed with the nutrient broth overnight cultures prior to the start of the experiment, and the results are evaluated before scoring the experiment. If the tester strain does not behave appropriately in the strain check, all test plates from that strain/culture should be discarded, and not scored.

3.7 Histidine and Biotin (his, bio) or Tryptophan (trp) Dependence

1. The *uvr*B deletion in the Salmonella strains also deletes the biotin (bio) genes so that the *uvr*B strains also require biotin for growth.
2. Streak a loop of the culture across a minimal agar plate supplemented with biotin alone, and with an excess of biotin and histidine (Salmonella strains) or tryptophan (*E. coli* strains).
3. Growth should be observed with all strains in the presence of histidine or tryptophan, but not on the plates containing only biotin. When using this technique, more than one strain can be streaked on the same plate, taking care not to have the individual streaks touch each other.

3.8 rfa (Deep Rough) Mutation

1. Streak a loop-full of the overnight culture across a nutrient agar plate supplemented with an excess of biotin and histidine or tryptophan.
2. Place a sterile filter paper disk in the center of the streak and apply 10 µL of a sterile 0.1 % crystal violet solution. The crystal violet disks can be prepared in advance and stored aseptically at room temperature [50].
3. All strains should show a zone of growth inhibition surrounding the disk.

3.9 Requirements for an Adequate Test

The minimum requirements for an adequate test are described in OECD Test guideline No. 471 [51].

1. Every experiment should contain a solvent control and a positive control using an appropriate control chemical (e.g., Table 4).
2. At least five doses of the test substance, at no greater than half-log concentrations should be used, with the highest dose selected by toxicity or solubility (*see* **Note 21**).
3. Typically, each dose, including the controls, should use triplicate plates.

4. A positive response in a single tester strain is sufficient for a showing of mutagenicity.

 5. Before a chemical is determined to be non-mutagenic it should be tested in a minimum of five tester strains, Salmonella strains TA97 (or TA1537), TA98, TA100, and TA1535, and either TA102 or one of the *E. coli* strains. Other strains can be used in addition to these strains; however the replacement of one of these strains with another strain needs to be justified in the test report.

 6. For regulatory purposes, testing is typically performed using a pre-incubation or a plate test procedure. These are described below. Other methods, including procedures for gasses and highly volatile substances [52–54], fluctuation tests [43, 45, 55, 56] and reductive metabolism procedures [57, 58] have been used, but are not described here (*see* **Note 22**).

3.10 Determination of Sample Test Concentrations for Use in the Test

 1. Before performing the test, it is necessary to determine the test dose range, which will be a function of the test substance's toxicity and solubility.

 2. A toxicity test is performed using a single tester strain, usually TA100, with and without S9 and only one or two plates per dose.

 3. In this procedure, a series of 5–10 doses at half-log intervals are used, with the top dose at 5 mg/plate.

 4. Chemicals that are poorly soluble should be tested at the lower precipitating levels but must be thoroughly mixed before adding to the top agar.

 5. The plates are incubated at 37 °C for 40–48 h, removed from the incubator, and the colonies are counted. It is important here, and in the performance of the mutagenicity test, that all plates in an experiment be incubated for the same length of time, and counted at the same time (*see* **Note 22**).

3.11 The Pre-incubation Procedure (See Note 23)

 1. To 13×100 mm sterile tubes maintained at room temperature, add in the following order, with mild mixing after each addition: 0.50 mL of S9 mix or buffer, 0.05 mL of the test chemical dilution, and 0.05–0.10 mL of overnight culture of the tester strain (about $1-2 \times 10^8$ cells per tube).

 2. Incubate the mixture at 37 °C for 20–30 min, with or without gentle shaking. If the test substance is known or suspected to be volatile, the tubes should be capped or sealed with parafilm prior to the incubation step.

 3. Add 2 mL of molten top agar maintained at 40–43 °C to each tube, mix the contents, and pour onto the surface of minimal agar plates. Gently swirl the plate to distribute the top agar evenly over its surface.

4. When the top agar has hardened (2–3 min), the plates are inverted and placed in a 37 °C incubator for 48 h.

5. The colonies are then counted and the results are expressed as number of revertant colonies per plate.

3.12 The Plate Test Procedure

1. The procedure in this test is the same as the pre-incubation procedure, Subheading 3.11, except that the pre-incubation step (**step 2** of Subheading 3.11) is eliminated.

2. The top agar is added to the tubes immediately after adding the S9 mix or buffer, test chemical, and cells. **Steps 4** and **5** of Subheading 3.11 are then carried out.

3.13 Testing Strategies (See Note 24)

1. The minimum recommended tester set for routine testing includes strains TA97 or TA1537, TA98, TA100, and TA1535, and either TA102 or one of the *E. coli* strains [51].

2. The predictivity of a positive response for carcinogenicity is the same regardless of which strain or the number of tester strains that are positive in the test. Therefore, if at least one strain produces a positive response, it may not be necessary to test additional strains unless one wants to determine the full mutagenic spectrum of the test substance.

3.14 Scoring the Plates

1. The plates are removed from the incubator after 40–48-h growth.

2. In situations where the colonies appear too small to count at this time, all the plates from the particular experiment should be incubated for an additional 24 h (*see* **Note 25**).

3.15 Data Evaluation

1. A clear positive response in any one of the tester strains is sufficient to demonstrate that a substance is mutagenic.

2. However, a substance should not be considered non-mutagenic unless it has been tested in, at least, the above five strains.

3. There are no formal, agreed-upon procedures for evaluating whether a response should be considered as mutagenic or non-mutagenic, and different laboratories use different procedures.

4. The most widely used is the so-called modified two-fold rule. In this, a test is considered positive if there is a dose-related response reaching at least two-fold over the control.

5. For tester strains with a low background mutant frequency, e.g., strains TA1535 or TA1537, a dose–response leading to a threefold increase is considered positive. Other labs require a reproducible increase for a positive result even if the increase does not reach two-fold.

6. It is important to determine the data evaluation procedures before the experiments are performed and to clearly describe them in the data reports.

7. Formal statistical procedures, if used, should be used with care because they may not be sufficiently conservative.

8. In addition, when using strains with high background plate counts, e.g., TA97 and TA100, the mutant count data may be hyper-poisson, which would affect the sensitivity of the statistical procedure if not taken into consideration [24].

9. In order for a test to be acceptable, the solvent control plate counts should be within the range previously established for the laboratory (*see* **Note 26**), and the positive control plates must show a positive response (*see* **Note 27**). If both of these do not occur, the experiment should be discarded and repeated. It may be appropriate to keep an experiment without an acceptable positive control response if the test chemical produces a clear positive response with a dose-related increase.

3.16 Data Interpretation

1. A positive response in the test in any one of the tester strains indicates that the test substance is a mutagen, and that it has a high probability of being carcinogenic in laboratory animals (typically rats and mice).

2. The probability of the chemical being a carcinogen is the same regardless of which strain, or the number of tester strains, in which the chemical is mutagenic, or the potency of the mutagenic response.

3. Similarly, the potency of the mutagenic response is no predictor of the potency of the rodent cancer response or the responses of other genetic toxicity tests; i.e., highly potent mutagens may be noncarcinogenic in laboratory animal tests [59, 60].

3.17 The Test Report

1. An example of the information that should be included in a test report can be found as part of the OECD Test Guideline [51].

2. At a minimum, the report should identify the type(s) of procedure used, e.g., plate test and pre-incubation test; the number of plates used per concentration and number of replicate tests performed; the solvent; rationale for selection of the highest test concentration; presence or absence of toxicity; the plate count means ± S.D. or S.E.M. for all tester strains used; laboratory criteria for determining a positive response; and an unequivocal statement as to whether the test substance is mutagenic, not mutagenic, or if the result cannot be clearly determined (*see* **Note 28**).

4 Notes

1. All solutions, media, test tubes, and pipettes must be sterile for use. All reagents can be stored, sterile, at room temperature or in the refrigerator unless otherwise noted.

2. All solutions and media should be sterile and labeled with the date prepared. All autoclaving should be at 121 °C. Filter sterilization should be with a 45 μm filter.

3. Let the agar cool to about 65 °C. Dispense 20–25 mL in sterile Petri plates (all plates must contain the same volume of agar). Allow the agar to set on a level surface, and then store upside down in sealed plastic bags at 4 °C.

4. Add each ingredient in the order indicated. Each substance should be thoroughly dissolved before adding the next. Adjust the volume to 1 L. Distribute in 20 mL aliquots and autoclave, loosely capped, for 30 min at 121 °C. When the solutions have cooled, tighten the caps and store at room temperature in the dark. Biotin is required for the growth of the Salmonella strains.

5. A precipitate may form when the VB salts are added, which will dissolve after thorough mixing. The agar should never be autoclaved with the VB salts and glucose because plates prepared this way will not allow proper growth of the bacterial strains.

6. These plates are used to perform the mutagenicity experiments, and also to test the bacterial strains for their identity. The plates contain VB salts, glucose as the energy source, 1.5 % agar, and biotin to support the growth of the Salmonella strains. Only cells that have reverted (mutated) to be able to synthesize histidine or tryptophan will be able to grow and form colonies.

7. These plates are used to purify and check the bacterial strains. In order to ensure that the strains are histidine-requiring (*his*⁻) (Salmonella) or tryptophan-requiring (*trp*⁻) *E. coli*, the strains are plated on biotin-supplemented minimal agar without histidine or tryptophan, and on the same agar supplemented with histidine and/or tryptophan. The strains should not be able to grow on the minimal agar without the amino acids, with the exception of a few mutant colonies, but should be able to grow on the supplemented minimal agar.

For growing the *his*⁻ or *trp*⁻ tester strains on minimal agar, the plates must be supplemented with biotin (needed by the Salmonella strains because of the *uvr*B deletion which causes a biotin requirement). The same plates can be used for Salmonella and *E. coli* if all three reagents are added. The added histidine and biotin will not affect the *E. coli* growth, and the tryptophan will not affect the Salmonella.

8. This soft agar, supplemented with minimal amounts of histidine and/or tryptophan, is used to suspend the test chemical, bacteria, and S9 mix for pouring onto the minimal agar plates. Minimal levels of histidine and/or tryptophan are added to enable the Salmonella and *E. coli* cells, respectively, to undergo

a few cell divisions, but not enough to allow them to form colonies. The few cell divisions make them more sensitive to mutagenesis, thereby increasing the response. This is done because dividing cells are more sensitive to mutation than cells that are not dividing.

9. Thaw sufficient cofactors and S9 fraction before each experiment. The percentage of S9 used is usually 10 or 30 %. A volume of 0.5 mL of the S9 mix is usually added per plate. Unused S9 and S9 mix should be discarded; it will lose potency even if stored frozen.

    The term "S9" refers to the $9,000 \times g$ supernatant fraction following homogenization of the tissue. Livers or other tissues from other animals, including humans, can be used depending upon the research question being studied. S9 can be prepared by the testing laboratory, but is available commercially from a number of reliable sources. Aroclor 1254- or Kanechlor 500-induced rat liver S9 is typically used for routine testing or screening of chemicals. Alternatively, to avoid the use of PCBs, the livers can be induced by phenobarbital and β-naphthoflavone [36]. The S9 must be stored at −70 to −80 °C for stability.

    S9 cofactors include the enzymes and cofactors, which are needed to support and maintain the enzyme activity of the S9. They can be prepared in advance and stored frozen until needed. The S9 mix contains either 10 or 30 % S9 in the mix for routine testing and should be prepared at the beginning of the experiment, and kept on ice until needed. Lower concentrations have been used, but they tend to provide less sensitivity for most chemicals.

10. Because the phosphate buffer components may chelate some divalent metal ions [61], an alternative buffer, such as HEPES, should be used when testing samples containing such ions.

11. Mix well and adjust pH to 7.4 using 0.1 M dibasic sodium phosphate solution. Dispense 100 mL aliquots in 250 mL screw-cap bottles. Autoclave it for 30 min. When cooled, tighten the caps and store at room temperature in the dark.

12. Add the glucose to the water in a 3 L flask. Stir on a magnetic stirrer until mixture is clear. Add water to bring the final volume to 1,000 mL. Dispense 50-mL aliquots into 250 mL screw-cap bottles. Autoclave for 20 min with caps on loosely. When cooled, tighten the caps and store at 4 °C.

    Freshly autoclaved glucose should not be used. It may contain free radicals produced by the heat which could affect the mutation response. Allow the autoclaved solution to age for at least 1–2 days before using it.

Glucose provides the energy source for the bacteria. Wild-type Salmonella and *E. coli* can grow with glucose as their only carbon source.

13. Histidine and tryptophan are needed for growth by the Salmonella and *E. coli* tester strains, respectively. Biotin is needed for growth of the Salmonella mutant and revertant strains because of the *uvr*B deletion, which also removed the biotin gene. If the *E. coli* strains will not be used, it is not necessary to add tryptophan to any of the solutions or agar.

14. This is used to test for the presence of the *rfa* mutation. Cells with the mutation will be more sensitive to the killing effects of the crystal violet.

15. This is used to test for the presence of the pKM101 plasmid, which carries a gene for ampicillin resistance.

16. This is used to test for the presence of the pAQ1 plasmid, which carries a gene for tetracycline resistance in strain TA102.

17. To 900 mL of distilled water add one ingredient at a time. Each ingredient must be dissolved before adding the next one. After all are dissolved, bring up to 1 liter using distilled water and filter sterilize the solution using a 0.45 μm filter. Dispense in sterile glass bottles in aliquots of 7 or 9 mL, or multiples of these volumes, for preparing 10 or 30 % S9, respectively, in the final S9 mix. This contains the enzymes and cofactors needed to maintain the enzymatic activity of the S9.

18. Solvent controls are included in all experiments. The solvent control value is used as the basis for judging whether a test response is positive or negative, and to ensure that the background mutant colony count is within the acceptable range for the tester strain and the testing laboratory. The solvent of choice is distilled water or normal saline. If the test sample is not soluble in water, the nonpolar solvent of choice is DMSO. Care should be taken when using solvents other than water to keep the solvent volume below 0.5 mL in the mixing tubes; otherwise it may be toxic to the tester cells or could adversely affect the enzymatic activity of the S9 mix. Other nonaqueous solvents can be used, including 70 % ethanol, acetone, and dimethylformamide. The use of solvents other than water or DMSO should be justified in the data report. When a nonaqueous solvent other than DMSO is used, the laboratory should have a negative control (i.e., buffer or water) in addition to the solvent control.

19. Positive control chemicals are included with all bacterial strains in all tests to show that the test bacteria were capable of being mutagenized, that the laboratory procedures used were correct, and that the S9 mix was enzymatically active. Control mutagens other than those in Table 4 may be used depending

on availability in the laboratory. When testing a series of structurally or functionally related chemicals it is advisable to use a related mutagenic chemical if one is available. It is recommended that, for routine testing, the laboratory use the same positive controls so that a database of positive control responses can be developed to better monitor the day-to-day performance of the assay. Separate positive control chemicals are used when testing without S9 and with S9 so that the enzymatic activity of the S9 can also be monitored.

Only a single dose level of the positive control is necessary. The concentration of positive control chemical selected should fall about one-quarter to one-half way up the dose–response curve for that substance. This will provide a concentration that will be most sensitive to changes in the mutagenic response. Positive control concentrations that produce very high numbers of revertants per plate should be avoided because at such high levels of response it is difficult to determine if the tester cell response or S9 activity is diminished for any reason.

20. As with all laboratory testing procedures, a number of factors can adversely affect the performance of the test. Although these bacterial tests are relatively easy to perform and do not require much technical knowledge, they are not foolproof. Because the test relies on the growth of specific strains of bacteria in culture and on Petri dishes, the researcher must be knowledgeable about the use of aseptic microbiological techniques. Many of the problems encountered when performing the assay are the result of mislabeling the tester strains, having contamination during the growth of the culture, or having contaminant colonies growing on the plates. For this reason, all instruments, including pipettes and test tubes, and solutions, must be sterilized and stored under sterile conditions.

Another aspect of the assay is that the chemicals used as positive controls are, by definition, mutagens, and many of them are also carcinogens. Similarly, many of the substances tested may also be mutagens. Thus, all handling of test chemicals, as well as the test itself, should be performed in a chemical or a biological safety cabinet. All test chemicals and positive control chemicals must be labeled and disposed of in such a manner that they do not contaminate the laboratory, laboratory personnel, trash handlers, or the environment. Laboratory personnel should protect themselves from chemical exposure by wearing gowns, eye glasses, and gloves. Mouth pipetting should never be practiced. Contact lens wearers should wear regular eye glasses because some chemicals might react with the lens material.

21. The presence of toxicity is indicated by a complete or a partial clearing of the background lawn, which is best seen by holding the plate up to a light, by inspection using a dissecting

microscope at 10–50× magnification, or by a decrease in the number of revertant colonies with increasing dose. If the chemical is not toxic, the highest tested dose should be 5 mg/plate unless otherwise justified. If the chemical is toxic or has poor solubility, the highest dose should show minimal toxicity, as determined by partial clearing of the background lawn or a decrease in revertant colony count, or be the dose just below the toxic dose. If precipitation is present, the highest dose chosen for the test should be one that shows some precipitation. The toxicity determination should be performed using the procedure and number of cells that will be used for the actual mutagenicity test because the effective concentration of the test substance is different when performing the plate and pre-incubation tests. If there is evidence that the substance would not be toxic at 5 mg/plate, it is not necessary to perform the toxicity test.

22. If the test chemical and/or solvent appears to be reacting with the plastic pipettes or Petri dishes, the test should be performed using only glass tubes, pipettes, and Petri dishes. If the test substance appears to be reacting with the reagents used, it may not be testable.

23. The pre-incubation assay is considered to be more sensitive than the plate incorporation assay because short-lived mutagenic metabolites may have a better chance of reacting with the tester strains in the small volume of pre-incubation mixture, and the concentration of S9 mix in the pre-incubation volume is higher than on the plate. As a result, the tester strains are exposed to the test chemical for 20–30 min in a small volume (0.5 mL) of either buffer or S9 mix at 37 °C prior to plating on minimal medium supplemented with a trace amount of histidine or tryptophan. Following this pre-incubation period, 2–3 mL of top agar (at 43–48 °C) is added to each tube, mixed well, and the contents poured onto the surface of the minimal agar plate.

24. Because TA98 and TA100 are the most efficient strains for detecting mutagens, a testing strategy that has been shown to be cost and labor effective is to test initially in TA98 and TA100, with and without S9. If the substance is mutagenic in one or both strains, it may not be necessary to test using additional Salmonella strains [62]. If the substance is not mutagenic in the two strains, the other strains should be used.

    Weak positive or equivocal results should be repeated using a modified protocol, e.g., a pre-incubation test instead of a plate test, or vice versa, 30 % S9 if the original test used 10 %, closer dose spacing around the response, and/or additional doses. If repeating a negative pre-incubation test, a plate test protocol should be used, and vice versa.

25. The colonies can be counted using an automated colony counter or by hand. It is important that all plates from the same experiment be scored at the same time using the same procedure because the colonies will continue growing and new mutant colonies will continue to appear with time. If the colony count appears to be greater than 2,000 or 3,000 per plate it is acceptable to record the count as >2,000 or >3,000, and not attempt to determine the exact number.

    Highly colored test chemicals, or the presence of precipitate on the plate, may interfere with the automated colony counts. In those cases, all the plates from that test should be counted by hand.

    If the colony growth is spread out across plates into large areas of confluent colony growth, the plates may have been too wet when used and water was present on the surface of agar while in the incubator. If the his revertant colonies are concentrated towards one end of plate it indicates that the plates were not level when bottom or top agar was poured, so that the cells were concentrated at the lower (deeper) end of the agar or that the top agar was not distributed evenly across the plate when allowed to set.

    A clear background lawn with or without microcolonies, and no standard mutant colonies, is an indication of high toxicity. The microcolonies are the result of the surviving his⁻ cells scavenging the histidine released from the dead cells and being able to undergo enough replication to form visible colonies. If lower test chemical concentrations also show such toxicity, it may be that the temperature of the top agar or the incubator is above 40 °C, which would kill a large proportion of the cells. Test samples that are highly acidic or basic can be cytotoxic. In this situation, the standard buffer may not be able to adjust the pH to near neutrality, and a stronger buffering solution may be needed. In rare cases, a new batch of agar has been found to be toxic to the cells.

    Contaminating bacterial colonies or mold may occasionally be present on plates. If present on only one or a few plates, it may have occurred during the performance of the test and is not a concern unless it reoccurs. Often this is caused by inadequate attention to sterile conditions when performing the test. If the contamination is present on all plates, one or more of the solutions, including the S9 mix components, were probably contaminated. In this case, check all solutions for sterility and replace any that are contaminated. Do not re-sterilize the solutions. Another possibility is that the minimal agar plates became contaminated when they were prepared or stored prior to use. If so, they will all have to be discarded and new plates prepared.

26. If the solvent control count is too high or low the strain may have been mislabeled. Alternative reasons are that if too high, it may have been contaminated by a mutagen, and if too low, there may be some condition that is killing or inhibiting cells (e.g., agar temperature is too high or the pH is too high or low).

27. If the positive control does not work, or responds too weakly, it is an indication that the wrong tester strain or the wrong positive control chemical may have been used, or that the strain has lost the deep rough mutation or the plasmid. If +S9-positive control does not work it may indicate that the S9 has lost activity; no or insufficient cofactors were added to the S9 mix.

28. In addition to the above, the report should include the laboratory's historical solvent control values for comparison with the current experiment's controls. A calculation of the mutagenic potency, typically the slope of the response, may be included, but only if it is in addition to the actual mean plate counts.

If the test data are for submission to regulatory authorities, the testing should be done, and the report written, in accordance with Good Laboratory Practice (GLP) guidelines [63–65].

Because there is a formal OECD Test Guideline for the test [51], data submitted to regulatory authorities of the OECD member countries, which include North America, most European countries, Japan, Korea, Australia, and New Zealand, should be obtained and reported following the Guideline requirements, and the Guideline should be referenced in the report. Data obtained from non-Guideline testing may be acceptable if it satisfies the minimum requirements of the Guideline, e.g., number of Salmonella strains and test doses.

References

1. Ames BN, Durston WE, Yamasaki E et al (1973) Carcinogens are mutagens: a simple test system combining liver homogenates for activation and bacteria for detection. Proc Natl Acad Sci USA 70:2281–2285
2. Ames BN, Lee FD, Durston WE (1973) An improved bacterial test system for the detection and classification of mutagens and carcinogens. Proc Natl Acad Sci USA 70:782–786
3. McCann J, Yamasaki E, Ames BN (1975) Detection of carcinogens in the Salmonella/microsome test: assay of 300 chemicals. Proc Natl Acad Sci USA 72:5135–5139
4. Sugimura T, Sato S, Nagao M et al (1976) Overlapping of carcinogens and mutagens. In: Magee PN, Takayama S, Sugimura T et al (eds) Fundamentals of cancer prevention. University Park Press, Baltimore, pp 191–215
5. Heddle JA, Bruce WR (1977) Comparison of tests for mutagenicity or carcinogenicity using assays for sperm abnormalities, formation of micronuclei, and mutations in Salmonella. In: Hiatt HH, Watson JD, Winsten JA (eds) Origins of human cancer, Book, C. Cold Spring Harbor, Cold Spring Harbor, pp 1549–1557
6. Purchase IFH, Longstaff E, Ashby J et al (1978) An evaluation of 6 short–term tests for detecting organic chemical carcinogens. Br J Cancer 37:873–959
7. Dunkel VC, Zeiger E, Brusick D et al (1985) Reproducibility of microbial mutagenicity assays: II. Testing of carcinogens and noncarcinogens in Salmonella typhimurium and Escherichia coli. Environ Mutagen 7(Suppl 5): 1–248

8. Tennant RW, Margolin BH, Shelby MD et al (1987) Prediction of chemical carcinogenicity in rodents from in vitro genetic toxicity assays. Science 236:933–941
9. Zeiger E, Haseman JK, Shelby MD et al (1990) Evaluation of four in vitro genetic toxicity tests for predicting rodent carcinogenicity: confirmation of earlier results with 41 additional chemicals. Environ Mol Mutagen 16(Suppl 18):1–14
10. Zeiger E (1998) Identification of rodent carcinogens and noncarcinogens using genetic toxicity tests: premises, promises and performance. Regul Toxicol Pharmacol 28:85–95
11. Kirkland D, Aardema M, Henderson L (2005) Evaluation of the ability of a battery of three in vitro genotoxicity tests to discriminate rodent carcinogens and non–carcinogens. I. Sensitivity, specificity and relative predictivity. Mutat Res 584:1–256
12. Margolin BH, Risko KJ, Shelby MD et al (1984) Sources of variability in Ames' *Salmonella typhimurium* tester strains: analysis of the International Collaborative Study on "Genetic Drift". Mutat Res 130:11–25
13. Piegorsch WW, Zeiger E (1991) Measuring intra–assay agreement for the Ames Salmonella assay. In: Hothorn L (ed) Statistical methods in toxicology, lecture notes in medical informatics, vol 43. Springer, Heidelberg, pp 35–41
14. Maron DM, Ames BN (1983) Revised methods for the Salmonella mutagenicity test. Mutat Res 113:173–215
15. Zeiger E, Mortelmans K (1999) The Salmonella (Ames) test for mutagenicity. In: Maines MD et al (eds) Current protocols in toxicology. Wiley, New York, pp 3.1.1–3.1.29
16. Mortelmans K, Zeiger E (2000) The Ames Salmonella/microsome mutagenicity assay. Mutat Res 455:29–60
17. Claxton LD, Allen J, Auletta A et al (1987) Guide for the *Salmonella typhimurium*/mammalian microsome tests for bacterial mutagenicity. Mutat Res 189:83–91
18. Claxton LD, Umbuzeiro G deA, DeMarini DM (2010) The Salmonella mutagenicity assay: the stethoscope of genetic toxicology for the 21st century. Environ Health Perspect 118:1515–1522
19. Dillon D, Combes R, Zeiger E (1998) The effectiveness of Salmonella strains TA100, TA102, and TA104 for detecting the mutagenicity of some aldehydes and peroxides. Mutagenesis 13:19–26
20. Gatehouse D, Haworth S, Cebula T et al (1994) Recommendations for the performance of bacterial mutation assays. Mutat Res 312:217–233
21. Kado NY, Langley D, Eisenstadt E (1983) A simple modification of the Salmonella liquid incubation assay: increased sensitivity for detecting mutagens in human urine. Mutat Res 122:25–32
22. Kamber M, Flückiger-Isler S, Engelhardt G et al (2009) Comparison of the Ames II and traditional Ames test responses with respect to mutagenicity, strain specificities, need for metabolism and correlation with rodent carcinogenicity. Mutagenesis 24:359–366
23. Ku WW, Bigger A, Brambilla G et al (2007) Strategy for genotoxicity testing – metabolic considerations. Mutat Res 627:59–77
24. Margolin BH, Kaplan N, Zeiger E (1981) Statistical analysis of the Ames Salmonella/microsome test. Proc Natl Acad Sci USA 78:3779–3783
25. Margolin BH, Kim BS, Smith MG et al (1994) Some comments on potency measures in mutagenicity research. Environ Health Perspect 102(Suppl 1):91–94
26. Maron D, Katzenellenbogen K, Ames BN (1981) Compatibility of organic solvents with the Salmonella/microsome test. Mutat Res 88:342–350
27. Pagano DA, Zeiger E (1985) The stability of mutagenic chemicals stored in solution. Environ Mutagen 7:293–302
28. Prival MJ, Zeiger E (1998) Chemicals mutagenic in *Salmonella typhimurium* strain TA1535 but not in TA100. Mutat Res 412:251–260
29. Waleh NS, Rapport SJ, Mortelmans KE (1982) Development of a toxicity test to be coupled to the Ames Salmonella assay and the method of construction of the required strains. Mutat Res 97:247–256
30. Yanofsky C (1971) Mutagenesis studies with Escherichia coli mutants with known amino acid (and base–pair) changes. In: Hollaender A (ed) Chemical mutagens: principles and methods for their detection, vol 1. Plenum, New York, pp 283–287
31. Zeiger E (2004) History and rational of genetic toxicity testing: an impersonal, and sometimes personal, view. Environ Mol Mutagen 44:363–371
32. Zeiger E (2007) Guest editorial. What is needed for an acceptable antimutagenicity manuscript? Mutat Res 626:1–3
33. Zeiger E (2010) Historical development of the genetic toxicity test battery in the United States. Environ Mol Mutagen 51:781–791
34. Ames BN, Hartman PE (1963) The histidine operon. Cold Spring Harbor Symp Quant Biol 28:349–356
35. Hartman PE, Hartman Z, Stahl RC et al (1971) Classification and mapping of spontaneous and

induced mutations in the histidine operon of Salmonella. Adv Genetics 16:1–34
36. Ong TM, Mukhtar H, Wolf CR et al (1980) Differential effects of cytochrome P450–inducers on promutagen activation capabilities and enzymatic activities of S–9 from rat liver. J Environ Pathol Toxicol 4:55–65
37. Beaune P, Lemestre-Cornet R, Kremers P et al (1985) The Salmonella/microsome mutagenicity test: comparison of human and rat livers as activating systems. Mutat Res 156:139–146
38. Hakura A, Suzuki S, Satoh T (1999) Advantage of the use of human liver S9 in the Ames test. Mutat Res 438:29–36
39. Hakura A, Shimada H, Nakajima M et al (2005) Salmonella/human S9 mutagenicity test: a collaborative study with 58 compounds. Mutagenesis 20:217–228
40. Mortelmans K, Riccio E (2000) The bacterial tryptophan reverse mutation assay with *Escherichia coli* WP2. Mutat Res 455:61–69
41. Ames BN (1971) The detection of chemical mutagens with enteric bacteria. In: Hollaender A (ed) Chemical mutagens: principles and methods for their detection, vol 1. Plenum, New York, pp 267–282
42. Levin DE, Yamasaki E, Ames BN (1982) A new Salmonella tester strain for the detection of frameshift mutagens: a run of cytosines as a mutational hot–spot. Mutat Res 94:315–330
43. Levin DE, Hollstein MC, Christman EA et al (1982) A new Salmonella tester strain (TA102) with A:T base pairs at the site of mutation detects oxidative mutagens. Proc Natl Acad Sci USA 79:7445–7449
44. Gee P, Maron DM, Ames BN (1994) Detection and classification of mutagens: a set of base-specific Salmonella tester strains. Proc Natl Acad Sci USA 91:11606–11610
45. Gee P, Sommers CH, Melick AS et al (1998) Comparison of responses of base-specific Salmonella tester strains with the traditional strains for identifying mutagens: the results of a validation study. Mutat Res 412:115–130
46. Hagiwara Y, Watanabe M, Oda Y et al (1993) Specificity and sensitivity of *Salmonella typhimurium* YG1041 and YG1042 strains possessing elevated levels of both nitroreductase and acetyltransferase activity. Mutat Res 291:171–180
47. Kushida H, Fujita K, Suzuki A et al (2000) Development of a Salmonella tester strain sensitive to promutagenic N–nitrosamines: expression of recombinant CYP2A6 and human NADPH–cytochrome P450 reductase in S. typhimurium YG7108. Mutat Res 471:135–143
48. Fujita K, Nakayama K, Yamazaki Y et al (2001) Construction of Salmonella typhimurium YG7108 strains, each co–expressing a form of human cytochrome P450 with NADPH–cytochrome P450 reductase. Environ Mol Mutagen 38:329–38
49. Matsui K, Yamada M, Imai M et al (2006) Specificity of replicative and SOS–inducible DNA polymerases in frameshift mutagenesis: mutability of Salmonella typhimurium strains overexpressing SOS–inducible DNA polymerases to 30 chemical mutagens. DNA Repair 5:465–478
50. Zeiger E, Pagano DA, Robertson IGC (1981) A rapid and simple scheme for confirmation of Salmonella tester strain phenotype. Environ Mutagen 3:205–209
51. Organization for Economic Co-operation and Development (OECD). (2012) OECD guideline for the testing of chemicals. No. 471. Bacterial reverse mutation test. http://www.oecd.org/document/55/0,3343,en_2649_34377_2349687_1_1_1_1,00.htmL
52. Araki A, Noguchi T, Kato F et al (1994) Improved method for mutagenicity testing of gaseous compounds by using a gas sampling bag. Mutat Res 307:335–344
53. Hughes TJ, Simmons DM, Monteith LG et al (1987) Vaporization technique to measure mutagenic activity of volatile organic chemicals in the Ames/*Salmonella* assay. Environ Mutagen 9:421–441
54. Zeiger E, Anderson B, Haworth S et al (1992) Salmonella mutagenicity tests: V. Results from the testing of 311 chemicals. Environ Mol Mutagen 19(Suppl 21):1–141
55. Gatehouse DG, Delow GF (1979) The development of a 'Microtitre' fluctuation test for the detection of indirect mutagens, and its use in the evaluation of mixed enzyme induction of the liver. Mutat Res 60:239–252
56. Green MHL, Muriel WJ, Bridges BA (1976) Use of a simplified fluctuation test to detect low levels of mutagens. Mutat Res 38:33–42
57. Prival MJ, Mitchell VD (1982) Analysis of a method for testing azo dyes for mutagenic activity in Salmonella typhimurium in the presence of flavin mononucleotide and hamster liver S9. Mutat Res 97:103–116
58. Reid TM, Morton KC, Wang CY et al (1984) Mutagenicity of azo dyes following metabolism by different reductive/oxidative systems. Environ Mutagen 6:705–717
59. Fetterman BA, Kim BS, Margolin BH et al (1997) Predicting rodent carcinogenicity from mutagenic potency measured in the Ames Salmonella assay. Environ Mol Mutagen 29:312–322

60. Zeiger E (2001) Mutagens that are not carcinogens: faulty theory or faulty tests? Mutat Res 492:29–38
61. Pagano DA, Zeiger E (1992) Conditions for detecting the mutagenicity of divalent metals in *Salmonella typhimurium*. Environ Mol Mutagen 19:139–146
62. Zeiger E, Risko KJ, Margolin BH (1985) Strategies to reduce the cost of mutagenicity screening using the Salmonella/microsome assay. Environ Mutagen 7:901–911
63. Food and Drug Administration (FDA). (2012) 21CFR58. Good laboratory practice for nonclinical laboratory studies. http://www.wrtolbert.com/Download%20Documents/21CFR58.pdf
64. Environmental Protection Agency (EPA). (2012) 40CFR792. Good laboratory practice standards. http://www.access.gpo.gov/nara/cfr/waisidx_99/40cfr792_99.htmL
65. Organization for Economic Co-operation and Development (OECD). (2012) OECD series on principles of good laboratory practice and compliance monitoring. http://www.oecd.org/document/63/0,3343,en_2649_34381_2346175_1_1_1_1,00.htmL

Chapter 2

In Vitro Mouse Lymphoma (L5178Y *Tk*⁺/⁻ -3.7.2C) Forward Mutation Assay

Melissa R. Schisler, Martha M. Moore, and B. Bhaskar Gollapudi

Abstract

The in vitro mouse lymphoma assay (MLA) is one of the most widely practiced assays in genetic toxicology. MLA detects forward mutations at the thymidine kinase (*Tk*) locus of the L5178Y (*Tk*⁺/⁻ -3.7.2C) cell line derived from a mouse thymic lymphoma. This assay is capable of detecting a wide range of genetic events including point mutations, deletions (intragenic) and multilocus, chromosomal rearrangements, mitotic recombination, and nondisjunction. There are two equally accepted versions of the assay, one using soft agar cloning and the second method using liquid media cloning in 96-microwell plates. There are two morphologically distinct types of mutant colonies recovered in the MLA: small- and large-colony mutants. The induction of small-colony mutants is associated with chemicals inducing gross chromosomal aberrations whereas the induction of large mutant colonies is generally associated with chemicals inducing point mutations. The source and karyotype of the cell line as well as the culture conditions are important variables that could influence the assay performance. The assay when performed according to the standards recommended by the International Workshops on Genotoxicity Testing is capable of providing valuable genotoxicity hazard information as part of the overall safety assessment process of various classes of test substances.

Key words Mouse lymphoma, Genetic toxicology, Thymidine kinase, Forward mutation assay, L5178Y (*Tk*⁺/⁻ -3.7.2C) cells

1 Introduction

Cell lines derived from humans and rodents have been extensively used for assessing the mutagenicity of test substances for more than four decades. The assay for forward mutations at the *Tk* locus of the mouse lymphoma L5178Y (*Tk*⁺/⁻ -3.7.2C) cells (referred to as the mouse lymphoma assay, MLA) was developed by Donald Clive and colleagues; [1–3] and is one of the most widely used tests in genetic toxicology, second only to the Ames bacterial reverse mutation assay.

Extensive research to understand the mechanistic basis and the types of mutations detected in the MLA reveals that the assay can detect a broad spectrum of genetic events including point mutations, deletions (intragenic), and multilocus chromosomal rearrangements, mitotic recombination, and nondisjunction [4–6]. Based on this ability, the MLA has been selected as the preferred in vitro mammalian gene mutation assay in a number of regulatory test batteries, including the US Environmental Protection Agency [7], the US Food and Drug Administration Center for Food Safety and Nutrition [8], and the International Conference for Harmonization [9] which provides international guidance for pharmaceuticals intended for human use.

1.1 Principle of the Test System

Thymidine is salvaged by cells from surrounding medium, utilizing the enzyme thymidine kinase (TK). This enzyme is essential for the initial phosphorylation of thymidine or thymidine analogs to thymidine monophosphate (TMP). The intracellular concentration of TMP regulates DNA synthesis and cell replication. In normal cells, TMP is synthesized de novo and hence the *Tk* gene is not essential for DNA replication. However, cells lacking a functional Tk gene ($Tk^{-/-}$ cells) are not capable of utilizing exogenous thymidine or its analogs. In the presence of toxic analogs of thymidine (and under conditions where de novo synthesis of TMP is not blocked), normal (i.e., TK competent) cells die because of the incorporation of the toxic nucleotide into DNA whereas cells lacking TK activity do not incorporate the toxic nucleotide into their DNA and therefore can multiply.

The mouse lymphoma forward gene mutation assay utilizes a strain ($Tk^{+/-}$ -3.7.2C clonal line) of L5178Y mouse lymphoma cells heterozygous at the *Tk* locus. Induced heritable loss of *TK* activity occurs because of a mutational event (from $Tk^{+/-}$ to $Tk^{-/-}$) that results from DNA damage by physical or chemical agents. $Tk^{-/-}$ mutants can be detected by their inherent resistance to toxic thymidine analogs such as trifluorothymidine (TFT) which is incorporated into the DNA of TK-competent ($Tk^{+/-}$) cells resulting in cytotoxicity. Forward mutations at the single functional *Tk* gene ($Tk^{+/-} \rightarrow Tk^{-/-}$) result in the loss of the salvage TK enzyme so that TFT is not incorporated into the cellular DNA, thereby enabling these cells to grow in the presence of TFT. This deficiency in the TK enzyme can result from a number of genetic events affecting the *Tk* locus including gene mutations (point mutations, frameshift mutations, small deletions, etc.) and chromosomal events (large deletions, chromosome rearrangements, and mitotic recombination). Chromosomal events lead to loss of heterozygosity (LOH), which is a commonly observed genetic event of tumor-suppressor genes in human tumorigenesis. Although there is some evidence that the MLA can detect the loss of the functional *Tk* locus as a result of whole chromosome loss, this assay is not generally recommended for the detection of aneugenic potential of test substances.

In the MLA, two phenotypically distinct classes of *Tk* mutant clones are generated: the large-colony (LC) mutants that grow at the same rate as wild-type cells, and the small-colony (SC) mutants that grow with prolonged cell cycle times. The biology underlying large- and small-colony mutants has been investigated in detail [4–6, 10, 11]. SC mutants harbor genetic damage involving putative growth-regulating gene(s) near the *Tk* locus that leads to prolonged cell doubling times and thus forming small colonies. Chemicals that induce gross structural chromosomal changes have been associated with the induction of SC mutants. Mutant cells with relatively less damage grow at rates similar to the wild-type cells and grow to become LC mutants. In general, the induction of primarily LC mutants is associated with substances that act primarily as point mutagens. Thus, it is essential to enumerate both SC and LC mutants in order to recover all of the mutants and to provide some insight into the type(s) of damage (mutagenesis vs. clastogenesis) induced by the test chemical.

There are two equally acceptable and widely practiced methodologies for the conduct of the MLA. The first method uses soft agar cloning and the second method uses liquid media cloning in 96-microwell plates. In this chapter, we describe in detail the conduct of the assay using both methodologies as practiced in one of the author's (MR Schisler) laboratory and explain some of the variables that should be taken into consideration in the interpretation of the results.

2 Materials

1. *Components of Growth Medium (F_{10P})*: 0.22 mg/mL sodium pyruvate, 2 mL of penicillin–streptomycin, 10 mL of Pluronic F68, 10 mL of L-glutamine, 877 mL of Fisher's medium, 100 mL of horse serum (horse serum must be heat inactivated at 56 °C for 30 min prior to use; *see* **Note 1**).

2. *Components of Treatment Medium (F_{0P})*: 0.22 mg/mL sodium pyruvate, 2 mL of penicillin–streptomycin, 10 mL of Pluronic F68, 10 mL of L-glutamine, and 977 mL of Fisher's medium.

3. *Components of Cloning Medium (R_{10P})*: 0.22 mg/mL sodium pyruvate, 2 mL of penicillin–streptomycin, 10 mL of Pluronic F68, 10 mL of L-glutamine, 877 mL of RPMI 1640 medium, 0.28 % Noble agar, 1 μg/mL TFT, 100 mL of horse serum (horse serum must be heat inactivated at 56 °C for 30 min prior to use; *see* **Note 1**).

4. *Cleansing Medium*
   (a) Thymidine–hypoxanthine–glycine (THG) stock: 60 mg Thymidine, 100 mg hypoxanthine, 150 mg glycine, 200 mL of treatment medium F_{0P}. This stock is added following the mutant purge to assist the culture with recovery of normal growth.

(b) Methotrexate (MTX) stock: 40 mg of methotrexate, 38.9 mL of Dulbecco's phosphate-buffered saline (DPBS), 0.7 mL 1N NaOH (used to dissolve MTX), 0.4 mL 1N HCl (used to adjust the solution to a neutral pH).

(c) Thymidine–hypoxanthine–methotrexate–glycine (THMG) stock: 99 mL of THG medium, 1 mL of MTX. This solution is used to purge a culture of existing $Tk^{-/-}$ mutants.

5. *0.28 % Nobel Agar:* Dissolve 3.75 g of Noble agar (USB Corporation/Affymetrix, a high-quality source of agar is critical for optimal growth of small-colony mutants) in 121 mL of distilled water and autoclave.

6. *S9 mix*: 4.07 mg/mL magnesium chloride, 2.82 mg/mL glucose-6-phosphate, 6.12 mg/mL nicotinamide adenine dinucleotide phosphate (NADP), 2.94 mg/mL calcium chloride, 4.47 mg/mL potassium chloride. Weigh, and dilute up the volume as desired with distilled water. This mix can be stored up to 1 year in -150°C or below.

7. *Phosphate Buffer (pH 8)*: Separately, prepare a 17.8 mg/mL sodium phosphate solution (dibasic) and a 17.3 mg/mL sodium phosphate soltuion (monobasic) in distilled water. Adjust the pH of sodium phosphate dibasic to a pH of 8 using sodium phosphate monobasic. These solution can be stored at 4–8 °C for 1 year.

8. Rat liver S9 homogenate.

9. 20-Methylcholanthrene (20-MCA).

10. Methyl Methanesulfonate (MMS).

11. Disposable 96-well plates.

12. Disposable Petri dishes (100 mm).

13. Disposable 125 mL Erlenmeyer flasks.

14. Disposable 50 mL centrifuge tubes.

15. Colony counting system (e.g., Sorcerer).

16. 37 °C incubator with CO_2.

3 Methods

3.1 Cells

The cell line $Tk^{+/-}$-3.7.2C clonal line of L5178Y mouse lymphoma cells (referred to as L5178Y cells) is the only acceptable cell line for this assay [12]. *See* **Notes 2** and **3** for cell line details and culture conditions.

3.2 Cell Cleansing

The L5178Y cells should be treated with THMG to cleanse the culture of spontaneous mutants prior to use in a mutation assay (*see* **Note 4**). The following outlines specific details of the cleansing process.

1. To 3×10^7 total cells in 100 mL of growth medium (F_{10P}), add 2.0 mL of THMG stock.
2. Gas cultures with 5 % CO_2 and incubate in an orbital shaker incubator at 37 °C for approximately 24 h.
3. After the 24-h incubation, determine the cell density (*see* **Note 5**), aliquot 3×10^7 cells, centrifuge at approximately $350 \times g$, and discard the supernatant containing THMG.
4. Resuspend the cell pellet in approximately 100 mL of fresh F_{10P} (final density of 3×10^5 cells/mL) and add 2.0 mL of THG stock solution.
5. Re-gas cultures with 5 % CO_2 and return cells to the orbital shaker incubator at 37 °C for approximately 24 h.
6. After the 24-h incubation determine the cell density (*see* **Note 5**), aliquot 3×10^7 cells, centrifuge at approximately $350 \times g$, and discard the supernatant containing the THG.
7. Resuspend the cell pellet in approximately 100 mL of fresh F_{10P} (final density of 3×10^5 cells/mL).
8. Cells are now ready to be frozen for long-term storage.

3.3 Cell Freezing

The L5178Y cells can be frozen for storage at -150°C or below using the freezing medium (growth medium (F_{10P}) containing 5 % DMSO). Cells should be stored in 2.0 mL cyrovials containing 2.25×10^7 total cells in 1.5 mL freezing medium.

3.4 Cell Thawing

1. Dispense 75 mL of prewarmed (37 °C) F_{10P} into a 125 mL disposable Erlenmeyer flask.
2. Remove cell vial from long-term storage storage and immediately thaw by placing in a 37 °C water bath.
3. Immediately after thawing remove vial of cells from the water bath, gently pipette cells, and slowly allow the cells to fall through the medium to the bottom of the flask containing F_{10P}.
4. Loosely tighten lid on flask and place in CO_2 incubator (approx. 5 % CO_2, 95 % air) maintained at approximately 37 °C.
5. Approximately 24–48 h later remove the cell flask from the incubator, swirl gently, tighten lid, and place in an orbital shaker incubator maintained at 37 °C.
6. Approximately 24 h later determine cell count and split the cells using the following guidance:

Type of cell split	Number of cells (cells/mL)
Routine split (1 day)	300,000
Two day split (2 days)	100,000
Weekend split (3 days)	12,000

7. Cells are ready to be used in a mutation assay after being thawed and maintained for 5–8 days, assuming that they are growing at a normal rate. Cells not growing at a normal rate should be discarded.

3.5 Treatment and Rinsing Procedures

The treatment procedures can either be of short duration without (Subheading 3.5.1) or with (Subheading 3.5.2) metabolic activation or continuous without metabolic activation (Subheading 3.5.3) as described in detail below (*see* **Note 6**).

3.5.1 Short Treatment in the Absence of Metabolic Activation and Rinsing of Cells

In short, prepare a stock cell culture of recently thawed L5178Y cells at a density of 1.0×10^6 cells/mL in 50:50 fresh F_{0P} medium: conditioned medium that the cells have been growing in (to minimize trauma to cells). This medium will now contain only 5 % horse serum (to be reduced later to 3 % in order to minimize serum interactions with test compounds and/or metabolites). The total number of cells used for each treatment will be 6×10^6 cells. Detailed steps are given below.

1. Determine the volume of cells needed by multiplying the number of cultures needed by 6.0 mL.
2. Determine the total cells needed for the assay by multiplying the total volume from **step 1** by the desired density of cells (i.e., 1×10^6 cells/mL).
3. Determine cell count from stock cultures.
4. Divide total cells needed for the assay (**step 2**) by the cell count of the stock cells (**step 3**) to determine the volume of stock cells needed for the assay.
5. Centrifuge the volume of stock needed for 5 min at approximately $350 \times g$.
6. Carefully decant the supernatant into a sterile bottle and save for later use (conditioned medium).
7. Transfer the cell pellet into a clean sterile culture flask along with sterile conditioned medium to equal half of the desired final volume.
8. Add an equal volume of fresh F_{0P} to the cells in the culture flask to provide fresh nutrients and to decrease the horse serum concentration from 10 to 5 %.
9. Gas the culture with 5 % CO_2 and shake in an orbital shaker incubator at 37 °C for at least 20 min to adapt the cells to the altered medium (treatment stock cell culture).
10. After incubation add 6.0 mL of the treatment stock cell culture (1×10^6 cells/mL) to each of pre-labeled, sterile 50 mL treatment centrifuge tube for a total of 6×10^6 cells per tube.
11. Add 4.0 mL of F_{0P} to each tube.

12. At this point the cell density is 0.6×10^6 cells/mL in medium containing 3 % horse serum.

13. Treat with an appropriate amount of test material concentration (1–10 %; *see* **Notes 7** and **8**), solvent control (*see* **Note 9**), or positive control (*see* **Note 10**). Carefully add the appropriate amount of test material stock solution to each treatment or positive control culture (for a final maximum in the test culture of 1 % DMSO or ethanol and 10 % for distilled water or cell culture medium). Each test culture should receive the same amount of solvent. Solvent controls should receive an equal (to the test cultures) amount of solvent. If used, negative controls receive nothing.

14. Re-gas each culture with 5 % CO_2 and vortex at low speed to rinse test material from the sides of the tube into the cell culture.

15. Incubate cultures in a roller drum (25–35 orbits) at 37 °C for approximately 4 h or other predetermined treatment period.

16. After the treatment period, centrifuge cell cultures for 5 min at approximately $350 \times g$.

17. Discard the supernatant, resuspend cell pellet in 10 mL of fresh F_{10P}, and gently vortex.

18. Centrifuge again at approximately $350 \times g$ for 5 min and decant the supernatant.

19. Resuspend pellet of cells in 20 mL of fresh F_{10P} with gentle vortexing and return to roller drum (25–35 orbits) at 37 °C for approximately 24 h.

3.5.2 Short Treatment in the Presence of Metabolic Activation and Rinsing of Cells

In brief, the preparation of cells for the short treatment in the presence of metabolic activation (*see* **Note 11**) is exactly the same as described in Subheading 3.5.1 above, except that instead of adding 4 mL of F_{0P} in **step 11** of Subheading 3.5.1, add 4.0 mL of the S9 components to each tube. The example detailed below will result in a final concentration of 2 % S9 in the treatment medium.

1. Determine the volume of cells needed by multiplying the number of cultures needed by 6.0 mL.

2. Determine total cells needed for the assay by multiplying the total volume from **step 1** by the desired density of cells (i.e., 1×10^6 cells/mL).

3. Determine cell count from stock cultures.

4. Divide total cells needed for the assay (**step 2**) by the cell count of the stock cells (**step 3**) to determine the volume of stock cells needed for the assay.

5. Centrifuge the volume of stock needed for 5 min at approximately $350 \times g$.

6. Carefully decant the supernatant into a sterile bottle and save for later use (conditioned medium).
7. Transfer the cell pellet into a clean sterile culture flask along with sterile conditioned medium to equal half of the desired final volume.
8. Add an equal volume of fresh F_{0P} to the cells in the culture flask to provide fresh nutrients and to decrease the horse serum concentration from 10 to 5 %.
9. Gas the culture with 5 % CO_2 and shake in orbital shaker at 37 °C for at least 20 min to adapt the cells to the altered medium (treatment stock cell culture).
10. After incubation, add 6.0 mL of the treatment stock cell culture (1×10^6 cells/mL) to each of pre-labeled, sterile 50 mL treatment centrifuge tubes for a total of 6×10^6 cells per tube.
11. Add 4.0 mL of S9 components to each tube in the activation series. For every S9 culture the following components (total volume of 4 mL) will be added to the 6 mL of stock cells as follows:

    S9 mix: 1 mL

    Phosphate buffer: 0.8 mL

    S9: 0.2 mL

    F_{0P}: 2 mL
12. At this point the cell density is 0.6×10^6 cells/mL in medium containing 3 % horse serum.
13. Treat with an appropriate amount of test material concentration (1–10 %; see **Notes 7** and **8**), solvent control (see **Note 9**), or positive control (see **Note 10**). Carefully add the appropriate amount of test material stock solution to each treatment or positive control culture (for a final maximum in the test culture of 1 % DMSO or ethanol and 10 % for distilled water or cell culture medium). Each test culture should receive the same amount of solvent. Solvent controls should receive an equal (to the test cultures) amount of solvent. If used, negative controls receive nothing.
14. Re-gas each culture with 5 % CO_2 and vortex at low speed to rinse test material from the sides of the tube into the cell culture.
15. Incubate cultures in a roller drum (25–35 orbits) at 37 °C for approximately 4 h or other predetermined treatment period.
16. After the treatment period, centrifuge cultures for 5 min at approximately $350 \times g$.
17. Discard the supernatant, resuspend cell pellet in 10 mL of fresh F_{10P}, and gently vortex.
18. Centrifuge again at approximately $350 \times g$ for 5 min.
19. Discard the supernatant, resuspend cell pellet in 10 mL fresh F_{10P}, and gently vortex.

20. Centrifuge at approximately $350 \times g$ for 5 min and discard the supernatant.

21. Resuspend the pellet of cells in 20 mL fresh F_{10P} with gentle vortexing and return to roller drum (25–35 orbits) at 37 °C for approximately 24 h.

3.5.3 Continuous (24 h) Treatment in the Absence of Metabolic Activation and Rinsing of Cells

In short prepare a stock cell culture of recently thawed L5178Y cells at a density of 5×10^5 cells/mL in part fresh F_{10P} medium and part conditioned medium that the cells have been growing in (to minimize trauma to cells). The total number of cells used for each treatment will be 6×10^6 cells.

1. Multiply the number of cultures needed by 12.0 mL of stock cells to determine the total volume of cells needed for the assay.

2. Determine total cells needed for the assay by multiplying the total volume from **step 1** by the desired density of cells (i.e., 5×10^5 cells/mL).

3. Determine cell count from stock cultures.

4. Divide total cells needed for the assay (**step 2**) by the cell count of the stock cells (**step 3**) to determine the volume of stock cells needed for the assay.

5. Centrifuge the volume of stock needed for 5 min at approximately $350 \times g$.

6. Carefully decant the supernatant into a sterile bottle and save for later use (conditioned medium).

7. Transfer the cell pellet into a clean sterile culture flask along with sterile conditioned medium to equal half of the desired final volume.

8. Add an equal volume of fresh F_{10P} to the cells in the culture flask to provide fresh nutrients.

9. Gas the culture with 5 % CO_2 and shake in orbital shaker at 37 °C for at least 20 min to adapt the cells to the altered medium (stock cell culture).

10. Add 12 mL of prepared stock cell culture (5×10^5 cells/mL) to each of the pre-labeled, sterile 50 mL treatment centrifuge tubes.

11. Add 8.0 mL of fresh F_{10P} to each centrifuge tube.

12. At this point the cell density is 0.3×10^6 cells/mL in medium containing 10 % horse serum.

13. Treat with an appropriate amount of test material concentration (1–10 %, *see* **Note 7** and **8**), solvent control (*see* **Note 9**), or positive control (*see* **Note 10**). Carefully add the appropriate amount of test material stock solution to each treatment or positive control culture (for a final maximum in the test culture of 1 % DMSO or ethanol and 10 % for distilled water or cell culture medium). Each test culture should receive the same amount of solvent. Solvent controls should receive an equal

(to the test cultures) amount of solvent. If used, negative controls receive nothing.

14. Re-gas each culture with 5 % CO_2 and vortex at low speed to rinse test material from the sides of the tube.
15. Incubate cultures in a roller drum (25–35 orbits) at 37 °C for approximately 24 h.
16. After the treatment period, centrifuge cultures for 5 min at approximately $350 \times g$.
17. Discard the supernatant, resuspend cell pellet in 10 mL fresh F_{10P}, and gently vortex.
18. Centrifuge at $350 \times g$ for 5 min., discard the supernatant, and resuspend cell pellet in 5 mL fresh F^{10P} with gentle vortexing.
19. Determine the cell count, adjust the density to 3×10^5 cells/mL with F^{10P} for a total of 20 mL, re-gas with 5 % CO2, and return to roller drum (25–35 orbits) at 37 °C for approximately 24 h.
20 After the 24-h incubation, proceed with the expression period (subheading 3.6.1; On Day 1 following treatment).

3.6 Expression Period

The expression period for the mouse lymphoma mutation assay has been established to be 2 days after treatment. On each day of the expression period, cultures are maintained by counting the cells and splitting them to a density of 3×10^5 cells/mL as described below.

3.6.1 On Day 1 After Treatment

1. Determine the cell count.
2. Dilute the cells to a density of 3×10^5 cells/mL by placing the appropriate amount of the cell culture into a new tube and adding fresh F_{10P} to give a final volume of 20 mL.
3. If the cells do not reach 400,000 cells/mL then cells should not be diluted; instead they are returned to the incubator until the next day. The cells may not reach this number due to toxicity.
4. Re-gas each culture with 5 % CO_2 and vortex at low speed to mix.
5. Incubate tubes in a roller drum (25–35 orbits) at 37 °C for approximately 24 h.
6. The following can be determined on day 1 of the expression period.

Suspension growth over first day = Day 1 SG

$$= \frac{\text{Number of cells per mL on day 1}}{3 \times 10^5 \text{ cells / mL}}$$

3.6.2 On Day 2 After Treatment

1. Determine the cell count.
2. Dilute the cells to a density of 3×10^5 cells/mL by placing the appropriate amount of the cell culture into a new tube and adding fresh F_{10P} to give a final volume of 20 mL.

Mouse Lymphoma Assay 37

3. Re-gas each culture with 5 % CO_2 and vortex at low speed to mix.

4. Incubate tubes in a roller drum (25–35 orbits) at 37 °C for approximately 1 h to allow cells to acclimate prior to mutant selection.

5. The following can be determined on day 2 of the expression period.

Suspension growth over the second day = Day 2 SG

$$= \frac{\text{Number of cells per mL on day 2}}{3 \times 10^5 \text{ cells / mL (or the previous day cell number if no adjustment was made on the previous day)}}$$

3.6.3 Relative Suspension Growth

Relative cumulative suspension growth is calculated over the 2-day expression period.

Relative suspension growth = RSG (%)

$$= \frac{\text{Cumulative suspension growth of culture during first 2 days} \times 100}{\text{Average cumulative suspension growth of solvent control during first 2 days}}$$

3.7 Mutant Selection and Viability Plates

All aspects up to this point in the assay are the same for both the soft agar cloning and the 96-well microwell cloning. Either cloning method is acceptable to carry out mutant selection and should be chosen based on the lab's capability or preference.

3.7.1 Soft Agar Cloning

After the cells have acclimated from the day-2 dilution, begin the soft agar cloning process by following the steps below.

1. Melt the prepared 0.28 % Noble agar in a microwave oven prior to use.

2. To each pre-labeled 125 mL Erlenmeyer flask containing 81 mL of cloning medium (R_{10P}), 9 mL of melted Noble agar, add 10 mL of vortexed cell suspension (total cells 3×10^6 cells) prepared in **step 4** of Subheading 3.6.2 above.

3. Place flasks in an orbital shaker incubator at 37 °C for 15–30 min to allow cells, agar, and medium to equilibrate.

4. After equilibration, transfer 1.0 mL of the above cell suspension (from **step 3**) and add to pre-labeled 50 mL tubes containing 9 mL of R_{10P}. Mix well (the total cells will be 3×10^4).

5. To the Erlenmeyer flasks from **step 3**, add 1 mL of 100 µg/mL stock TFT solution and return flasks to the orbital shaker

incubator until ready for plating. These flasks will be used for mutant selection (TFT).

6. Transfer 1.0 mL of cell suspension from **step 4** to a second 50 mL tube containing 4 mL of R_{10P}. This results in approximately 600 cells/mL.

7. Transfer 1.0 mL from **step 6** to a second pre-labeled 125 mL Erlenmeyer flask containing 9 mL of Noble agar and 90 mL of R_{10P}. Allow the cells, agar, and medium to equilibrate for 15–30 min in the orbital shaker incubator. These flasks will be used to plate cells for viable counts (VC). This results in approximately 600 cells per flask.

8. Remove the VC- and TFT-labeled flasks from the orbital shaker incubator one at a time and pour the mixture from each flask into each of the three disposable tissue culture Petri dishes labeled VC and three disposable tissue culture Petri dishes labeled TFT, respectively (approximately 33 mL per dish).

9. Place the Petri dishes containing agar, cells, and medium at approximately 4–8 °C for 15–20 min prior to incubation. This allows the agar medium to solidify. Care should be taken *not* to exceed this time at the low temperature as this will reduce the viability of the cells.

10. After cooling, incubate cultures at 37 °C in the presence of 5 % CO_2 for 10–14 days.

11. At the end of the incubation period, plates can be counted using an automated counting system (e.g., Perceptive Instruments; Loats Associates) by capturing digital images (Fig. 1). Colony sizing involves creating a histogram that displays the distribution of colony sizes. Generally the histogram is a bimodal distribution providing a distinction between the small and large colonies (Fig. 2). Two approaches can be taken. Either the laboratory should view each histogram and make a decision as to the cutoff between small and large colonies or the laboratory can set a standard cutoff based on the results of several experiments using chemicals that generate a bimodal colony size distribution.

12. The following will be determined from the data obtained for the VC and TFT plates.

    (a) *Cloning efficiency (CE)*
    $$= \frac{\text{Average \# of colonies in the TFT-free plates}}{\text{Average \# of cells seeded per plate}}$$

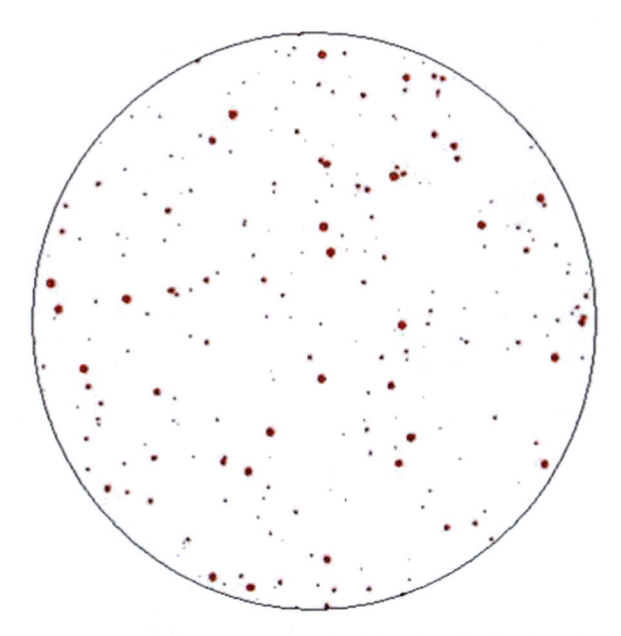

Fig. 1 Digital image of a soft agar plate with mutant colonies. The size distribution of these colonies is analyzed by a software (*see* Fig. 2)

(b) *% Relative cloning efficiency (RCE)*

$$= \frac{\text{Cloning efficiency of the treated at the time of mutant selection}}{\text{Cloning efficiency of the solvent control at the time of mutant selection}}$$

(c) *% Relative total growth (RTG)*

$$= \text{Day 2 RSG} \times \text{RCE} \times 100$$

(d) The *mutant frequency* per 10^6 clonable cells will be calculated as below:

$$= \frac{\text{Total no. of mutant colonies} \times 10^6}{\text{Total no. of cells plated on TFT plates} \times \text{CE}}$$

(e) *Mutant index*

$$= \frac{\text{Mutant frequency of the treated}}{\text{Average mutant frequency of the solvent control}}$$

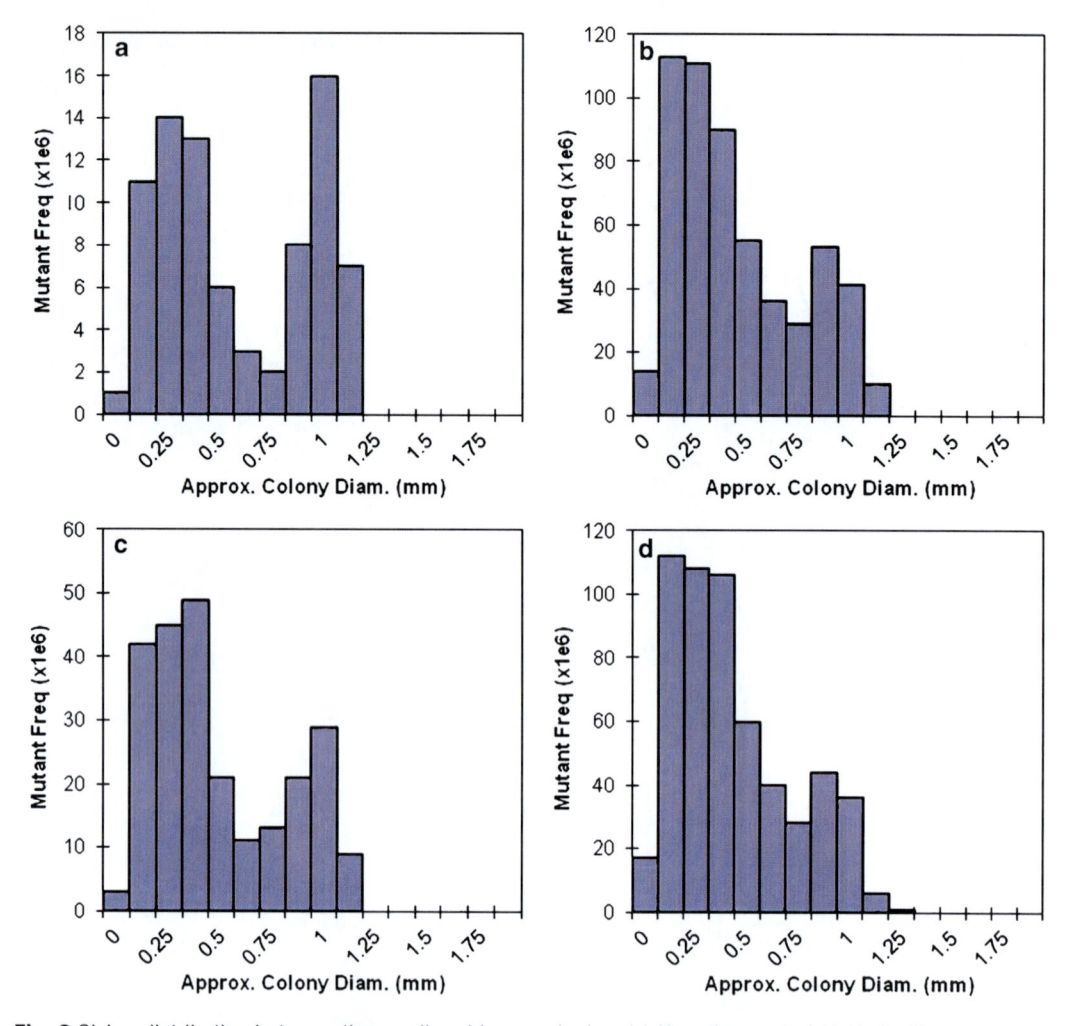

Fig. 2 Sizing distribution between the small and large colonies. (**a**) Negative control (1 % distilled water), MF 88×10^{-6}. (**b**) Positive control (10 µg/mL methyl methanesulfonate) MF 559×10^{-6}. (**c**) Test chemical (BIOBAN CS-1246, 5 µg/mL) MF 249×10^{-6}. (**d**) Test chemical (BIOBAN CS-1246, 6 µg/mL) MF 568×10^{-6} [29]

3.7.2 96-Microwell Cloning

After the cells have acclimated from the day-2 dilution, begin the microwell cloning process by following the steps below.

1. Vortex the acclimated cells from **step 4** of Subheading 3.6.2 and add 3.3 mL (total cells of approximately 1×10^6) to a 125 mL Erlenmeyer flask containing 96.7 mL of cloning medium (R_{10p}).

2. Place flasks in an orbital shaker incubator at 37 °C for 5–30 min to allow cells and medium to equilibrate.

3. After equilibration, transfer 0.5 mL of the above cell suspension (from **step 2**) to pre-labeled 50 mL tubes containing 9.5 mL of R_{10P} and mix well (approximately 500 cells).

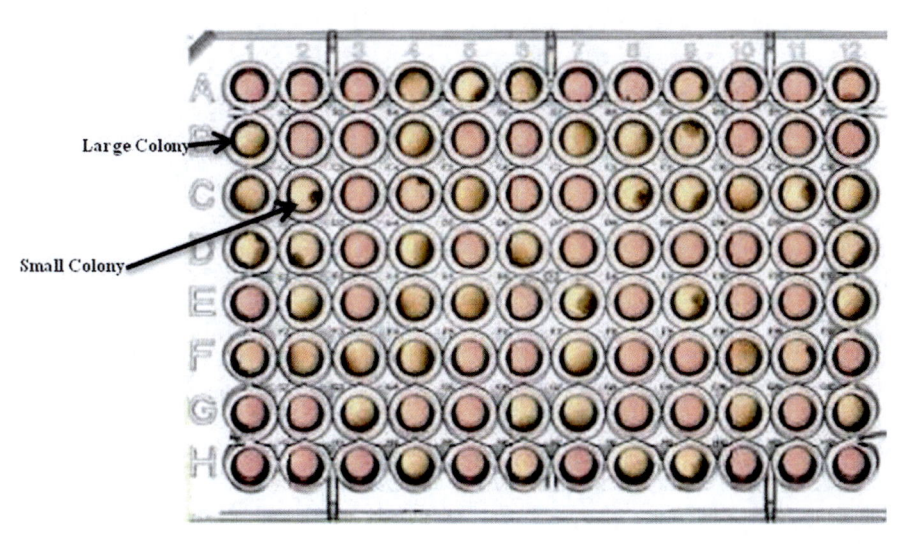

Fig. 3 A 96-well plate used for mutant selection. Note the change in the color of the medium in wells with colonies

4. To a second pre-labeled Erlenmeyer flask add 0.8 mL of cells from **step 3** to 49.2 mL of R_{10p} (final 8 cells/mL). These flasks will be used for viability counts (VC).

5. Place flasks in orbital shaker at 37 °C for 5–30 min to allow cells and medium to equilibrate.

6. To the original Erlenmeyer flask created in **step 2** add 3 mL of 100 µg/mL stock TFT solution. These flasks will be used to plate for mutant selection (TFT).

7. Plate two 96-well plates for each culture (labeled VC) with 0.2 mL per well and incubate (1.6 cells/well).

8. Plate four 96-well plates for each culture (labeled TFT) with 0.2 mL per well and incubate (2,000 cells/well).

9. Incubate plates for 10–12 days at 37 °C in the presence of 5 % CO_2.

10. After incubation, TFT plates are counted manually using the following guidance to determine small and large colonies (Fig. 3).

    (a) Small colonies are defined as those covering less than 25 % of the well's diameter.

    (b) Large colonies are defined as those covering more than 25 % of the well's diameter.

11. After incubation, VC plates are counted manually for the number of wells containing a colony.

12. The following endpoints will be determined from the data obtained for the VC and TFT plates.

    (a) *Cloning efficiency*

    $$= -\ln(\text{empty wells (VC)} / 192) / 1.6$$

    (b) *% RCE*

    $$= \frac{\text{Cloning efficiency of the treated at the time of mutant selection}}{\text{Cloning efficiency of the solvent control at the time of mutant selection}}$$

    (c) *% RTG*

    $$= \text{Day 2 RSG} \times \text{RCE} \times 100$$

    (d) The *mutant frequency* per 10^6 clonable cells is calculated as given below:

    $$\frac{\text{Average \# of mutants per TFT plate}}{\text{Cloning efficiency}}$$

    $$\frac{-\ln(\text{empty (TFT)} / 384) / 2000}{-\ln(\text{empty (VC)} / 192) / 1.6}$$

    (e) *Mutant index* is calculated as given below

    $$\frac{\text{Mutant frequency of the treated}}{\text{Average mutant frequency of the solvent control}}$$

3.8 Acceptance Criteria for a Valid Assay

Acceptance criteria for a valid assay were determined by the Mouse Lymphoma Assay Expert Workgroup of the International Workshop on Genotoxicity Testing (IWGT) [13, 14].

1. *Soft Agar Cloning*
   For an assay to be considered acceptable, the mutant frequency of the solvent controls should be within 35–140×10^{-6}. The solvent controls must also have an average absolute cloning efficiency between 65 and 120 % and a cumulative suspension growth over the 2-day expression period of between 8 and 32 for the short treatment and 32 and 180 for the continuous treatment.

2. *Microwell Cloning*
   For the microwell version of the assay the mutant frequency of the solvent control should fall between 50 and 170×10^{-6}. The solvent controls must also have an average absolute cloning efficiency between 65 and 120 % and a cumulative suspension growth over the 2-day expression period of between 8 and 32

for the short treatment and 32 and 180 for the continuous treatment.

3. *Positive Control*

There are two approaches to demonstrating an acceptable positive control: (1) An induced (i.e., treated minus the background) total mutant frequency of at least 300×10^{-6} with 40 % of the induced mutant frequency (IMF) reflected as small-colony mutant frequency, or (2) an induced small colony mutant frequency of at least 150×10^{-6}. In either situation, the upper limit of cytotoxicity should be >10 % RTG. In addition, the mutant frequency in the positive control should be within the acceptable range of the performing laboratory's historical values. It is sufficient to use a single dose in the absence of S9 and in the presence of S9 to demonstrate that the acceptance criteria for the positive control have been satisfied.

3.9 Interpretation of the Results

The criteria for positive and negative responses have evolved over time based on a better understanding of the assay and the more widespread use of the assay (*see* **Note 12**). Mutant frequencies are evaluated based upon the criteria developed by the IWGT MLA Expert Workgroup. These criteria were developed to provide more weight on biological significance rather than statistical significance [13].

1. The test chemical will be considered positive when the conditions listed below are met:

    (a) The average mutant frequency at one or more of the test concentrations resulting in $\geq$10 % RTG meets the global evaluation factor (GEF). The GEF is 90×10^{-6} or 126×10^{-6} mutants above the average mutant frequency of the concurrent solvent controls for the agar cloning and microwell versions, respectively (assuming that the concurrent solvent control mutant frequency is in the range of $35-140 \times 10^{-6}$ for the soft agar version and $50-170 \times 10^{-6}$ for the microwell version).

    (b) Once the above criterion is met, the mutant frequencies of the treated cultures may be evaluated by a trend test to determine whether there is a concentration-related response.

2. The test material is considered negative in this assay if there is no evidence of an increase in the mutant frequency that meets the GEF at RTG values $\geq$10 %.

3. The test material is considered equivocal in this assay if the following conditions are met:

    (a) There is a significant increase in mutant frequency only at RTG values >10 % and <20 %.

(b) There is no evidence of increase in mutant frequency at RTG values ≥20 %.

Equivocal responses can also occur when a positive response in one experiment is not corroborated by another experiment(s) and both experiments have approximately the same concentration range. Basically, one experiment is positive and the other one is not. This category also includes situations where the response barely meets the criteria for a positive response in one experiment and misses it barely in the repeat experiment. Generally, if the experiments have been conducted to optimally cover the appropriate dose–response range, repeating the experiment will not provide a clear outcome and in these situations, the definitive call should be "equivocal."

4 Notes

1. *Horse Serum*: It is critical that the horse serum be properly heat inactivated when using RPMI 1640 medium. Inadequate heat inactivation of horse serum can result in the recovery of clones that are not actually *Tk* mutants. Heat inactivation should occur with the horse serum at 56 °C for 30 min. It is essential that the horse serum is brought up to a temperature of 56 °C prior to starting the incubation.

2. *Cell Line*: The cell line $Tk^{+/-}$ -3.7.2C clonal line of L5178Y cells was derived from a 3-methylcholanthrene-induced thymic lymphoma from a DBA-2 mouse and it is the only acceptable cell line for this assay [12]. The karyotype for the cell line has been published [11, 15, 16]. The modal chromosome number is 40; the metacentric chromosome (t12;13) should be counted as one chromosome. The mouse *Tk* locus is located on the distal end of chromosome 11. The L5178Y cells has point mutations in both of the p53 alleles and produces mutant-p53 protein [17, 18]. Stock cell cultures can be stored at approximately -150 °C or below. The cultures should be periodically checked for *Mycoplasma* contamination either in house or outsourced to a commercial vender such as the American Type Culture Collection (ATCC). The method used to detect *Mycoplasma* contamination should be the Hoechst direct culture or DNA staining method. This suspension cell line can be grown either under static conditions or with gentle shaking (approximately 150 revolutions/min). The acceptable doubling time for this cell line is generally between 9 and 11 h. Population doubling time should be checked when master stocks are created and regularly (at each subculture) during cell maintenance. It is important that all laboratories conducting the MLA karyotype the cells and/or paint the chromosomes harboring the *Tk* locus when establishing a master stock.

3. *Culture Conditions*: It is extremely important that cell cultures be maintained under conditions that ensure log-phase cell growth during stock maintenance and preparation for testing. It is equally important that media and culture conditions ensure optimal growth of cells during the expression period and cloning including the optimal recovery of the small-colony *Tk* mutants. RPMI 1640 and Fischer's media have been successfully used with the MLA. In the protocol described in this chapter, Fischer's medium supplemented with horse serum was used for growth medium both for the stock cultures and for treatment and expression. RPMI 1640 medium was used during the mutant selection. Both horse serum and fetal bovine serum can be used in the assay. However, it is critical that the horse serum be properly heat inactivated when using RPMI 1640 medium (*see* **Note 1**). Once the medium is prepared it is recommended to use the prepared medium within 3 months.

4. *Cleansing of Cultures*: To reduce the frequency of spontaneous $Tk^{-/-}$ mutants, cell cultures are exposed to THMG for approximately 24 h to select against cells having the $Tk^{-/-}$ phenotype. Methotrexate blocks folate metabolism and thus thymidylate synthetase (TS)-mediated production of TMP is halted. $Tk^{-/-}$-deficient cells are devoid of other pathways leading to TMP production after TS is blocked and hence they die whereas *Tk*-competent cells are able to utilize THG in other pathways leading to TMP production. After 24-h treatment with THMG, the cells are spun down in a centrifuge and the cell pellet is placed in new growth medium containing THG for an additional 24-h period at which time they are returned to normal growth medium. While there are a number of strategies for maintaining cell stocks and cleansing cells prior to treatment, this laboratory recommends that cells be cleansed prior to freezing so that the cells are ready to be used after thawing. It is not recommended, as has often been past practice in a number of laboratories, to grow the cells for extended periods with weekly THMG cleansing.

5. *Cell Counting*: A properly grown stock culture is essential for proper conduct of the assay. This is achieved by counting and diluting cells to 3×10^5 cells/mL for optimal growth for an approximate 24-h culture. Cells can be counted manually using the standard hemocytometer method or automated cell counting systems such as a coulter counter. This laboratory generally uses a hemocytometer for cell counting to increase accuracy and precision.

6. *Treatment Duration*: Proliferating cells are treated with the test substance in the presence and absence of S9 for 3–4 h. In some instances (e.g., for poorly soluble, non-cytotoxic test substances), consideration can be given to extend the

treatment time without S9 to 24 h in an attempt to cover the recommended cytotoxicity range. The treatment in the presence of S9 should always be limited to 3–4 h because of the cytotoxicity of S9 itself during extended treatments.

7. *Concentration Levels of the Test Substance*: The concentration levels of the test substance in the MLA are based on the expected cytotoxicity as measured by the RTG (2). RTG includes the relative growth of the cells in suspension during the treatment and expression time and the relative cloning efficiency at the time of mutant selection (see the formula above). The highest concentration selected for the mutation assay should result in 80–90 % cytotoxicity (i.e., an RTG value between 20 and 10 %). Concentrations resulting in >90 % cytotoxicity (i.e., an RTG <10 %) should be avoided because of the potential for false-positive responses at these excessively cytotoxic concentrations. The other concentrations (at least four) for the initial mutation assay should be selected such that the cytotoxicity range covers the entire spectrum ranging from 90 to 0 % (i.e., 10–100 % RTG). This range of concentrations can be modified for the subsequent assays depending upon the results of this initial assay. For example, when a test substance induces a weak response in the initial assay, emphasis should be placed on concentrations resulting in higher cytotoxicity in subsequent assays in order to increase the likelihood of detecting a positive response if, in fact, the test chemical is mutagenic.

For freely soluble substances that are relatively nontoxic, the top concentration should be limited to a level that does not unduly increase the osmolality of the treatment medium or 2,000 μg/mL, 2 μL/mL, or 10 mM, whichever is lower; for studies intended for regulatory submission consult applicable guidance documents. For poorly soluble test substances that are not cytotoxic at soluble concentration, the highest concentration should be limited to one or two dose levels that produce turbidity/precipitate provided that the precipitate does not interfere with the conduct of the assay. Testing of multiple insoluble concentrations should be avoided because of the potential for artifacts.

In the MLA, either single or duplicate cultures can be set up at each concentration. In the initial mutation assay, the concentrations are usually spaced apart by a factor of 2–3. This spacing may be adjusted in subsequent assays to yield the desired levels of cytotoxicity.

Careful attention should be paid to the final pH and osmolality of the treatment medium following the addition of the test substance. It has been demonstrated that alterations in the pH and osmolality of the culture medium can result in false-positive responses in in vitro genotoxicity assays [19–22]. If the test substance causes a marked change (e.g., ±0.4 units) in the pH of the medium at the time of addition, the pH of the treatment medium should be adjusted by buffering. In general, the osmolality change following the addition of the test material should be limited to ±50 mOsm/kg H$_2$O.

8. *Preparation of Treatment Solutions in the Solvent*: Solutions used for treatment of the cell cultures should be prepared on the same day as the scheduled treatment and used within 3–4 h unless stability of the test material in the solvent has been established. Lower concentrations should be prepared carefully by serial dilutions of the highest concentration. Administration of the dilutions should occur under yellow light.

9. *Solvent Controls*: The solvent selected for dissolving the test substance should not have an adverse effect on the test system or the assay conduct. Special attention should be paid for any reaction of the solvent with culture vessels or the stability of the test substance. The solvent should not interfere with the normal growth of the cells or the functioning of the external metabolic activation system (S9; see below). The solvent used to dissolve the test material should also be added (in the same quantity) to an otherwise untreated culture to serve as the solvent control for the assay. Commonly used solvents in the order of preference are culture medium, distilled water, DMSO, and ethanol. Depending upon the solubility, the test material can be either dissolved directly in the treatment medium or dissolved first in any of the above solvents and further diluted (up to 1:10 for culture medium and water and 1:100 for other solvents) in the test culture. If a solvent is used for the first time in the MLA, it is necessary to include untreated cultures that can be used for comparison to the solvent control cultures within the same experiment.

10. *Positive Controls*: Positive controls are needed to demonstrate the adequacy of the experimental conditions to detect known mutagens and/or clastogens. Commonly used positive controls include methyl methanesulfonate (MMS) for the non-activation system (without S9) at concentrations ranging from 10 to 15 µg/mL with a short treatment and 1.0 to 3.0 µg/mL for the 24-h continuous treatment. In order to assure the capability of the S9 to metabolically activate the test substance, a promutagen is included as a positive control with each assay conducted in the presence of S9. The commonly used positive control for this purpose is 20-MCA (also referred to as 3-methylcholanthrene) at concentrations ranging from 1.0 to 7.5 µg/mL of culture medium. It is important that positive control chemicals be chosen based on their ability to induce small-colony mutants.

11. *Metabolic Activation*: Mouse lymphoma cells require the use of an exogenous source of metabolic activation in order to detect various classes of promutagens. Post-mitochondrial fraction (S9) of liver homogenates prepared from the livers of rats treated

with enzyme-inducing agents such as Aroclor 1254 or a combination of phenobarbitone and β-naphthoflavone is generally used as the external metabolic activation system. S9 can be prepared either in house or purchased from a commercial source and stored at approximately −80 °C or below. Thawed S9 needs to be reconstituted at a final concentration of approximately 10 % (v/v) in a "mix" [19]. The mix consists of the following cofactors: 10 mM $MgCl_2 \cdot 6H_2O$, 5 mM glucose-6-phosphate, 4 mM NADP, 10 mM $CaCl_2$, 30 mM KCl, and 50 mM sodium phosphate (pH 8.0) (*see* **item 6** of Subheading 2). The reconstituted mix is added to the culture medium to obtain the desired final concentration of S9 in the treatment medium (usually at 2 % v/v). Hence, the final concentration of the cofactors in the treatment medium is 1/5 of the concentrations stated above. However, lower or higher concentrations in the range of 0.5–5 % may be used based upon the activity of S9 in a given batch or the nature of the test material.

12. *Insight into the Interpretation of the MLA Databases and Published Literature*:
It is important to recognize, particularly when reviewing positive/negative calls in the published literature or in databases, that the criteria for positive and negative responses have evolved over time based on a better understanding of the assay and the more widespread use of the assay. Clive and colleagues published a validation paper for the assay in 1979 [23] and recommended that a twofold rule (twice the concurrent background mutant frequency) was an appropriate indication of a positive response. This criterion was widely applied by laboratories, particularly in the USA. The United Kingdom Environmental Mutagenesis Society developed statistical approaches for the evaluation of several genetic toxicology assays including the MLA. Their approach included using heterogeneity factors which determined the variability between duplicate cultures within a single experiment [24]. Based on experience over time with both of these approaches, MLA experts became concerned that these methods were calling responses positive that fell within the normal range of the background of the assay. The US Environmental Protection Agency's Gene-Tox MLA Expert Workgroup [25] recommended an approach that required a specific absolute increase in mutant frequency (100×10^6), a value that a group of individual MLA experts felt confident constituted a biologically relevant positive response. In the late 1990s, the International Workshop for Genotoxicity Procedures was initiated and the first meeting was held in Washington, DC in 2000. An expert workgroup for the MLA was established for this meeting and the group subsequently met a number of times to reach international consensus on acceptability criteria

(as described previously above) and on criteria for a positive response (also previously described above). These criteria were developed based on large primary datasets from the member laboratories [13, 14].

Unfortunately, there are large databases and a number of publications that use the older criteria for assay acceptance and for positive/negative response determination and these have been used in various exercises to evaluate the performance of the MLA in predicting cancer [26, 27]. Perhaps the largest of these databases is that of the National Toxicology Program (NTP). Recently, we have completed an evaluation of this database and found that a very large percentage of the experiments do not meet current acceptability criteria and that only a small subset of the chemicals called positive by NTP should be called positive by the current IWGT criteria [28].

We strongly encourage anyone using previously published MLA data or databases to review the primary data and to reevaluate it based on the current criteria. When the assay is properly conducted, as detailed in this chapter, and the data properly interpreted, the assay is robust and provides useful information that can be utilized in both hazard identification and as a part of the weight of the evidence assessment to inform mode of action evaluations, either for the induction of mutation or the induction of cancer.

Disclaimer

The views presented in this chapter do not necessarily reflect those of the Food and Drug Administration.

References

1. Clive D, Flamm GW, Patterson JB (1973) Specific-locus mutational assay systems for mouse lymphoma cells. In: Hollaender A (ed) Chemical mutagens: principles and methods for their detection, vol 3. Plenum, New York, pp 79–103
2. Clive D, Spector JFS (1975) Laboratory procedure for assessing specific locus mutations at the *Tk* locus in cultured L5178Y mouse lymphoma cells. Mutat Res 31:17–29
3. Clive D, Caspery W, Kirby PE et al (1987) Guide for performing the mouse lymphoma assay for mammalian cell mutagenicity. Mutat Res 189:143–156
4. Applegate ML, Moore MM, Broder CB et al (1990) Molecular dissection of mutations at the heterozygous thymidine kinase locus in mouse lymphoma cells. Proc Natl Acad Sci 87(1):51–55
5. Moore MM, Clive D, Hozier JC et al (1985) Analysis of trifluorothymidine-resistant (TFTr) mutants of L5178Y/TK$^{+/-}$ mouse lymphoma cells. Mutat Res 151(1):161–174
6. Wang J, Sawyer JR, Chen L et al (2009) The mouse lymphoma assay detects recombination, deletion, and aneuploidy. Toxicol Sci 109(1): 96–105

7. Dearfield KL, Auletta AE, Cimino MC et al (1991) Considerations in the U.S. Environmental protection agency's testing approach for mutagenicity. Mutat Res 258:259–283
8. FDA (2001) Mouse lymphoma thymidine kinase gene mutation assay. http://wwwcfsanfdagov/~redbook/redivc1chtmL. Accessed 2001
9. The international conference on harmonisation (ICH) of technical requirements for registration of pharmaceuticals for human use ICH S2 R1 (2011). Guidance on genotoxicity testing and data interpretation for pharmaceuticals intended for human use, 9 Nov 2011
10. Hozier J, Sawyer J, Moore M et al (1981) Cytogenetic analysis of the L5178Y/TK$^{+/-}$ leads to TK$^{-/-}$ mouse lymphoma mutagenesis assay system. Mutat Res 84(1):169–181
11. Sawyer J, Moore MM, Clive D et al (1985) Cytogenetic characterization of the L5178Y TK$^{+/-}$3.7.2C mouse lymphoma cell line. Mutat Res 147(5):243–253
12. Lorge E (2013) Recommendations for good cell culture practices in genotoxicity testing. Genetic Toxicology Technical Committee, ILSI-HESI, Washington, DC (Personal Communication)
13. Moore MM, Honma M, Clements J et al (2003) Mouse lymphoma thymidine kinase gene mutation assay: International Workshop on Genotoxicity Tests Workgroup Report-Plymouth, UK 2002. Mutat Res 540:127–140
14. Moore MM, Honma M, Clements J et al (2006) Mouse lymphoma thymidine kinase gene mutation assay: follow-up meeting of the International Workshop on Genotoxicity Testing–Aberdeen, Scotland, 2003–Assay acceptance criteria, positive controls, and data evaluation. Environ Mol Mutagen 47(1):1–5
15. Sawyer JR, Moore MM, Hozier JC (1989) High-resolution cytogenetic characterization of the L5178Y TK$^{+/-}$ mouse lymphoma cell line. Mutat Res 214(2):195–199
16. Sawyer JR, Binz RL, Wang J et al (2006) Multicolor spectral karyotyping of the L5178Y Tk$^{+/-}$ 3.7.2C mouse lymphoma cell line. Environ Mol Mutagen 47(2):127–131
17. Storer RD, Kraynak AR, McKelvey TW et al (1997) The mouse lymphoma L5178Y Tk$^{+/-}$ cell line is heterozygous for a codon 170 mutation in the p53 tumor suppressor gene. Mutat Res 373(2):157–165
18. Clark LS, Hart DW, Vojta PJ (1998) Identification and chromosomal assignment of two heterozygous mutations in the *Trp53* gene in L5178Y *Tk*$^{+/-}$ 3.7.2C mouse lymphoma cells. Mutagenesis 13:427–434
19. O'Neill JP, Machanoff R, San Sebastian JR et al (1982) Cytotoxicity and mutagenicity of dimethylnitrosamine in mammalian cells (CHO/HGPRT system): enhancement by calcium phosphate. Environ Mutagen 4:7–18
20. Thilagar AK, Kumaroo PV, Kott S (1984) Effects of low pH caused by glacial acetic acid and hydrochloric acid on chromosomal aberrations in CHO Cells. Toxicologist 4:51
21. Galloway SM, Bean CL, Armstrong MA et al (1985) False positive in vitro chromosome aberration tests with non mutagens at high concentrations and osmolalities. Environ Mutagen 7(3):48–49
22. Cifone MA (1985) Relationship between increases in the mutant frequency in L5178Y *Tk*$^{+/-}$ mouse lymphoma cells at low pH and metabolic activation. Environ Mutagen 7(3):27
23. Clive D, Johnson KO, Spector JFS et al (1979) Validation and characterization of the L5178Y/TK$^{+/-}$ mouse lymphoma mutagen assay system. Mutat Res 59:61–108
24. Robinson WD, Green MHL, Cole J et al (1989) Statistical evaluation of bacterial/mammalian fluctuation tests. In: Kirkland DJ (ed) Statistical evaluation of mutagenicity test data. Cambridge University Press, Cambridge, pp 102–140
25. Mitchell AD, Auletta AE, Clive DC et al (1997) The L5178Y/tk$^{+/-}$ mouse lymphoma specific gene and chromosomal mutation assay: a phase III report of the US Environmental Protection Agency Gene–Tox Program. Mutat Res 394:177–303
26. Kirkland D, Aardema M, Henderson L et al (2005) Evaluation of the ability of a battery of three in vitro genotoxicity tests to discriminate rodent carcinogens and non–carcinogens I. Sensitivity, specificity and relative predictivity. Mutat Res 584:1–256
27. Kirkland D, Aardema M, Müller L et al (2006) Evaluation of the ability of a battery of three in vitro genotoxicity tests to discriminate rodent carcinogens and non-carcinogens II. Further analysis of mammalian cell results, relative predictivity and tumour profiles. Mutat Res 608: 29–42
28. Schisler MR, Gollapudi BB, Moore MM (2010) Evaluation of the mouse lymphoma mutation assay (MLA) data of the U.S. National Toxicology Program (NTP) using International Workshop on Genotoxicity Testing (IWGT) criteria (2010). Environ Mol Mutagen 51:732
29. Charles GD, Spencer PJ, Schisler MR et al (2005) Mode of mutagenic action for the biocide Bioban CS–1246 in mouse lymphoma cells and implications for its in vivo mutagenic potential. Toxicol Sci 84:73–80

Chapter 3

Erythrocyte-Based *Pig-a* Gene Mutation Assay

Jeffrey C. Bemis, Nikki E. Hall, and Stephen D. Dertinger

Abstract

In addition to chromosomal damage, assessment of gene mutation is an important part of genotoxicity testing employed during preclinical safety testing. The *Pig-a* gene mutation assay is based on the loss of function of the *Pig-a* gene, which results in a lack of cell surface expression of specific proteins that are targeted to the surface by GPI anchors. This cell surface phenotype is readily assessed by flow cytometric analysis of red blood cells. This unit describes a procedure for the collection, processing, and analysis of peripheral blood samples using materials supplied in MutaFlow® kits and a common benchtop flow cytometer.

Key words Genotoxicity, Endogenous mutation, Reticulocytes, Erythrocytes, *Pig-a*, Glycosylphosphatidylinositol (GPI) anchor, Flow cytometry, Immunomagnetic separation, High throughput

1 Introduction

The *Pig-a* gene mutation assay is based on a cell surface phenotype characterized by the absence of glycosylphosphatidylinositol (GPI)-anchored proteins. Hematopoietic cells require GPI anchors to attach a host of proteins to their cell surface, e.g., CD24, CD59, and CD55 [1]. Importantly, of the genes required to form GPI anchors, only *Pig-a* is located on the X-chromosome. Thus, mutations in the *Pig-a* gene can prevent functional anchors from being produced, resulting in cells lacking these proteins on their surface. This phenomenon is illustrated in Fig. 1.

The *Pig-a* mutation assay protocol describes procedures for scoring the frequency of mutant phenotype erythrocytes (RBCs) and mutant phenotype immature erythrocytes (reticulocytes, or RETs) in the peripheral blood of rats using flow cytometry. The method is based on the endogenous *Pig-a* gene whose product is essential for the synthesis of GPI anchors [2, 3]. The cells without these cell surface markers can be differentiated from wild-type cells by using fluorescent antibodies and can represent a reliable phenotypic marker of *Pig-a* mutation [4–10].

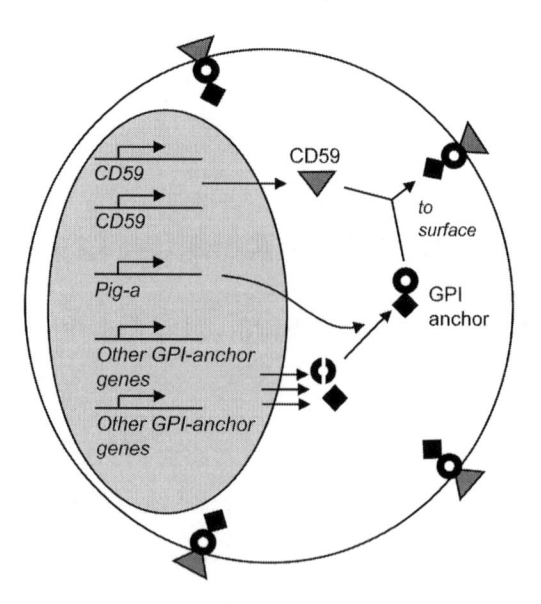

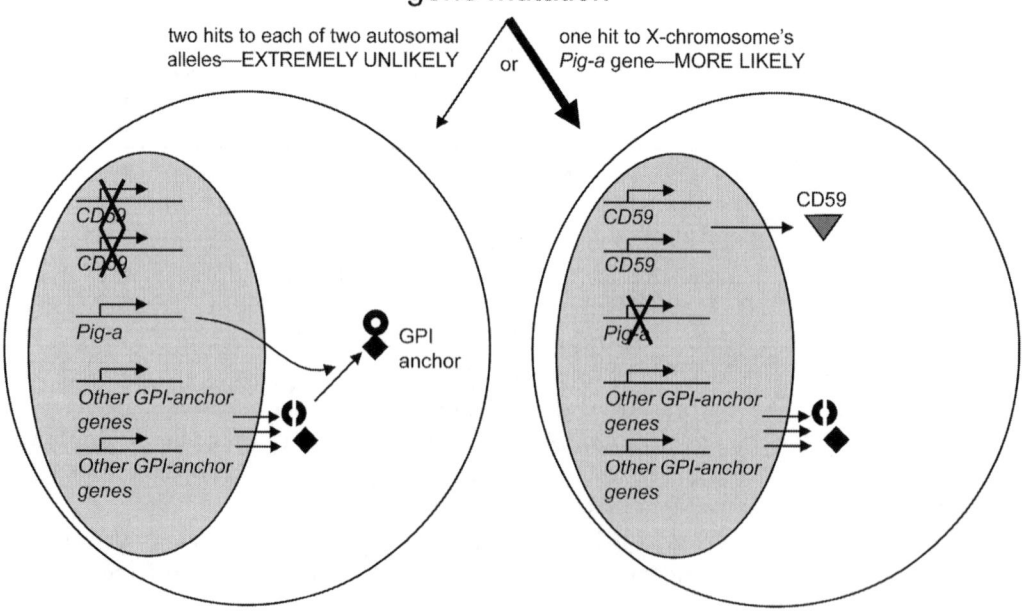

Fig. 1 Cartoon shows a normal, wild-type cell at the *top* and describes the acquisition of the *Pig-a* mutant phenotype

Litron Laboratories developed a sample processing method that incorporates antibody-labeling/nucleic acid staining of cells and immuno-magnetic separation followed by analysis by flow cytometry [11]. This methodology, commercially available as MutaFlow® kits, is illustrated in Fig. 2. Briefly, blood samples are processed through Lympholyte®-Mammal solution to remove the majority of leukocytes and platelets. Cells are then incubated with anti-CD59-PE (to

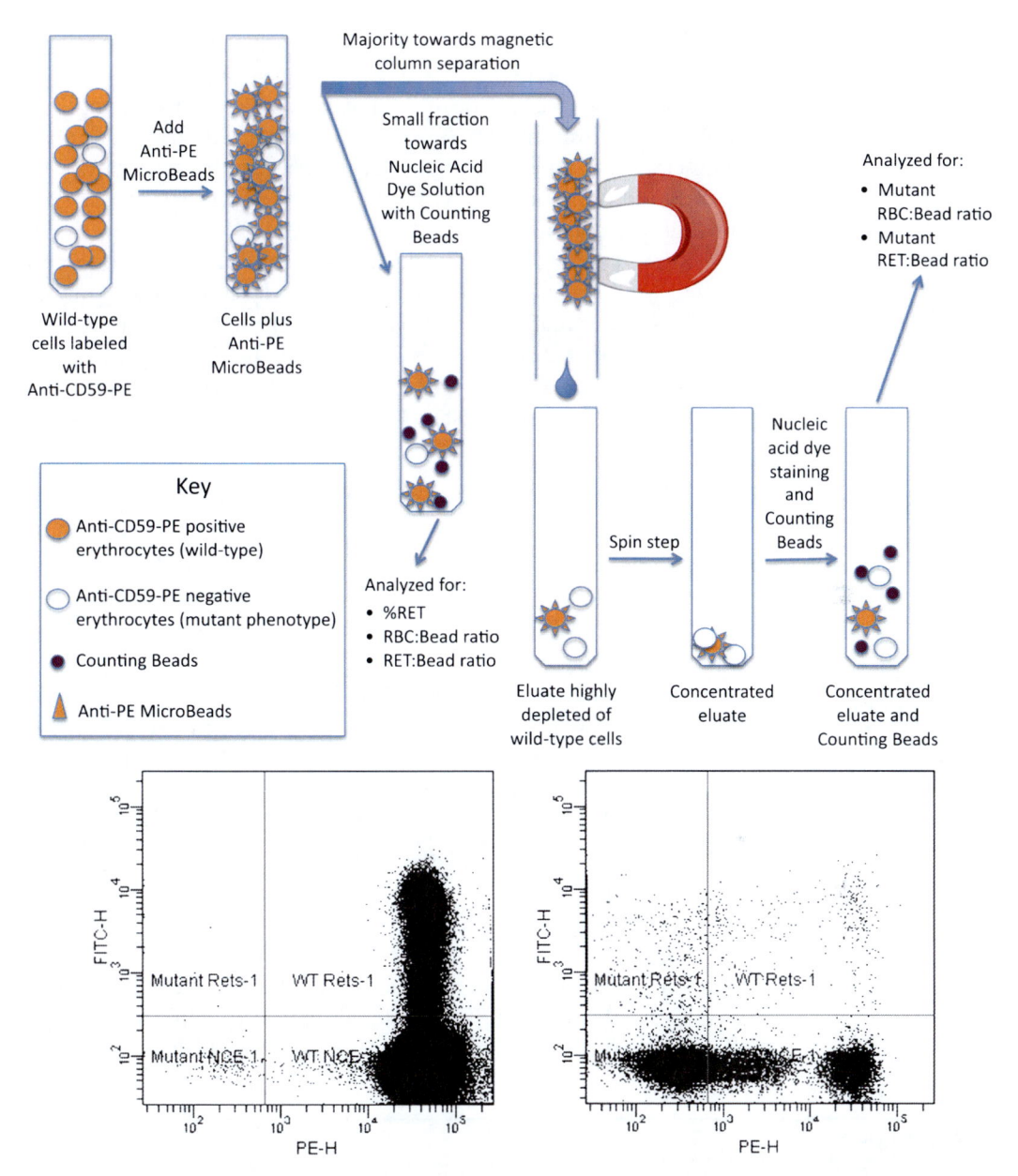

Fig. 2 The schematic at the *top* shows the strategy for immuno-magnetic separation and enumeration of the pre- and post-column samples for quantification of *Pig-a* mutant cell frequency. The *bottom plots* show data for a pre-column sample (*left*) and a post-column sample (*right*) from a blood sample obtained 25 days after 3 consecutive days of exposure to the mutagen ethyl nitrosourea. Notice how the post-column sample has become greatly enriched for mutant cells

label wild-type (wt) RBCs) and anti-CD61-PE (to label the remaining platelets). Antibody-labeled samples are incubated with anti-PE MicroBeads, which bind to these antibodies.

A small fraction of each sample is stained with a nucleic acid dye (to differentiate leukocytes and RETs from mature RBCs). This dye solution also includes fluorescent counting beads and these "pre-column" samples are analyzed on a flow cytometer to capture cell:bead ratios.

The remaining portion (majority) of the blood sample is applied to a Miltenyi LS Column that has been suspended in a magnetic field. These columns selectively retain the wild-type cells, whereas *Pig-a* mutants (without CD59 on their surface) pass through the columns.

Eluates are collected, concentrated by centrifugation, and stained with a nucleic acid dye in order to differentiate leukocytes and RETs from mature RBCs. This dye solution also includes fluorescent counting beads and these "post-column" samples are analyzed on a flow cytometer to capture mutant cell:bead ratios.

From the pre- and post-column analyses, the following values are calculated:

1. RET percentage, an index of bone marrow toxicity.

2. Frequency of mutant-phenotype RBCs.
   Note that upon acute mutagen exposure this index of genotoxicity is not expected to reach a maximal response until the entire cohort of circulating RBCs has turned over (approximately 42–65 days for rats; [12–14]).

3. Frequency of mutant-phenotype RETs.
   Note that upon acute mutagen exposure this index of genotoxicity reaches a maximal value faster than the frequency of mutant phenotype RBCs (often 2–3 weeks), since RETs are turned over at a much faster rate than the total RBC pool.

A comprehensive collection of investigations on the *Pig-a* assay, including results from an international validation trial and discussion of the role of this assay for hazard identification and risk assessment [15], can be found in a special issue of Environmental and Molecular Mutagenesis dedicated to the *Pig-a* assay (Vol. 52, issue 9, December 2011). Additional information and video clips of specific processing steps can be accessed at www.litronlabs.com.

2 Materials

As noted, certain reagents and materials described below are from the commercially available MutaFlow® kits (Litron Laboratories, Rochester, NY). Additional materials and supplies that are required, but not supplied with the kits, are specifically noted.

2.1 Supplies and Specialized Equipment

1. Centrifuge with swinging bucket rotor.
2. 15 mL polypropylene centrifuge tubes.
3. Microcentrifuge tubes.
4. −10 to −30 °C freezer.
5. 2–8 °C refrigerator.

6. 37 °C incubator or water bath.
7. Flaked/chipped ice.
8. Flow cytometer capable of 488 nm excitation.
9. Flow cytometry tubes.
10. 0.2 μm filters—various types: syringe, flask, etc.
11. Heparin-coated capillary tubes (optional).
12. LS Columns (Miltenyi Biotech).
13. MidiMACS™ or QuadroMACS™ Separator (Miltenyi Biotech).
14. Aspiration device (*see* **item 6** of Subheading 2.2).
15. Potassium EDTA microtainer tubes (e.g., Becton Dickinson, optional—for storage of blood samples and/or shipping samples off-site).
16. Exakt-Pak® shipping containers (optional—for shipping samples off-site).
17. Icepacks (optional—for shipping samples off-site).
18. Lympholyte®-Mammal (Cedarlane Laboratories).
19. CountBrite™ Absolute Counting Beads (Life Technologies).
20. Anti-PE MicroBeads (Miltenyi Biotech).
21. Heat-inactivated Fetal Bovine Serum.

2.2 Solution and Material Preparation

The following solutions are made up fresh daily. The formulas are based on the volumes required for *one sample* and should be scaled up and prepared in bulk as appropriate to the number of samples being processed (*see* **Note 1**).

1. Buffered salt solution + 2 % fetal bovine serum (FBS): Combine 29.4 mL of buffered salt solution (MutaFlow^{PLUS} kit) with 0.6 mL of FBS. Filter sterilize with 0.2 μm filter and store at 2–8 °C.

2. Working nucleic acid dye plus counting bead solution: Add 75 μL of stock nucleic acid dye solution (MutaFlow^{PLUS} kit) and 75 μL of CountBright™ absolute counting beads to 2.35 mL of buffered salt solution + 2 % FBS. Pipette to mix and store at room temperature, protected from light (*see* **Note 2**).

3. Working antibody solution: Combine 65 μL of buffered salt solution + 2 % FBS with 30 μL of stock anti-CD59-PE solution (MutaFlow^{PLUS} kit) and 5 μL stock anti-CD61-PE solution (MutaFlow^{PLUS} kit). Pipette to mix and store protected from light at 2–8 °C (*see* **Note 3**).

4. Working anti-PE MicroBead suspension: Add 25 μL of Anti-PE MicroBeads to 75 μL of buffered salt solution + 2 % FBS. Pipette to mix and store at 2–8 °C, protected from light.

5. Preliminary material preparation: Label all necessary tubes and vials with sample IDs. At least two flow cytometry tubes, three

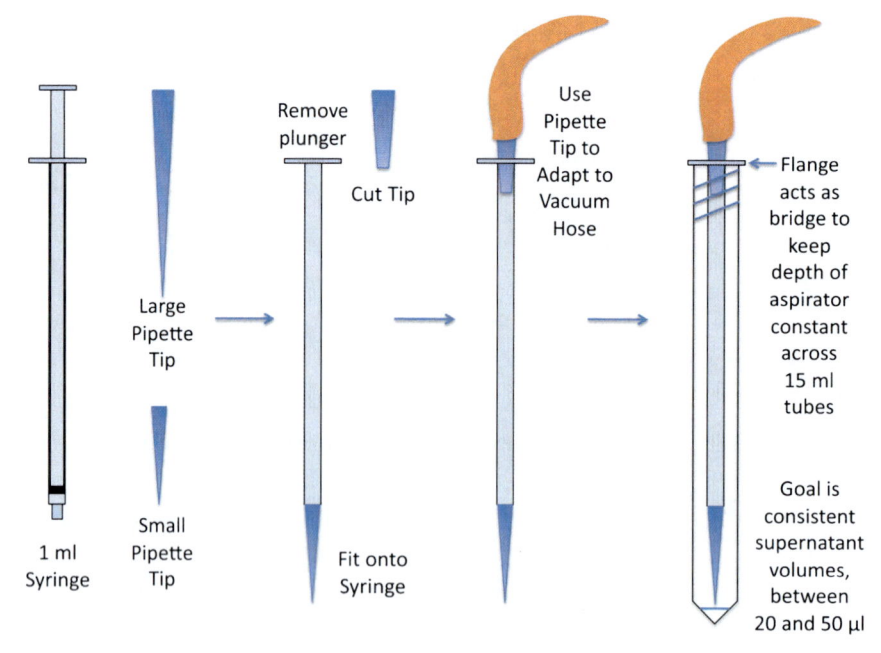

Fig. 3 Schematic for assembly of the aspiration device described in **item 6** of Subheading 2.2

centrifuge tubes (15 mL), and two microcentrifuge tubes are required per sample.

6. Aspiration device preparation: It is very important to carefully control and standardize the aspirations, especially the last aspiration (**step 2** of Subheading 3.9). To achieve this, fashion an aspirator with a bridge that controls the depth to which the tip can reach when aspirating from a standard 15 mL centrifuge tube (Fig. 3). With this bridge, the aspirator will leave a consistent and low volume of supernatant across all tubes and not contact the bottom of the tube or disturb the cell pellet. The goal for the volume of supernatant left after the final aspiration is a consistent volume within the range of 20–50 µL. This value is required to make mutant cell frequency calculations (*see* Subheading 3.12).

7. Template preparation: Data acquisition template files are available from Litron (download from www.litronlabs.com/support.html or e-mail pigatechsupport@litronlabs.com for information), but are specific to CellQuest™ Pro or FACSDiva™ software. Figure 4 shows actual screen images of the CellQuest™ Pro and FACSDiva™ template graphs. Flow cytometry operators who are not using CellQuest™ Pro or FACSDiva™ software should find the following instructions and graphics valuable for constructing their own data acquisition and analysis template.

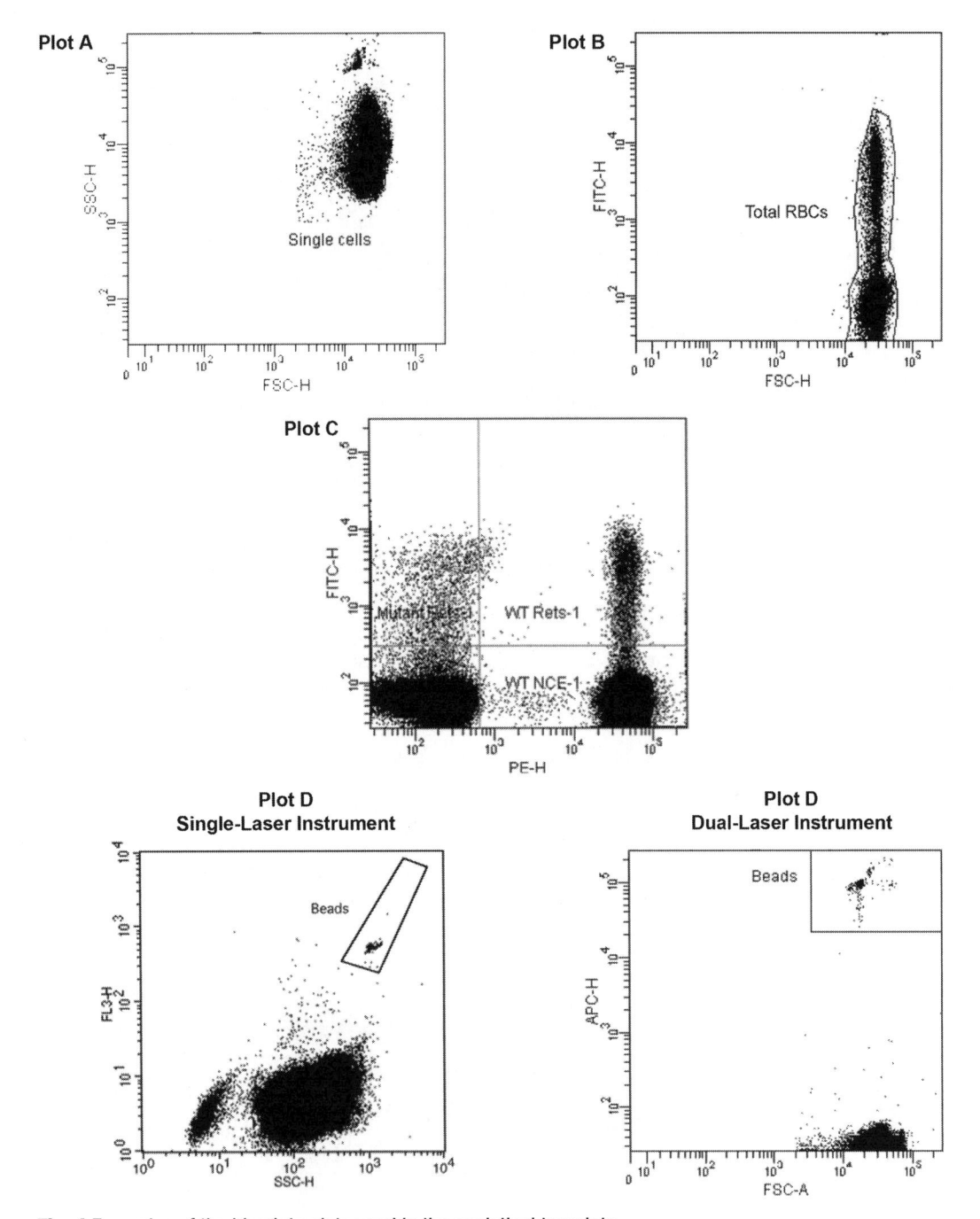

Fig. 4 Examples of the bivariate plots used in the analytical template

We recommend that if you are using FACSDiva™ software, set the fluorescence parameter to "Height" rather than "Area." The Internal Calibration Standard (ICS) may be run using single-color compensation controls and auto-compensation if available with your software package.

1. *Defining gates*:
   (a) G1 = R1 = "Single Cells"
   (b) G2 = R2 = "Total RBCs"
   (c) G3 = R3 = "Beads"
   (d) G4 = R1 and R2 and R3 = "Single cells" and "Total RBCs" and NOT "Beads" (this gating actively instructs the software that any plots that use that gate should purposefully take out anything defined in the R3 region as beads).
2. *Gate and parameters for each plot*:

Plot A	No gate	SSC-H vs. FSC-H
Plot B	G1	FL1-H vs. FSC-H
Plot C	G4	FL1-H vs. FL2-H
Plot D*	No gate	FL4-H vs. FSC-A *or* FL3-H vs. SSC-H

*If you have a second, red diode laser, use FL4 and either FSC or SSC for Plot D. Otherwise, use FL3. SSC is needed for single-laser analysis to provide optimal resolution when not using a red diode laser (and FL4).

3. *Quadrant key for Plot C*:

   UL = mutant RETs.

   UR = wild-type RETs.

   LL = mutant mature RBCs (i.e., mutant normochromatic erythrocytes [NCEs]).

   LR = wild-type mature RBCs (i.e., wild-type NCEs).

4. *Alternate names for detectors*:

Green	FL1	FITC
Orange	FL2	PE
Red	FL3	PerCP–Cy5.5
Far red	FL4	APC

5. *Save the template file*: This template file should be suitable for all analyses. To ensure consistency of data, once the regions, compensation, etc. have been set based on the ICS, it is preferable that no changes are made to the location and size of the regions between samples that are analyzed on the same day.

3 Methods

3.1 Blood Collection

1. Aliquot 100 µL kit-supplied (MutaFlow[PLUS]) anticoagulant solution into labeled microcentrifuge tubes, one for each blood sample. Refrigerate until use. This can be done before the day of blood collection.

2. On the morning of leukodepletion, gently shake the Lympholyte®-Mammal bottle and allow for air bubbles to disappear. Aliquot 3 mL of the solution into labeled 15 mL polypropylene centrifuge tubes, one for each blood sample. Protect from light and allow the aliquots to equilibrate to room temperature before use.

3. Obtain approximately 100 µL of blood per animal, taking care to collect free-flowing blood (*see* **Notes 4–6**).

4. Immediately upon collection of each blood sample, transfer 80 µL of blood into the labeled microcentrifuge tube containing 100 µL of anticoagulant solution.

5. Refrigerate blood samples in anticoagulant solution as soon as possible, although they can be stored at room temperature for up to 4 h before refrigeration. Process samples through Lympholyte®-Mammal (*see* Subheading 3.2) within 8 h of collection.

3.2 Leukodepletion and Platelet Removal

1. Using a pipettor, remove the entire contents of the microcentrifuge tube (80 µL of blood plus 100 µL of anticoagulant solution) and gently layer on top of the pre-aliquoted, room-temperature Lympholyte®-Mammal (Fig. 5). Repeat this step for the remaining samples (*see* **Note 7**).

2. Centrifuge the samples at $800 \times g$ for 20 min at room temperature. Maintain tubes at room temperature (*see* **Note 8**).

3. Holding the tube upright, aspirate supernatants, taking care not to disturb the pellets. Leave approximately 50 µL of supernatant above the pellet.

4. Gently add 300 µL of cold buffered salt solution (that does NOT contain FBS) to each supernatant just above the pellet to rinse the surface of the pellets and remove the remaining Lympholyte®-Mammal (*see* **Note 9**).

5. Holding the tube upright, carefully aspirate supernatants, leaving approximately 50 µL of supernatant on the pellet. Rinse the surface of each pellet a second time by repeating **step 4** of Subheading 3.2 (*see* **Note 10**).

6. Holding the tube upright, carefully aspirate the supernatants, but this time remove as much of the supernatant as possible without disturbing the loosely packed pellets (*see* **Note 11**).

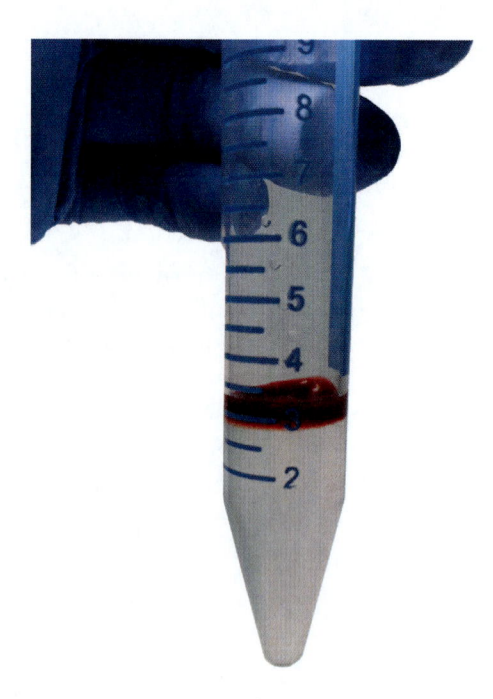

Fig. 5 Blood layered onto Lympholyte®-Mammal

7. Add 150 μL of cold buffered salt solution (without FBS) directly to each pellet (*see* **Note 12**).
8. Proceed immediately to Subheading 3.3. Otherwise, resuspend the cells with gentle pipetting and transfer each to a new polypropylene tube (*see* **Note 13**).

3.3 Sample Labeling

Once samples are fully labeled, it is advisable to analyze them within 3 h. Therefore, it is important to proceed through **step 1** of Subheading 3.3 to **step 7** of Subheading 3.9 in sequence; so careful staging of the samples and of the workday is critical. See work flow diagram at the end of the protocol for advice on personnel and time requirements for the processing steps in order to aid in experimental planning. At least one of the 15 mL tubes from **step 1** of Subheading 3.3, below, will need to be saved. The cells remaining in this tube will be used in **step 1** of Subheading 3.7 to make the Part A portion of the ICS. These unstained cells will become the "mutant mimicking cells."

1. Set a pipettor to 160 μL and pipette the first pellet (and buffered salt solution previously added) up and down to resuspend the cells (*see* **Note 14**).
2. Carefully transfer 160 μL of the resuspended cells directly into the working antibody solution in the labeled microcentrifuge tube, carefully pipetting up and down to mix. See the images in Fig. 6 for an example (*see* **Note 15**).

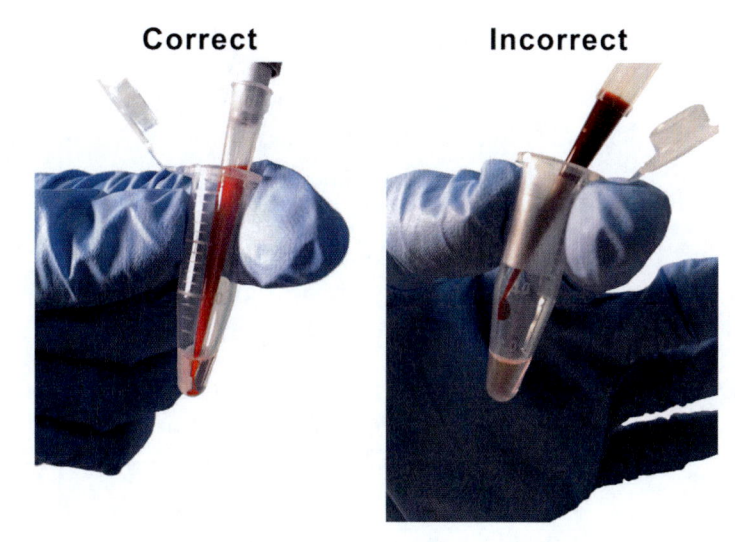

Fig. 6 Correct versus incorrect introduction of sample to working antibody solution

3. Repeat **steps 1** and **2** above of Subheading 3.3 for the remaining samples. Use a new pipette tip for each sample.
4. Incubate cells with working antibody solution for 30 min at 2–8 °C, covered to protect from light.
5. Save at least one 15 mL tube and its remaining contents (preferably from a vehicle control) at 2–8 °C (*see* **Note 16**).
6. During the incubation, aliquot 10 mL of cold buffered salt solution + 2 % FBS to labeled 15 mL centrifuge tubes, one for each sample. Store at 2–8 °C until needed in Subheading 3.4.

3.4 Wash Labeled Cells Out of Working Antibody Solution

1. After the incubation, resuspend the cells by gently pipetting the contents up and down and transfer directly into the cold, pre-aliquoted buffered salt solution + 2 % FBS prepared in **step 6** of Subheading 3.3 (*see* **Note 17**).
2. Centrifuge at 340 × *g* for 5 min at room temperature. After centrifugation, maintain at room temperature.
3. Holding the tube upright, aspirate supernatants, removing as much of the supernatant as possible without aspirating the pellet (*see* **Note 18**).

3.5 Incubate with Working Anti-PE MicroBead Suspension

1. Pipette the working anti-PE MicroBead suspension to mix.
2. Add 100 µL of MicroBead suspension to a sample tube by washing down the inside of the tube starting at about 0.5 cm above the pellet, carefully pipetting up and down to mix. Change the pipette tip and repeat for the remaining samples (*see* **Note 19**).
3. Incubate cells in a non-insulating rack for 30 min at 2–8 °C, covered to protect from light.

4. After incubation, add 10 mL of cold buffered salt solution + 2 % FBS to each tube. Cap tightly, invert each tube, and while inverted gently tap the bottoms to dislodge any settled cells. Repeat for all samples.

5. Centrifuge at $340 \times g$ for 5 min at room temperature. After centrifugation, keep at room temperature.

6. Holding the tube upright, aspirate supernatants, removing as much of the supernatant as possible without aspirating the pellet.

3.6 Stain Pre-column Samples

1. Add 1.0 mL of cold buffered salt solution + 2 % FBS directly to a pellet, carefully pipetting up and down to mix (*see* **Note 20**).

2. Transfer exactly 20 μL of this suspension (*see* **Note 21**) to the labeled flow cytometry tube containing 1980 μL of working nucleic acid dye plus counting bead solution kept at room temperature. Shake gently to mix (this is a pre-column sample).

3. Repeat **steps 1** and **2** of Subheading 3.6 for the remaining samples. Maintain them at room temperature until each sample has been processed to this point (*see* **Note 22**).

4. Maintain the cells remaining in the centrifuge tubes (~980 μL) in the dark at 2–8 °C or on ice until proceeding to Subheading 3.8 (*see* **Note 23**).

3.7 Prepare Instrument Calibration Standard and Incubate Pre-column Samples

1. Prepare Part A of the ICS. Do this by retrieving the leukodepleted sample that was stored in **step 5** of Subheading 3.3 and gently resuspend cells by pipetting up and down. Transfer 5 μL of this sample to the labeled flow cytometry tube containing 500 μL of working nucleic acid dye plus counting bead solution maintained at room temperature (prepared in **item 2** of Subheading 2.2). Shake gently to mix.

2. Incubate the pre-column samples (including Part A of the ICS) in working nucleic acid dye plus counting bead solution for 30 min at 37 °C in the dark.

3. After incubation, transfer the samples to ice and protect from light. Ensure that the tubes are surrounded by flaked/chipped ice, not resting on top (*see* **Note 24**).

3.8 Column Separation

Perform this section (**steps 1** through **5**) in batches of up to 8 samples (*see* **Note 25**).

1. Insert LS columns into either a MidiMACS™ or QuadroMACS™ Separator. Place a suitable vessel(s) under the columns to collect the eluate for further processing (*see* **Note 26**).

2. Gently add 3 mL cold buffered salt solution + 2 % FBS to each column reservoir to pre-wet it (*see* **Note 27**).

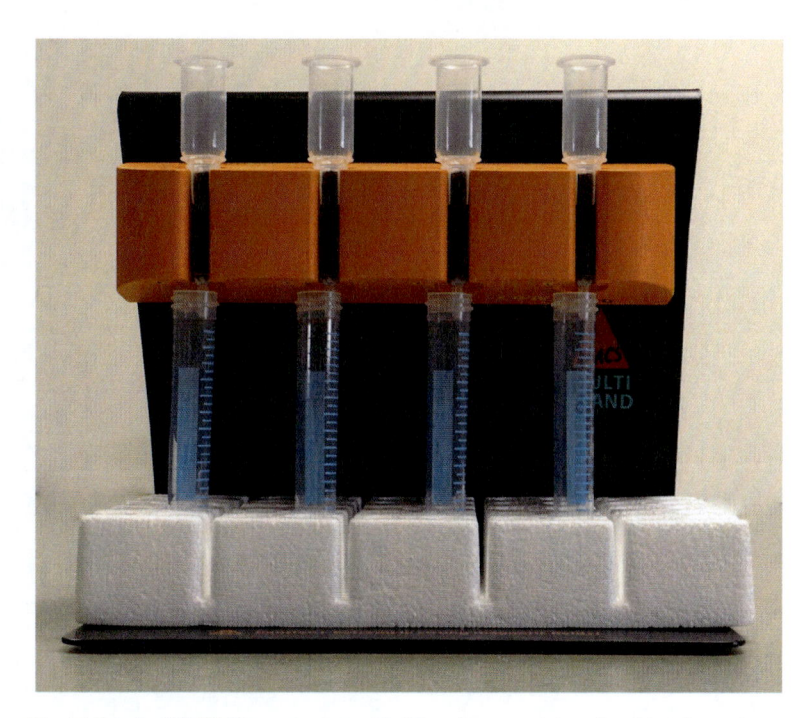

Fig. 7 QuadroMACS™ separator with LS columns and rack with tubes placed directly underneath

3. Place a clean, labeled 15 ml centrifuge tube under each column to collect the eluate for further processing (Fig. 7). Take the remainder of the pre-column sample (**step 4** of Subheading 3.6) and carefully pipette up and down to resuspend the cells and MicroBeads without creating bubbles. When the buffered salt solution + 2 % FBS has nearly completely entered the column matrix, gently add the entire remaining volume of sample (~980 µL) into the appropriate pre-wet LS column reservoir. Repeat for additional column reservoirs (*see* **Note 28**).

4. When the sample has fully entered the column (e.g., when the sample cannot be seen above the column matrix), slowly add 5 mL cold buffered salt solution + 2 % FBS to the column reservoir as a column wash. Repeat for additional column reservoirs (*see* **Note 29**).

5. The eluates will appear clear, as the vast majority of the cells will be trapped in the column (Fig. 8). Once an eluate has been collected, store it in the dark at 2–8 °C. Discard each LS column after use—DO NOT reuse the columns.

6. Repeat the above **steps 1–5** until all pre-column samples in the batch of 8 have been added to the columns—the eluates are now called post-column samples. Once this first batch is done, process the remaining samples in batches as described above.

64   Jeffrey C. Bemis et al.

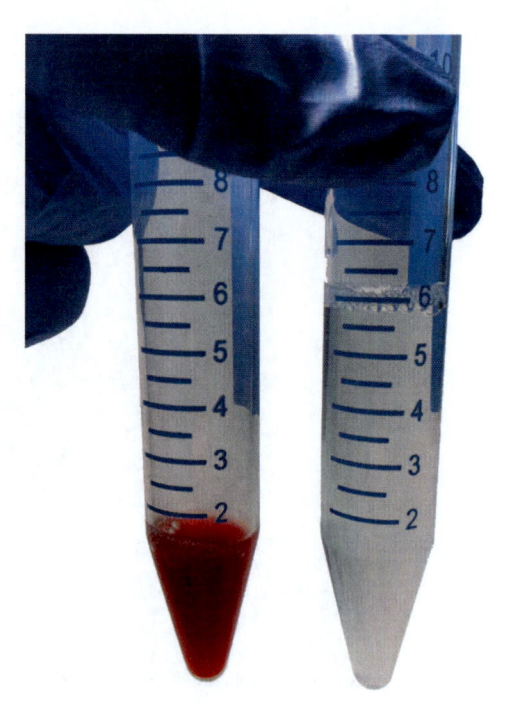

Fig. 8 Pre-column sample on *left*, post-column eluate on *right*

3.9 Centrifuge and Stain Post-column Samples

1. Centrifuge a set of 8 eluate tubes from **step 5** of Subheading 3.8 at $800 \times g$ for 5 min (*see* **Note 30**).
2. Holding the tube upright, use the aspiration device constructed in **item 6** of Subheading 2.2 to carefully aspirate supernatant starting at the top (meniscus) and working downwards to prevent disturbing the pellet (*see* **Note 31**).
3. Gently tap the pellets loose (*see* **Note 32**).
4. Add 300 μL of working nucleic acid dye plus counting bead solution at room temperature to each post-column sample (*see* **Note 33**).
5. Once a sample has been resuspended, transfer it to a flow cytometer tube. Incubate all the samples in the dark at 37 °C for 15 min.
6. After incubation, transfer the tubes to ice and protect from light. Store on ice for at least 5 min, but no more than 3 h, before flow cytometric analysis (*see* **Note 34**).
7. Repeat Subheadings 3.8 and 3.9 with the remaining batches of up to eight samples.

3.10 Flow Cytometric Analysis: Instrument Calibration

1. Before analyzing samples, ensure that the flow cytometer is working properly. Follow the manufacturer's instructions for the appropriate setup and quality control procedures.

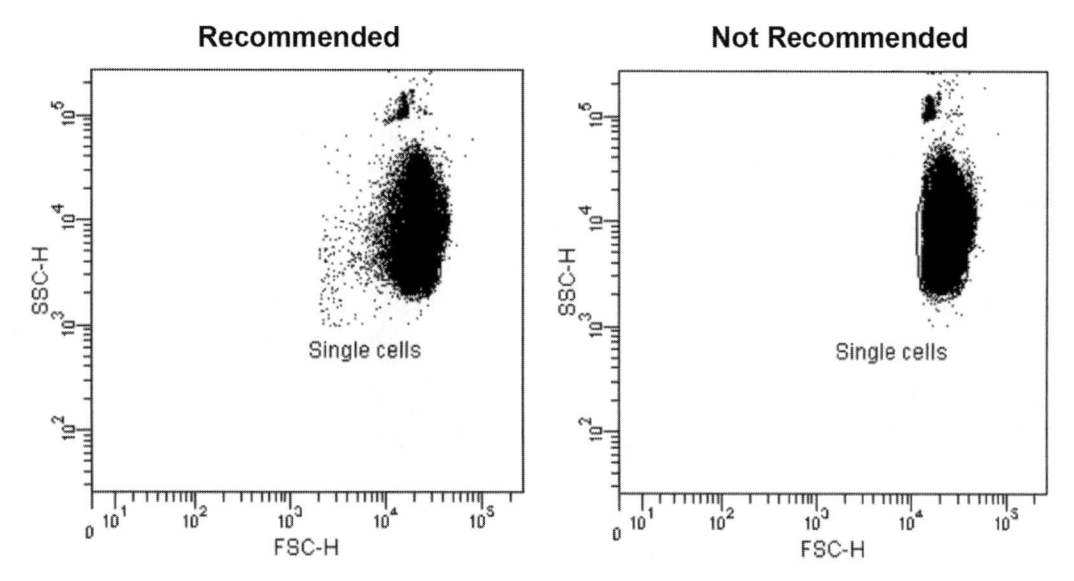

Fig. 9 Plot A: Recommended versus not recommended threshold setting

Download the data acquisition template file from www.litron-labs.com/support.html or create your own based on instruction in **item 7** of Subheading 2.2 (*see* **Note 35**).

2. Prepare the ICS by vigorously resuspending (pipetting) both Part A from **step 1** of Subheading 3.7 and Part B, which is obtained from any fully stained, pre-column vehicle control sample. Combine equal volumes of each into a flow cytometry tube (e.g., 200 μL of Part A and 200 μL of Part B) (*see* **Note 36**).

3. Immediately after creating the ICS, place it on the flow cytometer (*see* **Note 37**).

4. Threshold is set on FSC so that the remaining platelets and other subcellular debris are eliminated. If your instrument is capable, threshold should be set on both FSC and SSC, but be careful not to set the values so high that counting beads are thresholded out. In Plot A, adjust the "Single Cells" region so that it closely defines the major population of single, unaggregated erythrocytes. The resulting plot should look similar to the plot in Fig. 9.

   Viewing Plot B, adjust the "Total RBCs" region to eliminate contaminating leukocytes (those cells with high nucleic acid dye fluorescence). Together with the "Single Cells" region, this region is used to eliminate leukocytes from RBC-based measurements. The resulting plot should look similar to the plot in Fig. 10.

5. Viewing Plot D, adjust the FL4 (or FL3) PMT voltage so that counting beads fall within the "Beads" region. Adjust the position and size of the region as necessary. The resulting plot should look similar to one of the plots in Fig. 11.

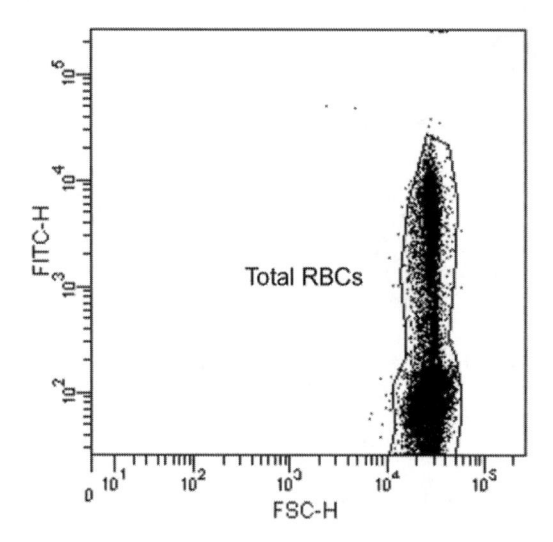

Fig. 10 Plot B: Placement of the "Total RBCs" gate

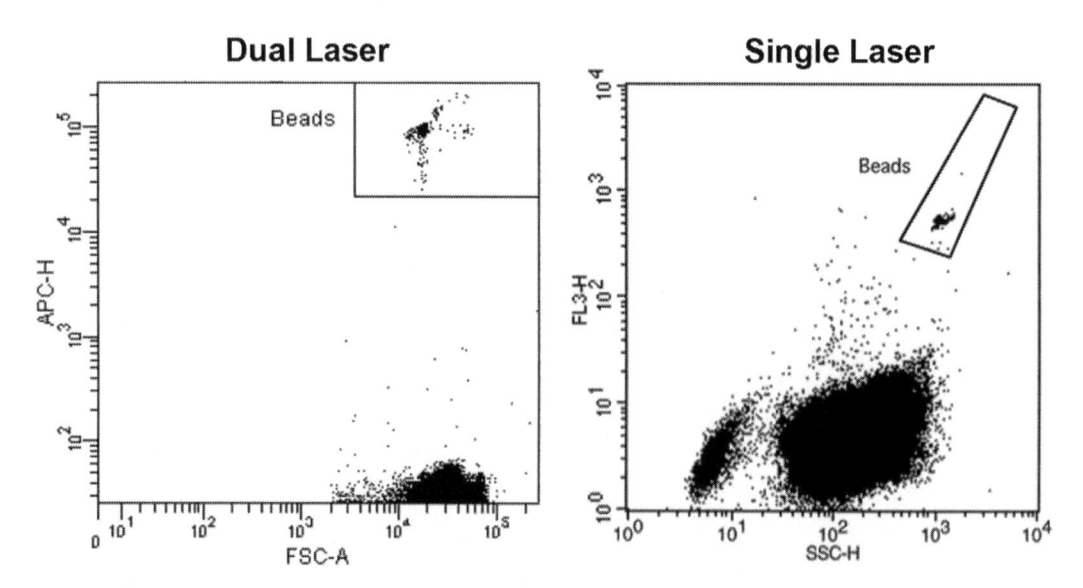

Fig. 11 Plot D: Capturing counting beads on a dual-laser versus single-laser instrument

6. Viewing Plot C, adjust PMT voltages so that mutant phenotype, mature RBCs (lower left quadrant; LL) are in the first to second decade of FITC and PE fluorescence. The resulting plot should look similar to the plot in Fig. 12.

7. Viewing Plot C, adjust compensation so that the green (FITC) component of the PE label is eliminated. This is evident when the wild-type phenotype, mature RBCs (lower right quadrant; LR) are at the same FITC fluorescence intensity as the mutant phenotype, mature RBCs (LL). See the before and after plots in Fig. 13 (*see* **Note 38**).

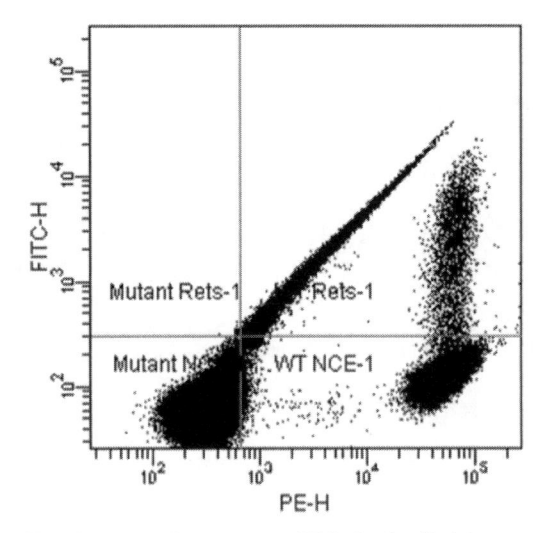

Fig. 12 Plot C: Adjusting PMT voltage to place mature RBCs in the first to second decade of FITC and PE fluorescence

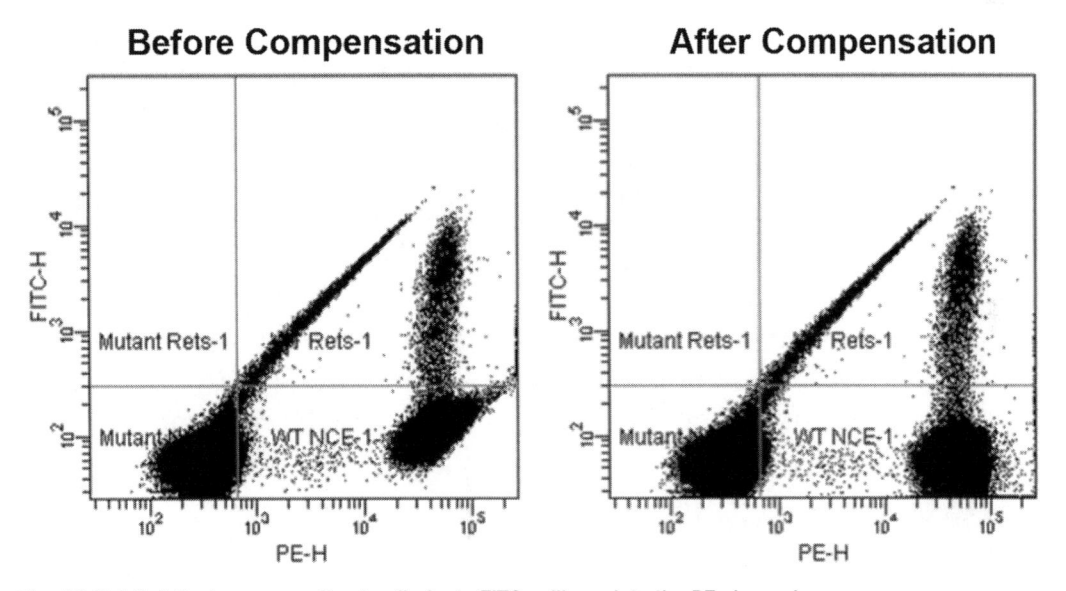

Fig. 13 Plot C: Adjust compensation to eliminate FITC spillover into the PE channel

8. Viewing Plot C, adjust compensation so that the orange (PE) component of the nucleic acid dye is eliminated. This is evident when the mutant-phenotype RETs (upper left quadrant, UL) are positioned directly above the mutant phenotype, mature RBCs (LL). It is appropriate for the cells with the highest FITC fluorescence to lean over to the right, as shown in the center plot in Fig. 14. If using a digital instrument capable of biexponential scaling, it can be useful to temporarily view the PE fluorescence with biexponential scaling. This view can

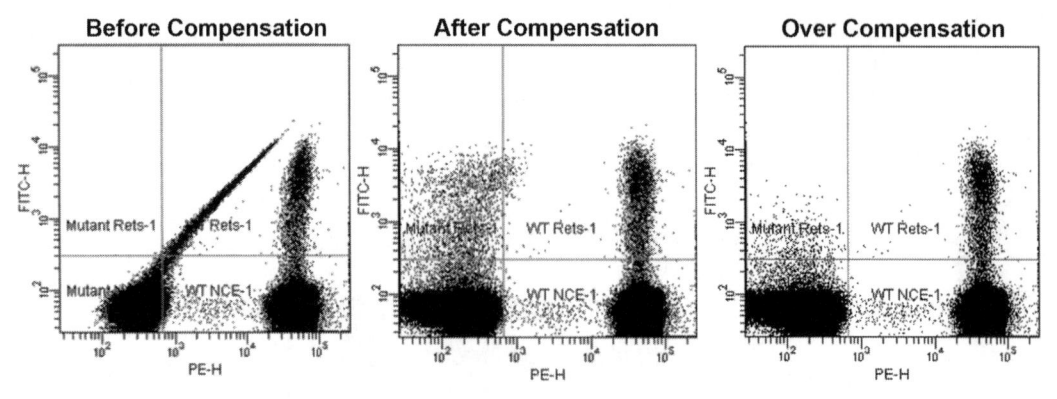

Fig. 14 Plot C: Adjust compensation to eliminate PE spillover into the FITC channel

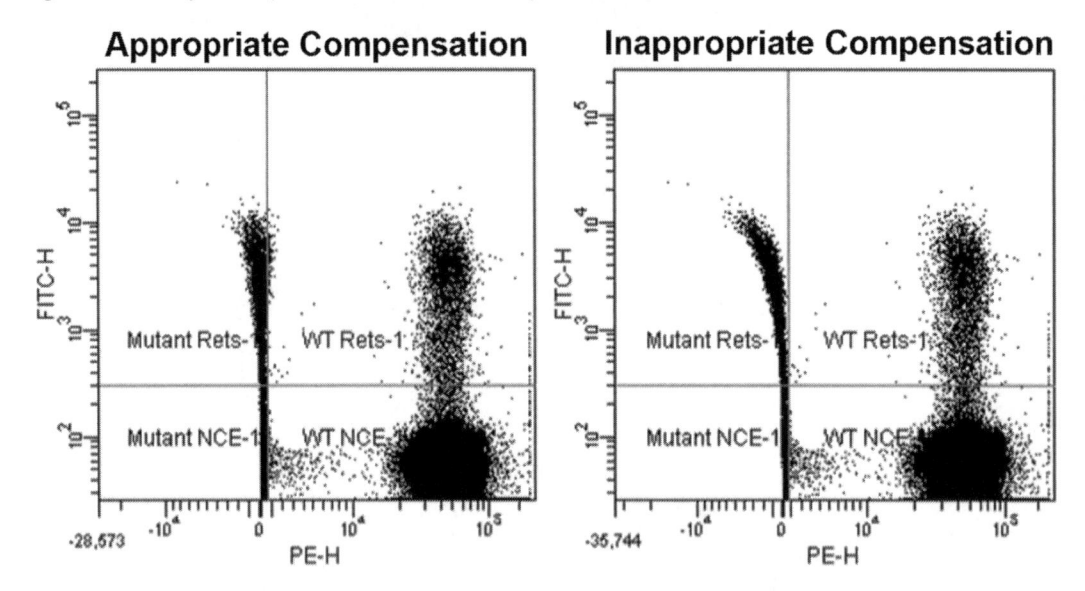

Fig. 15 Plot C: Appropriate versus inappropriate compensation on a digital machine

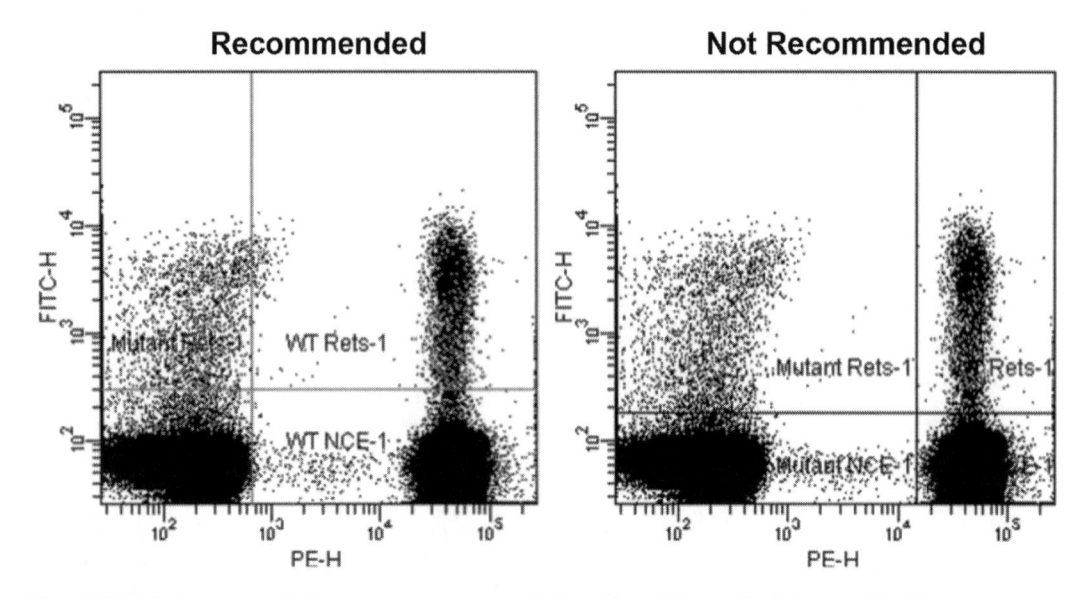

Fig. 16 Plot C: Recommended versus not recommended position of the vertical demarcation line

highlight overcompensation that may not be evident otherwise. The resulting plot should look similar to the one shown in Fig. 15.

9. Viewing Plot C, adjust the quadrant's position to ensure that it is appropriate. Use a conservative approach for scoring cells as mutant phenotype RBCs (i.e., these cells need to exhibit very low PE fluorescence, similar to that of the mutant mimics). The resulting plot should look similar to the plot on the left, in Fig. 16 (*see* **Note 39**).

3.11 Flow Cytometric Analysis: Experimental Samples

1. After analyzing the ICS sample and before analyzing experimental samples, place a tube of water on the flow cytometer and run for approximately 5 min to clear the lines of mutant-mimicking cells.

2. Having determined the appropriate PMT voltages and compensation settings with the ICS, keep these parameters constant when proceeding to the analysis of experimental samples. To ensure consistency of data, it is preferable that no changes be made to the location and size of the regions/quadrants between samples.

3. Remove the first pre-column sample from ice and pipette up and down until the cells and counting beads are well resuspended. Immediately place on the flow cytometer and begin acquiring data. Repeat until all pre-column samples have been analyzed (*see* **Note 40**).

4. Remove the first post-column sample from ice and pipette vigorously up and down until well suspended. Place on the flow cytometer, and begin acquiring data. Repeat until all post-column samples have been analyzed (*see* **Note 41**).

3.12 Mutant Cell Frequency Calculations

The data used to calculate % RET and mutant phenotype cell frequencies are derived from both pre-column and post-column analyses (*see* **Note 42**).

1. Abbreviations:
   (a) RETs = reticulocytes, RNA-positive fraction of total erythrocytes.
   (b) Mature RBCs = RNA-negative fraction of total erythrocytes.
   (c) RBCs = total erythrocytes, include both RNA-positive and -negative fractions.
   (d) UL = number of gated events occurring in Plot C's upper left quadrant, defined as mutant RETs.
   (e) UR = number of gated events occurring in Plot C's upper right quadrant, defined as wild-type RETs.
   (f) LL = number of gated events occurring in Plot C's lower left quadrant, defined as mature mutant RBCs.
   (g) LR = number of gated events occurring in Plot C's lower right quadrant, defined as mature wild-type RBCs.

(h) Counting beads = number of events occurring in Plot D's Counting Bead region.

2. Variables related to sample volumes:
   (a) a = Starting volume of antibody-labeled blood (μL), **step 1** of Subheading 3.6; usually 1,000 μL.
   (b) b = Volume of antibody-labeled blood added to working nucleic acid dye plus counting bead solution (μL), **step 2** of Subheading 3.6; usually 20 μL.
   (c) c = Volume of working nucleic acid dye plus counting bead solution used to prepare pre-column samples (μL), **step 2** of Subheading 3.6; usually 1,980 μL.
   (d) d = *LAB-SPECIFIC* value: The supernatant volume remaining in post-column samples following the final centrifugation and aspiration (μL), **step 2** of Subheading 3.9; should be between 20 and 50 μL.
   (e) e = Volume of working nucleic acid dye plus counting bead solution added to each post-column sample (μL), **step 4** of Subheading 3.9; usually 300 μL.

3. Calculations based on sample volume and dilution variables:
   (a) f = Cell dilution factor = $(b+c)/b$.
   (b) g = Cell concentration factor = $(a-b)/(d+e)$.
   (c) h = Bead dilution factor = $(e \times 100)/(d+e)$.

4. Pre-column data:
   (a) i = UL
   (b) j = UR
   (c) k = LL
   (d) l = LR
   (e) m = Counting beads

5. Calculations based on pre-column data:
   (a) n = Pre-column RBC-to-counting bead ratio = $(i+j+k+l)/m$
   (b) o = Pre-column RET-to-counting bead ratio = $(i+j)/m$
   (c) p = %RET = $(i+j)/(i+j+k+l) \times 100$

6. Post-column data:
   (a) q = UL
   (b) r = LL
   (c) s = Counting beads

7. Calculations based on pre- and post-column data:
   (a) t = Total RBC equivalents = $n \times s \times f \times g \times 100/h$
   (b) u = Total RET equivalents = $o \times s \times f \times g \times 100/h$

(c) $v =$ Number of mutant RBCs per 10^6 total RBCs $= (q+r)/t \times 10^6$

(d) $w =$ Number of mutant RETs per 10^6 total RETs $= q/u \times 10^6$

4 Notes

1. All sample preparation should be performed under sterile conditions in order to keep reagent vials from becoming contaminated. When working with any "bead" solution, be sure to sufficiently resuspend the particles before use and *do not sonicate* the stock or working solutions that contain beads.

2. Aseptically transfer 1,980 µL of working nucleic acid dye plus counting bead solution to one labeled flow cytometry tube for each pre-column sample. Periodically pipette the working nucleic acid dye plus counting bead solution up and down to ensure that the counting beads do not settle. Aseptically transfer 500 µL of working nucleic acid dye plus counting bead solution to a flow cytometry tube for the ICS sample (Part A). Cover the vessel containing unused working nucleic acid dye plus counting bead solution with foil and store at room temperature until needed again.

3. Aseptically transfer 100 µL of the working antibody solution to labeled microcentrifuge tubes, one for each sample. The working antibody solution is light sensitive; therefore cover these tubes with foil and store at 2–8 °C until needed.

4. Use an IACUC-approved method to collect blood. To prevent platelet activation and cellular aggregation, it is important that the blood is free flowing. For instance, if planning to collect blood by nicking the tail vein with a surgical blade, it is important to warm the animals under a heat lamp for several minutes. Once the blood starts flowing, use a heparin-coated capillary tube to collect approximately 100 µL. If planning to collect blood with a small-gauge needle and syringe, it is important to first coat the inside of the needle/syringe with a small amount of anticoagulant solution. Do not overfill the needle/syringe with excessive anticoagulant solution such that blood is overly diluted at this point. Rather, a ratio of one part anticoagulant solution to 9 parts whole blood is ideal. If not planning to label and analyze blood on the same day it is collected, or if you plan to ship blood samples to an off-site facility, it is preferable to transfer whole blood into EDTA Microtainer tubes.

5. Blood storage guidance: It is possible to store blood samples for up to 3 days after blood collection. If storing for subsequent labeling and analysis, collect blood as described in **step 3** of Subheading 3.1 and immediately transfer each whole blood sample (approximately 100 µL) into potassium EDTA

Microtainer tubes (e.g., BD). Store at 2–8 °C until blood dilution, leukodepletion, and platelet removal.

6. Sample shipping guidance: It is also possible to transport blood for off-site labeling and analysis. Blood samples collected into potassium EDTA Microtainer tubes can be shipped overnight, but they must be kept cold, not frozen. Litron recommends using Exakt-Pak® shipping containers and icepacks (e.g., Cold Chain Technologies) frozen at −20 °C. To prevent freezing of the blood samples, make sure that icepacks do not come into direct contact with the sample tubes.

7. Perform leukodepletion and subsequent washing steps at room temperature, and process all samples through **steps 1–8** of Subheading 3.2.

8. When removing from centrifuge, be aware that the resulting cell pellets will not be hard packed; therefore avoid tapping or bumping the tubes in a way that will cause the pellets to loosen or become dislodged from the bottom of the tube.

9. Do not add to the side of the tube, but directly into each supernatant.

10. Take care not to disturb the pellets.

11. The goal is to leave a minimal amount of supernatant behind.

12. Do not let the solution run down the side of the tube, as that can reintroduce cells removed by the Lympholyte®-Mammal.

13. These transferred cells can be stored at 2–8 °C for up to 24 h before proceeding with Subheading 3.3.

14. Continue as necessary until there is no visual evidence of aggregation—ten times is usually sufficient. **Steps 1** and **2** of Subheading 3.3 can be performed at room temperature, as long as processing of all samples occurs in less than 10 min. If it will take longer, maintain tubes on ice during processing.

15. Take care not to splash cells onto the side of the tubes. Continue pipetting up and down to ensure adequate mixing—ten times is usually sufficient. Ensure that all cells come into full contact with the working antibody solution. Working antibody solution is light sensitive. Keep the microcentrifuge tubes, both before and after addition of cells, protected from light.

16. This will become Part A of the ICS sample. There should be approximately 20–30 μL remaining in the tube after **step 2** of Subheading 3.3.

17. Be sure to only transfer cells that have been in contact with working antibody solution for the entire incubation period. For instance, do not transfer blood that may have been on the side of the microcentrifuge tube. Cap the centrifuge tubes and invert to mix.

18. The goal is to leave a minimal amount of supernatant behind.
19. Make sure not to splash cells high onto the sides of the tube. Continue mixing as necessary until there is no visual evidence of aggregation—ten times is usually sufficient.
20. Be sure not to splash cells high onto the sides of the tube. Continue mixing as necessary until there is no visual evidence of aggregation—four times is usually sufficient.
21. This dilution is critical in determining the final mutation frequencies, so be careful to transfer this exact amount.
22. These samples represent the *pre-column* samples. They will be incubated after preparing the ICS (Subheading 3.7).
23. These samples will be processed further (Subheading 3.8) and ultimately become *post-column* samples.
24. Store on ice for at least 5 min, but no more than 3 h, before flow cytometric analysis.
25. Avoid creating bubbles when adding samples or buffered salt solution + 2 % FBS to the column. Air bubbles can block the column and prevent the eluate from passing through. For this same reason, once wet, the column should not be allowed to dry. That is why it is important to add the next solution/suspension as soon as the previous one has fully entered the column (i.e., is no longer in the column reservoir).
26. Ensure that the bottoms of the columns are inside the open tops of the centrifuge tubes.
27. Be careful to avoid creating bubbles, and avoid disturbing the column matrix. Discard the pre-wetting solution once it stops dripping from the bottom of the columns.
28. Be careful not to disturb the top of the column matrix.
29. Be careful to avoid creating bubbles, and avoid disturbing the column matrix. The entire elution process should occur by the force of gravity only; DO NOT force buffer through the column with a plunger or other device. It takes approximately 5 min for the sample(s) and wash(es) to pass through the column and for the eluate(s) to collect in the centrifuge tube(s).
30. After centrifugation, the pellet will be small and difficult to see. This is normal.
31. It is critical that all samples have the same volume, and it is important to understand the average volume left in tubes. *See* **item 6** of Subheading 2.2 for a description of an apparatus for consistent aspiration that guards against disturbing the cell pellets.
32. Be careful that supernatants are not splashed high onto the sides of the tubes, as this may result in cells that do not come into contact with working nucleic acid dye plus counting bead solution.

33. Pipette to mix working nucleic acid dye plus counting bead solution prior to adding to the first sample, and to ensure a homogenous suspension of counting beads, pipette to mix after adding to every 4 or 5 samples. Carefully pipette up and down to resuspend the cells and counting beads, taking care not to splash onto the side of the tubes.

34. Ensure that the tubes are buried in the flaked/chipped ice, not resting on top.

35. It is advisable to ensure the appropriate template is running on your flow cytometer prior to processing samples. Do not wait until samples are ready for analysis before downloading or creating the template.

36. Place the remaining Part A sample back on ice in case you need to prepare another ICS sample and return the tube Part B to where the other pre-column samples are stored. This ICS now consists of adequate numbers of anti-CD59-PE-positive and -negative events to guide selection of PMT voltages and compensation settings.

37. A "Medium" flow rate is usually appropriate for digital instruments such as FACSCanto™ II as well as for analog instruments such as FACScan™ and FACSCalibur™.

38. One way that you can determine this is by looking at the "Y Geo Mean" values for the LL and LR quadrants. When these two values are approximately equal, compensation has been set correctly.

39. In the right-hand plot, below, the horizontal demarcation line that distinguishes mature RBCs from RETs is too low. This can lead to subtle variations in staining intensity causing greatly overestimated % RET values. Additionally, the vertical demarcation line is positioned too far right. This can lead to subtle variations in staining intensity causing greatly overestimated frequencies of mutant phenotype cells.

40. It is VERY IMPORTANT that each sample is pipetted IMMEDIATELY before analysis; ten times is usually sufficient. When analyzing pre-column samples, a medium fluidics rate is recommended in order to ensure that the event rate does not exceed approximately 2,000 events per second. Use a stop mode based on the length of time needed to acquire at least 1,000 counting beads. Some initial experimentation may be required to determine the specific time for each flow cytometer, but 2 min is usually sufficient when using a medium fluidics rate.

41. It is VERY IMPORTANT that each sample is pipetted IMMEDIATELY before analysis; ten times is usually sufficient. Use the same instrument settings as the pre-column

It is helpful to have 2 or 3 people available when many samples (≥ 16) will be collected and analyzed. Below is a recommended outline for a day of collecting and analyzing samples.

Day(s) before blood collection	Label all necessary tubes and vials with sample IDs. At least 2 flow cytometry tubes, 3 centrifuge tubes (15 ml), and 2 microcentrifuge tubes are required per sample (**step 5** of Subheading 2.2). Aliquot the anticoagulant into labeled vials prior to blood collection (**step 1** of Subheading 3.1).
Day of collection, early morning (3 people)	Two people collect blood samples (Subheading 3.1). / A third person prepares reagents and aliquots into pre-labeled tubes and vials (Subheading 2.2).
Day of collection, early morning (2 or 3 people)	Two people process all samples through Lympholyte, rinsing cell pellets, and adding to Working Antibody Solution. (Subheadings 3.2 & 3.3). If there are more than about 24 samples, it can be helpful for one person to aspirate after centrifugation (Subheading 3.2.3), while one or two people add the first rinse (Subheading 3.2.4) for all samples before moving to the next step. Repeat this process for the additional rinses.
Day of collection, late morning (2 or 3 people)	Two people add all samples to Buffered Salt Solution + 2 % FBS. (**step 1** of Subheading 3.4). After centrifugation (**step 2** of Subheading 3.4), one person aspirates (**step 3** of Subheading 3.4) while another adds Working Anti-PE MicroBead Suspension (**steps 1** and **2** of Subheading 3.5). If working up more than 24 samples at a time, have one person aspirate (**step 3** of Subheading 3.4), while one or two people add Working Anti-PE MicroBead Suspension (**step 1** and **2** of Subheading 3.5).
Day of collection, late morning (2 or 3 people)	After aspiration (**step 6** of Subheading 3.5), one person adds Buffered Salt Solution + 2 % FBS and resuspends cells (**step 1** of Subheading 3.6) while another removes 20 μl for Pre-Column samples (**step 2** of Subheading 3.6). If working up more than 24 samples at a time, have one person aspirate (**step 6** of Subheading 3.5), while one person adds Buffered Salt Solution + 2 % FBS and resuspends cells (**step 1** of Subheading 3.6). A third person could be removing 20 μl for Pre-Column samples (**step 2** of Subheading 3.6).
Day of collection, late morning to early afternoon (2 people)	After incubation, one person performs instrument calibration and analyzes Pre-Column samples on the flow cytometer (Subheadings 3.10 through 3.11). / One person performs column separation (in batches of up to 8 samples) and cell staining. (Subheadings 3.8 & 3.9).
Day of collection, early to late afternoon (1 person)	One person analyzes Post-Column samples on the flow cytometer (Subheading 3.11).

Fig. 17 A typical workday flowsheet diagram

samples. Do not adjust the location or the size of regions. When analyzing post-column samples, use the same fluidics rate that was used to acquire the pre-column data. Therefore, a medium fluidics rate is recommended for both digital instruments such as FACSCanto™ II, and analog instruments such as FACScan™ and FACSCalibur™. Use a stop mode based on the length of time needed to analyze nearly the entire volume of cells and counting beads. Some initial experimentation may be required to determine the specific time for each flow cytometer, but 5–6 min is usually sufficient when using a medium fluidics rate.

42. An Excel spreadsheet can be obtained from Litron (www.litronlabs.com/support.html or e-mail pigatechsupport@litronlabs.com more information). This spreadsheet can be used to make these calculations, and also provides examples of actual flow cytometric data. A typical work day flowsheet is shown in Fig. 17, to help in the planning of the experiments.

Acknowledgements

The authors would like to recognize the contributions made to this work by Dorothea Torous, Svetlana Avlasevich, and Souk Phonethepswath. This work was supported by a grant from the NIH-NIEHS to Stephen Dertinger (R44ES018017).

References

1. Hernández-Campo PM, Almeida J, Matarraz S et al (2007) Quantitative analysis of the expression of glycosylphosphatidylinositol-anchored proteins during the maturation of different hematopoietic cell compartments of normal bone marrow. Cytometry B Clin Cytom 72:34–42
2. Takahashi M, Takeda J, Hirose S, Hyman R et al (1993) Deficient biosynthesis of N-acetylglucosaminyl–phosphatidylinositol, the first intermediate of glycosyl phosphatidylinositol anchor biosynthesis, in cell lines established from patients with paroxysmal nocturnal hemoglobinuria. J Exp Med 177:517–521
3. Kawagoe K, Takeda J, Endo Y et al (1994) Molecular cloning of murine pig–a, a gene for GPI–anchor biosynthesis, and demonstration of interspecies conservation of its structure, function, and genetic locus. Genomics 23:566–574
4. Bryce SM, Bemis JC, Dertinger SD (2008) In vivo mutation assay based on the endogenous Pig–a locus. Environ Mol Mutagen 49:256–264
5. Miura D, Dobrovolsky VN, Kasahara Y et al (2008) Development of an in vivo gene mutation assay using the endogenous pig–a gene: I. Flow cytometric detection of CD59–negative peripheral red blood cells and CD48–negative spleen t-cells from the rat. Environ Mol Mutagen 49:614–621
6. Miura D, Dobrovolsky VN, Mittelstaedt RA et al (2008) Development of an in vivo gene mutation assay using the endogenous pig–a gene: II. Selection of pig-a mutant rat spleen t-cells with proaerolysin and sequencing pig-a cdna from the mutants. Environ Mol Mutagen 49:622–630
7. Kimoto T, Suzuki K, Kobayashi X et al (2011) Manifestation of Pig—a mutant bone marrow erythroids and peripheral blood erythrocytes in mice treated with N-ethyl-N-nitrosourea: direct sequencing of Pig–a cDNA from bone marrow cells negative for GPI–anchor protein expression. Mutat Res 723:36–42
8. Dobrovolsky VN, Miura D, Heflich RH et al (2010) The in vivo Pig—a gene mutation assay, a potential tool for regulatory safety assessment. Environ Mol Mutagen 51:825–835
9. Phonethepswath S, Bryce SM, Bemis JC et al (2008) Erythrocyte-based Pig—a gene mutation assay: demonstration of cross-species potential. Mutat Res 657:122–126
10. Dertinger SD, Phonethepswath S, Franklin D et al (2010) Integration of mutation and chromosomal damage endpoints into 28-day repeat dose toxicology studies. Toxicol Sci 115:401–411
11. Dertinger SD, Bryce SM, Phonethepswath S et al (2011) When pigs fly: immunomagnetic separation facilitates rapid determination of Pig-a mutant frequency by flow cytometric analysis. Mutat Res 721:163–170
12. Everds N (2007) Hematology of the laboratory mouse. In: Fox JG, Barthold SW, Davisson MT, Newcomer CE, Quimby FW, Smith AL (eds) The mouse in biomedical research, vol 3, 2nd edn. Academic, Burlington, MA
13. Koch M (2006) Experimental modeling and research methodology. In: Suckow MA,

Weisbroth SH, Franklin CL (eds) The laboratory rat, 2nd edn. Elsevier Academic, Burlington, MA
14. Car BD, Eng VM, Everds NE et al (2006) Clinical pathology of the rat. In: Suckow MA, Weisbroth SH, Franklin CL (eds) The laboratory rat, 2nd edn. Elsevier Academic, Burlington, MA
15. Schuler M, Gollapudi BB, Thybaud V et al (2011) On the need and potential value of the Pig–a in vivo mutation assay-a HESI perspective. Environ Mol Mutagen 52:685–689

Chapter 4

Detection of In Vivo Mutation in the *Hprt* and *Pig-a* Genes of Rat Lymphocytes

Vasily N. Dobrovolsky, Joseph G. Shaddock, Roberta A. Mittelstaedt, Daishiro Miura, and Robert H. Heflich

Abstract

Assays for in vivo mutation are used to identify genotoxic hazards and phenotypes prone to genomic instability and cancer. The hypoxanthine guanine phosphoribosyl transferase (*Hprt*) gene and the phosphatidyl inositol glycan, class A (*Pig-a*) gene are endogenous X-linked genes that can be used as reporters of mutation in peripheral blood lymphocytes from most mammals. Here we describe methodology for measuring *Hprt* and *Pig-a* mutation in rat T-lymphocytes. The identification and selective expansion of mutant lymphocytes are based upon the phenotypic properties of *Hprt*- and *Pig-a*-deficient cells, i.e., resistance to the purine analog, 6-thioguanine, or to the bacterial toxin, proaerolysin. Expanded mutants can be further analyzed by sequencing cDNA from the target transcripts for identification of small sequence alterations and by multiplex PCR analysis of genomic DNA for the detection of deletions.

Key words *Hprt*, *Pig-a*, alamarBlue™, Mutation, 6-Thioguanine, Proaerolysin

1 Introduction

Mutations, heritable changes in DNA sequence, are involved in many human diseases, including cancer [1]. Therefore, agents that induce mutations (i.e., mutagens) are considered to be risk factors for causing disease. Because it is often difficult to detect and measure mutations in disease-relevant genes (e.g., in proto-oncogenes, tumor supressors, or cell cycle regulators), proxy targets (or reporter genes) are often used to identify mutation. The reporter genes have characteristic mutant phenotypes that can be revealed under appropriate selective conditions. Although there are a number of in vitro assays developed to identify mutagens, the kinetics and specificity of metabolic processing and DNA repair in vitro may not be the same as in vivo. Thus, in vivo mutagenicity data are often important for evaluating human risk. The endogenous hypoxanthine guanine phosphoribosyl transferase (*Hprt*) gene was among the first reporter genes used for the detection of mutation in vitro [2]. Later, methods were developed for measuring in vivo mutation in *Hprt* and other endogenous reporter genes as well as in transgenic targets in genetically manipulated laboratory rodents (extensively reviewed in Lambert et al. [3] and in references therein; also discussed in other chapters of this book).

The endogenous X-linked *Hprt* gene is involved in the purine nucleotide salvage pathway. Both male and female somatic cells of placental mammals have a single functional copy of the *Hprt* gene (there is a single X-chromosome in male cells; in female cells one copy of the *Hprt* gene is nonfunctional due to transcriptional silencing). Cells having an inactivating mutation in the functional copy of the *Hprt* gene can be grown in the presence of the toxic purine analogue, 6-thioguanine (6-TG). In wild-type cells, Hprt enzyme converts 6-TG into a product that interferes with DNA synthesis and kills the cells.

The product of another reporter gene—the endogenous X-linked phosphatidyl inositol glycan, class A (*Pig-a*) gene—is involved in the synthesis of glycosyl phosphatidylinositol (GPI) anchors that tether multiple protein markers to the exterior surface of the cytoplasmic membrane [4]. *Pig-a* mutant cells lack GPI anchors and GPI-anchored markers. Marker-deficient mutants can be identified and enumerated with the help of immunofluorescent staining and flow cytometric analysis (discussed in another chapter in this book). GPI anchor-deficient cells also can be identified by their ability to expand in the presence of proaerolysin (proAER), a cytolytic toxin produced by the bacterium *Aeromonas hydrophila*. ProAER binds GPI anchors present on wild-type cells, and subsequently undergoes proteolytic conversion into aerolysin; aerolysin monomers, in turn, aggregate into multimers that form perforations in the cytoplasmic membrane and eventually kill the wild-type cells [5]. ProAER does not bind to GPI-deficient *Pig-a* mutant cells, and the mutants can be selectively expanded in a growth medium containing proAER.

The metabolic pathways for purine nucleotide metabolism and GPI anchor synthesis are conserved across mammalian species so that the *Hprt* and *Pig-a* genes potentially can serve as reporters of mutation in a variety of mammalian species, including laboratory rodents (transgenic and non-transgenic) and humans. Measurement of *Pig-a* and *Hprt* mutation in vivo, however, generally has been limited to blood cells. While *Pig-a* mutations can be detected in both erythrocytes and white blood cells by flow cytometry, (as also detailed in Chapter 3) limiting-dilution cloning of T-lymphocytes in the presence of proAER or 6-TG has the advantage of not only providing information on mutant frequency but also serving as a source of expanded mutant cells whose mutations can be further characterized.

The general procedures for determining the frequency and types of mutations in the *Hprt* and *Pig-a* genes of spleen lymphocytes from rats consist of the following steps:

1. Mononuclear peripheral white blood cells are released from the spleen, purified by density gradient centrifugation, and stimulated to proliferate by the mitogen concanavalin-A (Con-A; for *Hprt* mutants) or a combination of phorbol-12-myristate-13-acetate (PMA) and ionomycin (for *Pig-a* mutants).

2. Limiting-dilution cultures of lymphocytes are established in multi-well plates (usually in 96-well plates) in a medium that promotes T-lymphocyte growth in the presence of appropriate selecting agents (for expansion of mutants) and in a medium without selecting agents (for determining cloning efficiencies).

3. After 12–14 days of incubation, clones formed in individual wells of the plates are scored either by visual examination on an inverted microscope or by using a fluorescence plate reader and computer-assisted semi-automated counting [6], and the frequencies of mutant cells are calculated.

4. The 6-TG-resistant and proAER-resistant mutant clones can be further expanded and mutations in the target genes characterized by sequencing of *Pig-a* and *Hprt* cDNAs to identify point mutation and small deletions/insertions [7] or by multiplex PCR amplification of *Hprt* exons to identify large deletions within the gene [8].

5. When using the *Hprt* or *Pig-a* in vivo mutation detection assays for identification of potential genotoxic hazards, it is recommended that the mutant frequencies be determined in a vehicle-control group and three treatment groups, each group consisting of at least 6 animals.

2 Materials

2.1 Isolation and Priming of T-Lymphocytes

1. RPMI1640-based growth medium: 20 % HL-1™ medium (Lonza, Walkersville, MD), 13 % fetal bovine serum, 10 % T-STIM™ (Beckton Dickinson, Franklin Lakes, MA), 25 mM HEPES, 5 mM L-glutamine, 1× MEM nonessential amino acids, 1× sodium pyruvate, 0.2× penicillin–streptomycin, 55 µM 2-mercaptoethanol, 10 U/mL mouse interleukin-2 (IL-2; Roche, Indianapolis, IN), and RPMI1640 to final volume (*see* **Note 1**). The growth medium has been formulated after Dobrovolsky et al. [9].

2. Humidified CO_2 incubator (37 °C, 95 % humidity, 5 % CO_2).

3. Surgical instruments: One 6″ dissecting sharp scissors, one 6″ dressing forceps, two 4¾″ fine iris or dressing forceps, one 4″ micro dissecting curved scissors, disposable sterile razor blades.

4. 10 mL individually packaged disposable syringes with plungers that have serrated thumb rests, 25-gauge syringe needles.

5. 12-well plastic tissue culture plates (12WPs).

6. 75 cm² plastic tissue culture flasks (T75s).
7. Disposable 50 mL polypropylene tubes.
8. Disposable 15 mL polystyrene tubes.
9. 100 mm plastic Petri dishes (P100s).
10. Spray bottle with 70 % ethanol.
11. Lympholyte®-R (Cedarlane, Burlington, NC).
12. Con-A (250×; Worthington Biochemical, Lakewood, NJ): 1 mg/mL stock solution in RPMI1640 medium, sterilize using 0.2-μm filter, store in 2 mL aliquots at –20 °C.
13. Ionomycin: 1 mg/mL stock solution in DMSO, store in 100 μL aliquots at –80 °C. This is a 4,000× ionomycin solution.
14. PMA: 400 μg/mL stock solution in DMSO, store in 250 μL aliquots at –80 °C. This is a 40,000× PMA solution.

2.2 Lymphocyte Primary Culture

1. 12-channel pipettor and matching tips, 100 mL reagent reservoirs.
2. Disposable 50 mL polypropylene tubes, disposable 15 mL polystyrene tubes, and disposable 125 mL plastic media bottles.
3. Coulter®Z1 cell counter (Coulter, Miami, FL), Zap-Oglobin®II reagent (Coulter) for lysis of red blood cells.
4. Isotonic diluent, e.g., Hematall®.
5. X-ray irradiator, e.g., RS-2000 Biological Irradiator (Rad Source, Suwanee, GA).
6. 96-well round-bottom plastic tissue culture plates (96WPs).
7. 6-TG: 2 mg/mL in RPMI1640 medium. Weigh approximately 10 mg of powdered 6-TG, dissolve in two to three drops of 5 N aqueous NaOH, and add 5 mL of RPMI1640 medium to reach final concentration. Sterilize using 0.2 μm filter and store in 1 mL aliquots at –20 °C. This is a 1,000× 6-TG solution.
8. ProAER (available from Dr. Peter Howard, University of Saskatchewan, Canada): 1 μM solution in RPMI1640 medium. Dissolve 100 μg of lyophilized proAER in 2 mL of RPMI1640 medium, sterilize using 0.2 μm filter, and store in 2 mL aliquots at –80 °C. This is a 500× proAER solution.

2.3 Scoring Clones, Determining Mutant Frequencies, and Expansion of Mutant Clones

1. Inverted microscope, 40–100× magnifications.
2. 12-channel pipettor, tips, and reagent reservoirs.
3. alamarBlue™ viability indicator (Trek Diagnostics, Cleveland, OH).
4. Fluorometer capable of reading 96WPs.
5. 24-well plastic tissue culture plates (24WPs).
6. Phosphate-buffered saline (PBS), pH 7.0.

2.4 Amplification of Hprt and Pig-a cDNA and Multiplex PCR of Hprt Exons

1. Thermocycler for PCR.
2. Primers for *Hprt* cDNA amplification: R1F and R5R. Primers for multiplex amplification of *Hprt* exons from genomic DNA: RP2F, R124R, R265F, R264R, R307F, R308R, R403F, R404R, R505F, R506R, R601F, R602R, R701F, R804R, R901F, and R906R (for primer sequences *see* **Note 2**).
3. Primers for *Pig-a* cDNA amplification: 1F and WR. Primers for *Pig-a* nested PCR: 1F, 1R, 2F, 2R, 3F, 3R, 4F, and 4R. Primers for *K-ras* exon 2 amplification: K2-A and K2-B (for primer sequences *see* **Note 2**).
4. RNA release cell lysis solution for synthesis of *Hprt* cDNA: 2.5 % Nonidet P40 and 0.4 U/μL RNasin® (ribonuclease inhibitor; Promega, Madison, WI) in nuclease-free water.
5. Access RT-PCR System (Promega), for synthesis and amplification of *Hprt* cDNA.
6. MasterPure RNA Purification Kit (Epicentre Technology, Madison, WI), for extraction and purification of *Pig-a* cDNA.
7. 70 % ethanol, 100 % isopropanol, for use with MasterPure RNA Purification kit.
8. Reagents for *Pig-a* cDNA synthesis: Oligo(dT) primer (Ambion, Austin, TX), 10 mM dNTP mix (2.5 mM each), RETROScript™ RT buffer (Ambion), RNaseOUT™ ribonuclease inhibitor (Invitrogen, Grand Island, NY), SuperScript™ II reverse transcriptase (Invitrogen).
9. HotStarTaq™ DNA polymerase kit (Qiagen, Valencia, CA), for *Pig-a* cDNA amplification.
10. DNA release cell lysis buffer for amplification of genomic DNA of *Hprt* mutant clones: 10 mM Tris–HCl (pH 7.5), 25 mM MgCl$_2$, 0.45 % Triton-X, 0.45 % Tween-20, 0.4 mg/mL Proteinase K in nuclease-free water.
11. Reagents for multiplex amplification of *Hprt* exons: 1 M KCl, 1M Tris-HCl (pH 8.3), 1 M MgCl$_2$, DMSO, primers, 10 mM dNTP mix (2.5 mM each), light mineral oil, AmpliTaq DNA polymerase (Invitrogen).
12. Disposable 1.5 mL tubes, disposable PCR tubes.
13. Electrophoresis-grade agarose.
14. Horizontal gel electrophoresis apparatus and electrophoresis power supply.

3 Methods

3.1 Isolation of Lymphocytes

1. Prior to necropsy, sterilize the surgical instruments and prepare sterile growth medium (at least 80 mL of medium for each animal if assayed only for *Hprt* mutation or 120 mL if assayed only for *Pig-a* mutation; *see* **Note 3**).

2. In the animal processing room, euthanize the animals using methods approved by your Institutional Animal Care and Use Committee (e.g., CO_2 asphyxiation). In an aseptic environment on the lab bench, place sacrificed animal on its right side and soak the left side with 70 % ethanol. Pinch and lift the skin below the rib cage on the left side with 6″ forceps and, using large scissors, slice off a 2–3 cm^2 piece of skin without cutting into the abdomen. Identify the spleen under the body wall by its characteristic dark purple color. Use separate sets of scissors and forceps for external and internal surgical procedures. Lift the peritoneum with small forceps and make a 10–15 mm incision over the area where the spleen is located. Gently pull the spleen from the abdomen through the incision with small forceps; cut out the intact spleen, trimming off as much connecting tissue and fat as possible. Place the spleen into a 50 mL tube containing 3 mL of RPMI1640 medium, cap the tube, and place on ice until all animals are processed.

3. In a biological safety cabinet (i.e., laminar flow tissue culture hood), pour a spleen from its tube into a P100; using a sterile razor blade, cut the spleen into four approximately equal pieces. One piece of the spleen is usually sufficient for performing a mutation detection assay for one gene. If mutation assays are performed for both the *Hprt* and the *Pig-a* genes, then two pieces of spleen should be processed separately.

4. Add 3 mL of Lympholyte®-R into a 15 mL tube (one tube for each piece of spleen to be processed). Using sterile forceps, place one piece of spleen into a well of a 12WP filled with 3 mL of RPMI1640 medium. Crush the piece of spleen with the serrated thumb rest end of a 10 mL syringe plunger using several squeeze-and-twist motions. Slowly aspirate the cloudy medium containing released lymphocytes into another sterile 10 mL syringe fitted with a 25-gauge needle. Holding the needle against the wall of the 15 mL tube, slowly layer the contents of the syringe on top of the 3 mL of Lympholyte®-R.

5. Centrifuge the gradient tubes for 20 min at $1,500 \times g$ at room temperature. White blood cells (including T-lymphocytes) will concentrate at the interface of the clear Lympholyte®-R and the pink RPMI1640 medium. Transfer the lymphocyte fraction into a new 15 mL tube, add 5 mL of RPMI1640 medium, mix the contents with a few gentle inversions, and centrifuge for 10 min at $800 \times g$ at room temperature.

6. Discard the supernatant, resuspend the cell pellet in 5 mL of complete growth medium, and transfer the entire cell suspension into a T75 flask containing 15 mL of growth medium. Add 4 μg/mL Con-A (final concentration) for the *Hprt* assay or the mixture of 0.25 μg/mL PMA and 10 ng/mL ionomycin (final concentrations) for the *Pig-a* assay.

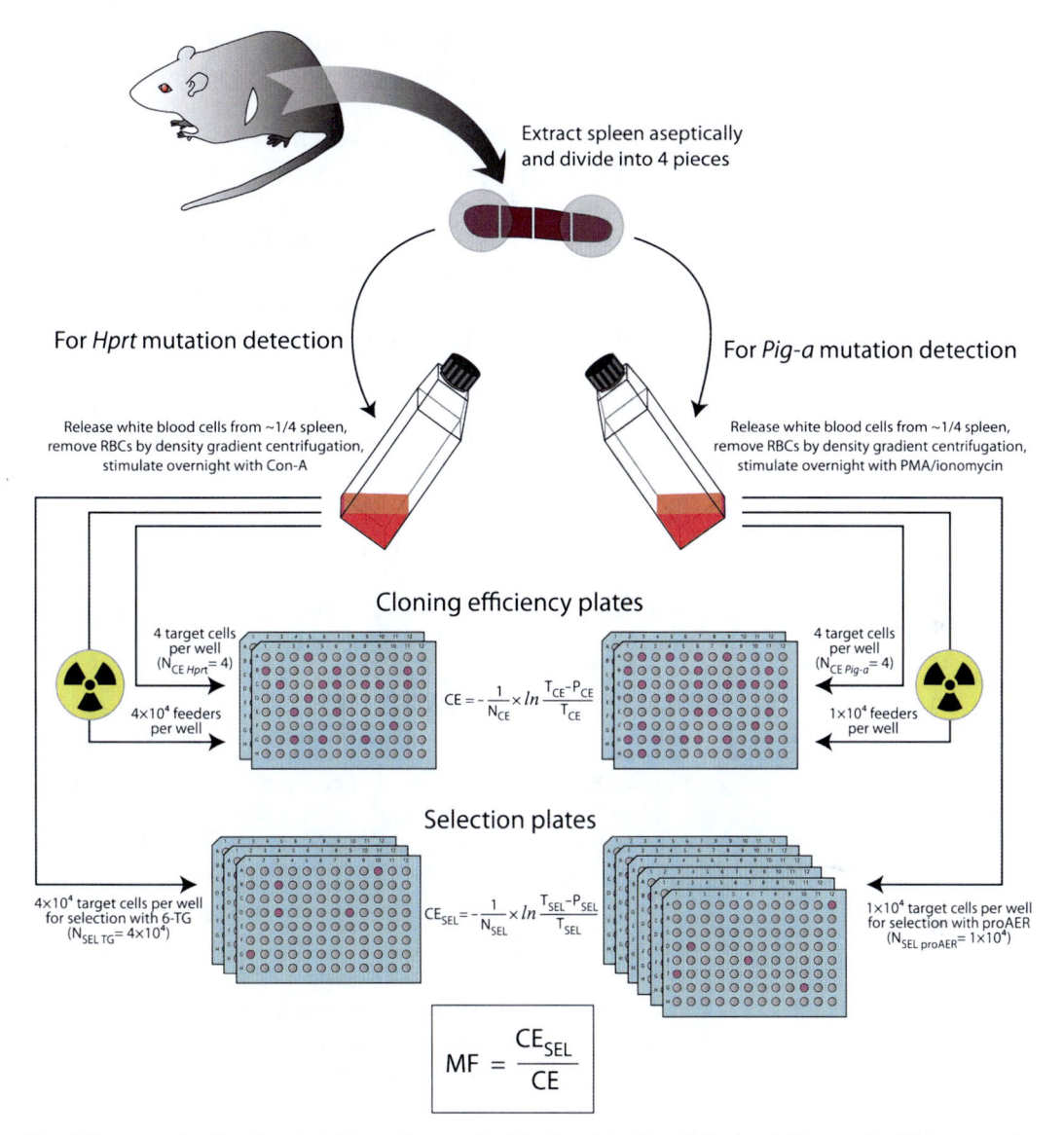

Fig. 1 Diagram of cell cultures and formulas required for the detection of *Hprt* and *Pig-a* mutant T-lymphocytes and the calculation of mutant frequencies using the Poisson distribution

7. Place the flasks into the CO_2 incubator, standing at a 45° angle to allow the cells to concentrate in the corner of the flask between the wall and bottom. Leave the screw caps loose to allow atmosphere exchange, and incubate at 37 °C overnight.

3.2 Limiting-Dilution Culture of Lymphocytes for Detection of Hprt Mutation

1. The next morning thaw a vial containing the stock solution of the selecting agent 6-TG; keep on ice until needed. Consult Fig. 1 for general information on the types of cell cultures required for the entire process.

2. Dispense 3 mL of growth medium into 15 mL tubes (one tube for each piece of spleen processed).

3. Remove the T75 flasks with Con-A-stimulated splenocyte cultures from the CO_2 incubator. Resuspend the settled cells by gentle agitation, and transfer 0.2 mL of the cell suspension into a counting vial filled with 9.8 mL of isotonic diluent. Add three drops of Zap-Oglobin®II to the vial, cap the vial, and mix the contents by vigorous inversion. Wait at least 30 s to allow Zap-Oglobin to completely lyse RBCs in the sample. Determine the concentration of mononuclear cells in the overnight cultures using the counts returned by the counter and the dilution factors (*see* **Note 4**).

4. Transfer 30 µL of the cell suspension from the T75 into the 15 mL tube containing 3 mL of growth medium to make a 1:100 dilution of cells for determining cloning efficiency (CE_{Hprt}) in the absence of the selection agent.

5. In 50 mL tubes (one tube for each piece of spleen), mix 13.2×10^6 cells from each overnight Con-A culture (from **step 7** of Subheading 3.1) with fresh growth medium to make up a cell suspension with a final concentration of 4×10^5 cells/mL in a final volume of 33 mL. Add 33 µL of the 1,000× 6-TG stock solution to the cell suspension to make a final concentration of 2 µg/mL 6-TG in the selection cultures. Cap the tubes and mix by inversion. These are the mixtures for determining $CE_{SEL\ TG}$.

6. Combine the unused cells from the overnight Con-A cultures from all animals (leftovers from the previous step) into one T75 and determine the cell concentration of the pooled cells as described above (these cells will be used as feeders for the CE_{Hprt} plates; alternatively, a few unused pieces of spleen from various rats can be used as a source of feeder cells). The unused cultures are pooled together for simplicity to make one large volume of feeders, irradiate them in one flask, and use for all CE cultures (without selecting agents). Alternatively, each CE culture can use feeders from the same animal, then multiple tubes with feeders should be irradiated, and calculation of volumes of the feeders and the medium for each CE culture will be more complicated. Irradiate these cells with 90 Gray in an RS-2000 irradiator or using another source of ionizing radiation. In a 50 mL tube, mix growth medium, the 1:100 dilution of cells (from **step 4** of Subheading 3.2), and the irradiated feeder cells to make up a final volume of 23 mL containing final concentrations of 40 nonirradiated target cells/mL, and 4×10^5 irradiated feeder cells/mL. The required volumes of feeder cells depend upon the concentration of cells in feeder cultures determined previously in this step; the required volumes of 1:100 dilutions of non-irradiated target cells depend upon the concentration of cells in Con-A-stimulated cultures determined in **step 3** of Subheading 3.2, divided by the factor of 100.

7. Pour the cell suspension for determining CE_{Hprt} (**step 6** of Subheading 3.2) from its 50 mL tube into a 100 mL reagent reservoir. Using the 12-channel pipettor, dispense 100 µL of cell suspension into each well of two 96WPs. Discard any leftovers. Pour the cells for determining $CE_{SEL\ TG}$ belonging to the same animal (from **step 5** of Subheading 3.2) into the emptied reservoir and dispense 100 µL of 6-TG-containing culture into each well of three 96WPs. Replace the reservoir and pipette tips and dispense the CE and 6-TG cell suspensions from the next animal. With the suggested concentrations of cells and dispensing volumes, each well in the CE_{Hprt} plates will contain four target cells ($N_{CE\ Hprt}$) and 4×10^4 irradiated feeder cells; each well in the $CE_{SEL\ TG}$ plates will contain 4×10^4 cells ($N_{SEL\ TG}$).

8. Load the 96WPs into the CO_2 incubator and incubate for 10–11 days at 37 °C.

3.3 Limiting-Dilution Culture of Lymphocytes for Detection of Pig-a Mutation

1. Thaw a vial containing the 1 µM (500×) proAER stock solution (selecting agent); keep on ice until needed.

2. Dispense 3 mL of growth medium into 15 mL tubes (one tube for each piece of spleen processed).

3. Remove the flasks with PMA/ionomycin-stimulated splenocyte cultures from the CO_2 incubator. Resuspend the settled cells by gentle agitation, and transfer 0.2 mL of the cell suspension into a counting vial filled with 9.8 mL of isotonic diluent. Add three drops of Zap-Oglobin®II to the vial, cap the vial, mix the contents by vigorous inversion, and determine the cell concentration using a Coulter counter (*see* **step 3** of Subheading 3.2 and **Note 4**).

4. Transfer 30 µL of the cell suspension from the T75 into a 15 mL tube containing 3 mL of growth medium to make a 1:100 dilution of cells for determining cloning efficiency in the absence of the selection agent (CE_{Pig-a}; same as in **step 4** of Subheading 3.2).

5. In 125 mL bottles (one bottle for each piece of spleen), mix 6.3×10^6 cells from each overnight PMA/ionomycin culture (from **step 7** of Subheading 3.1) with fresh growth medium to make up a cell suspension with a final concentration of 1×10^5 cells/mL in a final volume of 63 mL. Add 126 µL of the 500× stock of proAER to each bottle with 63 mL of cells to make a final concentration in the selection cultures of 2 nM proAER. Cap the bottle and mix by inversion. These are the mixtures for determining $CE_{SEL\ proAER}$.

6. Combine the unused cells from the overnight PMA/ionomycin cultures (from **step 5** of Subheading 3.3) into one T75 and determine the concentration of the pooled cells as described above (these cells will be used as feeders for the CE_{Pig-a} plates;

alternatively, a few unused pieces of spleen from several rats can be used as a source of feeder cells). Irradiate these cells with 90 Gray in an RS-2000 irradiator or using another source of ionizing radiation. In a 50 mL tube, mix growth medium, the 1:100 cell dilution (from **step 4** of Subheading 3.3), and irradiated feeder cells to make up a final volume of 23 mL containing final concentrations of 40 nonirradiated target cells/mL and 1×10^5 irradiated feeder cells/mL. The required volumes of feeder cells depend upon the concentration of cells in feeder cultures determined previously in this step; the required volumes of 1:100 dilutions of non-irradiated target cells depend upon the concentration of cells in PMA/ionomycin-stimulated cultures determined in **step 3** of Subheading 3.3, divided by the factor of 100.

7. Pour the cell suspension for determining CE_{Pig-a} (from **step 6** of Subheading 3.3) from its 50 mL tube into a 100 mL reagent reservoir. Using the 12-channel pipettor, dispense 100 μL of cell suspension into each well of two 96WPs. Discard any leftovers. Pour the cells for determining $CE_{SEL\ proAER}$ belonging to the same animal (from **step 5** of Subheading 3.3) into the emptied reservoir and dispense 100 μL of proAER-containing culture into each well of six 96WPs. Replace the reservoir and pipette tips and dispense the CE_{Pig-a} and $CE_{SEL\ proAER}$ cell suspensions from the next animal. With the suggested concentrations of cells and dispensing volumes, each well of the CE_{Pig-a} plates will contain four target cells ($N_{CE\ Pig-a}$) and 1×10^4 irradiated feeder cells; each well of the $CE_{SEL\ proAER}$ plates will contain 1×10^4 cells ($N_{SEL\ proAER}$).

8. Load the 96WPs into a CO_2 incubator and incubate for 10–11 days at 37 °C.

3.4 Scoring Lymphocyte Clones in 96WPs

1. *Manual method.* After 11 days of culture, inspect all wells of each plate using an inverted microscope at 40× magnification. Mark wells that contain expanded clones (positive wells). Expanded clones have common characteristics: the overall size of the cell mass in the well is relatively large; elongated or rounded cells are present on the periphery of the cell mass; and the majority of individual cells on the periphery have sharp refractive membranes. Dead cells are small, without a distinct refractive membrane; and the overall amount of the cell mass in a negative well is smaller. Switch the microscope to 100× magnification if needed for detailed examination of the cells on the periphery of the cell mass.

2. *Alternative automated method.* After 10 days of culture (*see* **Note 5**), make a 5 % solution of alamarBlue™ (v/v) in growth medium (2.5 mL for each 96WP). Using the 12-channel

pipettor, add 25 μL of alamarBlue™-containing medium to each well of all plates (*see* **Note 6**). Return plates to the CO_2 incubator for an additional overnight culture. The next day, read all plates with the fluorometer using a 530 nm excitation filter and a 590 nm emission filter, a gain of 47, and four flashes per well. Using the fluorescence data array generated by the reader, identify the well with the minimum fluorescence (MIN) for each plate. Determine the wells with fluorescence twofold or higher than the MIN; these are scored as positive wells for this plate. Determine a MIN value for each plate and use it to identify the number of positive wells on the plate using the 2× MIN criterion (*see* **Note 7**).

3. *Calculating mutant frequencies.* Calculate mutant frequencies for each gene for each individual animal using the set of cloning efficiency and selection plates for this specific animal. Count the total number of positive wells in CE plates without selection (P_{CE}), and the total number of positive wells in selection plates (P_{SEL}). Calculate the CE of cells without selection using the formula $CE = -1/N_{CE} \times ln([T_{CE} - P_{CE}]/T_{CE})$, where T_{CE} is the total number of wells seeded with target cells in the medium without selection (the number of CE plates multiplied by 96, or 2×96 in our case). Calculate the cloning efficiency of cells grown in selection media (CE_{SEL}) using the formula $CE_{SEL} = -1/N_{SEL} \times ln([T_{SEL} - P_{SEL}]/T_{SEL})$, where T_{SEL} is the total number of wells on selection plates (3×96 in the case of *Hprt* mutation detection and 6×96 in the case of *Pig-a* mutation detection). Determine the *Hprt* mutant frequency (MF_{Hprt}) using the formula $MF_{Hprt} = CE_{SEL\ TG}/CE_{Hprt}$. Determine the *Pig-a* mutant frequency (MF_{Pig-a}) using the formula $MF_{Pig-a} = CE_{SEL\ proAER}/CE_{Pig-a}$.

4. *Expansion of lymphocytes resistant to selecting agents beyond 96WPs.* Transfer the entire contents of a positive well into an individual well of a 24WP containing 0.5 mL of growth medium supplemented with Con-A or PMA/ionomycin at the final concentrations given in **step 6** of Subheading 3.1. After incubating the plate (angled at 30°) in a CO_2 incubator overnight, add another 0.5 mL of growth medium (without mitogens) and continue the incubation for an additional 2–5 days. Examine the wells daily for cell growth. For freezing expanded cells, resuspend the cells in the well medium by repeated pipetting and divide the cell suspension into two 1.5 mL microcentrifuge tubes containing 0.5 mL PBS. Spin the tubes for 10 min at 800×*g*, remove the supernatants, freeze the cell pellets on dry ice, and store the tubes at −80 °C.

3.5 cDNA Amplification and Analysis of Deletion Mutations

1. *RT-PCR amplification of Hprt cDNA from frozen cells*: Hprt cDNA synthesis and amplification are performed in a single step. Thaw the tubes containing cell pellets (from **step 4** of Subheading 3.4) on ice, resuspend each cell pellet in 50 µL of ice-cold RNA release cell lysis solution, and allow the tubes to remain on ice for 30 min. In fresh PCR tubes, combine the cell lysate and primers with components of the Access RT-PCR Kit System: 4 µL of reaction buffer, 0.4 µL of dNTP mix, 0.8 µL of MgSO$_4$ stock, 0.4 µL AMV reverse transcriptase, 0.4 µL *Tfl* DNA polymerase, 4 µL of cell lysate (containing released RNA), 1 µM of each primers R1F and R5R (final concentrations), and water to a final volume of 30 µL. Process the mixtures in a thermocycler using a temperature profile of 48 °C × 45 min + 94 °C × 2 min + (94 °C × 30 s + 60 °C × 1 min + 68 °C × 2 min) × 40 + 68 °C × 7 min. Analyze 5 µL of the RT-PCR products on a 1 % agarose gel. The full size of the amplified *Hprt* cDNA fragment is 867 bp. Perform sequencing of the amplified *Hprt* cDNA using your favorite protocol (*see* **Note 8**).

2. *Amplification of Pig-a cDNA from frozen cells*: *Pig-a* cDNA synthesis and PCR amplification are performed in separate steps. Extract total RNA from a cell pellet of an expanded clone using the MasterPure RNA purification kit and the manufacturer's suggested protocol. Reconstitute the total RNA in a final volume of 50 µL of kit-supplied RNA hydration solution. Synthesize total cDNA in a 20 µL reaction mixture containing 2 µL of total RNA, 1.25 µM oligo(dT), 0.2 mM of each dNTP, 1× RETROScript RT buffer, 40 U RNaseOUT ribonuclease inhibitor, 200 U SuperScript II reverse transcriptase, and nuclease-free water to final volume. Process the mixtures in a thermocycler using a temperature profile of 42 °C × 60 min + 97 °C × 3 min. Amplify *Pig-a* cDNA using an aliquot of total cDNA and components of the HotStarTaq kit. In a final volume of 20 µL, mix 2 µL of total cDNA, 1 µM of each primers 1F and WR (final concentrations), kit-supplied buffer to a final concentration of 1×, 2 mM dNTPs (final concentrations), 20 % kit-supplied Q-solution, and 1 U of HotStarTaq DNA polymerase. Process in a thermocycler using a temperature profile of 95 °C × 15 min + (94 °C × 1 min + 62 °C × 1 min + 72 °C × 3 min) × 30 + 72 °C × 7 min. The size of the full first-round PCR *Pig-a* amplicon is ~1.5 kbp, but it is rarely visible on an agarose gel. Nested PCR is required for amplification of material sufficient for subsequent sequencing.

3. *Pig-a cDNA nested PCR*: Dilute the product of the first-round RT-PCR 1:100 with water. The second (nested) PCR of the

Pig-a cDNA product is best done using either three or four separate amplification reactions that produce overlapping fragments. Amplify the fragments of *Pig-a* cDNA in final volumes of 20 μL using 2 μL of the 1:100 dilution and the components of HotStarTaq kit at concentrations similar to the first-round PCR. In every nested PCR reaction use one pair of primers at final concentrations of 1 μM each. Amplify nested products using the following temperature profiles: for primer pairs 1F/1R, 2F/2R, and 4F/4R use 95 °C × 15 min + (94 °C × 1 min + 60 °C × 1 min + 72 °C × 2 min) × 30 + 72 °C × 7 min (fragment sizes ~500 bp, ~460 bp, and ~450 bp, respectively); for primer pair 3F/3R use 95 °C × 15 min + (94 °C × 1 min + 66 °C × 1 min + 72 °C × 2 min) × 30 + 72 °C × 7 min (fragment size ~450 bp); and for primer pair 3F/4R use 95 °C × 15 min + (94 °C × 1 min + 64.5 °C × 1 min + 72 °C × 2 min) × 33 + 72 °C × 7 min (fragment size ~800 bp). Analyze 5 μL of each nested PCR product on a 1 % agarose gel. Perform sequencing of the cDNA fragments using your favorite protocol.

4. *Preparation of genomic DNA from expanded 6-TG-resistant Hprt mutant T-lymphocyte clones*: Resuspend the cells in 100 μL of DNA release cell lysis buffer. Incubate at 60 °C for 1 h; then inactivate Proteinase K by incubation at 95 °C for 15 min. Store the digest at −80 °C until ready for PCR analysis.

5. *Multiplex PCR amplification of Hprt exons for analysis of deletion mutations*: The final concentrations of individual components of multiplex PCR assume that the total volume of the reaction is 30 μL. In a 0.5 mL thin-walled PCR tube, combine in a final volume of 29 μL the following components: 2 μL of lymphocyte digest (from **step 4** of Subheading 3.5), nine pairs of primers to the final concentrations listed in Table 1, 125 mM KCl, 25 mM Tris–HCl (pH 8.3), 3.75 mM MgCl$_2$, 5 % DMSO, and 0.25 mM each dNTPs (final concentrations for all components are given). Overlay the reaction mixture with 50 μL of light mineral oil, denature DNA at 100 °C for 5 min, cool to 85 °C for 5 min, and add 1 μL (5 U) of AmpliTaq DNA polymerase to the tube. Immediately amplify nine DNA fragments with sizes listed in Table 1 using the temperature profile (94 °C × 1 min + 60 °C × 1 min + 72 °C × 1 min) × 35 + 72 °C × 10 min. Analyze the results of the amplification by electrophoresis on a 2 % agarose gel. If ambiguous results are produced for specific fragments, confirmation may be performed in duplex PCR reactions containing primers for amplification of *Hprt* exon(s) in question and for amplification of the internal control, exon 2 of the endogenous *K-ras* gene.

Table 1
Final concentrations of primers used for multiplex PCR of *Hprt* exons

Amplicon	Fragment size (bp)	Primer	Concentration in multiplex amplification (µM)
Hprt exon 1	296	RP2F	0.50
		R124R	0.50
Hprt exon 2	337	R265F	0.15
		R264R	0.15
Hprt exon 3	391	R307R	0.12
		R308R	0.12
Hprt exon 4	240	R403F	0.15
		R404R	0.15
Hprt exon 5	272	R505F	1.00
		R506R	1.00
Hprt exon 6	498	R601F	0.08
		R602R	0.08
Hprt exons 7 and 8	444	R701F	0.15
		R804R	0.15
Hprt exon 9	684	R901F	0.08
		R906R	0.08
K-ras exon 2	191	K2-A	0.20
		K2-B	0.20

4 Notes

1. For making growth medium, use the following tissue culture-grade concentrated stocks available from various reputable suppliers: 1 M HEPES (as 40× stock), 200 mM L-glutamine (40×), 10 mM MEM nonessential amino acids (100×), 100 mM sodium pyruvate (100×), 10,000 units/mL penicillin and 10,000 µg/mL streptomycin (penicillin–streptomycin mix, 100×), 55 mM 2-mercaptoethanol (1,000×), and 10,000 U/mL IL-2 (1,000×).

   Prepare growth medium from sterile components or sterilize by filtration through a 0.2 µM filter; keep at 4 °C for up to 2 weeks before use. For the detection of *Hprt* mutant lymphocytes T-STIM supplement may be omitted from the growth medium with a compensating increase in the volume of RPMI1640 medium. T-STIM supports better growth of proAER-resistant lymphocyte clones.

2. Sequences of primers used for PCR and RT-PCR.

Primer	5' → 3' sequence	Purpose
R1F	CTTCCTCCTCAGACCGCTTT	*Hprt* cDNA amplification
R5R	GGACGCAGCAACAGACATTTC	*Hprt* cDNA amplification
1F	CACATTTCTTAACAGGCTTGCT	*Pig-a* cDNA amplification
WR	AAAACCCAAAAACAATGAGCAA	*Pig-a* cDNA amplification
1R	GACTGTCTGAAGCCCCATTG	*Pig-a* cDNA nested PCR
2F	CGATAATCCATTCCCACAGTT	*Pig-a* cDNA nested PCR
2R	CCGGACTTCTTCCAAAATGA	*Pig-a* cDNA nested PCR
3F	AAGGGACTGATTTGCTTAGTGGTA	*Pig-a* cDNA nested PCR
3R	CGCCAGGTGTAGAAAGTCTTCA	*Pig-a* cDNA nested PCR
4F	TTTGTGAAGGGGTCTGGAAAAA	*Pig-a* cDNA nested PCR
4R	CAAATGCTAACAGGACTATTGCA	*Pig-a* cDNA nested PCR
RP2F	GAAGACGAGCGTTGGACTTACCTCA	*Hprt* exon 1 amplification
R124R	CAACCAGCCAGACTCCAGGAATGT	*Hprt* exon 1 amplification
R265F	GAATTAAATGTGGATATCT TATAATGCTT	*Hprt* exon 2 amplification
R264R	TATCAGTATGTATGCTCCAGGTTCA	*Hprt* exon 2 amplification
R307R	CAGGATGTGTCCTGTAGAAGTTT	*Hprt* exon 3 amplification
R308R	AATTGCACCATCATCAAGTAAGA	*Hprt* exon 3 amplification
R403F	TTGTAACCATTAGATAGCTAACT TCTTAGA	*Hprt* exon 4 amplification

(continued)

(continued)

Primer	5′ → 3′ sequence	Purpose
R404R	AGTTGAAAGCAGTTACTAATTGATTCTTGG	*Hprt* exon 4 amplification
R505F	CCCTTCTGAGTTCTAATAAGCCCAC	*Hprt* exon 5 amplification
R506R	GTCTCCCTGGCTTACCTTTATTAAT	*Hprt* exon 5 amplification
R601F	CCCACTGCTTGCTTAGAACCAGATA	*Hprt* exon 6 amplification
R602R	GGTAGGTAGGGAAGGGAGGACAGAT	*Hprt* exon 6 amplification
R701F	TCTGTAGTCTTCCCATATGTCCCATG	*Hprt* exons 7 and 8 amplification
R804R	TCAAATTACGAGGTGCTGGAAGGAGA	*Hprt* exons 7 and 8 amplification
R901F	GCTGGTGTTGTCTCTTGAGAATCCA	*Hprt* exon 9 amplification
R906R	GAAATGATAGGCATGGAGAATTCCT	*Hprt* exon 9 amplification
K2-A	TTCTCAGGACTCCTACAGGAAA	*K-ras* exon 2 amplification
K2-B	CCCACCTATAATGGTGAATATC	*K-ras* exon 2 amplification

3. The detection of mutant clones and calculation of mutant frequencies require establishing multiple cell cultures, containing precise numbers of target cells for each experimental animal. Since the yield of mononuclear cells from spleens of individual animals is variable, custom mixes of cells and media must be made for each experimental animal. In order to make consistent cell cultures, a properly designed computer spreadsheet template may significantly simplify the efforts in determining correct volumes of cells and media and adjusting these values in cases when fewer or more tissue culture plates are established or when the concentrations of cells are changed to reflect experimental modifications.

4. Most automated cell counters are configured to count cells in a 0.5 mL volume. The cell count for 0.5 mL should be multiplied by 2 and by the cell dilution factor (in our case, the dilution factor is 50, which is determined from mixing of 0.2 mL of cells from the overnight cultures with 9.8 mL of isotonic buffer). The resulting cell concentration is expressed in cells/

mL. A range of 1×10^6 to 3×10^6 cells/mL is typical for untreated rats. Cell counts may be lower in animals affected by a specific genotype or the experimental regimen.

5. If the animals are sacrificed on a Thursday, then cell plating occurs on Friday, alamarBlue™ is added on Monday (of the second week), and clone scoring is done on Tuesday.

6. Final concentration of alamarBlue™ in the wells is 1 %.

7. With computerized support, an 8×12 fluorescence data array for each plate can be either processed by software supplied with the plate reader or exported to a spreadsheet processor. The plate reading, finding MIN, calculating the cutoff value, and determining the total number of positive wells on each plate are achieved in one step.

8. Smaller fragments amplified in the first-round PCR may be indicative of deletions that eliminate all or parts of exons in genomic DNA, or they may be caused by aberrant processing of pre-mRNA due to mutations in splice sites (often called exon deletion mutants). Exon deletions may be verified and further characterized by multiplex PCR using genomic DNA from mutant cells (*see* **step 5** of Subheading 3.5).

Disclaimer

The views presented in this chapter do not necessarily reflect those of the US Food and Drug Administration.

References

1. Bertram JS (2000) The molecular biology of cancer. Mol Aspects Med 21:167–223
2. Adair GM, Carver JH, Wandres DL (1980) Mutagenicity testing in mammalian cells. I. Derivation of a Chinese hamster ovary cell line heterozygous for the adenine phosphoribosyl-transferase and thymidine kinase loci. Mutat Res 72:187–205
3. Lambert IB, Singer TM, Boucher SE et al (2005) Detailed review of transgenic rodent mutation assays. Mutat Res 590:1–280
4. Kinoshita T, Fujita M, Maeda Y (2008) Biosynthesis, remodelling and functions of mammalian GPI–anchored proteins: recent progress. J Biochem 144:287–294
5. Brodsky RA, Mukhina GL, Nelson KL et al (1999) Resistance of paroxysmal nocturnal hemoglobinuria cells to the glycosylphosphatidylinositol–binding toxin aerolysin. Blood 93:1749–1756
6. Dobrovolsky VN, Shaddock JG, Heflich RH (2000) 7,12–dimethylbenz[a]anthracene–induced mutation in the *Tk* gene of *Tk*[+/−] mice: automated scoring of lymphocyte clones using a fluorescent viability indicator. Environ Mol Mutagen 36:283–291
7. Miura D, Dobrovolsky VN, Mittelstaedt RA et al (2008) Development of an in vivo gene mutation assay using the endogenous *Pig-A* gene: II. Selection of *Pig-A* mutant rat spleen T-cells with proaerolysin and sequencing *Pig-A* cDNA from the mutants. Environ Mol Mutagen 49:622–630
8. Chen T, Aidoo A, Mittelstaedt RA et al (1999) *Hprt* mutant frequency and molecular analysis of *Hprt* mutations in Fischer 344 rats treated with thiotepa. Carcinogenesis 20: 269–277
9. Dobrovolsky VN, Shaddock JG, Heflich RH (2005) Analysis of in vivo mutation in the *Hprt* and *Tk* genes of mouse lymphocytes. In: Keohavong P, Grant SG (eds) Molecular Toxicology Protocols. Humana, Totowa, NJ, pp 133–144

Chapter 5

In Vivo *cII*, *gpt*, and Spi⁻ Gene Mutation Assays in Transgenic Mice and Rats

Mugimane G. Manjanatha, Xuefei Cao, Sharon D. Shelton, Roberta A. Mittelstaedt, and Robert H. Heflich

Abstract

Transgenic mutation assays are used to identify and characterize genotoxic hazards and for determining the mode of action for carcinogens. The three most popular transgenic mutational models are Big Blue® (rats or mice), Muta™ mouse (mice), and *gpt*-delta (rats or mice). The Big Blue® and Muta™ mouse models use the *cII* gene as a reporter of mutation whereas *gpt*-delta rodents use the *gpt* gene and the *red/gam* genes (Spi⁻ selection) as mutation reporter genes. Here we describe methodology for conducting mutation assays with these transgenes. Transgenes recovered from tissue DNA are packaged into infectious lambda phage, bacteria are infected with the phage, and *cII*-mutant and Spi⁻ plaques and *gpt*-mutant colonies are isolated using selective conditions and quantified. Selected mutants can be further analyzed for identification of small sequence alterations in the *cII* and *gpt* genes and large deletions at the Spi⁻ locus.

Key words Big Blue®, *gpt*-delta, *cII* mutants, *gpt* mutants, *cII*, *gpt* and Spi⁻ assays

1 Introduction

In vivo transgenic mutation assays have the unique ability to measure gene mutation in reporter genes engineered into the DNA of several commonly used mouse and rat animal models. The reporter genes are nonfunctional in vivo, but can be recovered from animal tissues as bacterial viruses or plasmids, and assayed for mutation in bacteria. As a result of the mechanistic relationship between somatic cell mutation and cancer, in vivo transgenic systems have found widespread application in research evaluating the mode of action of carcinogens. Although used less for this purpose, transgenic assays also are capable of measuring mutation in germ cells. In the past, the lack of a consensus protocol, their perceived regulatory need, and their cost limited the use of transgenic systems in regulatory science. These problems may have been overcome, at least partially, by the recent publication of an Organization for Economic Cooperation and Development (OECD) review on transgenic assays [1], an OECD Test guideline No. 488 for conducting the transgenic assays [2], and their recent application to a quantitative risk assessment of a drug contaminant [3].

The various transgenic mutation models have been reviewed recently [1, 4]. Although the models use both mice and rats and different transgene reporters, the considerations for animal treatment with test agents and the sampling of tissues for mutation are similar [5]. Because the transgenes are not expressed in vivo, they are essentially neutral when in the animal, meaning that transgene mutations can accumulate with repeat dosing. Mutant frequencies are a function of the amount and type of DNA damage in the target tissue, the rate of cell replication to fix mutations, and tissue kinetics. Liver is a relatively slow-growing tissue, so mutant frequencies tend to increase in liver slowly with time, while bone marrow is rapidly dividing, so mutant frequencies increase within a few days, and then (if dosing is not continued) fall rapidly as mutagenized cells transit to the periphery. As a compromise to account for different in vivo kinetics, the core treatment protocol, originally developed by the International Workshop on Genotoxicity Testing [6] and incorporated into OECD TG488 [2], calls for daily treatments with the test agent for 28 days, following by sampling tissues 3 days later.

In this chapter, we do not cover the in-life portions of the assay, but concentrate on processing of the tissues collected from the transgenic animals. We describe mutation detection by three commonly used assays, the *cII* assay conducted with Big Blue® mice and rats and Muta™ mouse, and the *gpt* and Spi⁻ assays conducted with *gpt*-delta mice and rats. In general, the *cII* and *gpt* assays detect smaller sequence changes, e.g., base pair substitutions and frameshifts, while the Spi⁻ assay, requiring the inactivation of both the *red* and *gam* genes, detects larger deletions. As shown in Fig. 1, the assays consist of (1) isolation of DNA from tissues of the experimental animals; (2) recovery of infectious lambda (λ) phage; (3) assaying for *cII*, *gpt*, and Spi⁻ mutant frequencies using indicator strains of bacteria; and (4) molecular analysis of target genes from the mutant transgenes.

2 Materials

Make all media components and solutions in distilled and autoclaved water.

2.1 Tissue Harvesting

1. Surgical tools: One 4″ micro-dissecting scissors, curved with sharp point; one 4″ micro-dissecting forceps, curved.
2. 1 mL disposable syringes with 25-gauge, 5/8″ needles.
3. 3 mL disposable syringes.

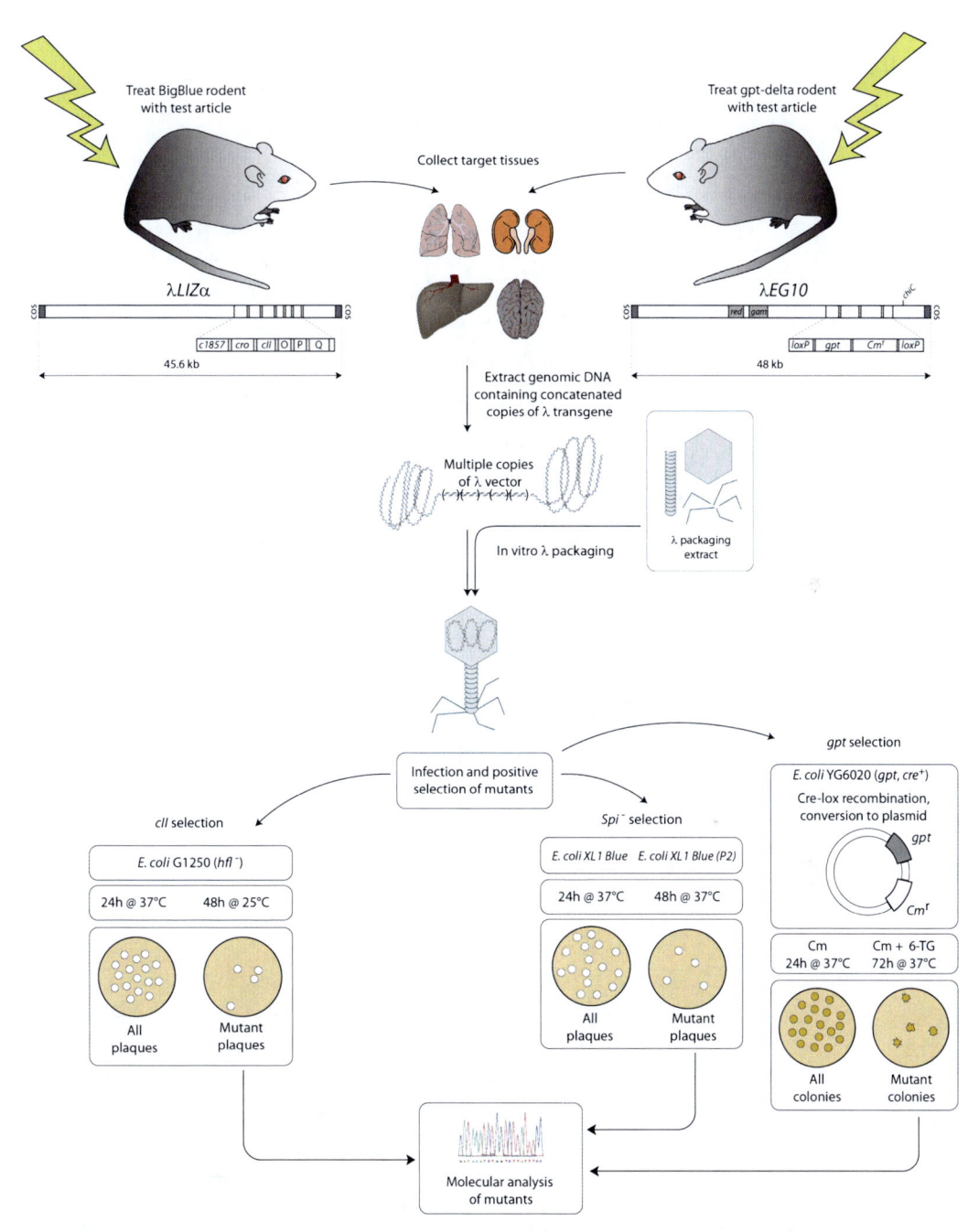

Fig. 1 Diagram illustrating different steps involved in transgenic assays using Big Blue® and *gpt*-delta mice and rats. Note that the *cII* assay in the Muta™ mouse model is similar to the *cII* assay with Big Blue® animals, but Muta™ mouse uses a different transgene vector. The following steps are shown: DNA isolation from target tissues; recovery of infectious λ phage; *cII*, *gpt*, and Spi− mutant selection using indicator bacteria; and molecular analysis of target genes from the mutant phages

4. 18-gauge syringe needles.
5. Precut aluminum foil, size 2″ × 2″, labeled with sample IDs.
6. Phosphate-buffered saline (PBS).
7. Spray bottle with 70 % ethanol.
8. Transgenic rodents: Big Blue® (Taconic Farms, German Town, NY); *gpt*-delta (Dr. T. Nohmi, Division of Genetics and Mutagenesis, National Institute of Health Sciences, Setagaya-ku, Tokyo 158, Japan); Muta™ mouse (Covance Research Products, Denver, PA).
9. Kimberly Clark Scott Multifold paper towel.
10. Absorbent underpads.
11. Micro-tube storage box.
12. 1.5 mL microcentrifuge tubes.

2.2 Genomic DNA Extraction

1. Kontes 15 mL Dounce tissue grinder with two pestles.
2. 50 mL conical polystyrene centrifuge tubes, sterile.
3. RecoverEase DNA Isolation Kit, 15 Preps, containing proteinase K, digestion buffer, and dialysis cups (Agilent Technologies, Cedar Creek, TX).
4. Centrifuge.
5. 1–200 µL specialty tips, large orifice.
6. Magnetic stir bar and stirrer.
7. Dialysis reservoir, e.g., a large plastic autoclave tray, clean and sterile.
8. 50 °C water bath.
9. Spectrophotometer.
10. Tris–HCl (pH 7.5): 1 M solution. Dissolve 60.55 g Trizma base in distilled water to a final volume of 500 mL. Adjust pH to 7.5 using 6 N HCl.
11. EDTA (pH 8.0): 0.5 M solution. Dissolve 186.1 g EDTA in distilled water to a final volume of 1 L. Use ~20 g NaOH to adjust pH to 8.0.
12. TE buffer: Combine 10 mL of 1 M Tris–HCl (pH 7.5) and 2 mL of 0.5 M EDTA (pH 8.0). Make up the volume to 1 L with distilled water.

2.3 In Vitro DNA Packaging

1. 1–200 µL specialty tips, large orifice.
2. In vitro Transpack Packaging Extract (Agilent Technologies) or packaging extract from Dr. Joseph Guttenplan (New York University, NY).
3. Mini vortex mixer.
4. 27 °C water bath.

5. MgCl₂: 40 mM solution. Dissolve 163 mg MgCl₂ in 20 mL distilled water.

6. SM buffer: 1 M solution. Combine and dissolve 5.8 g NaCl, 2.0 g MgSO₄·7H₂O, 50 mL of 1 M Tris–HCl (pH 7.5), and 5 mL of 2 % gelatin (w/v) in distilled water to a final volume of 1 L. Autoclave to sterilize. Store the solution at room temperature for up to 1 year.

2.4 cII Mutation Assay and Mutant Confirmation

1. Bacteriological Petri dishes, 100 × 15 mm.
2. Sterile inoculating loop.
3. Spectrophotometer.
4. Disposable 4.5 ml polystyrene 10 mm cuvettes.
5. Disposable borosilicate glass culture tubes, 15 × 100 mm.
6. Microwave oven.
7. Stationary 37 °C incubator.
8. Centrifuge, Sorvall RT 6000.
9. 37 °C bench top orbital shaker.
10. Disposable 0.45 μm filters.
11. *E. coli* host strain G1250 (Agilent Technologies).
12. Disposable sterile 15 mL centrifuge tubes.
13. 24 °C low-temperature incubator.
14. TB1 liquid medium: Dissolve and combine 5.0 g of NaCl, 10.0 g of casein peptone, and 0.1 % thiamine hydrochloride (1 mL/L) in distilled and autoclaved water to a final volume of 1 L. Adjust pH to 7.0 with either NaOH or HCl. Store the solution at room temperature for up to 3 months.
15. TB1 bottom agar: Add 12 g of Difco agar to TB1 liquid medium as prepared above, and make up the volume to 1 L with TB1 liquid medium after adjusting pH to 7.0 with either NaOH or HCl. 40 mL of agar is required per plate (*see* **Note 1**).
16. TB1–kanamycin agar: Add 50.0 mg kanamycin to 1 L of TB1 bottom agar. 40 mL of agar is required per plate (*see* **Note 2**).
17. TB1 top agar: Add 7.0 g of Difco agar to TB1 liquid medium as prepared above, and make up the volume to 1 L with TB1 liquid medium after adjusting pH to 7.0 with either NaOH or HCl. Prior to use, melt the TB1 top agar completely in a microwave or autoclave, mix well, and cool in 50 °C water bath.

2.5 Molecular Analysis of cII Mutants

1. Thermocycler for PCR.
2. Horizontal gel electrophoresis apparatus, Horizon® 58 and power supply, PowerPac 300.
3. Centrifuge with a 96-well plate rotor.
4. 96-well PCR plates.

5. *cII* PCR primers F1 and R1, and *cII* sequencing primers F2 and R2 (*see* **Note 3**).

6. Taq™ DNA polymerase and reaction buffer.

7. dNTP mix: 10 mM mix.

8. Electrophoresis-grade agarose: 1 % agarose.

9. Ethidium bromide: Dissolve 1 mg Ethidium bromide (EtBr) in 1 mL distilled and autoclaved water; use at 0.5× concentration.

10. DNA ladder: 100 bp.

11. Gel Pilot 5× loading buffer.

12. Multichannel Pipette C8 × 100.

13. MiniElute 96 UF PCR Purification Kit (Qiagen).

14. ThermalSeal RTS™ silicon adhesive sealing film (Sigma-Aldrich).

2.6 gpt Mutation Assay and Mutant Confirmation

1. *E. coli* YG6020 (Dr. T. Nohmi, Division of Genetics and Mutagenesis, National Institute of Health Sciences, Setagaya-ku, Tokyo 158, Japan).

2. Glass culture tubes, 16 × 125 mm and 12 × 75 mm, sterile.

3. 100 mm bacteriological Petri dishes, sterile.

4. 50 mL disposable polystyrene centrifuge tubes.

5. Round toothpicks, sterilize by autoclaving.

6. LB medium: Weigh 2.5 g LB broth into 100 mL distilled water and autoclave.

7. Maltose: 20 % (w/v) solution. Dissolve 20 g of maltose in 100 mL of distilled water.

8. Kanamycin: 25 mg/mL solution. Sterilize by filtration; store at −20 °C.

9. Glycerol: 50 % (v/v) solution. Mix 200 mL of glycerol with distilled water to a final volume of 400 mL. Sterilize by autoclaving; store at room temperature.

10. 10× M9: Dissolve 67.8 g of $Na_2HPO_4 \cdot 7H_2O$, 30 g of KH_2PO_4, 10 g of NH_4Cl, and 5 g of NaCl in distilled water to a final volume of 1 L; sterilize by autoclaving.

11. $MgSO_4$: 1 M solution. Dissolve 24.6 g of $MgSO_4 \cdot 7H_2O$ in distilled water to a final volume of 100 mL.

12. $CaCl_2$: 0.25 M solution. Dissolve 2.78 g of $CaCl_2$ in distilled water to a final volume of 100 mL.

13. Thiamine: 1 % (w/v) solution. Dissolve 1.0 g of thiamine in distilled water to a final volume of 100 mL; sterilize by filtration; store at 4 °C.

14. Amino acids: 10 mg/mL solution. Dissolve 50 mg of proline, 50 mg of leucine, and 50 mg of isoleucine together in distilled

water to a final volume of 500 mL. Sterilize by filtration; store at 4 °C.

15. Chloramphenicol: 25 mg/mL solution in ethanol. Dissolve 5 g of chloramphenicol in absolute ethanol to a final volume of 200 mL; store at −20 °C.

16. 6-thioguanine (6-TG): 25 mg/mL solution in DMSO. Make fresh every time.

17. *gpt* titration agar: Add 15.0 g of Difco agar to 900 mL of distilled water and autoclave. Once the agar cools to 50 °C, add 100 mL of 10×M9, 20 mL of 50 % (v/v) glycerol, 2 mL of 1 M $MgSO_4$, 400 μL of 0.25 M $CaCl_2$, 500 μL of 1 % (w/v) thiamine, 4 mL of 10 mg/mL amino acids, and 1 mL of 25 mg/mL chloramphenicol. Maintain the agar solution in 50 °C water bath.

18. *gpt* selection agar: Add 1 mL of 25 mg/mL 6-TG to 1 L molten *gpt* titration agar. Maintain the agar solution in 50 °C water bath.

19. Molten top agar: 0.6 % (w/v) solution. Add 0.6 g of Difco agar and 0.6 g of NaCl to 100 mL of distilled water and autoclave. Each plate needs 2.5 mL top agar.

20. Add 0.1 mL of 25 mg/mL 6-TG to *gpt* mutant selection/confirmation plates.

21. 1/15 Na-K: Dissolve 7.57 g of $Na_2HPO_4 \cdot 7H_2O$ and 1.82 g of KH_2PO_4 in distilled water to a final volume of 1 L; sterilize by autoclaving.

2.7 Spi− Selection and Mutant Confirmation

1. Glass culture tubes, 16×125 mm and 12×75 mm, sterile.
2. 1–200 μL specialty tips, large orifice.
3. 100 mm bacteriological Petri dishes, sterile.
4. 50 mL disposable polystyrene centrifuge tubes.
5. *E. coli* XL-1Blue MRA, *E. coli* XL-1Blue MRA(P2), and *E. coli* WL95 (P2) bacteria (Dr. T. Nohmi, Division of Genetics and Mutagenesis, National Institute of Health Sciences, Setagaya-ku, Tokyo 158, Japan).
6. $MgSO_4$: 1 M solution. *See* **item 11** of Subheading 2.6.
7. λ-Trypticase agar plates: Add 10.0 g of BBL trypticase peptone, 5.0 g of NaCl, and 10 g of Difco agar to 1 L of water and autoclave. Add 10 mL of 1 M sterile $MgSO_4$ after the agar solution is cooled to 50 °C in a water bath. Each plate needs 25 mL of the agar.
8. Molten λ-trypticase top agar: Add 10.0 g of BBL trypticase peptone, 5.0 g of NaCl, and 6.0 g of Difco agar to 1 L of distilled water and autoclave. Add 10 mL of 1 M sterile $MgSO_4$ after the agar solution is cooled to 50 °C in a water bath.

2.8 Molecular Analysis of gpt and Spi⁻ Mutant Phenotypes

1. Thermocycler for PCR.
2. Horizontal gel electrophoresis apparatus, Horizon® 58 and power supply, PowerPac 300.
3. Glass culture tubes, 16 × 125 mm, sterile.
4. 0.2 mL PCR tubes.
5. QIAquick PCR Purification Kit.
6. Taq™ DNA polymerase kit.
7. dNTP mix: 10 mM.
8. Electrophoresis-grade agarose: *See* **item 8** of Subheading 2.5.
9. Ethidium bromide: *See* **item 9** of Subheading 2.5.
10. DNA ladder: 100 bp.
11. Gel Pilot 5× loading buffer.

3 Methods

3.1 Tissue Harvesting

1. Prior to necropsy, soak the surgical tools in 70 % ethanol in a clean beaker. Place liquid nitrogen, dry ice, and wet ice within reach of the surgical area. Keep the microcentrifuge tube storage box on dry ice.
2. Sacrifice animals using methods approved by your Institutional Animal Care and Use Committee (IACUC), such as by CO_2 asphyxiation.
3. Place the sacrificed animals on their back over a disposable bench cover or multiple layers of paper towels. Spray the animals thoroughly with 70 % ethanol (*see* **Note 4**).
4. Pinch the skin anteriorly to the urethral opening and cut along the ventral midline from the groin to the rib cage. Make additional incisions from the end of the first incision to the axillae on both sides of the animal, ending up with a Y-shaped incision.
5. Expose the abdomen with forceps by pulling and holding the skin to the sides. Lift the peritoneum with a curved forceps and cut peritoneum along the ventral midline to expose the internal tissues.
6. Cut at least 1″ of small intestine from the desired segment. Withdraw 3 mL of PBS into a 3 mL disposable syringe with an 18-gauge syringe needle. Flush the intestinal lumen with PBS until feces are completely removed. Remove excess PBS by gently pressing the exterior of the small intestine. Quickly wrap the intestinal tissue in a piece of aluminum foil and snap freeze in liquid nitrogen.
7. If male animals are used, gently separate testes from their anchors of fat. Quickly wrap the testes in a piece of aluminum foil and snap freeze in liquid nitrogen.

8. Identify and collect the spleen, kidney, and liver in the order given. Gently pull and hold tissues with forceps; cut out intact tissues and trim off as much connective tissue and fat as possible. Quickly wrap these tissues in aluminum foil and snap freeze in liquid nitrogen.

9. Hold the center of the rib cage with forceps; punch a hole through the diaphragm and cut the rib cage open with a microdissecting scissors to expose the lung and heart. Gently pull and hold the lung with forceps and cut out the entire lung. Pinch the tip of the heart with forceps and cut out the heart. Quickly wrap the tissues in aluminum foil and snap freeze in liquid nitrogen.

10. Excise either the right or the left femur, and with a syringe, flush the marrow from the excised femur into a microcentrifuge tube with ~1 mL PBS with 5 % FBS, spin the tube at $300 \times g$ for 5 min, aspirate the supernate, and freeze the pellet in liquid nitrogen.

11. Remove snap-frozen tissues from liquid nitrogen and place them into the tube slots in the microcentrifuge tube storage box kept on dry ice. Proceed with the next animal.

12. At the end of the tissue collection, transfer the microcentrifuge tube storage box containing all the tissues to −80 °C for long-term storage.

13. Discard the animal carcasses in accordance with your institutional regulations.

3.2 Genomic DNA Extraction

1. Chill a clean 15 mL Kontes Dounce tissue grinder and cell lysis buffer on ice.

2. Add 5 mL ice-cold lysis buffer to the tissue grinder. Transfer an appropriate amount of the excised tissue sample to the tissue grinder according to the guideline given in Table 1.

3. Disaggregate the tissue samples with 15 strokes of the loose pestle until the samples appear completely homogenized (*see* **Note 5**).

4. Release the cell nuclei within the homogenates using 15 strokes of the tight pestle. Avoid twisting the pestle while raising and lowering it into the mortar of the tissue grinder as it may shear the nuclei and reduce the yield of high-molecular-weight DNA.

5. Pour the homogenate through a sterile cell strainer into a 50 mL sterile conical tube (*see* **Note 6**). Wash the tissue grinder with 3 mL of ice-cold lysis buffer; pour the wash through the cell strainer to bring the total volume in the conical tube to 8 mL.

6. Store the conical tube on ice until all samples are processed through **step 5**.

Table 1
Guideline for required amount of mouse tissues for genomic DNA extraction (adjust appropriately for rats)

Type of tissue	Approx. starting mass (mg)	DNA approx. yield (µg)
Liver	50–80	175
Kidney	100 (one-half of a whole mouse kidney)	200
Lung	175 (a whole mouse lung)	125
Heart	185 (a whole mouse heart)	75
Spleen	40 (one-half of a whole mouse spleen)	400
Bone marrow	From one whole mouse femur	75
Brain	100 (one-quarter of a whole mouse brain)	50
Testis	50 (one-half of a whole mouse testis)	50
Small intestine	175 (one inch of the small intestine)	100

7. Centrifuge the tubes at $1,100 \times g$ for 12 min at 4 °C. For bone marrow samples, centrifuge the conical tubes at $1,200 \times g$ for 15 min at 4 °C.

8. Uncap the tubes and discard the supernatant carefully to avoid losing the cell nuclei pellets. Special care needs to be taken with bone marrow pellets. Withdraw and keep the bone marrow pellet in a wide-bore pipette tip before decanting the supernatant; then release the cell pellets into the empty conical tubes.

9. Carefully invert the uncapped conical tubes on a paper towel for approximately 1 min to drain any excess liquid from the pellets. Note that the cell nuclei pellets may slide down the inner wall of the tubes.

10. Warm 100 µL of the proteinase K solution (from the RecoverEase kit) for each sample in a 50 °C water bath for 2–5 min in advance to activate the enzyme.

11. Prepare the complete digestion buffer by adding 20 µL of RNace-It™ ribonuclease cocktail to each mL of digestion buffer from the RecoverEase kit.

12. Add 100 µL of the complete digestion buffer to the nuclei pellets and rock the conical tubes gently to dislodge the pellets from the bottom of the tubes (*see* **Note 7**).

13. Place the conical tubes in a 50 °C water bath and add 100 µL of the activated proteinase K solution to the free-floating pellet. Swirl the tubes gently to mix.

14. Incubate the tubes in a 50 °C water bath for 1–2 h or until the nuclei pellets become dispersed and clear in color. Swirl the tubes gently every 30 min.

15. Pour 500 mL of TE buffer for each sample into a dialysis reservoir. Float the dialysis cups on the surface of the TE buffer, tilting the cups slightly to remove any large air bubbles from beneath the membrane surface.

16. Carefully transfer the viscous genomic DNA from the conical tubes to the center of the floating dialysis cups using a large orifice pipette tip.

17. Dialyze the genomic DNA at room temperature for 48 h while stirring the TE buffer gently with a magnetic stir bar.

18. Upon completion of the dialysis, remove the dialysis cups from the TE buffer. Immediately transfer the genomic DNA to a sterile microcentrifuge tube using a large orifice pipette tip. Measure DNA concentration by a spectrophotometer and store the genomic DNA at 4 °C.

3.3 In Vitro DNA Packaging

1. Quickly thaw one vial of packaging extract by holding the tube between the palms of your hands until it is completely thawed. Mix the contents by vortexing for 5 s. Centrifuge the tubes briefly in a microcentrifuge at the maximum speed to collect the contents from the sides and caps. Leave the tubes on ice until ready for use (*see* **Note 8**).

2. Using a large orifice pipette tip, transfer 8–14 μL of the genomic DNA samples (depending on the concentration of the genomic DNA as measured in **step 18** of Subheading 3.2) to 1.5 mL microcentrifuge tubes containing 12 μL of packaging extract, and place the tubes in a 27 °C water bath for 2 h with occasional shaking (*see* **Note 9**).

3. Boost the packaging reactions by adding an additional 12 μL of packaging extract to each tube containing the packaging reaction.

4. Place the tubes back in the 27 °C water bath for an additional 2 h with occasional shaking (*see* **Note 10**).

5. After incubation, bring the volume of each DNA sample to 1 mL (for *cII* assay) or 300 μL (for *gpt* and Spi⁻ assays) with SM buffer.

6. Vortex the tubes vigorously at the maximum speed for 1 min to completely disrupt the genomic DNA. Make sure that the tubes are capped tightly. The packaging reactions, diluted with SM buffer, are hereafter referred to as the "phage DNA." Keep the tubes on ice until use.

3.4 cII Mutation Assay and Mutant Confirmation

3.4.1 Preparing the G1250 Liquid Culture

1. At least 2 days before plating, use a sterile inoculating loop to streak splinters of solid ice from a bacterial glycerol stock containing the *E. coli* host strain G1250 onto a 100 mm TB1–kanamycin agar plate.

2. Incubate the bacterial streak plate overnight in a stationary 30 °C incubator (*see* **Note 11**).

3. On the day before plating, combine 10 mL of TB1 liquid medium with 100 μL of the 1 M $MgSO_4$ solution containing 20 % (w/v) maltose in a sterile 50 mL screw-cap conical tube.

4. Inoculate the liquid medium with a single colony from the bacterial streak plate (*see* **Note 12**).

5. Incubate the liquid culture overnight in a 30 °C shaking incubator with vigorous shaking (250–300 rpm).

3.4.2 Preparing the G1250 Plating Culture and cII Assay

1. Centrifuge the conical tube containing the G1250 liquid culture (from **step 5** of Subheading 3.4.1) at 1,500 ×*g* for 10 min to pellet the bacterial cells.

2. Discard the supernatant and gently resuspend the cell pellet in 10 mL of 10 mM $MgSO_4$.

   This stabilizes the bacteria in the absence of growth medium.

3. Measure the absorbance of 0.5 mL of the bacterial suspension at a wavelength of 600 nm and calculate the OD_{600} of the cell suspension.

4. Dilute the cell suspension to a final OD_{600} of 0.5 with 10 mM $MgSO_4$. The prepared suspension is referred to as the "G1250 plating culture."

5. Place the G1250 plating culture on ice until ready to use.

6. Prepare titer plates (to determine the total number of phages packaged) and mutant selection plates for the *cII* mutation assay. Each DNA sample requires 3 TB1 agar plates without kanamycin for titer and 5 TB1 agar plates without kanamycin for mutant selection (*see* **Notes 1** and **2**).

7. After completing the in vitro packaging reactions (from **step 6** of Subheading 3.3), transfer 1 mL of the packaged phage DNA to 50 mL tubes and add 2 mL of G1250 plating culture (from **step 5** of Subheading 3.4.2) to each tube. Agitate the tubes and then incubate for 30 min at room temperature without shaking for adsorption of phages to *E. coli* cells.

8. Dispense 200 μL of G1250 plating culture (from **step 5** of Subheading 3.4.2) into three small sterile glass tubes (15 × 100 mm) for titer plates (*see* **Note 13**) and add 1 μL of the adsorbed packaged DNA (from **step 7** of Subheading 3.4.2) to each titer tube. The final dilution of the adsorbed phage is 1:3,000.

9. Add 3 mL of molten TB1 top agar (55 °C) to each tube containing the 1:3,000 dilution for titer (from **step 8** of Subheading 3.4.2), mix, and immediately pour the contents onto a TB1 agar plate. Allow the top agar to solidify for at least 5 min at room temperature.

10. For mutant selection, add 20 mL of molten TB1 top agar to each of the 50 mL tubes containing the rest of the phage-adsorbed host cells (from **step 7** of Subheading 3.4.2), mix, and immediately pour the contents onto 5 TB1 agar plates (mutant selection plates), roughly dividing the ~23 mL among the 5 plates. Allow the top agar to solidify for at least 5 min at room temperature.

11. Invert and incubate the titer plates at 37 °C overnight; invert and incubate the mutant selection plates at 24 °C for 48 h.

3.4.3 Counting Titer Plates and Identifying Putative Mutant cII Plaques on Mutant Selection Plates

1. Following the 37 °C overnight incubation, examine the set of triplicate titer plates. Count the number of plaques on the titer plates. Determine the average number of plaques for the triplicate plates. The total number of plaques screened = Average number of plaques on the titer plates × 3,000 (dilution factor).

2. Examine the mutant selection plates immediately following the 48-h incubation at 24 °C. With the lid removed, hold each plate over a white-light box and look for any disturbance in the host lawn that may be a *cII* mutant plaque. To more easily identify the plaques, place a black object on a portion of the light box and move the plate across the contrasting background.

3. Using a marking pen, draw a circle on the bottom surface of the plate around each putative *cII* mutant plaque and give the putative plaque an identification number (*see* **Note 14**).

4. The plates can be stored at 4 °C until the putative *cII* mutant plaques are verified.

3.4.4 Verifying Putative cII Mutants and Calculating cII Mutant Frequency

1. For verification, core the mutant plaques on the selection plates (from **step 3** of Subheading 3.4.3) with a sterile large orifice pipette tip (or a sterile Pasteur pipette) and expel the core into a sterile microcentrifuge tube containing 500 μL of sterile SM buffer (*see* **Note 15**).

2. Incubate the tube for at least 2 h at room temperature, or overnight at 4 °C, to allow the phage particles to elute from the agar plug into the buffer.

3. In a sterile 15 × 100 mm glass tube, combine 0.2 mL of prepared G1250 plating culture (from **step 5** of Subheading 3.4.2) with 1 μL of the cored phage solution and incubate the tube at room temperature for 30 min.

4. Plate the sample on a TB1 agar plate using 2.5 mL of 55 °C molten TB1 top agar and incubate the plate at 24 °C for 48 h.

5. If the *cII* mutant phenotype is verified by plaque formation on the secondary verification plate, core an isolated plaque from the plate and store this secondary core in 100 μL of H_2O at

4 °C for DNA sequence analysis. For long-term storage, transfer an aliquot of the eluted phage to a new tube, add DMSO to a final concentration of 7 % (v/v), and store the aliquot at −80 °C (*see* **Note 16**).

6. Count all the verified *cII* mutant plaques observed on the selection plates. Calculate the *cII* mutant frequency as the ratio of verified *cII* mutant plaques to the total number of plaques screened from the titer plates (from **step 1** of Subheading 3.4.3) (*see* **Note 17**).

3.4.5 Molecular Analysis of *cII* Mutants

1. Use a sterile toothpick or a standard yellow pipette tip to carefully scrape the top agar, avoiding the bottom agar, from an individual mutant plaque and transfer the plaque to a microcentrifuge tube or a 96-well plate containing 100 μL H$_2$O per well. Alternatively, use material from a plaque stored in H$_2$O at 4 °C (*see* **step 5** of Subheading 3.4.4).

2. Cap the tube or the 96-well plate securely and place in thermocycler for 5 min at 99 °C.

3. Centrifuge the tube or the plate at 2,000 ×*g* for 3 min.

4. Immediately transfer 5 μL of the heated mutant phage supernatant (used as the template) to 15 μL of PCR master mix such that the final concentrations of the reagents are 1× Platinum *Taq* polymerase reaction buffer, 20 pmol each of primers *cII* F1 and *cII* R1 (*see* **Note 3**), 12.5 nmol of each dNTP, and 2.5 U of *Taq* DNA polymerase.

5. Amplify the template using the following cycling parameters: a 3-min denaturation at 95 °C, followed by 30 cycles of 30 s at 95 °C, 1 min at 60 °C, and 1 min at 72 °C, with a final extension of 10 min at 72 °C.

6. Analyze the resultant 432-bp fragment containing the *cII* gene and flanking regions on a 1 % agarose gel containing 0.5× ethidium bromide.

7. Remove unincorporated primers and dNTPs from the PCR products using the MinElute 96 UF Qiaquick PCR purification system as described in the manual. The purified PCR product is eluted into 20 μL sterile H$_2$O.

8. Sequence the PCR-purified *cII* fragment by your favorite method, using primers F2 or R2 as sequencing primers (*see* **Note 3**). Confirm all mutations at least once.

3.5 *gpt* Mutation Assay

3.5.1 Preparation of *E. coli* YG6020

1. Make an overnight culture of the *E. coli* YG6020 in 5 mL of LB medium supplemented with 0.2 % (v/v) maltose and 25 μL/mL kanamycin (*see* **Note 18**). Grow the bacteria overnight at 37 °C with shaking at 250 rpm. Preparation of the overnight culture should be done the day before the *gpt* mutation assay.

2. Subculture the overnight culture by making a 1:40 dilution in LB medium supplemented with 0.2 % (v/v) maltose and 25 μg/mL kanamycin. Grow the bacteria at 37 °C with vigorous shaking at 300 rpm until the OD_{600} reaches 1.0. This usually takes 3 h (*see* **Note 19**).

3. Recover the bacteria by centrifugation at 2,000 ×*g* for 10 min at 4 °C. Decant the supernatant and resuspend the pellet with half the volume of the LB broth supplemented with 10 mM $MgSO_4$ to make an OD_{600} of approximately 2.0.

4. Dispense 200 μL of the bacterial suspension into small sterile glass tubes (12×75 mm). Each phage DNA sample needs three glass tubes for titer and five glass tubes for 6-TG mutant selection.

3.5.2 Titration and 6-TG Mutant Selection

1. Prepare titer plates and mutant selection plates the day before the *gpt* mutation assay. Each sample requires three agar plates for titer and five agar plates for mutant selection (*see* **Notes 20 and 21**). Additional selection plates may be prepared and stored at 4 °C for up to 1 week for confirming the *gpt* mutant phenotype.

2. After completing the in vitro packaging reactions, mix 5 μL of the phage DNA (from **step 6** of Subheading 3.3) with 495 μL of LB medium supplemented with 10 mM $MgSO_4$ to make a 100-fold diluted suspension. Add 10 μL of the diluted phage DNA to the sterile glass tubes (12×75 mm) containing 200 μL of the twofold concentrated *E. coli* YG6020 suspension (from **step 4** of Subheading 3.5.1). Mix the contents by gently tapping the bottom of the glass tubes. These tubes are for the titer plates.

3. Add 60 μL phage DNA (from **step 6** of Subheading 3.3) to each of the five sterile glass tubes (12×75 mm) containing 200 μL of the twofold concentrated *E. coli* YG6020 suspension (from **step 4** of Subheading 3.5.1). Mix by gently tapping the bottom of the tubes. These tubes are for the mutant selection plates.

4. Incubate the titer and selection tubes at 37 °C for 20 min without shaking for adsorption of phages to *E. coli* cells.

5. Incubate the tubes at 37 °C with shaking at 300 rpm for 30 min to convert phage DNA into plasmids.

6. Remove the tubes from the shaker incubator. Add 2.5 mL of molten top agar at 55 °C to the glass tubes for titration. Mix the contents by gentle vortexing (avoid bubbles). Pour the contents immediately onto the titer plates. Allow the top agar to solidify for at least 5 min at room temperature.

7. Add 2.5 mL of molten top agar containing 25 μg/mL 6-TG to the tubes for *gpt* mutant selection, and mix by gentle vortexing.

Pour the contents immediately onto the mutant selection plates. Allow the top agar to solidify for at least 5 min at room temperature.

8. Invert the plates and incubate at 37 °C for 3 days (*see* **Note 22**).

9. Count the number of the chloramphenicol-resistant (Cm^r) colonies on each titer plate. The titer of the phage DNA is the product of the average number of Cm^r colonies and the dilution factor. The dilution factor in this protocol is 3,000 (*see* **Note 23**).

10. Mark the possible 6-thioguanine-resistant (6-TG^r) colonies that appear on the mutant selection plates after 72 h of incubation (*see* **Note 24**). Return the selection plates to the incubator for an additional 24 h (*see* **Note 25**).

11. Pick the mutant colonies on the selection plates with sterile toothpicks. Briefly rinse the tips of the toothpicks with 50 μL of 1/15 Na-K buffer. Streak the resuspended cells on a confirmation plate (*see* **Note 26**).

12. Invert the confirmation plates and incubate at 37 °C for 72 h.

13. Count the total number of confirmed 6-TG^r colonies.

14. The mutant frequency (MF) is calculated by dividing the total number of confirmed 6-TG^r colonies by the titer determined in **step 9** of Subheading 3.5.2.

3.5.3 Storage of 6-TG^r Mutant Clones

1. Inoculate the 6-TG^r colonies from the confirmation plates in 1 mL of LB medium containing 25 μg/mL chloramphenicol. Grow the bacteria overnight at 37 °C with shaking at 250 rpm.

2. Collect the bacteria by centrifugation at the maximum speed in a microcentrifuge for 1 min. Decant the supernatant. The bacterial pellets are stored at −80 °C for *gpt* gene sequencing analysis.

3.5.4 Sequencing Mutants

1. The primer pair for amplifying the 456-bp coding region of the *gpt* gene is as follows:

   Forward primer: 5′ TAC CAC TTT ATC CCG CGT CAG G 3′
   Reverse primer: 5′ ACA GGG TTT CGC TCA GGT TTG C 3′

2. Prepare a PCR master mix on ice by combining the reagents described in Table 2 and mix thoroughly by vortexing.

3. Withdraw ~1 μL of the bacterial pellet (from **step 2** of Subheading 3.5.3) and add to a PCR tube containing 29 μL of the master mix. The bacterial pellets of the mutant clones are used directly as the template for PCR reactions. Mix the contents by vortexing. Keep all samples on ice until the thermal cycler is programmed and ready to load.

4. Process samples in a thermal cycler as follows: 95 °C for 15 min, followed by 35 cycles of 30 s at 94 °C, 30 s at 58 °C, 1 min at 72 °C, and ending with 72 °C for 7 min.

Table 2
PCR master mix for amplifying *gpt* coding region

Reagent	Stock concentration	Volume added (µL)	Final concentration
dNTP mixture	2.5 mM	3	0.25 mM
Forward primer	10 µM	1	0.33 µM
Reverse primer	10 µM	1	0.33 µM
PCR buffer	10×	3	1×
HotStarTaq polymerase	5 U/µL	0.5	2.5 U/reaction
Total volume		29	

5. Check the PCR products by agarose gel electrophoresis. The size of the full-length PCR product is 456 bp. The PCR products are used as the templates for the sequencing reaction.

6. Purify the PCR products with a QIAquick PCR Purification Kit using the supplied reagents. The purified product is eluted in 30 µL of TE buffer.

7. Sequence the *gpt* mutants using an appropriate method and the following primers:

    Forward primer: 5′ GAG GCA GTG CGT AAA AAG AC 3′

    Reverse primer: 5′ CTA TTG TAA CCC GCC TGA AG 3′

3.6 Spi⁻ Selection Assay

3.6.1 Preparation of Bacterial Strains

1. Set up cultures of *E. coli* XL-1Blue MRA and XL-1Blue MRA (P2) by inoculating 2 mL of LB medium with 50 µL of bacterial stock (*see* **Note 18**). Grow the bacteria overnight at 37 °C with shaking at 250 rpm.

2. Subculture the overnight cultures by making a 100-fold dilution in LB medium supplemented with 0.2 % (v/v) maltose, e.g., add 100 µL of the overnight culture into 10 mL of LB broth with 0.2 % (v/v) maltose. Grow the bacteria at 37 °C with shaking at 300 rpm until the OD_{600} reaches 1.0.

3. Collect the bacteria by centrifugation at 2,000 ×*g* for 10 min at 4 °C. Discard the supernatant. Resuspend the bacterial pellets with an equal volume, e.g., 10 mL, of LB medium supplemented with 10 mM $MgSO_4$. Store the bacterial suspensions on ice until use (*see* **Note 27**).

3.6.2 Titration Using E. coli XL-1Blue MRA

1. After completing the in vitro packaging reactions, dilute phage DNA (from **step 6** of Subheading 3.3) by adding 1 µL of the phage DNA to 100 µL of LB medium.

2. Pipet 200 µL of *E. coli* XL-1Blue MRA suspension into 3 sterile glass tubes (12×75 mm). Add 5 µL of the diluted phage (from **step 1**) to the tubes containing the bacterial suspension.

Mix by gently tapping the bottom of the glass tubes. This is a 6,000-fold dilution.

3. Incubate the tubes at room temperature for 20 min.
4. Add 2.5 mL of molten λ-trypticase top agar to each tube. Vortex briefly to mix.
5. Pour the top agar immediately onto the λ-trypticase agar plates. Rock the plates back and forth to evenly distribute the soft agar (*see* **Note 28**), and allow the top agar to solidify on the bench for at least 5 min.
6. Invert the plates and incubate overnight at 37 °C.
7. Count the number of plaques on each titer plate. Calculate the number of plaque-forming units (p.f.u.) per packaging reaction by multiplying the average number of plaques by the dilution factor, which is 6,000 if the dilutions are made as instructed in this protocol (*see* **Note 23**).

3.6.3 Selection Using E. Coli XL-1Blue MRA (P2)

1. Add 150 µL of the phage DNA (from **step 6** of Subheading 3.3) to two sterile glass tubes containing 200 µL of the *E. coli* XL-1Blue MRA (P2) bacterial suspension (*see* **Note 20**). Mix gently by tapping the bottom of the tubes.
2. Incubate the tubes at room temperature for 20 min.
3. Add 2.5 mL of the molten λ-trypticase top agar to each tube. Vortex briefly to mix.
4. Pour the top agar immediately onto the λ-trypticase agar plates. Remove air bubbles trapped in the soft agar by exposing the plates to a gas flame for 1 s (*see* **Note 28**). Allow the top agar to solidify on the bench for at least 5 min.
5. Invert the plates and incubate overnight at 37 °C.
6. Count the total number of the clear plaques on the plates; these are the Spi⁻ candidate mutants.

3.6.4 Confirmation of the Phenotype

1. Prepare a sufficient number of λ-trypticase agar plates 1 day before the confirmation experiment. Each confirmation plate can be divided into 12 sections. The number of plates required for each bacterial strain is calculated by dividing the total number of clear plaques (as calculated in **step 6** of Subheading 3.6.3) by 12 (the number of sections on each plate). Prepare three sets of the λ-trypticase agar plates for the three bacterial strains that are used for confirmation.
2. Culture *E. coli* XL-1Blue MRA and XL-1Blue MRA (P2) bacteria as described in **steps 1–3** of Subheading 3.6.1 (*see* **Note 27**).
3. Prepare an overnight culture of *E. coli* WL95 (P2) by inoculating 10 mL of LB medium with 50 µL of stock bacteria. Grow the bacteria overnight at 37 °C with shaking at 250 rpm.

4. Centrifuge the overnight culture of *E. coli* WL95 (P2) at 2,000 ×*g* for 10 min at 4 °C. Discard the supernatant. Resuspend the bacterial pellet with 10 mL of LB medium (equal volume) supplemented with 10 mM MgSO$_4$.

5. For each plate, pipet 200 μL of the corresponding bacterial strain into a 5-mL sterile glass tube (12 × 75 mm). Add 2.5 mL of the molten λ-trypticase top agar into each tube. Mix by gentle vortexing and pour the contents immediately onto the agar plates. Dry the plates on the bench for 1 h or longer.

6. Punch out the clear plaques from the Spi⁻ selection plates (from **step 6** of Subheading 3.6.3) with sterile glass capillary tubes or wide-bore pipette tips. Resuspend the agar plugs in 60 μL of SM buffer.

7. Spot 15 μL of the plaque suspension onto the three λ-trypticase agar plates, each seeded with the XL-1Blue MRA, XL-1Blue MRA (P2), or WL95 (P2) strains (preceding **step 5**). Leave the plates on the bench for at least 1 h.

8. Invert the plates and incubate at 37 °C overnight.

9. Count mutants that make clear spots on all three strains of bacteria. These are confirmed Spi⁻ plaques.

10. Calculate the mutant frequency by dividing the total number of confirmed Spi⁻ plaques by the number of total plaques recovered (titer) as determined in **step 7** of Subheading 3.6.2.

3.6.5 Preparation of Mutant Plaques

1. Inoculate 1 mL of LB medium with *E. coli* LE392 and incubate overnight at 37 °C with shaking at 250 rpm.

2. Centrifuge the overnight culture at 2,000 ×*g* for 10 min at 4 °C. Discard the supernatant. Resuspend the bacterial pellet in an equal volume of LB medium supplemented with 10 mM MgSO$_4$. Dispense 50 μL of the *E. coli* LE392 suspension into sterile glass tubes.

3. Punch out the confirmed Spi⁻ mutant plaques from the confirmation plates inoculated with WL95 (P2) with capillary tubes or large orifice pipette tips. Add the agar plugs to the glass tubes containing the *E. coli* LE392 suspension (preceding **step**). Let stand for 5 min at room temperature.

4. Add 2.5 mL of LB medium supplemented with 10 mM MgSO$_4$ to the glass tubes. Incubate the tubes at 37 °C with shaking at 300 rpm until lysis occurs, which usually takes 7–8 h.

5. Add 100 μL of chloroform to the tubes when lysis is complete. Centrifuge at 12,000 ×*g* for 10 min at 4 °C.

6. Store the supernatant at 4 °C for molecular analysis.

Table 3
Primers used for molecular analysis of Spi⁻ mutants

Primer pair	Primer sequence (from 5′ to 3′)	Size of amplification product
1	Forward: CTC TCC TTT GAT GCG AAT GCC AGC Reverse: GGA GTA ATT ATG CGG AAC AGA ATC ATG C	4.8 kb
2	Forward: CGT GGT CTG AGT GTG TTA CAG AGG Reverse: CGG TCG AGG GAC CTA ATA ACT TCG	14 kb
3	Forward: CGT GGT CTG AGT GTG TTA CAG AGG Reverse: GTT ATG CGT TGT TCC ATA CAA CCT CC	21.9 kb

3.6.6 Molecular Identification of Spi⁻ Mutant Phenotypes

1. The primer pairs to detect different sized deletions are listed in Table 3.
2. Prepare a PCR master mix by combining the following reagents: 10 pmol of each primer, 200 μM of dNTP mix, and 2.5 U HotStar DNA polymerase in a 20 μL reaction. Transfer a small amount of phage DNA lysate (from **step 6** of Subheading 3.6.5) into the tubes as the templates for the PCR reaction.
3. Process the samples in a thermal cycler as follows: 98 °C for 3 min, followed by 25 cycles of 25 s at 98 °C and 1 min/Kb at 68 °C.
4. Determine the size of the PCR amplification products by agarose gel electrophoresis.
5. The size of the deletion is estimated by the difference between the actual size of the PCR products and the size of the products amplified from wild-type phage (Table 3).

4 Notes

1. Prior to use, melt the prepared TB1 top agar completely in a microwave or autoclave, mix well, and cool in a 50 °C water bath.
2. Prepare the plates at least 24 h prior to use. Add filter-sterilized kanamycin to TB1 bottom agar solution after cooling to 50 °C and pour the contents into bacteriological Petri dishes (100×15 mm) to prepare kanamycin–TB1 agar plates. These plates should be stored at 4 °C.
3. The primers used for PCR amplification and sequencing reactions of the *cII* gene are as follows:

PCR primers—F1: 5′-AAAAAGGGCATCAAATTAAAC-3′, and R1: 5′-CCGAAGTTGAGTATTTTTGCTG-3′

Sequencing primers—F2: 5′-CCACACCTATGGTGTATG-3′, and R2: 5′-GAGTATTTTTGCTGTATT-3′

4. Washing the animal carcass with 70 % ethanol ensures aseptic conditions for the incisions and reduces artifacts caused by hair dragging.

5. Highly vascularized tissues (e.g., heart or lung samples) typically require more than ten strokes to fully disaggregate the tissue samples. Do not proceed to the next step until the tissue sample is completely homogenized.

6. If the cell strainer becomes clogged, swirl a large orifice pipette tip over the surface to enable the liquid from the homogenized suspension to pass through the mesh.

7. The pellet should float freely in the solution.

8. Use thawed packaging extract as quickly as possible.

9. To pipette a small amount of genomic DNA, withdraw genomic DNA into a large orifice pipette tip and then press and twist the tip against the bottom of the tube.

10. Mix the contents by gently tapping the bottom of the tubes. DO NOT vortex. Centrifuge the tubes at 300 ×g for 3 min to collect the contents at the bottom of the tube.

11. The bacterial streak plates can be stored at 4 °C for up to 1 week.

12. Be sure to start with a sufficient number of 10 mL G1250 liquid cultures for the plating and titering requirements. Assume that each 10 mL of G1250 liquid culture will supply enough plating culture for ~40 plates.

13. Use the G1250 plating culture within 1–2 h.

14. Some *cII* plaques are large and very obvious while others may be difficult to distinguish from typical plating artifacts. Circle all possible mutants and verify.

15. Replate the putative mutant plaques at a low density under the selective conditions for two reasons: (a) Plating artifacts can sometimes look like tiny plaques. Therefore, verify the *cII* phenotype of each mutant before the mutant frequency is determined. (b) Agar cores taken from a screening plate may contain nonmutant phage along with the mutant phage. Secondary plaques from a low-density replating will provide an uncontaminated mutant template for PCR and subsequent sequencing.

16. The plates can be stored at 4 °C until the plaques are cored. Phages stored with DMSO should be replated to provide fresh plaques for molecular analysis of mutants.

17. The final mutant frequency calculation for a given tissue should be performed after screening a minimum of 200,000–300,000 rescued phages.

18. When preparing overnight bacterial cultures, a mock culture without bacteria should always be included in order to detect possible contamination of the LB medium.

19. The exact time depends on the concentrations of different batches of the bacterial stock. Bacterial stocks from the same batch usually take the same time to reach an OD_{600} of 1.0.

20. A large number of mutants are usually expected with potent mutagens. In this case, more selection plates should be used to reduce the number of 6-TG^r colonies or Spi$^-$ plaques on each selection plate.

21. The agar plates must be made 1 day before the *gpt* mutation and Spi$^-$ selection assays in order to obtain an appropriate level of moisture on the agar plates. Top agar cannot solidify when the surface of the agar plates is too moist.

22. Temperature and humidity determine the growth rate of the bacteria. Therefore, when incubating the plates in a 37 °C incubator, temperature and humidity should be kept constant throughout the entire experiment in order to obtain consistent results. Avoid areas in the incubator where the level of humidity is much lower.

23. The optimal number of Cm^r colonies or Spi$^-$ mutant plaques on the plates is between 30 and 100. To achieve this optimal number, dilutions can be adjusted depending on the quality of the genomic DNA and the efficiency of the in vitro packaging reactions.

24. When marking possible 6-TG^r colonies, hold the selection plates at a 45° angle towards a light source. Since the 6-TG^r colonies are usually tiny, holding the plates in this way will make the colonies more visible. Mark as many colonies as possible. Real 6-TG^r colonies will grow after 96 h of incubation.

25. Transgenic mutation assays can require a long period of time to complete. To achieve consistent results between replicate assays conducted over an extended period of time, the plates should be incubated for similar amounts of time throughout the entire experimental period.

26. At this point, the selection plates can be stored at 4 °C for up to 1 week. However, growth of the bacteria is not completely inhibited at 4 °C, and thus small colonies may become bigger after long storage times, leading to more false positives. Therefore, it is recommended that 6-TG^r colonies be confirmed as soon as possible.

27. XL-1Blue MRA and XL-1Blue MRA (P2) suspensions are used for titration, Spi$^-$ mutant selection, and mutant confirmation. They can be stored for up to 3 days at 4 °C if they are used for titration and can be used on the second day if they are used for selection and confirmation.

28. Avoid bubbles on the plates. It will be difficult to identify Spi⁻ mutant plaques after overnight incubation if there are small bubbles in the top agar. Small bubbles in the top agar can be removed by exposing the surface of freshly poured (top agar still molten) selection plates to a gas flame for 1 s.

Acknowledgement

We are grateful to our colleague, Vasily N. Dobrovolsky, for help in making the figure.

Disclaimer: The guidelines for conducting the transgenic assays described in this chapter are not necessarily endorsed by the US Food and Drug Administration.

References

1. Lambert IB, Singer TM, Boucher SE et al (2005) Detailed review of transgenic rodent mutation assays. Mutat Res 590:1–280
2. Organization for Economic Cooperation and Development (OECD) (2011). OECD guideline for the testing of chemicals; transgenic rodent somatic and germ cell gene mutation assay. Available at: http://www.oecd-ilibrary.org/docserver/download/fulltext/9748801e.pdf?expires=1342710912&id=id&accname=freeContent&checksum=06A7713CC9C6E1B7EEFCC61F2565CD7
3. Pozniak A, Muller L, Salgo M et al (2009) Elevated EMS in nelfinavir mesylate (Viracept, Roche): overview. AIDS Res Ther 6:18–23
4. Boverhof DR, Chamberlain MP, Elcombe CR et al (2011) Transgenic animal models in toxicology: historical perspectives and future outlook. Toxicol Sci 121:207–233
5. Heddle JA, Martus HJ, Douglas GR (2003) Treatment and sampling protocols for transgenic mutation assays. Environ Mol Mutagen 41:1–6
6. Thybaud V, Dean S, Nohmi T et al (2003) *In vivo* transgenic mutation assays. Mutat Res 540:141–151

Part II

Assays for Chromosomal Abnormalities

Chapter 6

In Vitro Cytogenetic Assays: Chromosomal Aberrations and Micronucleus Tests

Pasquale Mosesso, Serena Cinelli, Adyapalam T. Natarajan, and Fabrizio Palitti

Abstract

Chromosome damage is a very important indicator of genetic damage relevant to environmental and clinical studies. Detailed descriptions of the protocols used for detection of chromosomal aberrations induced by unknown agents in vitro both in the presence or the absence of rat liver-derived metabolizing systems are given. Structural chromosomal aberrations that can be observed and quantified at metaphases are described here. For the detection of chromosomal damage (fragments or whole chromosome) in interphase, the micronucleus test can be used and a description of this test is also presented. Criteria for determining a positive result using appropriate statistical methods are described.

Key words Chromosome aberrations, Micronucleus, Cytogenetic tests, Adherent cell lines (CHO, CHL, V79), Chromosome aberration assay protocols

1 Introduction

The presence of genotoxic agents in the environment may cause through different mechanisms chromosomal mutations, which are associated with the induction of both heritable defects as well as cancer in human population. Chromosomal mutations are distinguished under two major categories related to changes in the chromosome structure (chromosomal aberrations) and changes in the number of chromosomes (numerical aberrations). In the germ cells chromosomal mutations contribute to fetal mortality up to 50 % of spontaneous abortions (which account only for 15–20 % of identified pregnancies). The total frequency of chromosomal mutations at birth has been estimated to be approximately 6%: about 2% is caused by anomalies in the number of sex chromosomes, 1.4% by autosomal chromosomes, and 2.5% by anomalies in the structure of chromosomes. In somatic cells chromosomal mutations play a key role in processes leading to malignancies; if mutations occur in proto-oncogenes, tumor suppressor genes, and/or DNA damage response genes, they are responsible for a variety of genetic diseases [1]. Accumulation of DNA damage in somatic cells has also been proposed to play a role in degenerative conditions such as accelerated aging, immune dysfunction, and cardiovascular and neurodegenerative diseases [2–5].

Structural chromosome aberrations are induced by physical and chemical mutagenic agents through a variety of molecular mechanisms which are almost exclusively a consequence of a direct damage to DNA (e.g., DNA strand breaks, base damage, hydrolysis of bases, pyrimidine dimers, DNA cross-links) or indirect (e.g., inhibition of DNA topoisomerases I and II, nucleotide pool imbalance, generation of reactive oxygen species) left unrepaired, or mis-repaired to produce chromosome breaks or rearrangements [6]. However, numerical changes may not always be induced by the same genotoxic agent but could result from events, which involve disturbance or interference of the cell division spindle.

In this chapter we deal with cytogenetic tests, which are aimed to detect structural chromosomal aberrations by estimating different classes of chromosome changes scored either in metaphase or interphase (micronuclei) using the light microscope. For chromosomal aberrations in metaphase two major classes of chromosomal structural changes are identified. This is based on whether one or both chromatid (at the same locus) in a metaphase chromosome is involved in an aberration. The type of aberration observed in metaphase depends on the phase of cell cycle in which the treatment has been administered and the type of mutagenic agent used. Only few clastogenic agents are able to induce structural chromosomal aberrations in any phase of cell cycle: chromosome type in G_1, chromatid type in G_2, and a mixture of chromosome-type and chromatid-type aberrations if the treatment is made in S phase [7]. They include ionizing radiations and chemical agents such as bleomycin, cytosine arabinoside, and streptonigrin which induce DNA strand breaks directly. The majority of chemical agents induce essentially chromatid-type aberrations even when cells are treated in G_1 phase of cell cycle. On these bases, two classes of chromosome-breaking agents are usually identified: S-dependent and S-independent agents. Chromosome aberrations induced by S-independent agents are converted immediately after primary DNA damage irrespective of the cell stage at which the treatment is performed and these aberrations are assumed to arise as a consequence of "non-repair" or "misrepair" of the lesions. In contrast, aberrations induced by S-dependent agents are assumed to be formed due to "misreplication" [8, 9]. Three types of aberrations are generally scored in metaphase: gaps, breaks, and exchanges. A detailed description of them adapted from the work of Savage [10] is reported in **Note 1**.

For the analysis of chromosomal aberrations in interphase, micronuclei are scored. Micronuclei represent a proportion of fragments or whole chromosome which lag in anaphase and are not included in the main nucleus. In the subsequent interphase they condense to form small nuclei which are easily identified for scoring. Since it is essential for the cells to divide to generate micronuclei, the in vitro micronucleus assay has been improved by identification of cells which have completed cell division. This has been done by different methods which include 5-bromo-2'-deoxyuridine (BrdU) labeling [11], [^{3}H]thymidine labeling [12], or treatment of cells with cytochalasin B (Cyt B) which inhibits actin polymerization and blocks cytokinesis, and cells that have completed one cell cycle after treatment can be distinguished from undivided cells by their binucleate appearance [12]. This latter method is at present widely used because of its efficacy and the absence of any interfering effects on chromosomal structure and chromosome segregation. Micronuclei that have been formed as a consequence of chromosome breakage can be generally distinguished from those formed by whole chromosomes analyzing the size. However, this is not a completely reliable method and it is preferable to classify them on the presence or the absence of centromeres. Available methods include the immunochemical labeling of kinetochores [13–15], often called CREST method as this antibody is derived from the serum of patients with the autoimmune disease scleroderma, CREST syndrome, and the use of DNA pericentromeric sequence probes in combination with fluorescent in situ hybridization (FISH) techniques to mark centromeres [16, 17].

It is known that a large number of chemicals are metabolically transformed in vivo to avoid accumulation of lipophilic compounds in the organism. In some cases, the intermediates in xenobiotic metabolism can themselves be potentially reactive metabolites which are able to form DNA adducts. Since these processes are normally absent in cultured cells in vitro, attempts to mimic these events are achieved through the addition of rodent liver-derived metabolizing systems (S9 fraction) to the cell cultures [18]. Typically, S9 fraction contains Phase I enzymes which activate mutagens and also detoxifying enzymes such as glutathione-S-transferases, sulfo-transferases, and glucuronyl transferases (Phase II enzymes).

2 Material

2.1 Chinese Hamster Cell Lines

Chinese hamster ovary (CHO) cells, Chinese hamster V79 cells, and Chinese hamster lung (CHL) fibroblasts are established cell lines, adherent growing and widely used in cytogenetic tests because of their relatively small number of large chromosomes (diploid number, $2n=22$) and short cell cycle duration (12–14 h) with high plating efficiency.

- CHO cell line was derived from epithelial ovarian cells of female Chinese hamster originally isolated and described by Kao and Puck [19].

- V79 cell line was derived from an explant of the lung of the Chinese hamster and can be traced back directly to the original V79 isolate prepared by Ford and Yerganian [20].
- CHL cell line was derived from lung cells of the Chinese hamster and established by Utakoji in 1970 [21].

These cell lines are characterized by extensive rearrangement of chromosomal material which may change with the time [22] generating phenotypically different cell strains. As a consequence there is a widespread distribution of different cell lines with different phenotypes which may not respond in the same way to mutagens. To minimize the chances of "genetic drift" to generate possible "spurious" positive results, it is important to regularly check the cell lines for karyotype, generation time, and cloning efficiency (*see* **Note 2**).

2.2 Culture Medium

Cultured cells are normally cultivated in their usual growth media. The most common culture media for the above-mentioned cell lines are Eagle's Minimal Essential Medium (EMEM), Dulbecco's Minimal Essential Medium (DMEM), RPMI, Ham's F.10, and Ham's F.12 (*see* **Note 3**).

A description of preparation of two representative *complete culture media* is presented below:

1. *Ham's F.10 complete medium*: Ham's F.10 (424 mL), 50,000 IU/mL streptomycin sulfate and 50,000 IU/mL penicillin G (0.850 mL), newborn calf serum (75.0 mL). For preparing 500 mL of medium add and mix the respective volumes given in brackets.

2. *EMEM complete medium*: 10× EMEM (50.0 mL), 200 mM L-glutamine (5.0 mL), 7.5 % sodium bicarbonate (13.4 mL), 100× nonessential amino acids (5.0 mL), 50,000 IU/mL streptomycin sulfate and 50,000 IU/mL penicillin G (1.02 mL), 10 % fetal calf serum (50.0 mL), sterile distilled water (376 mL). For preparing 500 mL of medium add the respective volumes given in brackets.

2.3 Metabolic Activation

S9 fraction, 0.1 M nicotinamide adenine dinucleotide phosphate (NADP), 0.1 M glucose-6-phosphate (G-6-P), 0.33 M KCl, 0.1 M $MgCl_2$, 0.2 M HEPES, Hank's balanced salt solution (HBSS).

Post-mitochondrial S9 liver fraction from rodents (usually rats) pretreated with Aroclor 1254 or a combination of phenobarbitone and β-naphthoflavone for inducing mixed-function oxidases [23, 24] is combined with a mix of cofactors to form an NADPH-generating system (S9 mix) which is the most commonly used exogenous metabolic activation system with the capacity to metabolize chemicals in vitro. An example of a widely used S9 mix preparation is displayed below:

For preparing 10 mL of S9 mix, add 3.0 mL of S9, 0.4 mL of NADP, 0.5 mL of G-6-P, 1 mL each of KCl and Hepes, 0.4 mL of MgCl$_2$, and 3.7 mL of HBSS.

2.4 Solvents

1. Water and culture medium are the most commonly used solvents in case of solid test substances.
2. Add liquid test substances directly to the test systems and/or dilute prior to treatment.
3. Use organic solvents, such as dimethylsulfoxide (DMSO) and acetone, at a maximum concentration of 1 % (v/v), since it has been demonstrated that they do not interfere with cell proliferation, structural and numerical chromosomal changes, and metabolic activation.
4. Even in case of water as solvent, do not exceed a maximum concentration of 10 % (v/v) because of molarity changes of medium and dilution of nutrients.

2.5 Cell Culture Reagents

1. Phosphate-buffered saline (PBS), calcium and magnesium free, suitable for cell cultures.
2. Trypsin: 0.25 % trypsin–EDTA in PBS.
3. Colcemid: 10 μg/mL Colcemid aqueous solution.
4. Cytochalasin-B: 1.2 mg/mL Cytochalasin B prepared in DMSO.
5. Hypotonic solution: 1 % trisodium citrate.
6. Fixative solution for chromosome aberration test: 3:1 (v/v) methanol:glacial acetic acid.
7. Fixative solution for micronucleus test: Methanol.
8. Staining solution: 3 % Giemsa, v/v in tap water (for both chromosome aberration and micronucleus tests).
9. 0.0125 % Acridine orange, w/v in PBS (alternative staining solution for micronucleus test).

2.6 Positive Control Agents

The most commonly used positive control agents are the following:

1. Mitomycin C, clastogen active without metabolic activation, dissolved in culture medium.
2. Cyclophosphamide, clastogen active with metabolic activation, dissolved in sterile distilled water.
3. Colchicine, aneugen, dissolved in sterile distilled water.
4. Vinblastine, aneugen, dissolved in sterile distilled water.

3 Methods

As a general principle, cultured cells are treated with the test substance in the absence and presence of an exogenous metabolic activation system with adequate negative, solvent, and positive control agents and are harvested at one or more intervals after treatment. Before fixation, cells are exposed to a metaphase-blocking compound such as colcemid (analysis of chromosomal aberrations at metaphase) or cytochalasin B (analysis of micronuclei) and subsequently to an appropriate hypotonic solution to enlarge the cells for better cytogenetic analysis. They are then fixed in an acetic acid/methanol solution, dropped onto microscope slides, stained, coded according to a randomized system, and analyzed for the presence of chromosomal aberrations or micronuclei with a light or a fluorescence microscope.

3.1 Preparation of the Test Cultures

Approximately 20 h before treatment, prepare an appropriate number of flasks (F25) for the experiment from a single pool of cells. Each 25 cm^2 flask is seeded with 300,000 cells in complete culture medium. These cells should be in exponential growth phase at the time of treatment. Incubate all cell culture at 37 °C in a 5 % carbon dioxide atmosphere (100 % humidity nominal).

3.2 Selection of Dose Levels for Treatment

The highest dose level of the test substance to be assayed is determined according to the solubility in the solvent vehicle and culture medium. For relatively non-cytotoxic compounds the maximum concentration should not exceed 5 μL/mL, 5 mg/mL, or 10 mM, whichever is the lowest (*see* **Note 4**). Up to one dose level at which precipitation is observed may be included in the treatment series. Seven lower dose levels separated by a factor of two are used.

3.3 Treatment and Harvest of Cell Cultures

1. Prepare treatment medium for each test point as follows:

Without metabolic activation:	
Test substance or control solution	0.05 mL
Complete culture medium	4.95 mL
With metabolic activation:	
Test substance or control solution	0.05 mL
S9 mix	0.50 mL
Complete culture medium	4.45 mL

2. Include negative and positive controls in each treatment series.
3. Prepare duplicate cultures at each experimental test point, with the exception of the positive controls which are prepared as single cultures.
4. Both in the absence and presence of S9 metabolic activation, the cultures are incubated in this treatment medium for 3 h.

5. When negative results are obtained, a continuous treatment until sampling time is performed in the absence of metabolic activation only. In the presence of metabolic activation, due to toxicity and reactivity of S9 in culture, the exposure time is limited to the short treatment time.

3.4 Harvesting Cells for Chromosome Aberration Test

1. At the end of treatment, remove the medium containing test substance and wash the flasks twice with Ca^{2+}/Mg^{2+}-free PBS.
2. For the short treatment time (3 h), add fresh medium and incubate the cultures for a further 17 h until sampling at the time corresponding to 1.5 normal cell cycle lengths. (Perform continuous treatment in the absence of S9 metabolism for 20 h)
3. Add 0.2 µg/mL colcemid (final concentration) for the last 3 h of the recovery period, leading up to harvesting, as displayed below:

± S9	Treatment	Recovery period	Harvest: 1.5 normal cell cycles Colcemid	
	0 h	3 h	17 h	20 h

− S9	Treatment	Harvest: 1.5 normal cell cycles Colcemid	
	0 h	17 h	20 h

4. At the end of the recovery period, collect the medium from each F25 flask into appropriately labeled 15 mL centrifuge tube to avoid loss of detached mitotic cells.
5. Add 1 mL of trypsin/EDTA solution for 5–7 min at 37 °C to bring cell monolayers into suspension.
6. When monolayer cells are completely detached, add 1 mL of complete culture medium to block enzymatic digestion.
7. Add this cell suspension to the corresponding 15 mL centrifuge tube prepared as described in **step 4** above.
8. Centrifuge the cell suspensions for 10 min at $600 \times g$ and aspirate the supernatant with a Pasteur pipette.
9. Resuspend the cell pellet in 5 mL of hypotonic solution and leave for 5 min at room temperature.
10. Centrifuge for 8 min at $600 \times g$ and aspirate with a Pasteur pipette the supernatant.
11. Fix the cells by adding 5 mL of cold (at 4 °C), freshly prepared fixative solution and leave at 4 °C for at least 1 h.
12. Wash the cells two times with cold (4 °C), freshly prepared fixative and produce an appropriately concentrated cell suspension.

13. Drop a few drops of concentrated cell suspension onto clean, wet, grease-free glass slides and air-dry to produce metaphase chromosome spreads.

14. Prepare at least three slides for each experimental point, labeled with the identity of the culture.

3.5 Harvesting Cells for the Micronucleus Test

1. At the end of the treatment, remove the medium containing test substance and wash the flasks twice with Ca^{2+}/Mg^{2+}-free PBS.

2. For the short treatment time (3 h), add fresh medium and incubate the cultures until sampling at the time corresponding to 1.5–2.0 normal cell cycle lengths (24 h) (Perform continuous treatment in the absence of S9 metabolism for 24 h).

3. Add cytochalasin B at a final concentration of 3 µg/mL for the last 21 h, leading up to harvesting, as displayed below:

± S9	Treatment		Harvest: 1.5-2.0 normal cell cycles	
	0 h	3 h	cytochalasin B	24 h

- S9	Treatment		Harvest: 1.5-2.0 normal cell cycles	
	0 h	3 h	Treatment + cytochalasin B	24 h

The experiment can be performed also in the absence of cytochalasin B using the same treatment schedules without addition of cytochalasin B.

4. At the end of the recovery period, collect the medium from each F25 flask into appropriately labeled 15 mL centrifuge tube to avoid loss of detached cells.

5. Add 1 mL of trypsin/EDTA solution to each F25 culture flask and incubate for 5–7 min at 37 °C to bring cell monolayers into suspension.

6. When monolayer cells are completely detached, add 1 mL of complete culture medium to block enzymatic digestion.

7. Add this cell suspension to the corresponding 15 mL centrifuge prepared as described in **step 5**.

8. Centrifuge the cell suspensions for 10 min at 600 × *g* and aspirate the supernatant with a Pasteur pipette.

9. Resuspend the cell pellet in 5 mL of hypotonic solution and leave for 1 min at room temperature. The period of hypotonic treatment may slightly vary to improve the quality of interphase cells for scoring of micronuclei in the form of binucleated cells in the presence of cytochalasin B (*see* **Note 5**) and mononucleated cells in its absence.

10. Fix the cells by adding 5 mL of cold (4 °C) methanol and leave at 4 °C for at least 1 h.
11. Wash the cells two times with cold (4 °C) methanol and produce appropriately concentrated cell suspensions.
12. Drop a volume of approximately 20 µL of the concentrated cell suspensions onto clean, wet, grease-free glass slides and air-dry.
13. Prepare at least three slides for each experimental point, labeled with the identity of the culture.

3.6 Osmolality and pH Measurement

Measure the osmolality and pH of treatment media, both in the absence and presence of S9 metabolic activation at the end of the treatment time, to ensure that treatment has not been performed under extreme non-physiological conditions (*see* **Note 6**).

3.7 Staining of Slides

1. Allow the prepared slides from **step 14** of Subheading 3.4 and **step 13** of Subheading 3.5 to air-dry and keep at room temperature prior to staining for at least 1 day.
2. Stain the slides in 3 % aqueous solution of Giemsa for 8–10 min at room temperature and rinse stepwise in tap and distilled water.
3. Make the slides permanent with Eukitt.

For micronucleus test, alternative staining procedure can be performed as follows:

1. Allow the slides to air-dry and keep at room temperature prior to staining for at least 1 day.
2. Stain slides with a solution of acridine orange (0.0125 % w/v in PBS) for 10 s.
3. Rinse slides in PBS and air-dry.
4. Immediately before microscope analysis, place a coverslip using one drop of PBS.

3.8 Selection of Dose Levels for Scoring

1. In order to obtain appropriate sensitivity and specificity of in vitro mammalian cytogenetic assays, appropriate induced cytotoxicity levels should be selected (*see* **Note 7**).
2. To determine cytotoxicity for chromosome aberration assay and micronucleus test in the absence of cytochalasin B, calculate the population doubling (PD) according to the following formula:

$$PD = \frac{\left[\log(N/X_0)\right]}{\log 2}$$

where N is the cell count (*see* **Note 8**) at the time of harvesting and X_0 is the starting (baseline) count which is evaluated at the beginning of treatment. This is performed using two additional control cultures set up along with the other culture flasks for the experiment. Cells are brought into suspension with 1 % trypsin solution and counted approximately at time zero. The baseline count is the average of the total number of cells from the two flasks.

3. For the micronucleus test, when cytochalasin B is used, the most appropriate method to assess cytotoxicity is to calculate the cytokinesis-block proliferation index (CBPI) according to the following formula:

$$\text{CBPI} = \frac{\left[\begin{array}{c}(\text{No. of mononucleated cells}) + (2 \times \text{No. of binucleated cells})\\ + (3 \times \text{No. of multinucleated cells})\end{array}\right]}{(\text{Total number of cells})}$$

Evaluate the number of mononucleated, binucleated, and multinucleated cells by analyzing 1,000 cells per dose level (500 cells per replicate culture).

Determine the CBPI for each test substance treatment level and vehicle control cultures and then calculate the percentage of cytostasis as follows:

$$\%\ \text{cytostasis} = 100 - 100 \left[\frac{(\text{CBPI}_T - 1)}{(\text{CBPI}_C - 1)}\right]$$

where T = test substance treatment culture and C = vehicle control culture.

4. Select a dose which produces an adequate level of cytotoxicity compared with the solvent controls as the highest dose level for genotoxicity assessment.

5. For chromosome aberration test, PD value should be approximately 50 % of the control, while for micronucleus test this parameter should be between 40 and 50 % of the control value. In the presence of cytochalasin B, the % cytostasis, which is used to assess cytotoxicity, should be 50–60 %.

6. Select two lower dose levels for the scoring of chromosomal damage. The lowest dose level should be on the borderline of cytotoxicity and the intermediate dose level should be evenly spaced between the two.

7. If the test item does not induce toxicity at any dose level, then the highest treatment level is selected for scoring.

3.9 Cytogenetic Analysis

1. Let a person not subsequently involved in slide evaluation randomly assign code numbers to the slides, and conceal any other identification marks.

2. For chromosome aberration test, examine metaphase spreads, judged to be of sufficient quality to permit scoring, at high magnification (100×). Do not score metaphases that differ from the modal chromosomal complement by more than two centromeres.

3. Report polyploid and endoreduplicated cells encountered, but do not include them in the count of eligible metaphases. From 100 eligible metaphases per test culture, record the number of chromosomes, specific types, and number of aberrations. Classification of aberration is based on the criteria displayed in **Note 1**.

4. Record the Vernier readings of aberrant or equivocal metaphases.

5. For the micronucleus assay, score at least 2,000 cells per concentration (1,000 cells per replicate culture) to assess the frequency of micronucleated cells. When treatment is performed in the presence of cytochalasin B, only binucleated cells must be scored.

6. The criteria for identifying micronuclei are as follows:
    (a) The micronucleus diameter must be less than 1/3 of the nucleus diameter.
    (b) The micronucleus diameter must be greater than 1/16 of the nucleus diameter.
    (c) No overlapping with the nucleus must be observed.
    (d) The condensation status must be similar to the main nucleus.

3.10 Evaluation of Results

There are several criteria for determining a positive result including different appropriate statistical methods. Widely used criteria are described below:

1. The vehicle controls are used as the reference point for comparison in the statistical evaluation and the evaluation of the results. Positive control results should be statistically significant compared with the concurrent negative control and should fall within the historical positive control range values.

2. For chromosome aberration test, the number of cells bearing aberrations in the negative control and treated cultures are compared using Fisher's exact test. The comparison is performed both including and excluding gaps from the aberration counts. Since multiple comparisons are performed, the problem of Type I error (chance "positive" results) arises. Accordingly, significance levels for each treatment level after application of Bonferroni's correction are considered.

3. For a test substance to be considered clastogenic, four criteria must be met:
   (a) *Increases over the concurrent controls*: If any dose level shows a statistically significant increase in aberration-bearing cells, this is considered as evidence of a clastogenic effect.
   (b) *Increase over historical controls*: If the increases fall within the range of values normally observed in the negative control cultures, the test compound cannot be judged clastogenic. Any significant increase over the concurrent negative controls is therefore compared with historical control values derived from recent studies (*see* **Note 9**).
   (c) *Reproducibility*: Any increases observed must be present in both replicate cultures.
   (d) *Biological significance*: It is not required that an increased response is observed at increasing dose levels, but dose-related activity is taken as further evidence of a clastogenic effect. The types of aberrations observed should be also taken into consideration (e.g., chromatid exchanges).
4. For the micronucleus assay, the frequencies of cells with micronuclei (with one, two, and more than two micronuclei) are recorded.
5. The numbers of cells with micronuclei in the control and treated cultures are compared using a χ^2 calculation.
6. A test substance is considered as clearly positive in this assay if the following criteria are met:
   (a) *Increases over the concurrent controls*: Statistically significant increases in the proportion of micronucleated cells over the concurrent controls must be observed at one or more concentrations.
   (b) *Increase over historical controls*: If increases of micronucleated cells at such data points fall within the range of values normally observed in the negative control cultures, the test item cannot be classified as positive. Any significant increase over the concurrent negative controls is therefore compared with historical control values derived from recent studies (*see* **Note 9**).
   (c) *There is a significant dose–effect relationship*: The variability inherent in biological material may inevitably lead to the generation of equivocal data sets. In such cases it is strongly recommended that further evaluation should be undertaken in terms of additional scoring or alternatively in the form of replication of the relevant experimental treatment scheme previously used.

4 Notes

1. Classification of the most common induced structural chromosomal aberrations observed at metaphase. The examples given below have been adapted from the work of Savage [10].

   A. *Chromosome-type aberrations (involving both chromatids of a chromosome at identical loci)*

      a. Chromosome intra-changes

         i. *Deletions*: Deletions represent a discontinuity at the same locus in both chromatids giving an acentric fragment. The abnormal (shortened) monocentric chromosome may not be identified. They are distinguished in terminal and interstitial deletions. Terminal deletions appear as paired fragments resulting from a single break across the chromosome and are not associated with an obvious exchange aberration (Fig. 1). Interstitial deletions may appear as paired dots classified as "minutes" (Fig. 2) or as larger deletion without a centromere and which are joined to give an acentric ring and generate from double breaks across the chromosome (Fig. 3). The mechanisms leading to these aberrations are identical and the different classification is essentially an indication of the different sizes of interstitial deletions.

         ii. Asymmetrical intra-changes

             1. *Centric ring with fragment*: Centric rings generate from breakage of both chromosome arms and illegitimate rejoining of them. It can be easily

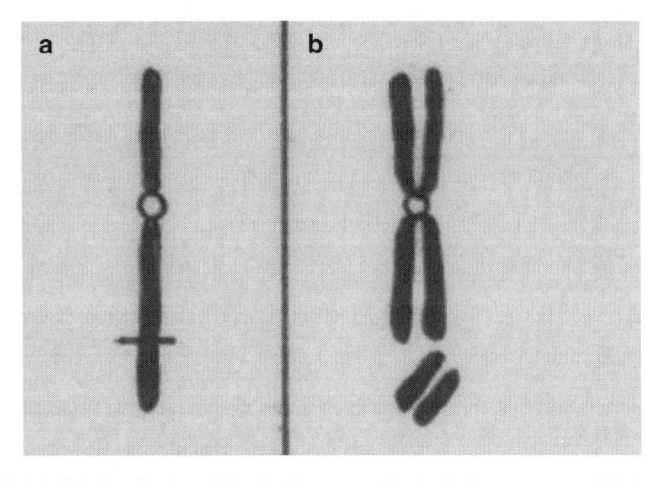

Fig. 1 (**a**) Origin of aberration in the unreplicated chromosome; (**b**) terminal deletion

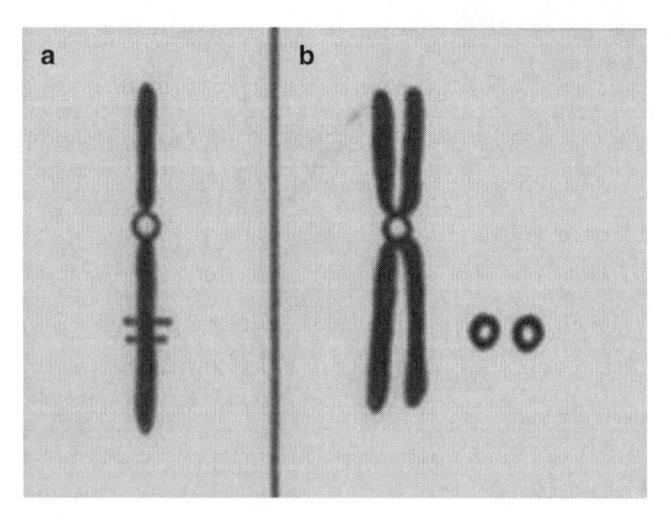

Fig. 2 Interstitial deletion: **(a)** Origin of aberration in the unreplicated chromosome; **(b)** minute

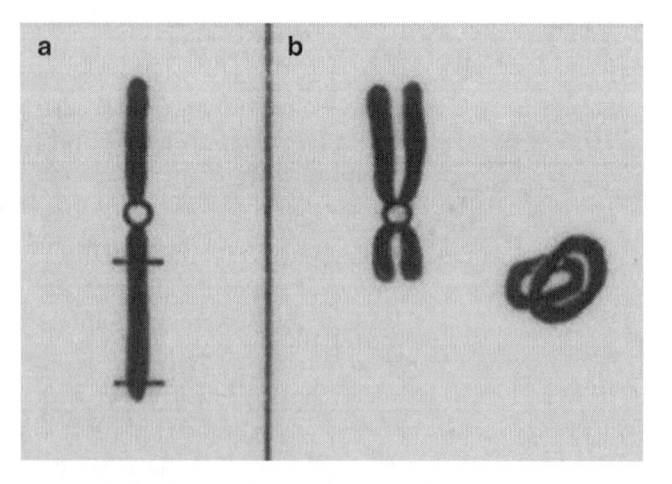

Fig. 3 Interstitial deletion: **(a)** Origin of aberration in the unreplicated chromosome; **(b)** acentric ring

distinguished by cytogenetic analysis from the acentric type and is generally accompanied by one acentric fragment (Fig. 4). The fragment is an element of the exchange process and should not be scored as a separate event.

2. *Inversions*: Chromosome inversions are classified into two categories: paracentric inversions when both breakage points and rejoining lie on the same arm of the chromosome and pericentric inversions when breakage points and inversion lie on different arms of the same chromosome (Fig. 5). In cells where the breakage

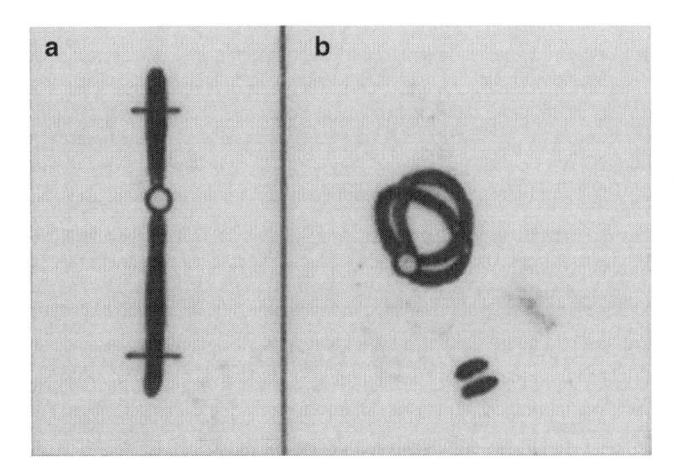

Fig. 4 (**a**) Origin of aberration in the unreplicated chromosome; (**b**) centric ring with fragment

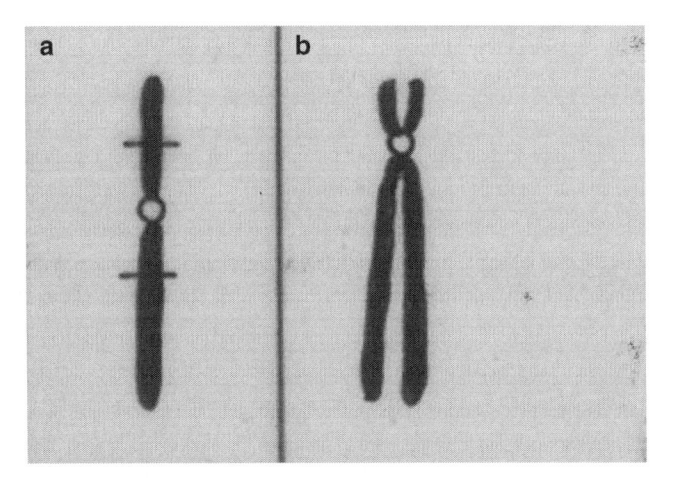

Fig. 5 (**a**) Origin of aberration in the unreplicated chromosome; (**b**) pericentric inversion

points are located at different distance from the centromere, the resulting abnormal chromosome can be morphologically distinguished by the rearranged position of the centromere. However, when breakage points are equally distant from the centromere and in the case of paracentric inversions, the resulting aberrations cannot be detected cytogenetically since the position of centromere is not changed. The detection of such inversions is only possible through the use of chromosome banding or FISH techniques.

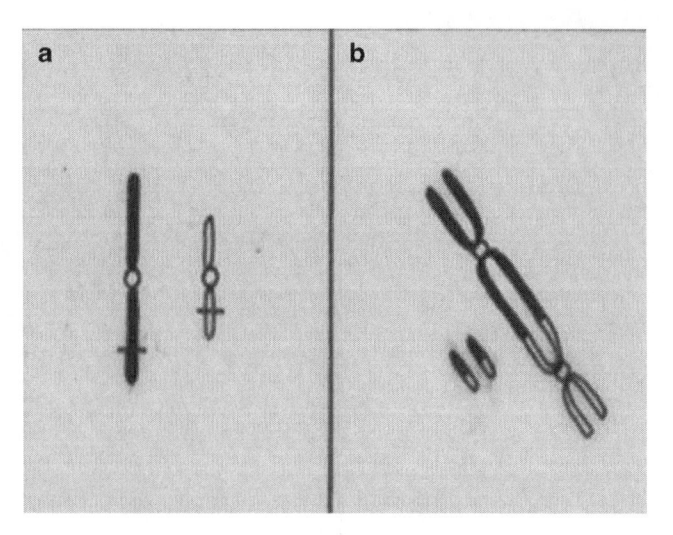

Fig. 6 Asymmetrical exchanges. **(a)** Origin of aberration in the unreplicated chromosome; **(b)** dicentric chromosome with fragment

b. Chromosome inter-changes
   i. *Asymmetrical exchanges (generally dicentrics)*: These involve two or more broken chromosomes which illegitimately rejoin to generate chromosomes with two (dicentric), three (tricentric), or more (polycentric) centromeres. It is widely recognized that each dicentric scored at 1st metaphase following treatment is accompanied by one acentric fragment and therefore it should not be scored as a separate event. On this basis, one cell bearing one dicentric and two acentric fragments would be scored as one dicentric plus fragment and one terminal deletion (Fig. 6). Correspondingly, a tricentric chromosome is assumed to bear two accompanying fragments.
   ii. *Symmetrical exchanges (reciprocal translocations)*: These involve breakage of two chromosomes and the reciprocal exchange of broken chromosome parts (Fig. 7). Conventional cytogenetic analysis do not allow detection of reciprocal translocations involving the exchange of two chromosome fragments of similar size, since the resulting rearranged chromosomes are only slightly different from the normal karyotype. They can be better detected by the use of chromosome banding or more efficiently by FISH techniques. However, symmetrical exchanges may also occur in the centromeric regions of two chromosomes involving whole arm exchange. In case of involvement of acrocentric

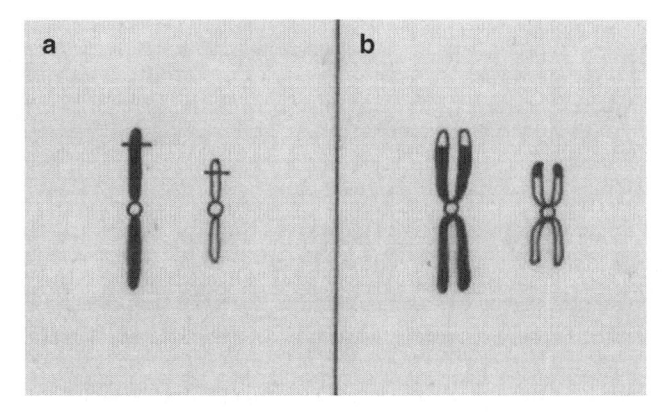

Fig. 7 Symmetrical exchanges. **(a)** Origin of aberration in the unreplicated chromosome; **(b)** reciprocal translocation

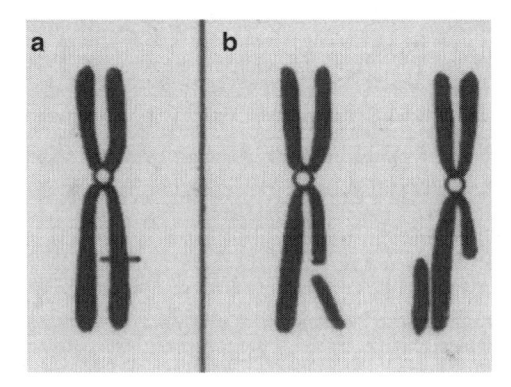

Fig. 8 (a) Origin of aberration in the replicated chromosome; **(b)** terminal deletions aligned and displaced

chromosomes they are morphologically identified as centric fusions (*Robertsonian translocations*).

B. Chromatid-type aberrations

  a. *Chromatid break or terminal deletion*: It is a discontinuity in which there is displacement of the chromatid fragment or alternatively a non-staining region with a width greater than that of a chromatid (Fig. 8).

  b. *Interstitial deletions*: They originate from two independent breaks on the same chromatid. The deleted fragment can often be separated from the deleted chromosome. It is generally small and classified as chromatid minutes (Fig. 9).

  c. *Isochromatid deletions*: These involve exchanges between sister chromatids and may be confused with chromosome-type terminal deletion since nonunion distal will lead to an acentric fragment indistinguishable

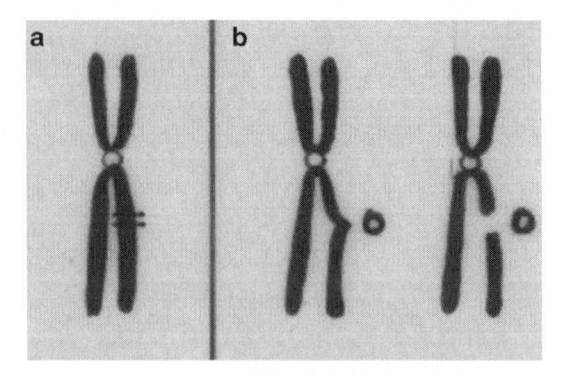

Fig. 9 Interstitial deletions. (**a**) Origin of aberration in the replicated chromosome; (**b**) minutes

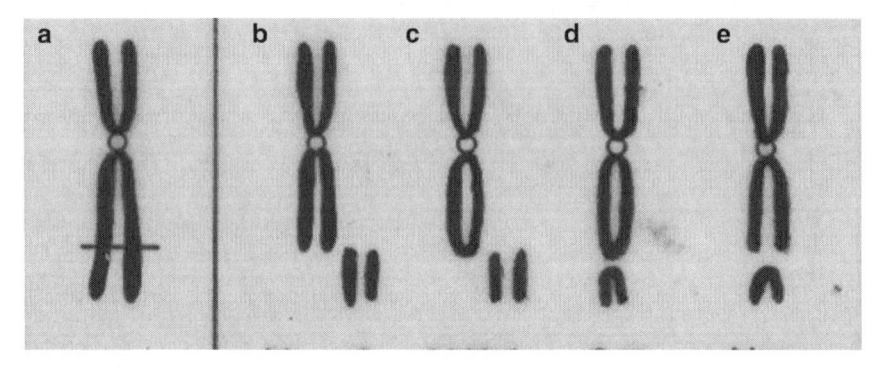

Fig. 10 Isochromatid deletions. (**a**) Origin of aberration in the replicated chromosome; (**b**) nonunion proximal and distal; (**c**) nonunion distal; (**d**) sister union; (**e**) nonunion proximal

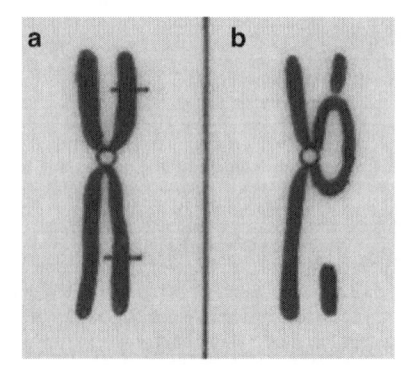

Fig. 11 Inter-arm intra-changes. (**a**) Origin of aberration in the replicated chromosome; (**b**) asymmetrical exchange

from a chromosome deletion. However, they are usually distinguished from chromosome-type aberrations since the majority of isochromatid breaks involve a union between sister chromatids either proximal or distal to the point of breakage (Fig. 10).

d. *Inter*-arm intra-changes: There are two forms of both asymmetrical (Fig. 11) and symmetrical intra-changes (Fig. 12) but conventional cytogenetic analysis allows

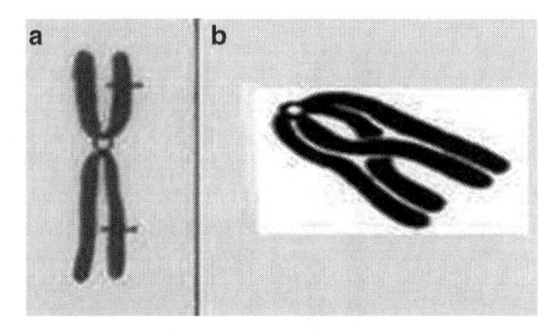

Fig. 12 Inter-arm intra-changes. **(a)** Origin of aberration in the replicated chromosome; **(b)** symmetrical exchange

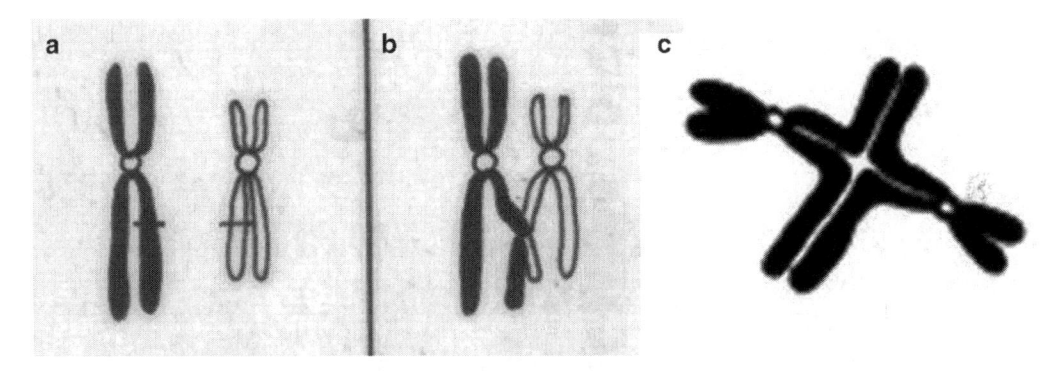

Fig. 13 Inter-arm inter-changes. **(a)** Origin of aberration in the replicated chromosome; **(b)** completion of symmetrical exchange; **(c)** cytological appearance at metaphase

distinguishing only one of each. The asymmetrical form is unstable and is normally lost at the next mitosis.

e. *Inter-arm inter-changes (symmetrical and asymmetrical)*: This type of aberration involves an exchange between two or more chromosomes. The symmetrical exchanges are the chromatid-type equivalent of reciprocal translocations while the asymmetrical ones are the equivalent of dicentric chromosomes (Figs. 13 and 14). In both cases, rearranged chromosomes result in a configuration having four arms and often called quadri-radials.

f. Tri*radials*: This aberration originates from the interaction between an isochromatid break and a chromatid break to give a wide range of configurations. The most common ones are displayed below (Fig. 15).

2. Cell lines should be obtained from reliable sources at known early passage and should be grown for a minimal number of passages in order to generate a sufficiently large permanent stock cell line checked for the absence of bacterial and mycoplasma contamination, cell viability, karyotype, and generation time. To keep variability at the minimum level, a vial from the

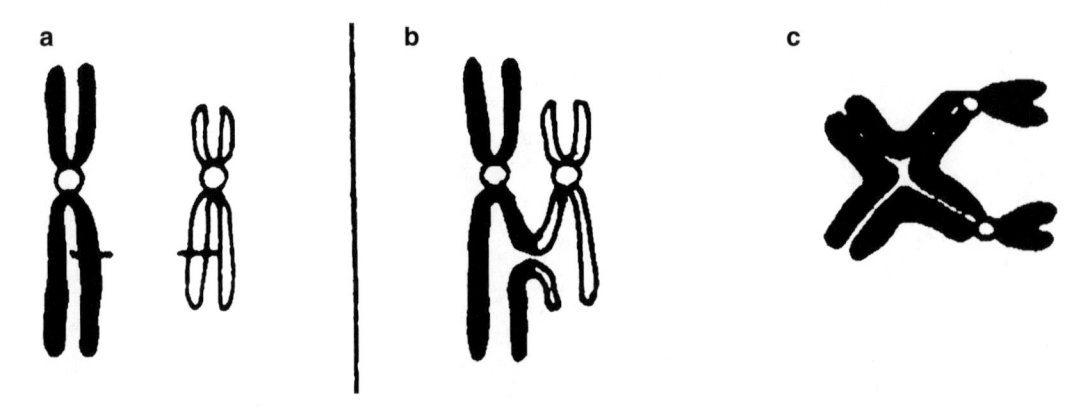

Fig. 14 Inter-arm inter-changes. (**a**) Origin of aberration in the replicated chromosome; (**b**) completion of asymmetrical exchange; (**c**) cytological appearance at metaphase

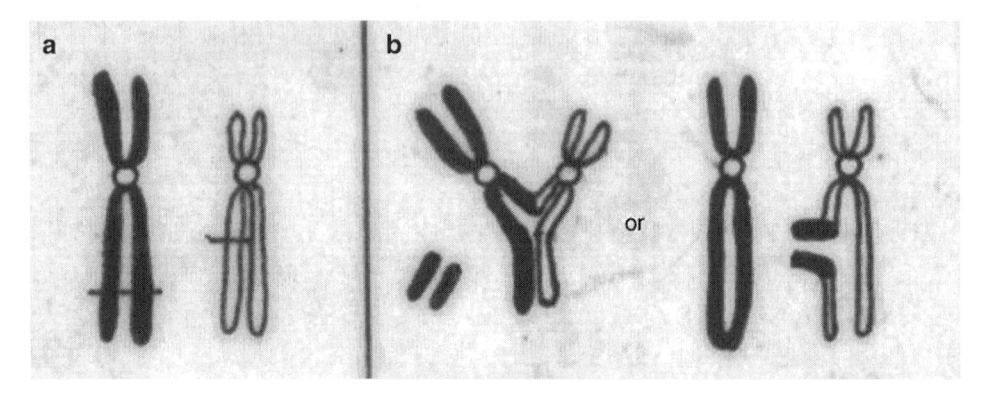

Fig. 15 Inter-arm inter-changes. (**a**) Origin of aberration in the replicated chromosome; (**b**) triradials

permanent stock should be thawed, grown for few passages, and frozen to generate a "working stock" cell line. For individual experiments, single vials are thawed and used up to completion of the study. On completion of "working stock" a new vial from the permanent stock should be thawed and grown for an appropriate number of passages to generate a new "working stock" with the same number of passages as the previously employed one.

3. It should be noted that some culture media have the potential to oxidize a wide range of chemicals including flavonoids and thiols to produce the clastogenic compound hydrogen peroxide [25, 26]. Results obtained for oxidation of ascorbic acid and epigallocatechin gallate showed that Ham's F-10 and F-12 generate much lower levels of peroxide compared to the other most commonly used culture media (i.e., DMEM, EMEM, RPMI) [27, 28].

4. The current testing limit of 10 mM or 5 mg/mL for in vitro mammalian cell assays is presently under discussion [29, 30] because of the concern of generation of false-positive results

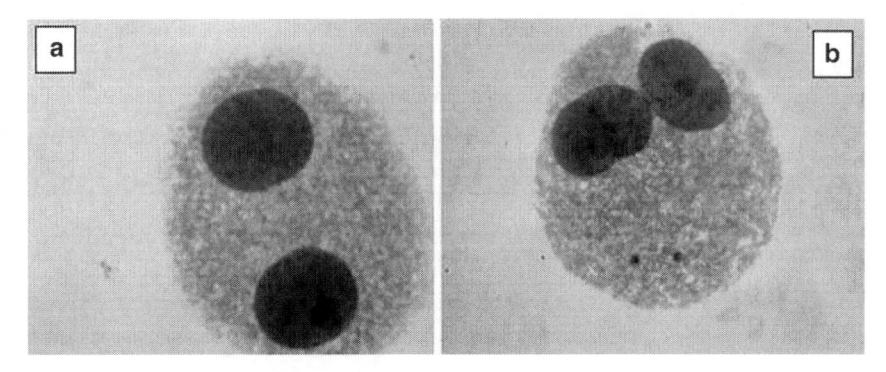

Fig. 16 Microphotograph (×1,000 magnification) of (**a**) normal and (**b**) micronucleated binucleated cells from untreated and mitomycin-C treated CHO cells, respectively

and it has been recognized that 1–2 mM could be used as suitable limit without significant loss of sensitivity of the assay. However, exceptions to use higher concentrations may be needed for different aspects such as elevated test substance volatility, presence of structural alerts for genotoxic carcinogenicity, inadequate in vitro metabolism, and high in vivo exposure expected.

5. Microphotograph of a micronucleated cell (Fig. 16).

6. It is widely recognized that treatments performed under extreme non-physiological conditions may induce false-positive results which are not relevant for assessing genotoxic risk [31–34]. It is therefore important that pH and osmolality are controlled and any observed deviation from physiological conditions, particularly low pH in the presence of S9 metabolism and high osmolality, should be corrected to avoid false-positive responses.

7. To ensure sensitivity of the cytogenetic assay it is necessary to evaluate test substance at high concentrations. However, it is presently recognized that excessive toxicity can generate false-negative results when cell growth is delayed and cells cannot reach the metaphase stage necessary for chromosome aberration analysis or progress into the next interphase where the micronuclei can be scored. Alternatively, under excessive cytotoxic conditions false-positive results can be generated through indirect non-genotoxic mechanisms which are not observed at lower concentrations. Preferred methods for measuring toxicity should take into account cell proliferation after the beginning of treatment like relative increase in cell count (RICC), relative population doubling (RPD), CBPI, and replicative index (RI). Parameters such as relative cell count (RCC) or relative percentage of confluence which do not provide information on proliferation can underestimate cytotoxicity [30]. On the other hand, mitotic index becomes a very inaccurate

measure of cytotoxicity in case of mitogenic stimulation induced by the test substances which interact with spindle structures and functions [32].

8. Cell count is performed using a hemocytometric chamber such as Burker or Neubauer chamber. Cells brought into suspension as described in Subheading 2.3 are completely monodispersed by repeated pipetting; a very small volume, adequate to filling in the chamber without floating with an excess of it, is added. The "charged" counting chamber is then placed on the microscope stage and the counting grid is brought into focus at low power.

9. The historical data set should be as large as possible and updated regularly. Data should rise from an unmodified experimental protocol unless any variations applied do not impact on the assay performance and results. No negative control values should be eliminated from the data set unless there is a scientifically justified reason. For assessment of experimental data the distribution of historical control values together with appropriate statistics (e.g., confidence intervals, 95–99 % percentiles) should be considered [35].

An example of historical control data set for untreated and positive control cultures in the chromosome aberration assay using CHO cells is given in Tables 1 and 2, respectively.

Table 1
Example of historical control data set for the incidence of chromosomal aberrations both including and excluding gaps in untreated control cultures in the absence and presence of rat S9 metabolism

	Absence of S9		Presence of S9	
	+ gaps	− gaps	+ gaps	− gaps
	Untreated controls			
Mean	1.8	0.5	2.4	0.6
SD (σ_{n-1})	1.5	0.6	1.9	0.7
n	145	145	113	113
Minimum	0.0	0.0	0.0	0.0
Maximum	7.0	3.0	9.5	3.0
99 % Percentile	5.8	2.5	8.4	2,5
1 % Percentile	0	0	0	0

Mean: Mean incidence (%) of aberration-bearing cells
SD (σ_{n-1}): Standard deviation
n: Number of experiments

Table 2
Example of historical control data set for the incidence of chromosomal aberrations both including and excluding gaps in positive control cultures in the absence and presence of rat S9 metabolism

	Absence of S9		Absence of S9		Presence of S9	
	3-h treatment time 20-h sampling time		20-h treatment time 20-h sampling time		3-h treatment time 20-h sampling time	
	+ gaps	− gaps	+ gaps	− gaps	+ gaps	− gaps
	Positive controls					
Mean	38.9	37.3	31.6	29.6	40.1	38.0
SD (σ_{n-1})	13.6	13.8	11.3	11.0	14.1	14.3
n	45	45	77	77	114	114
Minimum	20.0	18.5	14.0	12.0	15.5	15.5
Maximum	72.0	72.0	62.0	59.0	84.0	82.0
95 % Percentile	64.6	61.8	57.6	52.2	64.4	63.0
5 % Percentile	24.1	20.8	17.4	15.4	20.0	19.0

Mean: Mean incidence (%) of aberration-bearing cells
SD (σ_{n-1}): Standard deviation
n: Number of experiments

References

1. Erickson RP (2010) Somatic gene mutation and human disease other than cancer: an update. Mutat Res 705:96–106
2. De Flora S, Izzotti A (2007) Mutagenesis and cardiovascular diseases: molecular mechanisms, risk factors, and protective factors. Mutat Res 621:5–17
3. Hoeijmakers JH (2009) DNA damage, aging, and cancer. New Engl J Med 361:1475–1485
4. Frank SA (2010) Evolution in health and medicine Sackler colloquium: somatic evolutionary genomics: mutations during development cause highly variable genetic mosaicism with risk of cancer and neurodegeneration. Proc Natl Acad Sci USA 107:1725–1730
5. Slatter MA, Gennery AR (2010) Primary immunodeficiencies associated with DNA-repair disorders. Expert Rev Mol Med 685:146–165
6. Bender MA, Griggs HG, Bedford JS (1974) Mechanisms of chromosomal aberration production. III. Chemicals and ionising radiation. Mutat Res 23:197–212
7. Evans HJ (1961) Chromatid aberrations induced by gamma irradiation. I. The structure and frequency of chromatid interchanges in diploid and tetraploid cells of Vicia faba. Genetics 46:257–275
8. Evans HJ, Scott D (1969) The induction of chromosome aberrations by nitrogen mustard and its dependence on DNA synthesis. Proc R Soc Lond B Biol Sci 173:491–512
9. Kihlman BA (1977) 1,3,7,9-tetramethyluric acid, a chromosome-damaging agent occurring as a natural metabolite in certain caffeine-producing plants. Mutat Res 39:297–315
10. Savage JRK (1975) Classification and relationships of induced chromosomal structural changes. J Med Genet 13:103–122
11. Pincu M, Bass D, Norman A (1984) An improved micronuclear assay in lymphocytes. Mutat Res 139:61–65
12. Fenech M, Morley AA (1985) Measurement of micronuclei in lymphocytes. Mutat Res 147:29–36
13. Degrassi F, Tanzarella C (1988) Immunofluorescent staining of kinetochores in micronuclei: a new assay for the detection of aneuploidy. Mutat Res 203:339–345
14. Thomson EJ, Perry PE (1988) The identification of micronucleated chromosomes: a possible assay for aneuploidy. Mutagenesis 3:415–418
15. Eastmond DA, Tucker JD (1989) Identification of aneuploidy-inducing agents using

cytokinesis-blocked human lymphocytes and an anti-kinetochore antibody. Environ Mol Mutagen 13:34–43
16. Eastmond DA, Pinkel D (1990) Detection of aneuploidy-inducing agents in human lymphocytes using fluorescence *in situ* hibridization with chromosome specific DNA probes. Mutat Res 234:303–318
17. Marshall RR, Murphy M, Kirkland DJ et al (1996) Fluorescence *in situ* hybridisation (FISH) with chromosome-specific centromeric probes: a sensitive method to detect aneuploidy. Mutat Res 372:233–245
18. Natarajan AT, Tates AD, van Buul PPV et al (1976) Cytogenetic effects of mutagens/carcinogens after activation in a microsomal system *in vitro*. Mutat Res 37:83–90
19. Kao FT, Puck TT (1968) Genetics of somatic mammalian cells, VII. Induction and isolation of nutritional mutants in Chinese hamster cells. Proc Natl Acad Sci USA 60:1275–1281
20. Ford DK, Yerganian G (1958) Observations on the chromosomes of Chinese hamster cells in tissue culture. J Natl Cancer Inst 21: 393–425
21. Koyama H et al (1970) A new cell line derived from newborn Chinese hamster lung tissue. Gann 61:161–167
22. Xiao Y, Natarajan AT (1998) Development of arm-specific and subtelomeric region-specific painting probes for Chinese hamster chromosomes and their utility in chromosome identification of Chinese hamster cell lines. Cytogenet Cell Genet 83:8–13
23. Matsushima T, Sawamura M, Hara K et al (1976) A safe substitute for polychlorinated biphenyls as an inducer of metabolic activation system. In: De Serres FJ, Fouts JR, Bend JR, Philpot RM (eds) In vitro metabolic activation in mutagenesis testing. Elsevier/North-Holland, Amsterdam, pp 85–88
24. Elliot BM, Combes RD, Elcombe CR et al (1992) Report of the UK Environmental Mutagen Society working party. Alternatives to Aroclor 1254-induced S9 in *in vitro* genotoxicity assays. Mutagenesis 7:175–177
25. Halliwell B (2003) Oxidative stress in cell culture: an under-appreciated problem? FEBS Lett 540:3–6
26. Wee LM, Long LH, Whiteman M et al (2003) Factors affecting the ascorbate- and phenolic-dependent generation of hydrogen peroxide in Dulbecco's modified Eagles medium. Free Radic Res 37:1123–1130
27. Long LH, Kirkland D, Whitwell J et al (2007) Different cytotoxic and clastogenic effects of epigallocatechin gallate in various cell-culture media due to variable rates of its oxidation in the culture medium. Mutat Res 634:177–183
28. Santoro A, Lioi MB, Monfregola J et al (2005) L-Carnitine protects mammalian cells from chromosome aberrations but not from inhibition of cell proliferation induced by hydrogen peroxide. Mutat Res 587:16–25
29. Parry JM, Parry E, Phrakonkham P et al (2010) Analysis of published data for top concentration considerations in mammalian genotoxicity testing. Mutagenesis 25:531–538
30. Galloway SM, Lorge E, Aardema MJ et al (2011) Workshop summary: top concentration for in vitro mammalian cell genotoxicity assays; and report from working group on toxicity measures and top concentration for in vitro cytogenetics assays (chromosome aberrations and micronucleus). Mutat Res 723: 77–83
31. Brusick D (1987) Genotoxicity produced in cultured mammalian cell assays by treatment conditions. Mutat Res 189:1–80
32. Scott D, Galloway SM, Marshall RR et al (1991) International Commission for Protection Against Environmental Mutagens and Carcinogens. Genotoxicity under extreme culture conditions: a report from ICPEMC Task Group 9. Mutat Res 257:147–205
33. Seeberg AH, Mosesso P, Forster R (1988) High-dose-level effects in mutagenicity assays utilizing mammalian cells in culture. Mutagenesis 3:213–218
34. Kirkland D, Pfuhler S, Tweats D et al (2007) How to reduce false positive results when undertaking in vitro genotoxicity testing and thus avoid unnecessary follow-up animal tests: report of an ECVAM workshop. Mutat Res 628:31–55
35. Hayashi M, Dearfield K, Kasper P et al (2011) Compilation and use of genetic toxicity historical control data. Mutat Res 723:87–90

Chapter 7

Analysis of Chromosome Aberrations in Somatic and Germ Cells of the Mouse

Francesca Pacchierotti and Valentina Stocchi

Abstract

Chromosome aberration tests are used to evaluate the clastogenicity of chemical and physical agents, that is, the capacity of these agents to cause breaks in chromosomes and produce microscopically visible fragments or structural rearrangements. Aberrations are scored in metaphase chromosomes of dividing cells. In the mouse, bone marrow progenitors of erythrocytes and leukocytes provide abundant metaphases to study the effects on somatic cells, whereas the response of male germ cells to clastogenic agents can be visualized on metaphases of spermatogonia and primary spermatocytes. The techniques to prepare the slides for analyses are well standardized and internationally harmonized protocols for tests in bone marrow and spermatogonia provide the guidance necessary to obtain meaningful results. It is advisable to adhere as much as possible to these recommendations. Not all tests are suitable to score the same kind of aberrations. Here an overview of the application domains of these tests is provided with warnings on the scoring criteria and statistical analysis.

Key words Chromosome aberrations, Bone marrow, Germ cells, Mouse, Analysis and classification

1 Introduction

The term chromosome aberrations usually means damage of chromosome structure as opposed to genome mutations, which are changes in the number of chromosomes. Structural chromosome aberrations are produced by chromosome breaks with or without a subsequent exchange between two or more broken ends. Chromosome aberrations can be scored at metaphase, when distinct chromosomes are visible under the optical microscope. More recently, methods have been standardized to artificially force chromatin to condense during interphase (premature chromosome condensation) and make any aberration visible also at that stage [1], before selection and repair have occurred; however, these will not be discussed in this chapter. The analysis of chromosome aberrations is used to detect and quantify the genotoxic effects of clastogenic agents, i.e., agents able to induce chromosome breaks.

Other genotoxicity tests are also available that indirectly score the outcome of chromosome breaks, such as the micronucleus test [2].

In the mouse, metaphases can be sampled from bone marrow cells, spermatogonia, spermatocytes, oocytes, and zygotes, and can be also obtained from peripheral blood and spleen lymphocytes after a short-term in vitro culture. In this chapter, methods to prepare and especially to analyze metaphases from bone marrow and spermatogenic cells are presented. In fact, these are the most abundant source of metaphases that can be collected in vivo and, as such, can be used to test the induction of chromosome aberrations in the somatic and germ cell compartments, respectively, of exposed mice. Preparation and analysis of metaphases from mouse oocytes, zygotes, and lymphocytes can be referred to in the relative literature [3–5]. Two official, internationally accepted, guidelines on the chromosome aberration test in mammalian bone marrow (TG 475) and spermatogonia (TG 483) have been issued by the Organization for Economic Co-operation and Development [6, 7]. These should be regarded as a reference for conducting experiments with these tests.

Chromosome aberration tests are based on the microscopic analysis of metaphases, stained with appropriate dyes, most commonly Giemsa. An alternative approach for the detection of chromosomal aberrations consists in labelling the metaphases with fluorescent whole-chromosome painting probes [8]. This approach is most suitable to detect stable chromosome translocations due to the interaction between a centric and an acentric broken end (symmetrical exchanges) and multi-chromosome complex aberrations, which are difficult to score in Giemsa-stained slides. However, the technique requires more expensive materials and instruments. A detailed protocol for the application of chromosome painting has been already published in this book series [9], while, here, the general-purpose classical cytogenetics approach is described. Special emphasis will be given to scoring criteria and classification of aberrations.

Aberrations must be distinguished into chromatid and chromosome type, and breaks and exchanges need to be separately reported [10]. Gaps, which are achromatic lesions but do not correspond to true discontinuities, are always recorded, but reported separately from chromosomal aberrations. Several different aberrations have been classified [10]. In the mouse, whose karyotype contains 40 acrocentric chromosomes, all of similar length, staining metaphases with Giemsa allows reliable detecting of chromatid breaks, fragments, and exchanges, provided that sister chromatids are aligned. Chromosome fragments can also be easily scored together with dicentric and metacentric (Robertsonian) chromosomes produced by asymmetrical exchanges. However, the rare true metacentrics due to centromeric fusions must be distinguished from acrocentrics that are only in close proximity, and a technique for differential centromere staining (c-banding) might help in this respect.

1.1 Mouse Bone Marrow Chromosome Aberration Test

In adult mice, bone marrow is the principal site of erythropoiesis and myelopoiesis. It contains mitotically proliferating cells, which are the precursors of blood erythrocytes and leukocytes. For instance, about seven cell division cycles occur during the transition from pro-erythroblast to erythrocyte, each one lasting 10–11 h [11]. A percentage of mitotic cells (mitotic index, MI) of about 1.5 in unexposed animals, which is increased by a factor of about 3–4 following an ad hoc colchicine treatment, allows scoring of 200 well-spread metaphases in each mouse, as recommended by international guidelines [12]. In bone marrow cells, chromatid-type aberrations can be scored in the first metaphase after treatment. Dicentric chromosomes, metacentric chromosomes, and chromosome fragments can be detected as well. Possible changes of chromosome number, like hyperploidies and polyploidies, can be analyzed at the second cycle after a prospective aneugenic treatment. In Fig. 1, examples of aberrations in bone marrow metaphases are shown.

1.2 Mouse Spermatogonial Chromosome Aberration Test

In rodent testes, the germ cells mature from spermatogonia to elongated spermatids. Before entering into meiotic prophase, differentiating spermatogonia undergo a series of clonal mitotic divisions. These provide the metaphases for scoring chromosome aberrations. The duration of the cell cycle is about 26 h, much longer than that of bone marrow cells and, accordingly, the mitotic index is lower. This makes the collection of a sufficient number of scorable metaphases a bit more difficult. In addition, since the germinal tissue is finely structured and the spermatogonia layer is confined at the periphery of the seminiferous tubules, embedded in the cytoplasm of Sertoli cells, a mechanical treatment is not sufficient to obtain a suspension of these cells to prepare metaphases, and an enzymatic treatment needs to be used. Chromatid-type aberrations and asymmetrical chromosome exchanges can be scored as in bone marrow cells. Hyperploidies can be scored as in bone marrow cells, but detection of polyploidies is not reliable because, in the testis, spermatogonia form chains of cells connected by cytoplasmic bridges that allow them to mature synchronously, and two metaphases in close proximity, even with identical degree of chromosome condensation, may be erroneously interpreted as a polyploid one.

1.3 Mouse Spermatocyte Chromosome Aberration Test

After completion of the meiotic prophase, rodent germ cells enter into the first meiotic metaphase. At this stage, the homologous chromosomes are still kept together by chiasmata and form the so-called tetrads or bivalents. Among them, the X and Y chromosomes, paired at their short region of homology, are morphologically distinguishable. The first meiotic metaphase is amenable to the analysis of reciprocal translocations between nonhomologous chromosomes, occurred before chromosome pair at the zygotene stage.

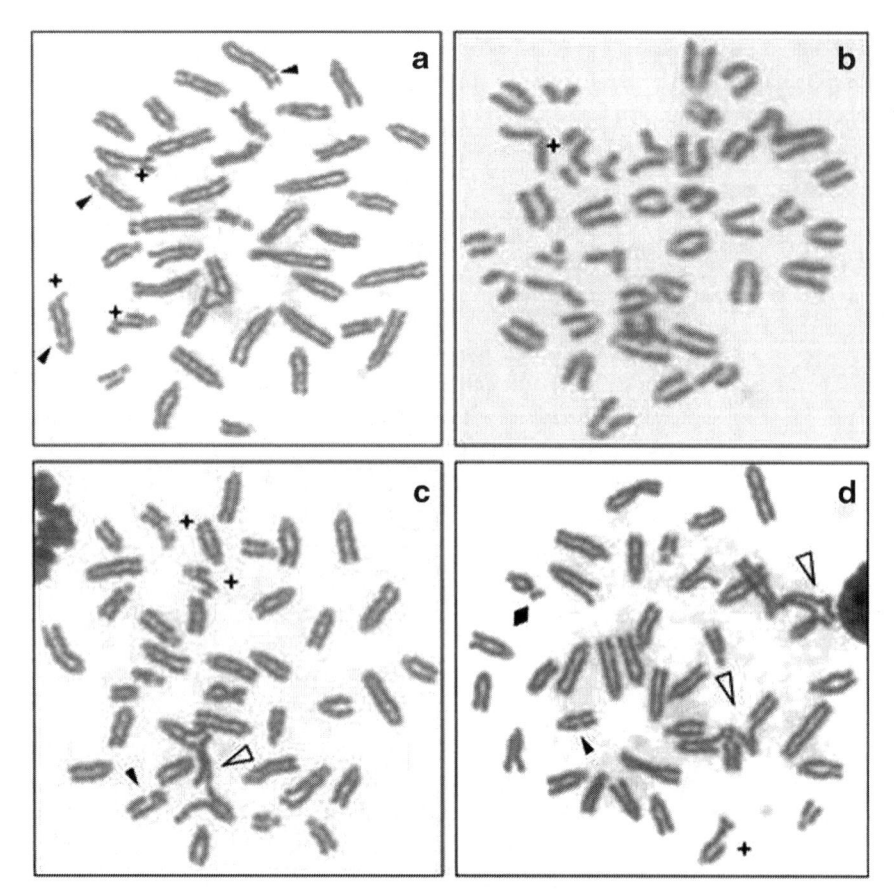

Fig. 1 Examples of chromatid-type aberrations in bone marrow metaphases. Note that the same kind of aberrations will look very similar in spermatogonial metaphases. (**a**) Three *arrowheads* mark as many gaps, three stars mark as many breaks. (**b**) One star marks one chromatid break. Note that chromosomes in this metaphase are a little too short for an optimal screening of aberrations, and a couple of chromosomes show unaligned chromatids. (**c**) One *arrowhead* marks a gap, two stars mark two chromatid breaks, one empty *arrowhead* marks a chromatid exchange. (**d**) One *arrowhead* marks a gap, one star marks one chromatid break, two empty *arrowheads* mark two chromatid exchanges, one diamond marks one isochromatid deletion with distal union

In fact, such rearrangements determine the formation of multivalent configurations, which are classified as "chains" or "rings" according to their morphology. Chromatid-type breaks and exchanges can be detected, but the metaphase must be well fixed and stained to preserve as much as possible the chromatin structure, since bivalents and their chiasmata may interfere with detection of chromatid aberrations. Chromosome-type, otherwise defined isochromatid, fragments can be detected as well. One rather common observation at the first meiotic metaphase consists in unpaired homologous chromosomes, or the so-called univalents, especially, but not only, in the case of sex chromosomes. These events do not correspond to structural aberrations and, by no means, can be summed up with reciprocal translocations and chromatid-type aberrations to assess the clastogenic potential of a given treatment. Sometimes, they have been proposed to represent a marker of subsequent chromosome mis-segregation at anaphase I and, consequently, of aneuploidies in secondary spermatocytes. However, direct comparative analyses between metaphase I univalency and metaphase II aneuploidies have disproved this hypothesis [13]. In Fig. 2, examples of multivalent chains and rings, and various chromatid-type aberrations in spermatocytes, are shown.

1.4 Application Domains of Chromosome Aberration Tests in Mouse Somatic and Germ Cells

In this paragraph, the purpose of chromosome aberration tests in mouse somatic and germ cells and their limits of application are discussed. All these tests can be used to detect the clastogenic effects potentially induced by a physical or a chemical agent (or combination of agents) in mice. However, the mechanisms of aberration induction are different among various agents. This means that not all tests might be right or, at least, sensitive enough to detect the effects of a given agent. Most chemicals produce chromosome aberrations only, when the primary alterations they have induced on the DNA double helix pass through the S-phase. For this reason these chemicals are known as S-phase-dependent clastogens. Their effects are best detected by scoring chromatid-type aberrations in the first metaphase after treatment. Ionizing radiation and a few chemicals can directly produce DNA breaks in any phase of the cell cycle and, for this reason, will produce chromosome-type aberrations when treatment occurs in G1 phase, and chromatid-type aberrations when treatment occurs after the S-phase. These considerations account for the choice of the optimal sampling time after treatment that, according to OECD Guidelines, is between 18 and 24 h in the case of bone marrow and between 24 and 48 h in the case of spermatogonia. The analysis of aberrations in metaphase I spermatocytes is the method of choice for detecting reciprocal translocations induced in proliferating and stem-cell spermatogonia. Such aberrations are effectively induced by ionizing radiation, whereas chemical clastogens have been much less effective for various reasons extensively discussed by Adler [14]. Repeated treatments with a chemical agent might increase the sensitivity of the test. According to the well-known duration of spermatogenesis in the mouse, to score reciprocal translocations in spermatocytes, at least 3 weeks should pass between treatment and sacrifice. To specifically test the effects induced in stem-cell spermatogonia, at least 7 weeks should elapse between treatment and sacrifice, and even more when a sterile transitory period is expected due to cytotoxic effect. Chromatid-type aberrations can be scored in metaphase I spermatocytes to evaluate the effects of ionizing radiation during meiotic prophase, with sampling times fixed by the duration of the various stages of prophase, from preleptotene to diplotene [14]. As an example, this approach might be applied

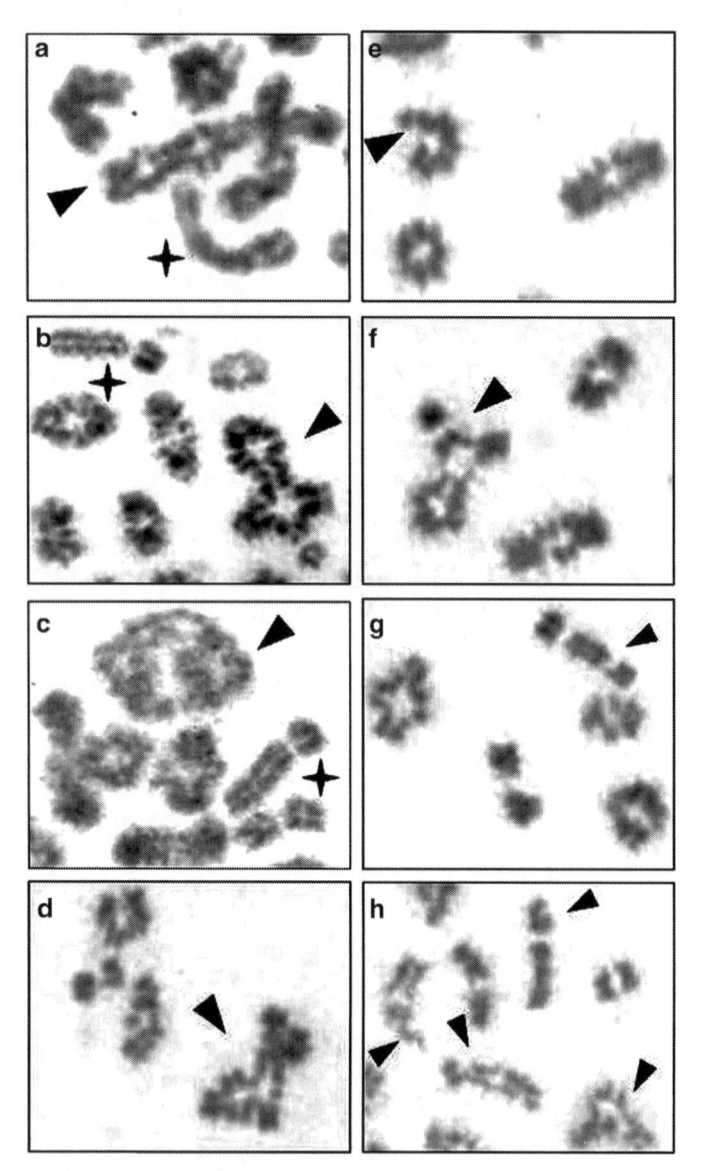

Fig. 2 Examples of aberrations in metaphases of primary spermatocytes. (**a**) The *arrowhead* marks one reciprocal translocation shaped as a chain quadrivalent. The star marks the XY bivalent. (**b**) The *arrowhead* marks one reciprocal translocation shaped as a chain quadrivalent. The star marks the XY bivalent. (**c**) The *arrowhead* marks one reciprocal translocation shaped as a ring quadrivalent. The star marks the XY bivalent. (**d–h**) The morphology of meiotic metaphase chromosomes makes it hard to distinguish chromatid gaps from chromatid breaks: in these metaphases all staining discontinuities on single chromatids have been interpreted as breaks. (**d**) One *arrowhead* marks one chromatid exchange between two nonhomologous autosomes. (**e**) One *arrowhead* marks one chromatid break. (**f**) One *arrowhead* marks one chromatid exchange between the X chromosome and an autosome. (**g**) One *arrowhead* marks one chromatid break on the X chromosome (with a possible second break on the opposite chromatid). (**h**) From *left* to *right* the four *arrowheads* mark: one chromatid break, another chromatid break (with a possible second break on the same bivalent), one chromatid break on the Y chromosome, one chromatid exchange

in DNA repair-defective mice, to assess the role of specific repair pathways during prophase, but induction of translocation multivalents should not be expected. Conversely, S-phase-dependent clastogens are not expected to produce chromatid-type aberrations in metaphase I if treatment occurs after the spermatocytes have gone through the preleptotene stage, during which the last replicative DNA synthesis occurs.

2 Materials

2.1 Mouse Bone Marrow Chromosome Aberration Test

1. Colchicine: 10^{-3} M (0.4 mg/mL) solution in sterile distilled water.
2. 1 mL syringes (three for each mouse) with 26 gauge needles.
3. A water bath.
4. Hank's balanced salt solution (HBSS).
5. Glass round-bottom 10 mL centrifuge tubes.
6. Dissecting instruments.
7. Cotton gauze.
8. A laboratory centrifuge.
9. Glass Pasteur pipettes and rubber bulbs.
10. Hypotonic solution: 0.9 % w/v trisodium citrate dihydrate in water.
11. Fixative: 3:1 methanol–glacial acetic acid.
12. 21 or 22 gauge needles (one for each mouse).
13. Pre-cleaned microscope slides (*see* **Note 1**).
14. Staining solution: 8 % Giemsa solution in phosphate buffer (pH 6.8; *see* **Note 2**).
15. 24 × 35 mm coverslips.
16. Mounting medium.
17. A research microscope with 20 and 100× objectives.

2.2 Mouse Spermatogonial Chromosome Aberration Test

1. Testis isolation medium (TIM): For 5 L of medium prepare two separate solutions:
   (a) Weigh 30.25 g of NaCl, 16.90 g of KCl, 4.25 g of Na_2HPO_4, 0.45 g of KH_2PO_4, 5 g of glucose and dissolve in a final volume of 4 L of purified distilled water.
   (b) 0.9 g of $CaCl_2.2 H_2O$ and 1.5 g of $MgSO_4.7 H_2O$ are dissolved in 0.9 L of purified distilled water.

   Dropwise add solution (b) to solution (a) under slow agitation; then add 0.025 g of phenol red, adjust the pH to 7.2–7.3, and bring to final volume of 5 mL.
2. Collagenase: Type I or collagenase A.
3. Isotonic solution: 2.2 % w/v trisodium citrate dihydrate.

4. Hypotonic solution: 0.9 % w/v trisodium citrate dihydrate.
5. Fixative: 3:1 ethanol–glacial acetic acid.
6. Colchicine: 10^{-3} M (0.4 mg/mL) colchicine solution in sterile distilled water.
7. Staining solution: 8 % Giemsa solution in phosphate buffer (pH 6.8) (see **Note 2**).
8. 1 mL syringe with 26 gauge needle.
9. A shaking water bath.
10. Dissecting instruments, including straight and curved forceps and scissors.
11. 100 mL glass flasks.
12. 60 mm diameter Petri dishes.
13. 10 mL round-bottom centrifuge tubes.
14. Glass Pasteur pipettes and rubber bulbs.
15. A laboratory centrifuge.
16. Pre-cleaned microscope slides (see **Note 1**).
17. 24×35 mm coverslips.
18. Mounting medium.
19. A research microscope with 20 and 100× objectives.

2.3 Mouse Spermatocyte Chromosome Aberration Test

1. Isotonic solution: 2.2 % w/v trisodium citrate dihydrate.
2. Hypotonic solution: 0.9 % w/v trisodium citrate dihydrate.
3. Fixative: 3:1 ethanol–glacial acetic acid.
4. Colchicine: 1 mg/mL colchicine solution in sterile distilled water (optional) (see **Note 3**).
5. Staining solution: 2 % acetic Orcein solution (see **Note 4**).
6. Absolute ethanol.
7. Xylene.
8. Mounting medium.
9. 1 mL syringe with 26 gauge needle.
10. Dissecting instruments, including straight and curved forceps and scissors.
11. 60 mm diameter Petri dishes.
12. 10 mL round-bottom centrifuge tubes.
13. Glass Pasteur pipettes and rubber bulbs.
14. Pre-cleaned microscope slides (see **Note 1**).
15. 24×35 mm coverslips.
16. A phase contrast research microscope with 20 and 100× objectives.

3 Methods

3.1 Mouse Bone Marrow Chromosome Aberration Test

1. An hour and a half before sacrifice, inject the mice with 0.01 mL/g b.w. of the colchicine solution (*see* **Note 5**). This corresponds to a treatment dose of 4 mg/kg b.w.

2. Fill round-bottom centrifuge tubes with 1 mL of HBSS and pre-warm them in the water bath at 37 °C.

3. Prepare the hypotonic solution and keep it at 37 °C in the water bath until use.

4. Sacrifice the animals according to regulatory guidelines applied in your country. Isolate both femurs and clean away the muscular tissue with the help of cotton gauze.

5. Aspirate the HBSS from the centrifuge tube into a 1 mL syringe.

6. Cut open the femur at the distal (knee) end and insert the needle.

7. Flush the content of the syringe to force the marrow out through the open end into the centrifuge tube. Repeat aspiration of the cell suspension and flushing until the bone is visibly empty of marrow. This will also allow obtaining a well-dispersed cell suspension. With the same syringe repeat the procedure on the second femur to add its marrow to the individual mouse cell suspension.

8. With a Pasteur pipette add dropwise 2 mL of the warm HBSS to the cell suspension and gently resuspend.

9. Centrifuge at $100 \times g$ for 10 min.

10. Discard the supernatant completely and add dropwise 3 mL of the pre-warmed hypotonic solution, while gently resuspending (*see* **Note 6**). Keep the tubes in the water bath for 13 min. During this time discard possible fat residues that float on the surface.

11. At the end of the hypotonic treatment, centrifuge the tubes for 10 min at $160 \times g$.

12. Discard the supernatant and very gently add 3 mL of the fixative maintained at room temperature without resuspending the pellet. Keep the tubes at room temperature for 10 min. Fixation of the pellet as such allows a gentle and progressive cell fixation that is especially suitable for this type of cells.

13. Repeat the former step (**step 12**) twice, always without resuspending the pellet.

14. Replace the fixative over the pellet with 1 mL of fresh fixative. Fit a 21 or 22 gauge needle to a 1 mL syringe and use it to gently scrape completely the pellet from the bottom of the tube. Then aspirate the pellet into the syringe several times to disperse the cells. Change the 21/22 gauge needle into a

26 gauge needle and aspirate again a few times to complete the process of cell dispersion.

15. Add dropwise 4 mL more of the fixative to each tube and keep it at 4 °C overnight.

16. The morning after, centrifuge the tubes at $160 \times g$ for 10 min, discard the supernatant, and add dropwise about 0.5 mL of freshly prepared fixative at room temperature.

17. Spot few drops of the cell suspension on a perfectly clean slide (*see* **Note 1**) and air-dry. Score the slide under the microscope at low magnification. Adjust the cell density if necessary (*see* **Note 7**). Check the degree of chromosome spreading: if, in spite of using perfectly cleaned slides, the spreading is insufficient, a few drops of acetic acid can be added to the cell suspension, gently mixing it with a Pasteur pipette, and, after a few minutes, more slides can be prepared.

18. Proceed to staining slides with Giemsa for 10 min in a Coplin jar (*see* **Note 2**), not less than 2 h, and possibly 24 h after air-drying is complete.

19. Rinse the slides in tap water and distilled water for a few seconds.

20. Mount the slides with coverslips.

21. Proceed to slide scoring. It is advisable to firstly determine the mitotic index in a total of 500 cells per mouse to check the efficacy of the colchicine treatment and any possible toxic effects of the test item. Then score the slides for chromosome aberrations. Under low magnification select well-spread metaphases excluding those with excessive chromosome scattering or isolated chromosomes in the proximity. Under high magnification, count the chromosomes and proceed to scoring aberrations only in metaphases with at least 40 centromeres. Exclude from the analysis metaphases with very short and/or unaligned chromatids (*see* **Note 8**). In each metaphase record the following observations: gaps, chromatid breaks with or without a corresponding identifiable fragment, chromatid fragments, chromosome fragments, chromatid exchanges, and chromosome exchanges. Stop the scoring when you have reached the set number of metaphases/mouse (*see* **step 22**).

22. Proceed to the statistical analysis. Remember that the single mouse is the smallest unit of analysis allowed. This means that one has to:

    (a) Score enough cells in each mouse to reliably represent its "true" frequency of cells carrying aberrations (*see* **Note 9**).

    (b) Include enough animals in each experimental group to reach a set detection power of the test.

International guidelines have been published that indicate these numbers on the basis of the mean and variance of historical control data [12]. With a mean percent of 1.0 ± 1.0 aberrant cells, a minimum of 200 cells per mouse should be scored to avoid too many zero counts. A minimum number of 5 mice per group is recommended to detect at least 80 % of treatments which induce a twofold increase in aberrant cells over the historical control level of 1.0 % at the significance level of 0.05.

(a) For each animal, record the following data: number of metaphases scored; number of each type of aberration; frequency of each type of aberration per cell; and number and percent of metaphases with aberrations (including and excluding gaps).

(b) For each experimental group calculate the mean percent of cells with aberrations and its standard deviation; the mean frequency of each kind of aberration per cell; and the mean frequency and standard deviation of total aberrations per cell.

A typical chromosome aberration test consists of a solvent control group, multiple dose groups, and a positive control group. Test results should be interpreted after statistical analysis of the dose–response relationship by a trend test and pair-wise comparisons between each dose group and the solvent control group by chi-square analysis or G statistics. Remember that preliminary chi-square analysis should be done to verify possible interindividual variability within a same experimental group, and in such a case, an F-test should be applied for pair-wise comparisons. The use of these tests can be found in Sokal and Rohlf [15]. Since the spontaneous frequency of some kind of aberrations, like exchanges, is very low (around 1/1,000 cells), the statistical analysis should be only applied to the percent of cells with aberrations and frequencies of total aberrations per cell, possibly divided into chromatid- and chromosome-type ones. A significant positive trend test plus significant differences between one or more dose groups and the solvent control one obviously indicate that the test item was effective in chromosome aberration induction. If none of the individual treated groups shows a significant difference but significance was shown in the trend test the interpretation requires biological judgement. The same applies to data that show no trend but a significant difference between an individual treatment group and the historical negative control, provided the pair-wise evaluation was corrected for multiplicity of comparisons.

3.2 Mouse Spermatogonial Chromosome Aberration Test

1. Four to five hours before sacrifice, inject the mice with 0.01 mL/g b.w. of the colchicine solution (*see* **Note 5**). This corresponds to a treatment dose of 4 mg/kg b.w.

2. Sacrifice the animals according to regulatory guidelines applied in your country. Dissect and isolate both testes in a Petri dish containing a small volume of TIM (37 °C) (*see* **Note 10**).

3. Use curved forceps and fine scissors to make an incision of the tunica albuginea and release the seminiferous tubules with a gentle pressure on the surface of the testis (you can use curved forceps for this operation).

4. Remove the tunica and gently tease apart the tubules.

5. Transfer the seminiferous tubules into a 100 mL flask containing 10 mL of TIM supplemented with 0.5 mg/mL collagenase right before use (*see* **Note 11**). Shake the flask in a water bath for 15 min at 37 °C.

6. At the end of the incubation time, release the germ cells into suspension by gentle pipetting using a Pasteur pipette. Filter through a 90 μm nylon membrane into a 10 mL centrifuge tube and centrifuge at $100 \times g$ for 10 min.

7. Wash the cell suspension by adding dropwise 2 mL of 2.2 % sodium citrate solution at room temperature. Resuspend the pellet in the solution and right after centrifuge for 10 min at $100 \times g$.

8. Discard the supernatant and resuspend the pellet by adding 3 mL of hypotonic citrate solution dropwise. Keep the cells in the hypotonic solution for at least 15 min at room temperature (invert the tubes once during this step to avoid cell sedimentation by gravity and help osmotic exchange between medium and cytosol).

9. Centrifuge for 10 min at $100 \times g$.

10. Discard the supernatant and resuspend the pellet in 5 mL freshly prepared, cold fixative. Incubate for 10 min on ice.

11. Centrifuge for 10 min at $190 \times g$.

12. Repeat **steps 10** and **11** twice.

13. Finally, discard the supernatant and resuspend the pellet in an appropriate volume of fresh fixative (*see* **Note 7**). Spot a few drops of the cell suspension on perfectly clean slides (*see* **Note 1**) and air-dry. Adjust the cell density if necessary.

14. Stain the slides with the Giemsa solution for 10 min (*see* **Note 2**).

15. Rinse the slides in tap water and distilled water for a few seconds.

16. Mount the slides with coverslips.

17. Score the slides for chromosome aberrations. Under low magnification, select well-spread metaphases excluding those with excessive chromosome scattering or isolated chromosomes in the proximity. Under high magnification, count the chromosomes and proceed to scoring aberrations only in metaphases with at least 40 centromeres. Exclude from the analysis metaphases with very short and/or nonaligned chromatids (*see* **Note 8**). In each metaphase record the following observations: gaps, chromatid fragments, chromatid breaks, chromosome fragments, chromatid exchanges, and chromosome exchanges. Stop the scoring when you have reached the set number of metaphases/mouse.

18. Proceed to the statistical analysis as described in **step 22** of Subheading 3.1.

3.3 Mouse Spermatocyte Chromosome Aberration Test

1. Three hours before sacrifice, inject the mice with 0.01 mL/g b.w. of the colchicine solution (*see* **Notes 3** and **5**). This corresponds to a treatment dose of 10 mg/kg b.w.

2. Sacrifice the animals according to the regulatory guidelines of your country. Isolate both testes in a 60 mm Petri dish containing 2.2 % sodium citrate (isotonic solution) at room temperature. Pay attention to clean the testes from fat residues as much as possible.

3. Transfer the testes in a second Petri dish containing fresh isotonic solution. Isolate the seminiferous tubules from the tunica albuginea as described in **step 3** of Subheading 3.2.

4. Cut through the tubular mass and repeatedly squeeze out the tubules with curved forceps to achieve their mechanical disruption. The increasing turbidity of the medium is an indicator of the dissociation of meiotic cells from the tubules and their release into suspension.

5. Transfer the suspension into a 10 mL centrifuge tube, together with the tubular debris. The latter will settle in a few minutes at the bottom of the tube. Now transfer the supernatant into a new tube.

6. Centrifuge for 10 min at $60–100 \times g$ (*see* **Note 12**).

7. Discard the supernatant and resuspend the pellet by dropwise adding 3 mL of 0.9 % trisodium citrate (hypotonic solution). The cell suspension must be incubated for 12 min at room temperature (invert the tubes once during this step).

8. Centrifuge as in **step 6** above and then go through **steps 10–13** of Subheading 3.2.

9. Stain the slides with the Orcein solution in a Coplin jar for 30 min.

10. Rinse the slides in 70 % ethanol (30 s), 90 % ethanol (2 × 30 s), 100 % ethanol (2 × 2.5 min), and then finally xylene (2 × 2.5 min) for differentiation and clearing.

11. Mount the slides with coverslips.

12. Score the slides for chromosome aberrations. Under low magnification select well-spread metaphases. At high magnification, count the number of bivalents, identify the X–Y bivalent, and proceed to scoring of aberrations only in metaphases with 20 bivalents. Record the following observations: univalents of sex chromosomes and autosomes; chromosome (iso-chromatid) fragments; chromatid breaks; chromatid exchanges; and multivalent chains and rings. Stop the scoring when you have reached a minimum of 100 metaphases per mouse (*see* **Note 13**).

13. Proceed to the statistical analysis as indicated in **step 22** of Subheading 3.1.

4 Notes

1. The slides must be perfectly degreased to obtain the highest yield of well-spread analyzable metaphases. To clean slides several procedures can be followed, for example: keep them in absolute ethanol for 1 week, then dry, and transfer them to hot water with soap for 2 days. The day before use wipe them with gauze and soap water until they form a homogeneous water film when put under tap water. Then rinse them with tap water and wash in distilled water. Keep them in distilled water at 4 °C and, when ready, drop the cell suspension onto the thin water film covering the slide surface. Alternatively, after degreasing with ethanol and soap, the slides can be air dried and used as such, or used after air-drying and chilling in the refrigerator for 1 h.

2. The quality of Giemsa staining may vary with different brands and even with different bottles. You may need to slightly adjust the concentration and time of staining to your specific condition. A concentration of 5 % is sometimes suggested with a staining time longer than 10 min.

3. In the case of metaphase I spermatocytes, the antimitotic treatment is optional because you will easily collect enough metaphases even without it. However, a treatment like the one suggested in this protocol may improve the quality of preparations and allow obtaining less fluffy chromosomes and more easily detectable chromatid aberrations.

4. For observation of the delicate features of meiotic chromosomes, Orcein staining is much preferred to Giemsa staining. The Orcein solution suggested in this protocol can be prepared as follows: mix 50 mL of glacial acetic acid with 50 mL

of distilled or deionized water. Slowly bring it to a boil. It need not be a "rolling boil." Slowly add to it 2 g of Orcein. Mix as you add it, and let it mix for about 60 min with a magnetic mixer. Place some kind of glass cover over the beaker. The heat of the acid will condense on the glass and then move back down into the mixture. This serves to help in the mixing. Filter when still warm. Once cooled, the solution is ready for use and can be kept at room temperature for years. Filter before each use.

5. A precise relation to the body weight is not necessary. With 2–3-month-old mice of most strains you may inject a fixed volume of 0.3 mL/mouse. Do not increase the time because, even if this might increase the mitotic index, you will end up with too many metaphases with short chromosomes and/or unaligned chromatids, unsuitable for chromosome aberration detection.

6. One has to be gentle during the resuspension of cells into the 37 °C hypotonic solution because otherwise undesirable clumping could occur due to collagens and fat in the bone marrow. You may alternatively decide to make the hypotonic treatment at room temperature for a longer incubation time. Remember that the optimal hypotonic conditions also depend on the temperature and humidity conditions of the laboratory and may need to be adjusted to your specific environment.

7. The cell density is correct when the suspension has an opalescent color. In most cases this corresponds to a volume of 0.5 mL, but remember that it is easier to dilute than to reconcentrate your suspension (this would obviously require a new centrifugation), so it is advisable to keep the initial test drops a little over-concentrated.

8. Whatever your colchicine protocol, some metaphases in your slides will display very short chromosomes and/or unaligned chromatids because the cell was caught in the mitotic phase right after colchicine injection and, as such, underwent a long mitotic arrest. These metaphases should be excluded from the analysis because the loss of chromosome alignment can contribute to failures in recognizing aberrations of the chromatid type.

9. One rather common mistake in the design of in vivo genotoxicity tests is to consider the analyzed cell as the smallest experimental unit, thereby aiming at scoring a set number of cells per group, irrespectively of the number of analyzed mice and analyzed cells per mouse. In other words, you can neither compensate for bad preparations by increasing the number of animals and scoring only a few metaphases in each of them nor can you score all the metaphases from a single or a few mice.

10. During testis isolation, it is important to avoid fat tissue residues. To obtain high-quality preparations, clean as quickly as

possible the fat residues from the testis before releasing the seminiferous tubules.

11. In order to detach spermatogonia from the basal layer of the seminiferous tubules, a quicker technique, not based on an enzymatic treatment, has been also proposed [16]. It is based on inverting the steps of cell isolation and hypotonic treatment/fixation: the seminiferous tubules are treated with the hypotonic solution and fixed with methanol/acetic acid in bulk and, only after these steps, they are digested into a 50 % acetic acid solution to release the spermatogonia with the aid of mechanical pipetting. After eliminating the acetic acid by centrifugation at increased speed ($400 \times g$ for 5 min), the cells released from the tubules are refixed and the suspension can be dropped onto the slides as described here.

12. Do not increase the centrifugation speed to avoid sedimentation of the many spermatozoa present in the suspension.

13. You can score 200 cells per animal to improve the statistical power of your analysis, if you suspect that the agent being tested is not highly effective and/or interindividual variability could affect your data.

References

1. Gotoh E, Durante M (2006) Chromosome condensation outside of mitosis: mechanisms and new tools. J Cell Physiol 209:297–304
2. Hayashi M, MacGregor JT, Gatehouse DG et al (2000) In vivo rodent erythrocyte micronucleus assay. II. Some aspects of protocol design including repeated treatments, integration with toxicity testing, and automated scoring. Environ Mol Mutagen 35:234–252
3. Russo A (2000) In vivo cytogenetics: mammalian germ cells. Mutat Res 455:167–189
4. Adler ID, Pacchierotti F, Russo A (2012) The measurement of induced genetic change in mammalian germ cells. Methods Mol Biol 817:335–375
5. Darroudi F, Farooqi Z, Benova D et al (1992) The mouse splenocyte assay, an in vivo/in vitro system for biological monitoring: studies with X-rays, fission neutrons and bleomycin. Mutat Res 272:237–248
6. Organization for Economic Co-operation and Development, OECD (1997a) Guideline for testing chemicals 475: mammalian bone marrow chromosome aberration test. http://www.oecd-ilibrary.org/environment/test-no-475-mammalian-bone-marrow-chromosome-aberration-test_9789264071308-en. Accessed 31 July 2012
7. Organization for Economic Co-operation and Development, OECD (1997b) Guideline for testing chemicals 483: mammalian spermatogonial chromosome aberration test. http://www.oecd-ilibrary.org/environment/test-no-483-mammalian-spermatogonial-chromosome-aberration-test_9789264071469-en. Accessed 31 July 2012
8. Stronati L, Farris A, Pacchierotti F (2004) Evaluation of chromosome painting to assess the induction and persistence of chromosome aberrations in bone marrow cells of mice treated with benzene. Mutat Res 545:1–9
9. Pacchierotti F, Sgura A (2008) Fluorescence in situ hybridization for the detection of chromosome aberrations and aneuploidy induced by environmental toxicants. Methods Mol Biol 410:217–239
10. Savage JR (1975) Classification and relationships of induced chromosomal structural changes. J Med Genet 13:103–122
11. Cole RJ, Taylor N, Cole J et al (1981) Short-term tests for transplacentally active carcinogens. I. Micronucleus formation in fetal and maternal mouse erythroblasts. Mutat Res 80:141–157
12. Adler ID, Bootman J, Favor J et al (1998) Recommendations for statistical designs of in vivo mutagenicity tests with regard to

subsequent statistical analysis. Mutat Res 417: 19–30
13. Liang JC, Pacchierotti F (1988) Cytogenetic investigation of chemically-induced aneuploidy in mouse spermatocytes. Mutat Res 201:325–335
14. Adler ID (1982) Male germ cell cytogenetics. In: Hsu T (ed) Cytogenetic assays of environmental mutagens. Allenheld, Osmun Publishers, Totowa, USA, pp 249–276
15. Sokal RR, Rohlf FJ (2012) Biometry, 4th edn. W.H. Freeman & Co, New York
16. Adler ID, Venitt S, Parry JM (1984) Cytogenetic tests in mammals. In: Mutagenicity testing. A practical approach. IRL, Oxford, pp 275–306

Chapter 8

Chromosomal Aberration Test in Human Lymphocytes

Christian Johannes and Guenter Obe

Abstract

Human peripheral lymphocytes (HPL) are non-cycling primary cells (G0 cells). They are easily collectable by venipuncture. In the presence of suitable culture media and stimulants in vitro HPL enter the cell cycle and divide mitotically. Metaphase-like stages can be arrested using the spindle fiber poison colcemid and prepared on microscopic slides. Following appropriate staining, chromosomal aberrations can be analyzed in the microscope. These aberrations may be induced either in vivo by environmental or occupational influences or in vitro after experimentally controlled manipulations in order to detect or to test the mutagenic potency of various agents.

Key words Human peripheral lymphocytes, Culture media, Stimulation, Chromosomal aberrations, Mutagenicity testing, Biological dosimetry, Human primary cells

1 Introduction

Human peripheral lymphocytes (HPL) are ideal to analyze chromosomal damage in human primary cells. HPL arise from pluripotent hematopoietic stem cells in the bone marrow and give rise to B- and T-lymphocytes in the bone marrow (B) and thymus (T), from where they migrate to the peripheral lymphoid organs [1, 2]. In the peripheral blood, there are about 2 % of the 500×10^9 lymphocytes in the body, but there is a permanent circulation of lymphocytes between the lymphoid tissues and the blood [3]. The circulation of HPL between blood and the lymphoid organs is very important when analyzing chromosomal damage in blood lymphocytes after in vivo exposure to DNA-damaging agents, because HPL which were damaged at any site in the body will eventually appear in the peripheral blood and can be analyzed after in vitro cultivation. Almost all HPL are in the G0 state of the cell cycle; thus they are nondividing.

With mitogens, HPL can be stimulated to enter the cell cycle and eventually to divide mitotically. T cells can be stimulated with phytohemagglutinin (PHA) and with concanavalin A (ConA), and both T and B cells with pokeweed mitogen (PWM; [2, 4]). Various types of blood cells as well as lymphocytes and their morphologic stages after mitogenic stimulation are beautifully figured by Bessis [5] and Petrzilka and Schroeder [6].

Petrzilka and Schroeder [6] performed a careful kinetic and stereological study of the stimulation of T-lymphocytes with PHA and some of their results are presented as follows: Cellular/nuclear/nucleolar volume changed from 103.8/47.9/0.5 in non-stimulated to 518/169.9/13.7 µm^3 after 48 h in culture. The volume of the heterochromatin diminished from 31.8 to 21.7 µm^3 and the volume of euchromatin elevated from 15.6 to 134.5 µm^3. In un-stimulated cells 0.9 % and at 48 h culture with PHA, 98.4 % cells were activated. These and other data given by the authors clearly highlight the stimulation process of HPL in the presence of PHA in culture. This is underlined by the morphological features of stimulated as compared to un-stimulated HPL [6]. G0 lymphocytes have very small nucleoli, but during stimulation a considerable increase of nucleolar material occurs [7]. Biochemical changes in PHA-stimulated HPL such as DNA polymerase activity and uptake of thymidine, uridine, and leucine were analyzed by Loeb et al. [8].

The characteristics of how HPL enter the cell cycle and undergo mitotic divisions depend on several factors, such as the stimulating agent and the choice of the culture medium which in the studies discussed in the following contained 10 % fetal bovine serum. In cultures with HAM's F-10 medium and PHA stimulation two peaks of DNA synthesis measured as (^{3}H)TdR labeled interphases were found, one at about 34 h after culture start (40 % labeled interphases) and the other at about 40 h (49 % labeled interphases). First mitoses occurred at about 36 h; the mitotic index showed two peaks, one at about 44 h (1.3 % mitoses) and a second at about 48 h (2.8 % mitoses) [9]. Corresponding to the two waves of DNA synthesis and mitoses, two peaks of RNA synthesis were found with maxima of labeled cells at about 14 and 19 h in PHA/HAM's F-10 medium. Up to 12 h only few labeled cells were found [10]. The data on DNA and RNA synthesis indicate that there are two subpopulations of cells, which, upon stimulation with PHA in culture, undergo RNA synthesis with maxima at about 14 and 19 h, and DNA synthesis with maxima of about 34 and 40 h, and with two maxima of mitoses at about 44 and 48 h [10]. In cultures with PHA and TC 199 medium the mitotic activity is low up to about 50 h and after that it slowly raises but does not reach the maximal values obtained with HAM's F-10. DNA synthesis, too, starts later than in cultures with HAM's F-10 [11]. DNA synthesis and mitoses of HPL stimulated with PWM and cultured in HAM's F-10 are similar to cultures set up with TC medium M199 [12]. Eagle's MEM medium in PHA-stimulated cells leads to a higher degree of asynchrony when compared to HAM's medium [13].

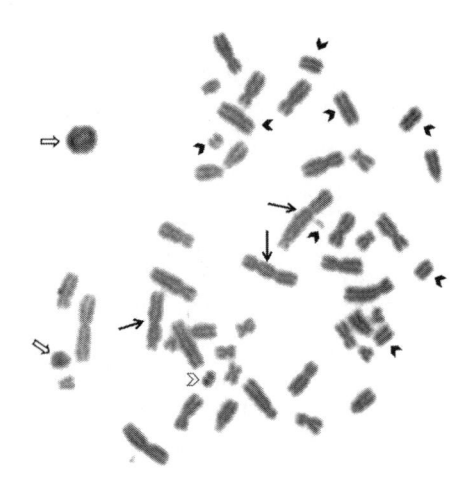

Fig. 1 A heavily damaged human lymphocyte revealing several chromosomal aberrations. Indicated are dicentric chromosomes (*solid arrows*), ring chromosomes (the *open arrow at top* shows a ring with centromere and the *lower* one a ring lacking a centromere), acentric fragments (*filled arrowheads*), and interstitial deletion (*open arrowhead*)

Labeling cells with 5-bromodeoxyuridine (BrdU) during cultivation and subsequent differential staining allows determination of the frequencies of first (M1), second (M2), and third and further (M3+) metaphases. In the HAM's F-10/PHA culture system we found the following percentages of M1, M2, and M3+ at culture times ranging from 41 to 102 h. At 42 h: 100.0 M1; 54 h: 96.5 M1 and 3.5 M2; at 66 h: 73.5 M1, 24.0 M2, and 2.5 M3+; at 78 h: 25 M1, 55.5 M2, and 19.5 M3+; at 90 h: 17.0 M1, 37.0 M2, and 46.0 M3+; and at 102 h: 4.0 M1, 24.0 M2, and 72.0 M3+ [14].

In both in vivo as well as in vitro studies with HPL it is mandatory to analyze only M1 cells, because there is a selection against damaged cells during subsequent cell cycles. In cultured HPL (PHA/HAM's F-10) from 100 persons, we found the following frequencies of M2 metaphases after a culture time of 48 h: 0–5 % in 63 blood donors (BD), 5–10 % in 13 BD, 10–15 % in 7 BD, 15–20 % in 5 BD, 20–25 % in 7 BD, 25–30 % in 2 BD, 30–35 % in 2 BD, and 35–40 % in 1 BD (data from Obe and Schmidt, *see* [14]). These data underline the necessity to check the cell cycle.

HPL are used to analyze chromosome breaking activities in vitro and in vivo. Metaphases can be made visible in the microscope by using staining procedure such as Giemsa block staining (Fig. 1), Giemsa banding, and various types of chromosome painting (fluorescence in situ hybridization, FISH) [15–18]. FISH with respective DNA probes allows analysis of chromosomal aberrations which are not detectable with Giemsa block staining, e.g., reciprocal translocations and insertions.

The problems to properly quantify structural chromosomal aberrations (CA) obtained in vitro are exhaustively discussed by Savage and Papworth [19]. In addition to CA other end points can be analyzed in cultured HPL, such as Comet assay [20], micronuclei [21], and SCE [22].

HPL are an integral part of genetic toxicology testing [23–32]. A problem in testing unknown agents for possible CA inducing activities in vitro is that in case of a negative result it cannot be excluded that the agent is mutagenic when metabolically activated. Metabolic activities are nearly absent in HPL, but this drawback can be partly overcome by adding metabolic systems to the HPL culture [33–35].

Several agents have been tested for their mutagenic potency in vivo and in the following results are shortly described which were obtained from studies concerning the mutagenic activity of alcohol and smoking in man. Alcohol drinking leads to elevated frequencies of CA in HPL and this is mainly caused by acetaldehyde which is the first metabolite in vivo [7, 36–41]. In vitro studies showed that nonvolatile residues of alcoholic beverages (whisky, brandy, rum) induce SCE in HPL and this apart from acetaldehyde may add to the in vivo findings [42]. Another widespread habit, namely, cigarette smoking, also leads to an elevation of CA in HPL [43–45].

The spontaneous frequencies of CA in human peripheral blood lymphocytes were extensively analyzed. A compilation of the data obtained is given in a recent review by Obe et al. [46]. In one set of data the following numbers of spontaneous aberrations were found in 10,000 metaphases: achromatic lesions (AL) = 536.24, chromatid breaks (B′) = 59.81, isochromatid breaks (B″) = 41.07, dicentric chromosomes (DIC) = 8.00; chromatid interchanges (RB′) = 5.13, and centric rings (RC) = 1.77. The frequencies of aberrations result in the following sequence: AL > B′ > B″ > DIC > RB′ > RC.

Dose–effect relationships of DIC (Fig. 1) induced by ionizing radiations in G0 HPL have a linear-quadratic (low LET radiation) or linear (high LET radiation) characteristic with similar parameters after radiation exposure in vivo or in vitro [46, 47]. Comparing a given frequency of DIC obtained from a person who was exposed to radiation of unknown dose with a respective dose–effect curve obtained in vitro allows a dose reconstruction. The methodological aspects of this biological dosimetry are exhaustively described in a recent IAEA report [47]. The lowest radiation dose resulting in a significantly elevated DIC frequency in HPL exposed in vitro when compared to no-irradiated controls is about 10 mGy for X-rays and 20 mGy for Co-60 γ-rays [48, 49].

2 Materials

2.1 Culture Components

Prepare all solutions using deionized water. All solutions and all other materials (plasticware) for culturing HPL must be sterile to avoid contaminations.

1. Heparinized tubes should be used to collect venous blood (*see* **Note 1**). Venipuncture should be performed by a physician.
2. Plastic tubes (for a 5 mL culture) or flasks (for total volumes of 10 mL or more) for culturing.
3. Culture media of different composition which are suitable for culturing HPL, are commercially available, such as RPMI 1640, Ham's F10, or McCoy's 5A medium. They are offered as ready for use or in powdered form (*see* **Notes 2** and **3**).
4. Fetal bovine serum (FBS).
5. Phytohemagglutinin M (PHA-M): Available in frozen or lyophilized form. Defreeze or reconstitute (*see* **Note 4**).
6. 1 mM BrdU: Dissolve 30.71 mg in 100 mL of water and sterilize by filtering. Store at 4 °C and wrap with aluminum foil (*see* **Note 5**). BrdU is not an essential compound of the culture medium, but has to be added at the start of the culture, if the cell cycle progression should be controlled.
7. 2 µg/mL Colcemid (Demecolcin): Dissolve or dilute to a concentration of 2 µg/mL in culture medium. Store at 4 °C in the dark.

2.2 Metaphase Preparation

1. Hypotonic solution: 75 mM KCl (5.5 g/L). Mix and store at 4 °C. Before use, warm the solution to 37 °C.
2. Fixation solution: 3:1 ratio of methanol and acetic acid, prepare fresh for each use (*see* **Note 6**).
3. Glass microscope slides: Rinse several times in deionized water. Cool and store the slides at 4 °C in deionized water in a refrigerator (*see* **Note 7**).

2.3 Giemsa Staining

1. Phosphate buffer, pH 6.8.
2. Giemsa staining solution: Mix 5 mL Giemsa stain with 20 mL phosphate buffer and 75 mL deionized water. Filter the solution through a paper filter to retain crystals (*see* **Note 8**).

2.4 Sister Chromatid Differentiation Staining

1. Bisbenzimide (H33258, Hoechst 33258): Dissolve 0.45 mg bisbenzimide in 100 mL of deionized water. Store at 4 °C in the dark.
2. Phosphate-buffered saline (PBS).
3. Giemsa staining solution (same as in Subheading 2.3).

3 Methods

Until specified, all procedures should be carried out at room temperature.

During handling human blood and blood cultures, wear protective gloves and ensure that all material which came in contact with the specimens is disposed adequately. Avoid usage of needles during lab work wherever possible to avoid puncture hazard. If needles are used, handle and dispose them with great care. When handling BrdU, Giemsa, and bisbenzimide wear protective gloves.

3.1 Blood Sampling and Whole Blood Cell Culturing

1. Select participants of a population study by preparing a questionnaire, as the significance of results depends on a careful matching of exposed and control persons. Before the study starts participants give information about age, sex, socioeconomic status, type of work, and workplace. Exposure to organic solvents, chemotherapeutic agents, drugs, or recent medical radiation should be taken into consideration. Habits like smoking and alcohol consumption (and if appropriate other habits known to influence chromosomal aberration frequency, e.g., chewing of betel) should also be recorded. Donors with malignancies, chronic diseases requiring regular therapies, exposure to radioactive compounds in diagnosis of therapy, or more than one diagnostic X-ray exposure per year should be excluded from a cytogenetic study. The blood sampling should be done at the same time in exposed and control persons. For in vitro studies blood samples should be taken from healthy persons, preferably nonsmokers.

2. Obtain a blood sample of approximately 10 mL by venipuncture in a heparinized tube (*see* **Note 9**). Immediately swing the tube gently to mix blood and heparin. Set up blood cultures as soon as possible (*see* **Note 10**).

3. For each culture, mix 0.5 mL of heparinized blood with 4 mL of culture medium, 0.5 mL of FBS, and 0.2 mL of phytohemagglutinin under sterile conditions (in laminar flow) in a tube (*see* **Notes 11–14**).

4. Incubate the cultures at 37 °C in 5 % of CO_2 in a humidified atmosphere for 48 h. Add 200 μL of demecolcin (colcemid) to 5 mL cultures for the last 3 h, in order to arrest metaphase cells with sufficiently condensed chromosomes.

5. At any time of an in vitro experiment, treatments can be performed such as irradiations, chemical treatments, or other modifications, which are addressed in the experiment. Often the treatments are performed at the beginning of the cell culturing, i.e., in the G0 phase of the cell cycle. The agents tested can be washed out after a certain time of incubation or can remain inside the culture up to fixation according to the experiment.

3.2 Harvest of Cells

1. Spin down the cells at $120 \times g$ for 5 min (*see* **Note 15**). Remove the supernatant.
2. Add 5 mL of pre-warmed (37 °C) hypotonic solution to the cell sediment and mix gently. Let the cells swell for 10 min at 37 °C.
3. Spin down the cells as in **step 1** above.
4. Remove the supernatant and add slowly 5 mL of freshly prepared fixation solution on a "Vortex" (at the beginning add drop by drop until the color of the cell suspension changes from red to brown) (*see* **Note 16**).
5. Spin down the cells as in **step 1** above. The volume of the cell sediment decreases dramatically, because the erythrocytes burst during the fixation step and the cell fragments can be washed out (*see* **Note 17**).
6. Repeat **step 3** two or three times until the supernatant is clear and all erythrocyte debris is removed. Leave about 0.5 mL of the supernatant after the last fixation and centrifugation. Resuspend the cells with a Pasteur pipette. For each culture tube use a separate pipette.
7. Drop 3–4 drops of the cell suspension, from about 20 cm height onto clean, grease-free, water-wet, refrigerated glass microscopic slides with frosted ends. If a phase contrast microscope is available control the number of cells and the quality of chromosome spreading (*see* **Note 18**). If necessary adjust the number of drops.
8. Make necessary notes on the slides using a pencil. Let the preparations air-dry overnight before staining them.

3.3 Giemsa Block Staining and Mounting

1. Stain the slides in a chamber for 8–10 min in Giemsa staining solution.
2. Rinse with deionized water and let air-dry.
3. For routine scoring of chromosomal aberrations a uniform staining is appropriate. If the distribution of first, second, and further in vitro mitosis should be controlled, a variation of the staining has to be performed (Fig. 2).
4. Mount the slides using an appropriate mounting agent with a coverslip (*see* **Note 19**), if the slides should be available for multiple microscope analyses or kept for a longer duration.

3.4 Sister Chromatid Differentiation Staining

1. Stain the prepared slides (from **step 8** of Subheading 3.2) in H33258 for 20 min.
2. Rinse in deionized water for few seconds.
3. Rinse in PBS for few seconds and place a second clean microscope slide (also rinsed in PBS) on the slide to make it a "sandwich."

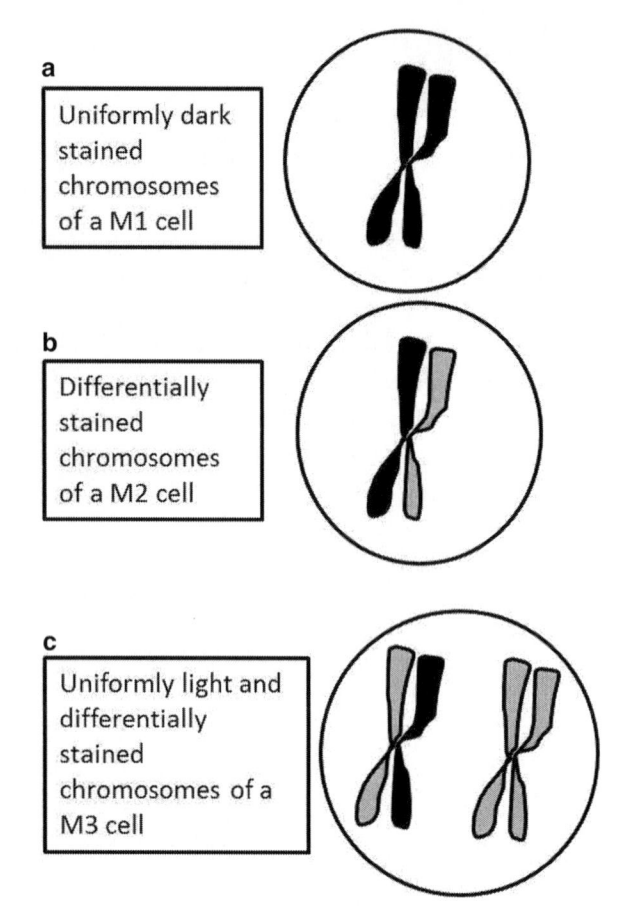

Fig. 2 If BrdU is added to the culture medium in adequate amounts cells will incorporate the base analogue during each round of DNA replication. Following appropriate staining the results will be different for first (M1), second (M2), and third (M3) mitosis. The staining pattern of the chromosomes gives information on the type of mitosis you look at. (**a**) M1 cells show uniformly darkly stained chromosomes. (**b**) Second divisions have chromosomes with a lighter and a darker stained chromatid. (**c**) Third and further metaphases consist of chromosomes with differentially stained chromatids and of chromosomes which show only light chromatids. The frequency of the latter type is 50 % in M3 cells and increases with each further in vitro division

4. Place the "sandwich" onto a heating plate at 60 °C. Irradiate the "sandwich" from above with UV for 20 min (254 nm, e.g., Osram HNS, 30W). The distance should be 150 mm.
5. Rinse the "sandwich" in PBS and remove the cover slide carefully so as not to scrape and thereby damage the cells on the slide.
6. Rinse in deionized water.
7. Stain in Giemsa solution for 8 min and let air-dry.
8. Mount the slides using an appropriate mounting agent with a coverslip (*see* **Note 19**).

3.5 Scoring Chromosomal Aberrations in Giemsa-Stained Preparations

1. Before analyzing the slides for chromosomal aberrations a second person should code them in order to ensure that the person analyzing the slides does not know about the particular experimental point (or blood donor).

2. Scoring must be performed at 500- to 1,000-fold magnification using oil-immersion lens, to get an appropriate resolution for the detection of all types of aberrations. Cells which show 46 uniformly stained chromosomes with no or little overlapping are suitable for analysis of chromosomal aberrations (Fig. 1). It should be ensured that the cells scored are not selected by any criteria and that no cell is repeatedly analyzed.

3. The number of cells scored is subject to the type of study performed. Usually the number has to be higher in in vivo studies than in in vitro experiments where exposures are performed under controlled conditions. If high yields of aberrations are expected to occur due to a high mutagenic activity 100 cells may be sufficient. In other cases 1,000 cells or more should be analyzed to avoid false-negative results.

4 Notes

1. Human blood is potentially infectious (e.g., from hepatitis virus). It must be handled with great care and protecting gloves should be worn throughout the blood collection, cultivation of cells, and chromosome preparation. All materials which have been in contact with the blood sample or lymphocyte cultures have to be sterilized (in an autoclave) before giving it to the waste. After fixing the cells the infection risk is eliminated.

2. Ready-for-use liquid media are very comfortable. If greater amounts of medium are needed, it is cost saving to purchase powdered media. This has then to be solved according to the manufacturer's protocol and sterile filtered. Liquid media usually are offered without L-glutamine, because once solved it is unstable at physiological pH. L-glutamine has then to be added to liquid media at the time of use in concentrations as indicated by the manufacturer.

3. Although not essential for successful HPL cultivation, it is recommended to add antibiotics to the culture medium prior to usage. Penicillin (active against Gram-positive bacteria) and streptomycin (active against Gram-negative bacteria) are available as a lyophilisate or in frozen form. 100× concentrated mixtures are very comfortable to use. The final concentration for penicillin is 100 U/mL, and for streptomycin 100 µg/mL.

4. Reconstituted or thawed PHA should be used within 2 weeks, because it loses its mitogenic activity.

5. BrdU is very light sensitive and should always be kept in the dark. During usage (cultivation and chromosome preparation) the ambient light should be dimmed.

6. Fixation solution should not be stored for more than one day, because esterification will spoil the fixative due to water generation.

7. We find a better spreading of metaphase chromosomes when using water-wet slides. We remove most of the water using a film wiper before dropping the cells onto them. A film wiper is an item which was formerly used to remove liquids from a photographic film during developing and fixing procedures.

8. Giemsa solution can be used for several staining passages (bathes). A prolongation of the staining time is necessary with each passage, because the Giemsa solution loses its staining property. We add 1 min per passage, starting with 8 min for the first dipping bath.

    After some hours Giemsa crystals may accumulate on the surface of the solution. Then (or on the next day) a second filtering of the Giemsa solution through a paper filter avoids crystal precipitation on the stained slides.

9. Heparin avoids cell agglutination. Mostly lithium heparin is used, but sodium or ammonium heparin is also suitable. If heparin is not already present in the tube it should be added immediately. Use of other anticoagulants than heparin (e.g., ETDA) blocks mitogenic stimulation nearly entirely and must be avoided. In case any other anticoagulant was used, it can be washed out with balanced salt solutions or culture medium. Centrifuge the sample, fill up with washing solution, mix, and centrifuge again. Repeat two times and add culture medium containing 10 % FBS after the last centrifugation. Proceed as described in **step 2** of Subheading 3.1.

10. Cell viability is optimal if the cell cultures are set up without any delay after blood collection. The mitotic activity decreases with time between blood collection and culture setup. We found viable cells up to 5 days after collection, but such a long period should be avoided, if possible. If blood specimens have to be transported from other places to the cytogenetic laboratory or cannot be set up immediately for any other reasons the ideal temperature to maintain growth ability of the lymphocytes is between 18 and 24 °C. Freezing or heating of the samples should be avoided during the intervening period. Ensure that the vials cannot break during transportation by packing the samples carefully. For sending blood samples by courier services use a container inside a second container to avoid any leakage. Label the container properly with addresses and telephone numbers of responsible persons, both from sender and receiver. Follow the national and international regulations for

transportations of biological material. Add information (like no freezing–no heating–no X-irradiation) on the package. A temperature logger may be enclosed to control for proper conditions. It may be convenient to employ a courier service for longer or international shipments as these companies are used to deal with customs and the necessary formalities.

11. Adjust respective amounts to greater culture volumes (e.g., 10 mL flask cultures) accordingly.

12. Usually the quantity of a human blood sample is rather small (e.g., 10 mL). In these cases we perform whole-blood lymphocyte cultures to obtain a maximum number of metaphase cells out of it. If greater volumes of blood are available, lymphocytes can be isolated prior to culturing. Separation media are commercially available (Ficoll, Percoll, etc.). After counting a number of 10^6 to 2×10^6 leukocytes should be set up per 5 mL culture. Nonetheless, a loss of lymphocytes during isolation is inevitable.

13. Some authors report about better metaphase yields when the culture contains 15 or 20 % FBS instead of 10 %.

14. To control the cell cycle (*see* also Subheading 1), 100 µL BrdU should be added to the cultures at the beginning. BrdU substitutes for thymidine into the replicating DNA of the HPL during each S-phase. Following two rounds of DNA replication in BrdU-containing medium a differential loss of stainable material can finally be visualized by a modified Giemsa staining. Chromatin containing BrdU and stained with HOECHST 33258 is very photosensitive. Following irradiation with UV light at 60 °C the DNA degrades and can be washed out of the fixed chromatids on the slides. This results in a darker and a lighter stained chromatid if two replication cycles have been completed (*see* Fig. 2). Sister chromatid exchanges are also detectable.

15. If other containers than tubes have been used for culturing (e.g., flasks), the cells have to be transferred to a centrifugation tube by use of a Pasteur pipette in a first step. Make sure that you rinse the bottom of the flask's surface gently with the supernatant to get all cells out of the flask.

16. If the first fixation solution is added too quickly the cells tend to clump together. This can be avoided by firstly adding 5–10 drops. In the following fixation steps the solution can be added quickly.

17. After the first addition of the fixative and following centrifugation the supernatant appears dark brown (colored by denatured hemoglobin) and makes it difficult to detect the cell sediment. Just leave about 0.5 mL of supernatant. After the second fixation and centrifugation the cell sediment appears as a very small white pellet.

18. If the chromosome spreading as controlled in the phase contrast is not satisfactory it may help to put the slides onto a heating plate after dropping the cells. This accelerates the evaporation of the fixative and may improve the chromosome separation of the metaphases (always control in phase contrast).

19. We use Entellan (Merck) for mounting, but other appropriate glue for microscopy is available. The mounting should be done under a hood, because solvents (xylene, toluol, etc.) evaporate during the procedure. Use only two small drops of the glue per slide; otherwise the distance between the cells on the slide and the oil-immersion lens on the coverslip will become too big for a sharp image. The mounting agent dissolves much better when both the slide and the coverslip are dipped into xylene before mounting.

References

1. Alberts B, Johnson A, Lewis J et al (2008) Molecular biology of the cell, 5th edn. Garland, New York
2. Murphy K (2012) Janeway's immunobiology. Garland Science, Taylor & Francis Group, New York
3. Trepel F (1975) Kinetik lymphatischer Zellen. In: Theml H, Begemann E (eds) Lymphocyt und klinische Immunologie. Springer, Berlin, pp 16–26
4. Ling NR, Kay JE (1975) Lymphocyte stimulation. North-Holland
5. Bessis M (1973) Living blood cells and their ultrastructure. Springer, Berlin
6. Petrzilka GE, Schroeder HE (1979) Activation of human T–lymphocytes, a kinetic study. Cell Tissue Res 201:101–127
7. Obe G, Beek B (1982) The human leukocyte system. In: de Serres FJ, Hollaender A (eds) Chemical mutagens, principles and methods for their detection, vol 7. Plenum Press, New York, pp 337–400
8. Loeb LA, Ewald JL, Agarwal SS (1970) DNA polymerase and DNA replication during lymphocyte transformation. Cancer Res 30:2514–2520
9. Dudin G, Beek B, Obe G (1974) The human leukocyte test system I. DNA synthesis and mitoses in PHA-stimulated 2-day cultures. Mutat Res 23:279–281
10. Beek B, Obe G (1975) The human leukocyte test system IV. The RNA-synthesis pattern in the first G1-phase of the cell cycle after stimulation with PHA. Mutat Res 29:165–168
11. Obe G, Beek B, Dudin G (1975) The human leukocyte test system V. DNA synthesis and mitoses in PHA-stimulated 3-day cultures. Humangenetik 28:295–302
12. Dudin G, Beek B, Obe G (1976) The human lymphocyte test system VIII. DNA synthesis and mitoses in 3-day cultures stimulated with pokeweed mitogen. Hum Genet 32:323–327
13. Obe G, Brandt K, Beek B (1976) The human leukocyte test system IX. DNA synthesis and mitoses in PHA-stimulated 2-day cultures set up with Eagle's minimal essential medium (MEM). Hum Genet 33:263–268
14. Beek B, Obe G (1979) Sister chromatid exchanges in human leukocyte chromosomes: spontaneous and induced frequencies in early- and late-proliferating cells in vitro. Hum Genet 49:51–61
15. Chudoba I (2007) Fluorescence in situ hybridization. In: Obe G, Vijayalaxmi (eds) Chromosomal alterations: methods, results and importance in human health. Springer, Berlin, pp 285–299
16. Johannes C, Horstmann M, Durante M et al (2004) Chromosome intrachanges and interchanges detected by multicolor banding in lymphocytes: searching for clastogen signatures in the human genome. Radiat Res 161: 540–548
17. Johannes C, Chudoba I, Obe G (1999) Analysis of X–ray induced aberrations in human chromosome 5 using high resolution multi-colour banding FISH (mBAND). Chromosome Res 7:625–633
18. Obe G, Pfeiffer P, Savage JRK et al (2002) Chromosomal aberrations: formation, identification and distribution. Mutat Res 504:17–36
19. Savage JRK, Papworth DG (1991) Excogitations about the quantification of structural chromosomal aberrations. In: Obe G (ed) Advances in mutagenesis research. Springer, Berlin, pp 162–189

20. Müller WU (2007) Comet assay. In: Obe G, Vijayalaxmi (eds) Chromosomal alterations: methods, results and importance in human health. Springer, Berlin, pp 161–176.
21. Fenech M (2007) Cytokinesis-block micronucleus assay: a comprehensive "cytome" approach for measuring chromosomal instability, mitotic dysfunction and cell death simultaneously in one assay. In: Obe G, Vijayaylaxmi (eds) Chromosomal alterations: methods, results and importance in human health. Springer, Berlin, pp 241–255
22. Wojcik A, Bruckmann E, Obe G (2004) Insights into the mechanisms of sister chromatid exchange formation. Cytogenet Genome Res 104:304–309
23. Albertini RJ, Anderson D, Douglas GR et al (2000) IPCS guidelines for the monitoring of genotoxic effects of carcinogens in humans. Mutat Res 463:111–172
24. Carrano AV, Natarajan AT (1988) Considerations for population monitoring using cytogenetic techniques. Mutat Res 204:379–406
25. Hsu C–H, Stedeford T (2010) Cancer risk assessment, chemical carcinogenesis, hazard evaluation, and risk quantification. John Wiley & Sons, Hoboken, New Jersey
26. Kirkland D, Gatehouse D (2010) In vitro genotox assays. In: Hsu C-H, Stedeford T (eds) Cancer risk assessment, chemical carcinogenesis, hazard evaluation, and risk quantification. Wiley, New Jersey, pp 272–288
27. Müller L, Martus HJ (2010) Genetic toxicology testing guidelines and regulations. In: Hsu C-H, Stadeford T (eds) Cancer risk assessment, chemical carcinogenesis, hazard evaluation, and risk quantification. John Wiley & Sons, New Jersey, pp 238–271
28. Marshall R, Obe G (1998) Application of chromosome painting to clastogenicity testing *in vitro*. Environ Mol Mutagenesis 32:212–2228
29. Natarajan AT, Obe G (1980) Screening of human populations for mutations induced by environmental pollutants: use of human lymphocyte system. Ecotoxicol Environ Saf 4:468–481
30. Obe G, Sperling K, Belitz HJ (1971) Some aspects of chemical mutagenesis in man and in Drosophila. Angew Chem, Internat Edn 10:302–314
31. Thybaud V (2010) *In vivo* genotoxicity assays. In: Hsu C–H, Stedeford T (eds) Cancer risk assessment, chemical carcinogenesis, hazard evaluation, and risk quantification. Wiley, New Jersey, pp 289–359
32. Zeiger E (2010) Development of genetic toxicology testing and its incorporation into regulatory health effects test requirements. In: Hsu C-H, Stedeford T (eds) Cancer risk assessment, chemical carcinogenesis, hazard evaluation, and risk quantification. Wiley, New Jersey, pp 225–237
33. Madle S, Obe G (1980) Methods for analysis of the mutagenicity of indirect mutagens/carcinogens in eukaryotic cells. Hum Genet 56:7–20
34. Madle S, Westphal D, Hilbig V et al (1978) Testing *in vitro* of an indirect mutagen (cyclophosphamide) with human leukocyte cultures, activation by liver perfusion and by incubation with crude liver homogenate. Mutat Res 54:95–99
35. Obe G, Jonas R, Schmidt S (1986) Metabolism of ethanol *in vitro* produces a compound which induces sister-chromatid exchanges in human peripheral lymphocytes *in vitro*: acetaldehyde not ethanol is mutagenic. Mutat Res 174:47–51
36. IARC (1988) Vol 44, Alcohol drinking. International Agency for Research on Cancer, Lyon
37. IARC (2010) Vol 95, Alcohol consumption and ethyl carbamate. International Agency for Research on Cancer, Lyon
38. Obe G, Anderson D (1987) Genetic effects of ethanol. Mutat Res 186:177–200
39. Obe G, Beek B (1979) Mutagenic activity of aldehydes. Drug and Alcohol Dependence 4:91–94
40. Obe G, Ristow H (1979) Mutagenic, cancerogenic and teratogenic effects of alcohol. Mutat Res 65:229–259
41. Obe G, Ristow H, Herha J (1979) Effect of ethanol on chromosomal structure and function. In: Majchrowicz E, Noble EP (eds) Biochemistry and pharmacology of ethanol, vol 1. Plenum, New York, pp 659–676
42. Hoeft H, Obe G (1983) SCE-inducing congeners in alcoholic beverages. Mutat Res 121:247–251
43. IARC (1986) Vol 38, Tobacco smoking. International Agency for Research on Cancer, Lyon
44. Obe G, Vogt HJ, Madle S et al (1982) Double-blind study on the effect of cigarette smoking on the chromosomes of human peripheral blood lymphocytes *in vivo*. Mutat Res 92:309–319
45. Obe G, Heller WD, Vogt HJ (1984) Mutagenic activity of cigarette smoke. In: Obe G (ed) Mutations in man. Springer, Berlin, pp 223–246
46. Obe G, Lloyd DC, Durante M (2011) Chromosomal aberrations in human populations and cancer. In: Obe G, Jandrig B, Marchant GE et al (eds) Cancer risk evaluation, methods and trends. Wiley-VCH, Weinheim, pp 139–161

47. IAEA (2011) Cytogenetic dosimetry: applications in preparedness for and response to radiation emergencies. International Atomic Energy Agency, Vienna
48. Iwasaki T, Takashima Y, Suzuki T et al (2011) The dose response of chromosome aberrations in human lymphocytes induced *in vitro* by very low-dose γ-rays. Radiat Res 175:208–213
49. Lloyd DC, Edwards AA, Leonard A et al (1992) Chromosomal aberrations in human lymphocytes induced *in vitro* by very low doses of X-rays. Int J Radiat Biol 61:335–343

Chapter 9

In Vivo Micronucleus Assay in Mouse Bone Marrow and Peripheral Blood

Sawako Kasamoto, Shoji Masumori, and Makoto Hayashi

Abstract

The rodent micronucleus assay has been most widely and frequently used as a representative in vivo assay system to assess mutagenicity of chemicals, regardless of endpoint of mutagenicity. The micronucleus has been developed to assess induction of structural and numerical chromosomal aberrations of target chemical. In this chapter, we describe the standard protocols of the assay using mouse bone marrow and peripheral blood. These methods are basically applicable to other rodents. The methodology of the micronucleus assay is rapidly developing, especially automatic analysis by flow cytometry (see also Chapter 11). Also we have to pay attention to the animal welfare, for example integration into repeat dose toxicity assay, combination of the micronucleus assay and Comet assay, and also omission of concurrent positive control group. Therefore, modification of the standard protocol is necessary for the actual assay on a case-by-case basis.

Key words Chromosomal aberration, Micronucleus, In vivo assay, Bone marrow, Peripheral blood

1 Introduction

Chromosomal aberration is one of the main endpoints of genotoxicity as well as gene mutation to assess the mutagenicity of chemicals. To assess chromosomal aberration in vitro, established cell lines (e.g., CHL/IU, CHO) or primary culture of human lymphocytes are mainly used for the assay. For the in vivo assay, which is more important for the risk characterization, the rodent erythrocyte micronucleus assay is most widely and frequently used to assess induction ability of chromosomal aberration of chemicals in the body of animals.

The micronucleus assay was developed as an assay system in the 1970s by Heddle [1] and Schmid [2] using mouse bone marrow erythropoietic cells. A good correlation was shown between chromosomal aberration induction and micronucleus induction [3]. The Collaborative Study Group of Micronucleus Test (CSGMT) of Mammalian Mutagenicity Study (MMS) group which belongs to the Japanese Environmental Mutagen Society (JEMS) has widely studied the many factors that might affect the result of the assay, e.g., sex difference [4], strain difference [5], dose administration route [6], number of treatments [7], and evaluation of the assay sensitivity [8, 9]. These outcomes were reflected in the internationally accepted guidelines such as OECD test guideline TG474 [10] and ICH guideline [ICH S2(R1); [11]]. The micronucleus assay has been harmonized mainly because of methodological aspects reviewed by the International Workshop on Genotoxicity Testing (IWGT) [12–14].

The main target tissue of the micronucleus assay has been the bone marrow of mouse and rat. The use of peripheral blood instead of bone marrow erythropoietic cells was introduced by MacGregor et al. [15]. The Collaborative Study Group for the Micronucleus Test [8] and Hayashi et al. [16] using acridine orange supravital staining. There are several advantages of using peripheral blood instead of bone marrow, for example, only tiny amount of blood is enough for analysis and then there is no need to sacrifice the animals, as well as samples can be periodically obtained from the same animals. On the other hand, the most important disadvantage exists, which is the elimination of micronucleated erythrocytes by spleen of many species, e.g., human, monkey, and rat [17], except in the mouse. Therefore, we have to select the assay system on a case-by-case basis according to the purpose of the study.

Here, we describe the standard practical methods of the micronucleus assay. Of course these are examples that can be modified according to the purpose of the study. Based on the animal welfare, several attempts have been started, which are trying to reduce animal usage: (1) integration of micronucleus endpoint into repeat dose toxicological study [18], (2) combination of study of micronucleus and Comet assays as one study [19], and (3) omitting the concurrent positive control group.

2 Materials

To perform the micronucleus assay, we need the following materials and working solutions.

2.1 Controls

1. Negative control: The solvent/vehicle that is used to prepare dosing formulations is used as the negative control (*see* **Note 1**). Examples of negative control include:

    (a) Water.

    (b) Physiological saline.

    (c) Methylcellulose solution.

    (d) Carboxymethyl cellulose sodium salt solution.

    (e) Olive oil.

2. Positive control: Positive controls should produce micronuclei in vivo at exposure levels expected to give a detectable increase over the background. Examples of positive control substances include:
   (a) Ethyl methanesulfonate.
   (b) Ethyl nitrosourea.
   (c) Mitomycin C.
   (d) Cyclophosphamide (monohydrate).
   (e) Triethylenemelamine.

2.2 Micronucleus Assay in Mouse Bone Marrow

1. Fetal calf serum.
2. Methanol.
3. 1/100 mol/L sodium phosphate buffer (pH 6.8).
4. Staining solution: 3 % Giemsa's solution prepared in 1/100 mol/L sodium phosphate buffer.
5. 0.001 % citric acid solution.

2.3 Micronucleus Assay in Mouse Peripheral Blood

1. Fetal calf serum.
2. Acridine orange (AO)-coated slide: 20 µL of 0.1 % acridine orange in ethanol is placed on a cleaned glass slide (end frosted slides can also be used) and covered with another slide. Move these slides right and left, separate, and then air-dry. The AO-coated slides are stored in a refrigerator or a freezer. The preparation of AO-coated slide refers to the original method described in the literature (*see* also ref. 16).
3. Sealing agent, e.g., Permount.

3 Methods

3.1 Animal Maintenance

1. Commonly used laboratory strains of young healthy animals (6–10 weeks old at the start of the treatment schedule) should be employed.
2. In general, only one sex, usually male, is used (*see* **Note 2**). However, if there is evidence indicating a relevant difference in toxicity or metabolism between males and females, then both sexes of the animals should be used.
3. Animals are housed in a temperature- and humidity-controlled room. Lighting should be artificial, the sequence being 12-h light, and 12-h dark. For feeding, conventional laboratory diets can be used with an unlimited supply of drinking water.
4. Animals are housed individually, or caged in small groups of the same sex.

5. Following arrival at the testing facility, each animal should be monitored for at least 5 days to ensure that it is healthy (no abnormalities) and growing normally; animals are acclimated to the laboratory environment.

6. Animals are randomly assigned to the control and treatment groups.

7. The animals are identified uniquely.

3.2 Study Design and Chemical Treatment

1. Each treated and control group must include at least 5 analyzable animals per sex.

2. The study may be performed in two ways:

    (a) Animals are treated with the test substance once. Samples of bone marrow are taken at least twice, starting not earlier than 24 h after treatment, but not extending beyond 48 h after treatment with appropriate interval(s) between samples. The use of sampling times earlier than 24 h after treatment should be justified. Samples of peripheral blood are taken at least twice, starting not earlier than 36 h after treatment, with appropriate intervals following the first sample, but not extending beyond 72 h. When a positive response is recognized at one sampling time, additional sampling is not required.

    (b) If two or more daily treatments are used (e.g., two or more treatments at 24-h intervals), samples should be collected once between 18 and 24 h following the final treatment for the bone marrow and once between 36 and 48 h following the final treatment for the peripheral blood.

3. A dose around the maximal tolerated dose or 2,000 mg/kg, whichever is lower, will be selected as the high dosage level, and two additional lower doses separated by a factor less than square root of 10 will be selected as the middle and lower dosage levels. The highest dose may also be defined as a dose that produces some indication of toxicity of the bone marrow (e.g., a reduction of immature erythrocytes among total erythrocytes in the bone marrow or the peripheral blood).

4. The test substance is usually administered by gavage using a stomach tube or a suitable intubation cannula, or by intraperitoneal injection. Other routes of exposure may be acceptable where they can be justified. The maximum volume of liquid that can be administered by gavage or injection at one time depends on the size of the test animal. The volume should not exceed 2 mL/100 g body weight.

5. The body weight of each animal is measured at the time of assignment to groups and just before sampling. Animals are assigned randomly considering keeping similar average of body weight.

6. Clinical signs of animals are examined at appropriate period after the dosing and just before sampling.

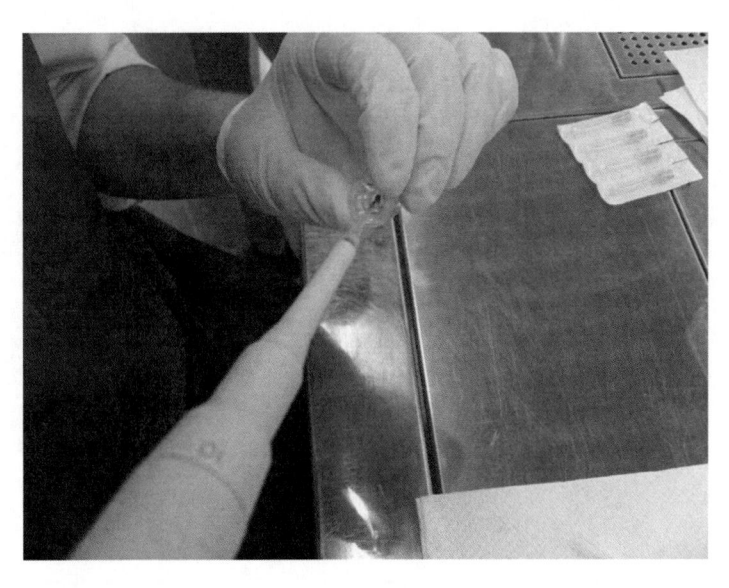

Fig. 1 Approximately 5 μL of peripheral blood are mixed well with the same volume of fetal bovine serum by pipetting

3.3 Preparation of Bone Marrow Smears

1. Euthanize the animals before sampling.
2. Remove one femur, trim the muscle around it, and cut both ends of the femur to collect the bone marrow cells.
3. Flush the bone marrow cells out with 0.5 mL of fetal calf serum into centrifuge tubes using a 1-mL syringe fitted with a 22 G needle.
4. Centrifuge the tubes at approximately $200 \times g$ in a microcentrifuge for 5 min.
5. Remove the supernatant, and resuspend the cells with the remaining serum.
6. Drop a small amount of the cell suspension on the end of a glass slide and spread it by pulling the material behind a polished cover glass held at an angle of approximately 45° (*see* **Note 3**).
7. Allow the slides to air-dry and fix them in methanol for 3 min.
8. Stain the slides with 3 % Giemsa solution diluted and prepared with 1/100 mol/L sodium phosphate buffer (pH 6.8) for 30 min.
9. Rinse the slides with 1/100 mol/L sodium phosphate buffer and then with purified water, and dry them.
10. Rinse the slides with 0.001 % citric acid solution (*see* **Note 4**), and then with purified water, and dry again.

3.4 Preparation of Peripheral Blood Smears

1. At the time of sampling, pierce a tail blood vessel and collect about 5 μL of peripheral blood with a micropipette.
2. Mix the blood with 5 μL of fetal bovine serum and slightly stir (Fig. 1).

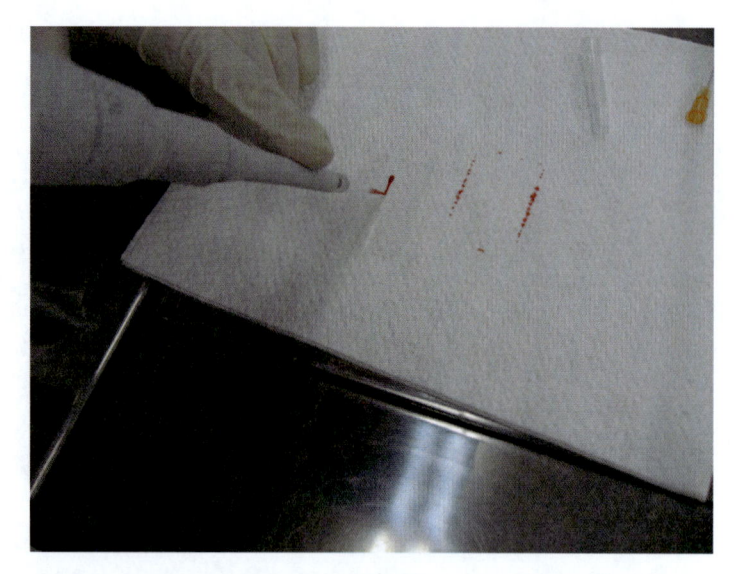

Fig. 2 Drop the diluted peripheral blood onto coverslip (approximately 7 μL)

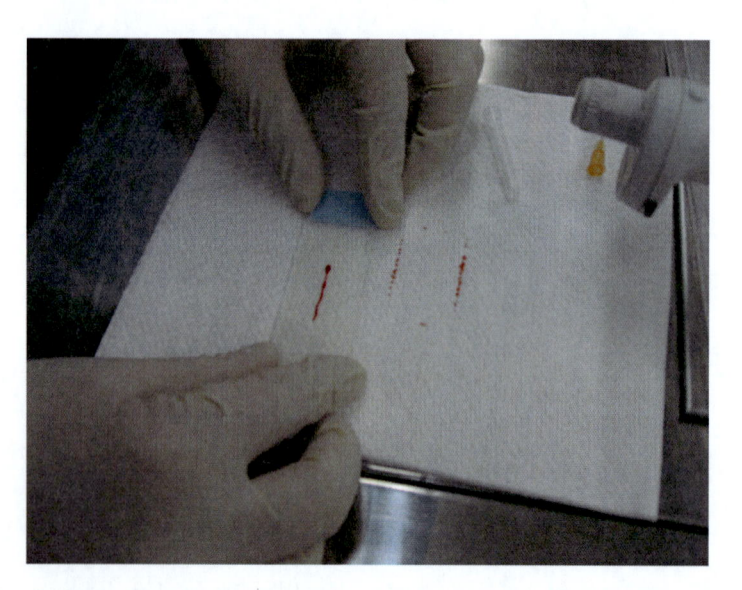

Fig. 3 Cover with glass slide that was pre-coated with acridine orange

3. Drop 5–7 μL of the serum-mixed blood on a 24 × 40 mm coverslip and place on it an AO-coated slide (*see* preparation in Subheading 2.3; Figs. 2, 3 and 4).
4. Seal the peripheral blood smear with a sealing, e.g., Permount (*see* **Note 5**).
5. Slides should be allowed to stand for a few hours or overnight to allow cells to settle and to maximize staining, and examined with a microscope within 5 days after preparation. The preparations can be stored for about a week in a refrigerator or a couple of months in a deep freezer.

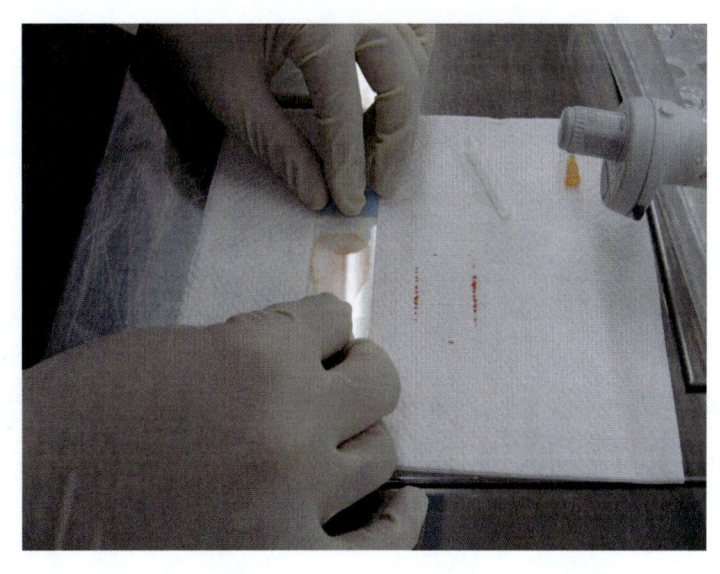

Fig. 4 Spread blood film and let stay for staining before microscopy

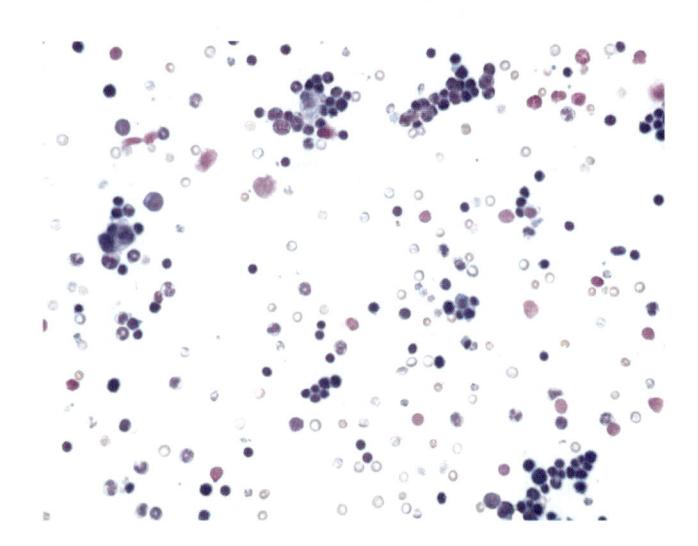

Fig. 5 Bone marrow slide (×100: low magnification)

3.5 Microscopic Observation of Bone Marrow Smears

1. The slides are coded so as not to reveal the treatment groups to the scorer.
2. First the slides are scanned at low or medium magnification looking for region of suitable technical quality, where the cells are well spread without overlapping, undamaged, and clearly stained to distinguish young and mature erythrocytes.
3. The erythrocytes should be well spread, neither globular nor having slurred contours.
4. Their staining has to be vigorous, pink in mature erythrocytes (normochromatic erythrocytes: NCE), and with a bluish tint in the immature forms (polychromatic erythrocytes: PCE) (Fig. 5).

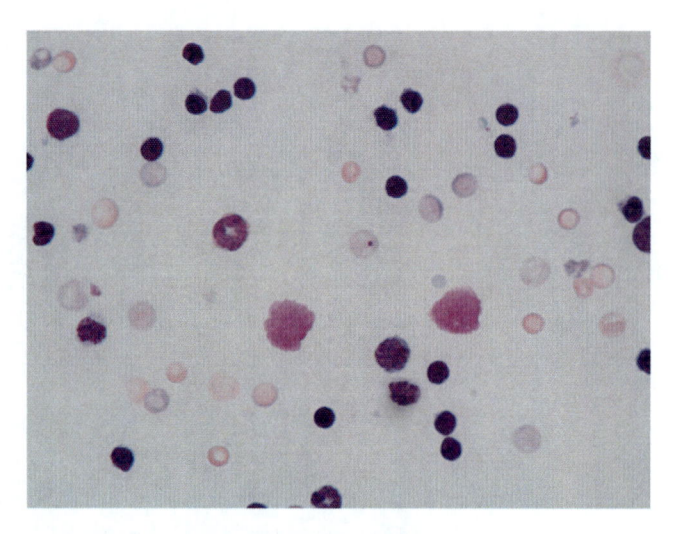

Fig. 6 Bone marrow slide (×1,000: high magnification)

5. As an indicator of cell toxicity in bone marrow, the number of PCEs is counted by examining at least 200 erythrocytes (PCE + NCE) per animal under a microscope (e.g., ×1,000) with immersion oil. Then, the ratio of PCE to total erythrocytes is calculated as a percent (*see* **step 3** of Subheading 3.7).

6. At least 2,000 PCEs per animal are examined and the number of micronucleated polychromatic erythrocytes (MNPCE) is recorded. Micronuclei are identified according to the criteria established by Schmid [2] and are darkly stained (purple) and generally round or almond shaped, although lightly stained, ring-shaped micronuclei occasionally occur (Fig. 6). Micronuclei have sharp borders and are generally 5–20 % the size of the PCEs and may occur in either PCE or NCE. However, only MNPCE is counted (*see* **Note 6**). Then, the incidence of MNPCE to total PCEs is calculated as a percent.

3.6 Microscopic Observation of Peripheral Blood Smears

1. The slides are coded to avoid the bias of the scorer who is aware of the treatment.

2. First the slides are scanned at low or medium magnification looking for region of suitable technical quality, where the cells are well spread, undamaged, and clearly stained (Fig. 7).

3. Cells should be intact and the nuclei of nucleated cells and the reticulum structure of reticulocytes (RET; immature cell) should fluoresce strongly green-yellow and red, respectively, using a fluorescent microscope (e.g., ×800) equipped with a blue excitation filter and a green barrier filter (*see* **Note 7**).

4. As an indicator of cell toxicity in bone marrow, the number of RET is determined by examining at least 1,000 erythrocytes (RET + mature cells) per animal using a fluorescent microscope

In Vivo Mouse Micronucleus Assay 187

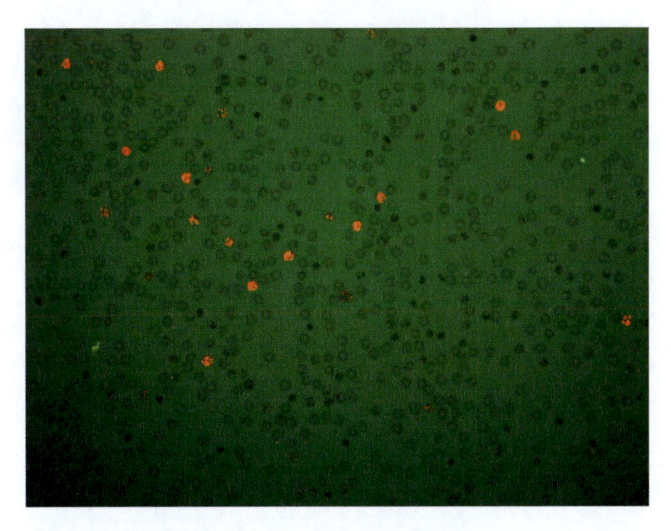

Fig. 7 Peripheral blood slide (×100: low magnification)

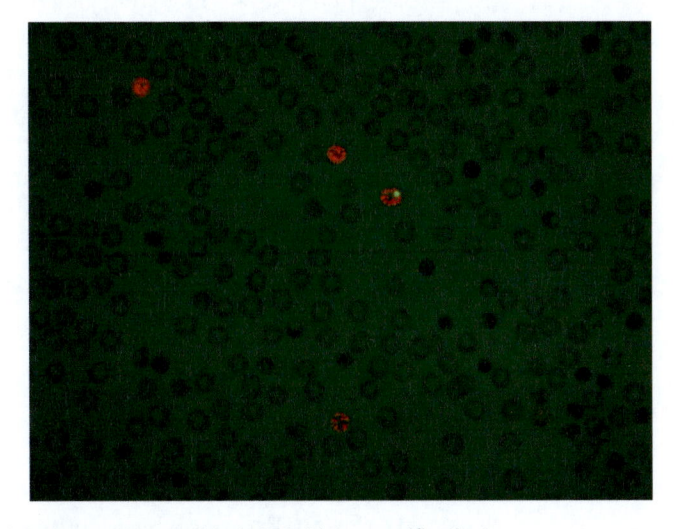

Fig. 8 Peripheral blood slide (×800: high magnification)

       (e.g., ×800) equipped with a blue excitation filter and a green barrier filter. Then, the ratio of RET to total erythrocytes is calculated as a percent.

5. At least 2,000 RET per animal are analyzed, and the number of micronucleated reticulocytes (MNRET) is recorded. Micronuclei are round in shape and fluoresce green-yellow (*see* **Notes 6** and **8**; Fig. 8). Then, the incidence of MNRET to total RET is calculated as a percent.

3.7 Evaluation of Results

1. The definition of a positive call should be predetermined, for example, increment of incidence of micronucleated immature erythrocytes is statistically significant compared to that

of the negative control group. The biological relevance, of course, should be included, namely, the response is dose dependent and reproducible among animals within the treatment group.

2. The negative call can be defined as not positive, i.e., no significant increase over negative control. If the result is not clear and equivocal, we recommend engaging a repeat or a confirmatory test.

3. Statistically significant reduction of the ratio of PCE or RET as compared with the negative control is used as an indication of toxicity (inhibition of cell proliferation) in bone marrow.

4 Notes

1. If peripheral blood is used, a pretreatment sample may also be acceptable as a concurrent negative control to reduce the total number of animals to be used.

2. Extensive studies of the activity of known clastogens in the mouse bone marrow micronucleus assay have shown that, in general, male mice are more sensitive than female mice for micronucleus induction [4]. Quantitative differences in micronucleus induction have been identified between the sexes, but no qualitative differences have been described.

3. To prepare good smears of the bone marrow, all favorable conditions including concentration and volume of the cell suspension, angle of the cover glass, and speed of pulling the cover glass are required.

4. Slides are immersed in 0.001 % citric acid solution for a few seconds to bleach the Giemsa stain slightly. This procedure allows us to make easy differentiation of the PCE and NCE. If the NCE cannot be identified, slides are immersed in the citric acid solution again for another few more seconds. On the other hand, if the PCE cannot be identified, which would be caused by overbleaching, they should be re-stained after fully removing Giemsa stain by immersing in the ethanol.

5. If the slides are not sealed, air comes in between a slide and cover glass and makes it hard to clearly observe the slides.

6. Only typical micronuclei should be counted. To confirm that micronuclei are present inside of a cell, change the microscopic focus up and down. If white-shining granules can be seen, then these should not be counted as micronuclei because they are artificial particles attached on the cells.

7. Acridine orange stains both RNA and DNA fluorescing differently, i.e., RNA fluoresces red and DNA appears green-yellow

when viewed under florescent light. RET contain RNA in their cytoplasm and can be distinguished easily from mature erythrocytes, which do not fluoresce because they lack RNA.

8. Bacterial contamination is suspected when moving green-fluorescing particles are observed in the slides.

References

1. Heddle JA (1993) A rapid in vivo test for chromosomal damage. Mutat Res 18:187–190
2. Schmid W (1975) The micronucleus test. Mutat Res 31:9–15
3. Hayashi M, Sofuni T, Ishidate M Jr (1984) Kinetics of micronucleus formation in relation to chromosomal aberrations in mouse bone marrow. Mutat Res 127:129–137
4. Sutou S, Hayashi M, Nishi Y et al (1986) Sex difference in the micronucleus test. The collaborative study group for the micronucleus test. Mutat Res 172:151–163
5. Sutou S, Hayashi M, Shimada H et al (1988) Strain difference in the micronucleus test. The Collaborative Study Group for the Micronucleus Test. Mutat Res 204:307–316
6. Hayashi M, Sutou S, Shimada H et al (1989) Difference between intraperitoneal and oral gavage application in the micronucleus test: the 3rd collaborative study by CSGMT/JEMS.MMS. Mutat Res 223:329–344
7. Collaborative Study Group for the Micronucleus Test, the Mammalian Mutagenesis Study group of the Environmental Mutagen Society, Japan (CSGMT/JEMS·MMS) (1990) Single versus multiple dosing in the micronucleus test: the summary of the fourth collaborative study by CSGMT/JEMS·MMS. Mutat Res 234:205–222
8. The Collaborative Study Group for the Micronucleus Test (1992) Micronucleus test with mouse peripheral blood erythrocytes by acridine orange supravital staining: the summary report of the 5th collaborative study by CSGMT/JEMS—MMS. Mutat Res 278:83–98
9. Morita T, Asano N, Awogi T et al (1997) Evaluation of the rodent micronucleus assay in the screening of IARC carcinogens (Groups 1, 2A, and 2B). The summary report of the 6th collaborative study by CSGMT/JEMS·MMS. Mutat Res 389:3–122
10. Organization for Economic Co-operation and Development (OECD) (1997) OECD Guideline for the Testing of Chemicals 474 Mammalian Erythrocyte Micronucleus Test
11. International Conference on Harmonisation; guidance on S2(R1) (2011) Genotoxicity testing and data interpretation for pharmaceuticals intended for human use. http://www.ich.org/fileadmin/Public_Web_Site/ICH_Products/Guidelines/Safety/S2_R1/Step4/S2R1_Step4.pdf
12. Hayashi M, Tice RR, MacGregor JT et al (1994) In vivo rodent erythrocyte micronucleus assay. Mutat Res 312:293–304
13. Hayashi M, MacGregor JT, Gatehouse DG et al (2000) In vivo rodent erythrocyte micronucleus assay. II. Some aspects of protocol design including repeated treatments, integration with toxicity testing, and automated scoring. Environ Mol Mutagen 35:234–252
14. Hayashi M, MacGregor JT, Gatehouse DG et al (2007) In vivo erythrocyte micronucleus assay III. Validation and regulatory acceptance of automated scoring and the use of rat peripheral blood reticulocytes, with discussion of non-hematopoietic target cells and a single dose-level limit test. Mutat Res 627:10–30
15. MacGregor JT, Wehr CM, Gould DH (1980) Clastogen-induced micronuclei in peripheral blood erythrocytes: the basis of an improved micronucleus test. Environ Mutagen 2:509–514
16. Hayashi M, Morita T, Kodama Y et al (1990) The micronucleus assay with mouse peripheral blood reticulocytes using acridine orange-coated slides. Mutat Res 245:245–249
17. Schlegel R, MacGregor JT (1984) The persistence of micronucleated erythrocytes in the peripheral circulation of normal and splenectomized Fischer 344 rats: implications for cytogenetic screening. Mutat Res 127:169–174
18. Hamada S, Sutou S, Morita T et al (2001) Evaluation of the rodent micronucleus assay by a 28-day-treatment protocol: summary of the 13th collaborative study by CSGMT/JEMS·MMS. Environ Mol Mutagen 37:93–110
19. Pfuhler S, Kirkland D, Kasper P et al (2009) Reduction of use of animals in regulatory genotoxicity testing: identification and implementation opportunities—report from an ECVAM workshop. Mutat Res 680:31–42

Chapter 10

Micronucleus Assay in Human Cells: Lymphocytes and Buccal Cells

Claudia Bolognesi and Michael Fenech

Abstract

The micronucleus (MN) assay, applied in different surrogate tissues, is one of the best validated cytogenetic techniques for evaluating chromosomal damage in humans. The cytokinesis-block micronucleus cytome assay (CBMNcyt) in peripheral blood lymphocytes is the most frequent method in biomonitoring human populations to evaluate exposure to genotoxic agents, micronutrient deficiency, or excess and genetic instability. Furthermore recent scientific evidence suggests an association between an increased MN frequency in lymphocytes and risk of cancer and other age-related degenerative diseases. The micronucleus cytome assay applied in buccal exfoliated cells (BMNCyt) provides a complementary method for measuring DNA damage and cytotoxic effects in an easily accessible tissue not requiring in vitro culture. The protocol for CBMNcyt described here refers to the use of ex vivo whole blood involving 72 h of culture with the block of cytokinesis at 44 h. BMNCyt protocol reports the established method for sample processing, slide preparation, and scoring.

Key words Micronucleus, Chromosomal damage, DNA damage, Lymphocyte, Buccal cell, Cytome, CBMNcyt, BMCyt, Bud, Nucleoplasmic bridge

1 Introduction

The micronucleus cytome assay, applied in different surrogate tissues, is a comprehensive approach for evaluating genomic damage, cell death, and cytostasis in human populations. The cytokinesis-block micronucleus (CBMN) is the preferred assay to evaluate genomic damage in peripheral blood lymphocytes. The CBMN test, developed in the early 1980s [1] involving the block of cytokinesis with cytochalasin B (Cyt-B), allowed evaluation of the micronuclei frequency in once divided cells accumulated in the binucleated stage, overcoming the confounding effect associated with variations in in vitro cell division kinetics [2]. The CBMN assay more recently evolved in a comprehensive cytome assay (CBMNCyt) involving the cytological analysis of every scored cell for its viability, mitotic status, and chromosomal damage [3].

This assay, applied simply as MN counting in binucleated cells or as comprehensive cytome mode, in isolated lymphocytes or whole-blood cultures, is widely used in human populations to assess the exposure to genotoxic agents, micronutrient deficiency or excess [4–6], and genetic instability [7, 8]. Recent studies report evidence that MN frequency in peripheral lymphocytes is associated with an increased risk of cancer and other age-related degenerative diseases [9–13] further justifying the potential application of CBMN assay in screening programs.

The buccal micronucleus cytome (BMNCyt) assay, as a minimally invasive application of the MN assay in exfoliated buccal mucosa cells, was successfully used in biomonitoring inhalation environmental and/or occupational exposure to genotoxic agents and in the evaluation of the impact of nutrition and lifestyle factors [14]. The complete potentiality of this assay is not known: important gaps yet subsist on the biological meaning of different cell types and nuclear alterations, other than micronuclei in buccal samples.

An international collaborative project on human micronucleus (HUMN) assays (www.humn.org) was launched in 1997, to better understand the significance of this biomarker and to validate its use in human population studies. The first phase of this project focused on CBMN in peripheral blood lymphocytes. The effects of methodological and demographic variables on MN frequency were defined through the analysis of a large database [15–17]. An international slide scoring exercise using standardized scoring criteria was carried out and allowed to evaluate inter-laboratory and intra-laboratory variability affecting the analysis of data from multicenter studies [18–21]. Furthermore, the HUMN project also led to a prospective study showing that a high MN frequency in lymphocytes predicts an increased risk of cancer [9].

More recently a new project (HUMNxl project) was launched to validate the MN assay in exfoliated buccal cells following the same steps experienced for the CBMN assay in human peripheral blood lymphocytes [22, 23].

Detailed standardized protocols for CBMNCyt and BMCyt assays were established by the HUMN consortium including sampling, slide preparation, and staining and scoring criteria by taking into account the available procedures, confounding factors, and staining artifacts [24, 25].

A special issue, recently published in *Mutagenesis* (January 2011), provides an overview of the potential applications of the MN test in different surrogate tissues in different fields and of the latest advances in understanding the molecular mechanisms underlying the MN induction and kinetics.

The aim of this chapter is to provide guidelines for the application of micronucleus cytome protocols in peripheral lymphocytes and exfoliated buccal cells in biomonitoring of human populations.

The experimental protocol for CBMNCyt assay described below refers to the use of ex vivo whole blood culture. The method applied to isolated lymphocyte culture was described in detail in a previous paper [24]. The assay involves the block of cytokinesis with cytochalasin at 44 h and cell harvesting at 72 h of culture. A comprehensive approach is described involving the evaluation of preexisting MN (frequency of MN in mononucleated cells), other different nuclear anomalies (nuclear buds and nucleoplasmic bridges), and cytotoxicity biomarkers (necrotic and apoptotic cells).

The protocol for BMCyt assay reports the established method for sample processing, slide preparation, and scoring. The procedure involves repeated washing with a buffer solution to remove bacteria and cell debris and a filtration process to obtain single-cell suspension. Criteria for scoring micronuclei and other nuclear anomalies in buccal cells are also described.

2 Materials

2.1 Laboratory Equipment

1. Biological safety cabinet (Class II biosafety cabinet to provide personnel, environment, and product protection).
2. CO_2 cell culture incubator.
3. Centrifuge: Benchtop centrifuge speed 0–3,000 rpm operating at −5 to 25 °C.
4. Cytocentrifuge is recommended for slide preparation, but not mandatory. Cytocentrifuge cups are supplied with the instrument and need to be assembled following the instructions.
5. Microscope with excellent optics for bright-field and fluorescence examination of stained slides at 400× and 1,000× magnification.

2.2 Materials for CBMN Cyt Protocol

1. Test tubes for culturing cells.
2. Conical 15 mL polypropylene test tubes.
3. Microscope slides pre-cleaned/ready to use (76×26 mm and 1 mm thick) wiped with alcohol and allowed to dry. The slides are stored at −20 °C before use.
4. Coverslips—22×50 mm.
5. Filter or Whatman filter paper.
6. Multichannel cell counter.

2.3 Reagents for CBMN Cyt Protocol

1. Culture medium: RPMI 1640 medium with 2 mM glutamine and 25 mM Hepes sterile liquid. Store at 4 °C. Use at 37 °C when preparing whole blood culture.
2. Fetal bovine serum (FBS): Heat-inactivated, sterile. Store frozen at −20 °C. Thaw in a 37 °C water bath before adding to the culture medium. Once thawed, FBS will remain stable at 4 °C for 3–4 weeks (*see* **Note 1**).

3. Antibiotic/antimycotic solution: 10,000 U/mL penicillin, 10 mg/mL streptomycin, and 25 μg/mL amphotericin.
4. Phytohemagglutinin (PHA) M form: 1 mg/mL, liquid. PHA should be aliquoted into a volume convenient for use in sterile tubes. The aliquots may be stored at −20 °C for up to 6 months. If lyophilized powder form is available reconstitute to prepare 1 mg/mL solution.
5. Lymphocyte culture medium: RPMI-1640 medium, 10 % FBS, 1 % antibiotic/antimycotic solution. Mix 500 mL of RPMI-1640 medium, 50.6 mL FBS, and 6.14 mL of antibiotic/antimycotic solution. Prepare the culture medium in sterile culture-grade glass or plastic bottles. Prepare aliquots of culture tubes containing 4.7 mL of the culture medium. Culture medium can be stored for 1 week at 4 °C before use. 75 μg of PHA is added to each test tube immediately before preparing the cultures.
6. Dimethyl sulfoxide (DMSO).
7. Cyt-B: 300 μg/mL Cyt-B prepared in DMSO and isotonic saline. Stored at −20 °C (*see* **Note 2**).
8. Hypotonic solution: 0.075 M Potassium chloride (KCl). Prepare 500 mL of solution by dissolving 2.8 g of KCl in 500 mL of distilled water. Do not store. Use only freshly prepared solution.
9. 100 % Methanol.
10. Prefixing solution: 3:5 ratio of methanol:glacial acetic acid. Mix and put at −20 °C before use. This procedure should be performed in a well-ventilated fume hood with appropriate safety precaution.
11. Fixing solution: 5:1 ratio of methanol:glacial acetic acid. Prepare an adequate volume of fixing solution. The fixative should be freshly prepared each time and used at 4 °C. This procedure should be performed in a well-ventilated fume hood with appropriate safety precaution.
12. Sorensen buffer: 9.07 g/L of potassium dihydrogen phosphate (KH_2PO_4), 11.87 g/L of disodium hydrogen phosphate dehydrate (Na_2HPO_4), pH 6.8. To obtain 100 mL of Sorensen buffer solution (pH 6.8), mix 53.4 mL of KH_2PO_4 with 46.6 mL of Na_2HPO_4. Utilize this final solution to prepare Giemsa staining solution.
13. Staining solution: 2 % (v/v) Giemsa in Sorensen buffer. Filter 50 mL of Giemsa's azur-eosin-methylene blue solution with filter paper. Protect from light. Prepare 200 mL of Giemsa staining solution (2 % v/v) by adding 192 mL of distilled water with 4 mL of Sorensen buffer at room temperature.
14. Eukitt or DePex mounting medium.

2.4 Materials for BMN Cyt Protocol

1. Small-headed toothbrushes.
2. 30–50 mL polystyrene containers or test tubes.
3. Centrifuge tubes.
4. Swinnex filter holders.
5. 100 μm Nylon net filters.
6. 10 mL Syringes.
7. Filtercards.
8. Microscope slides superfrosted (76×26 mm 1 mm thick) wiped with alcohol and allowed to dry.
9. 22×50 mm Coverslips.
10. Eukitt or DePex mounting medium.

2.5 Reagents for BMN Cyt Protocol

1. Buccal cell buffer: 1.6 g of Tris–HCl, 38.0 g of ethylenediaminetetraacetic acid (EDTA) tetra sodium salt, and 1.2 g of NaCl. Weigh and dissolve in 600 mL of Milli-Q water. Make up the volume to 1,000 mL. Adjust pH to 7.0 using 5 M HCl and autoclave at 121 °C for 30 min. The buffer will last for up to 3 months when stored at room temperature.
2. Saccomanno's fixative.
3. Fixing solution: 3:1 (v/v) ethanol:glacial acetic acid. Prepare an adequate volume of fixing solution. The fixative should be freshly prepared each time and used at 4 °C. This procedure should be performed in a well-ventilated fume hood with appropriate safety precaution.
4. DMSO.
5. Light green cytoplasmic stain: Dissolve 1 g of light green in 450 mL of Milli-Q water. When dissolved, make up to 500 mL and filter through Whatman No. 1 filter paper. Store in the dark at room temperature where it should remain active for 3 years.
6. Ethanol.
7. 5 M HCl.
8. Schiff's reagent.

3 Methods

3.1 CBMN Cyt Protocol

3.1.1 Sampling and Whole Blood Cell Culture

1. Collect fresh blood (2 mL is enough to prepare duplicate cultures for baseline and in vitro-challenged lymphocytes) by venepuncture into vacutainer blood tubes with sodium or lithium heparin anticoagulant. All the procedures of cell culturing and treatment must be performed in a Class II biosafety cabinet.
2. Cultures are prepared within 24 h of phlebotomy (*see* **Note 3**).

3. 0.3 mL heparinized blood is added to 4.7 mL of complete medium RPMI-1640 supplemented with 10 % FBS, antibiotics, and antimycotics in round-bottom culture test tubes. Set up duplicate cultures per subject and/or treatment.

4. Add 75 µL of PHA solution to each test tube.

5. Incubate the test tubes at 37 °C with lids loose in a humidified atmosphere containing 5 % CO_2.

6. At 44 h, add 100 µL of the 300 µg/mL Cyt-B solution.

7. Incubate the test tubes at 37 °C.

8. At 28 h after the addition of Cyt-B, harvest the cells for slide preparation.

3.1.2 Harvesting of Cells

1. Transfer the content of each culture to centrifuge tubes.
2. Centrifuge the cell suspension for 10 min at $146 \times g$.
3. Discard the supernatant and replace with 5 mL of 0.075 M KCl hypotonic solution.
4. Mix the cells gently into suspension and keep the test tubes at room temperature for 5 min to allow red blood cell lysis to occur (*see* **Note 4**).
5. Add 400 µL of prefixing solution to each test tube and gently mix (*see* **Note 5**).
6. Centrifuge the cell suspension for 10 min at $146 \times g$.
7. Discard the supernatant and replace with 5 mL of cold methanol (−20 °C).
8. Gently mix the cell suspension. Samples in methanol can be stored for months (*see* **Note 6**).
9. Centrifuge the cell suspensions in methanol for 10 min at $228 \times g$.
10. Discard the supernatant and replace with 5 mL of fixing solution.
11. Gently mix the cell suspension and centrifuge for 10 min at $228 \times g$.
12. Repeat **steps 9** and **10** twice.
13. Discard the supernatant and finally resuspend the pellet in 500 µL of fixing solution.

3.1.3 Slide Preparation and Staining

1. Drop the cell suspension directly onto clean iced slides.
2. Dry the slides in air for at least 20 min.
3. Immerse the microscope slides for 5 min at room temperature in staining dishes containing 2 % Giemsa solution.
4. Rinse the slides for 5 min in deionized water.
5. Place the slides to be coverslipped on tissue paper.
6. Put two large drops of Eukitt or DePex (use a plastic dropper) on each coverslip.

7. Invert the slide and place on the coverslip. Allow the DePex to spread. Turn the slide over so that the coverslip is on top, and press the coverslip gently to expel any excess DePex and air bubbles.
8. Place the slides on a tray and leave overnight in the fume hood to dry.
9. Store the slides in the slide boxes at room temperature.

3.1.4 Slide Scoring

1. Slides should be coded before scoring. Slide scoring should be performed by a person not aware of the experiment conditions using transmitted light microscopy at 400× magnification (*see* **Note 7**).
2. As a first step, 500 cells are scored for viability and mitotic status. Frequencies of viable (with intact cytoplasm and normal nuclear morphology) mono-, bi-, and pluri-nucleated cells are evaluated to calculate the nuclear division index:

$$\text{i.e. NDI} = [N(\text{Mono}) + 2(\text{Bi}) + 3(\text{Multi})] / 500$$

Necrotic and apoptotic cells are identified and enumerated, but not included among the viable cells scored. The photomicrographs of the different cell types considered in the CBMN Cyt are shown in Figs. 1, 2, 3, 4, and 5. Necrotic cells are identified by the presence of numerous vacuoles in the cytoplasm and sometimes in nuclei and damaged cytoplasmic membranes (Fig. 4). Apoptotic cells are identified by the presence of chromatin condensation and in the late stages by nuclear fragmentation in intact cytoplasm (Fig. 5).

3. Only binucleated cells are scored for DNA damage biomarkers which include micronuclei (MNi), buds, and nucleoplasmic bridges (NPBs). A minimum of 1,000 binucleated cells for each culture needs to be scored (*see* **Note 8**).

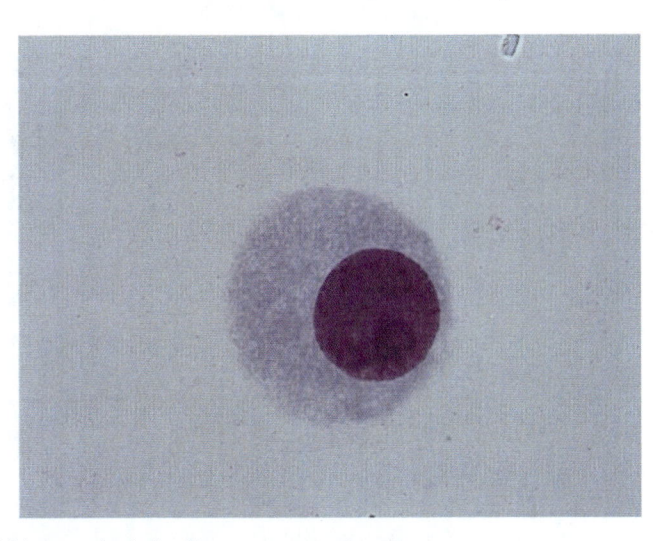

Fig. 1 Mononucleated cell

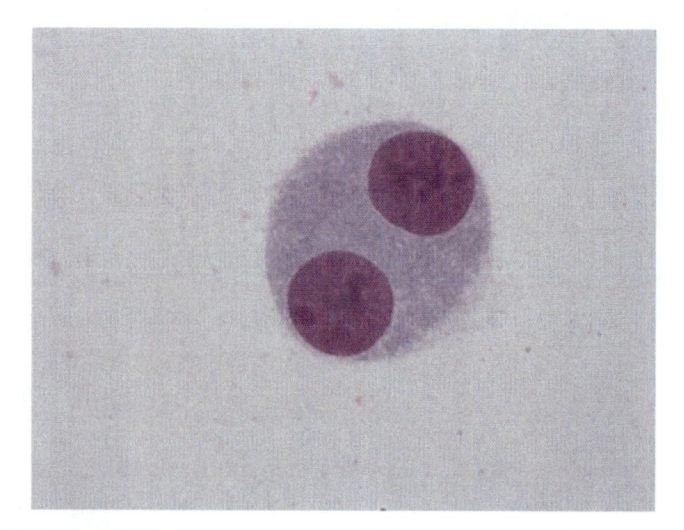

Fig. 2 Binucleated cell

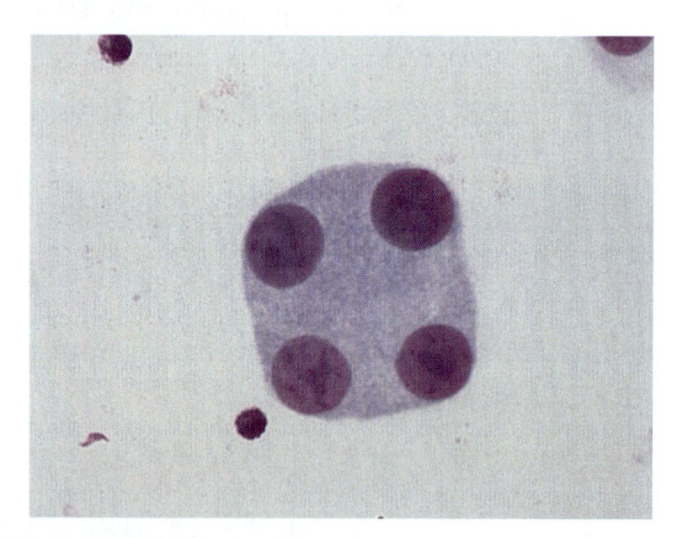

Fig. 3 Multinucleated cell

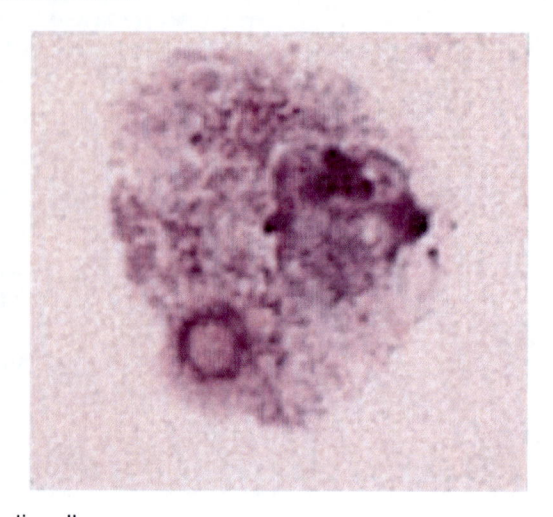

Fig. 4 Necrotic cell

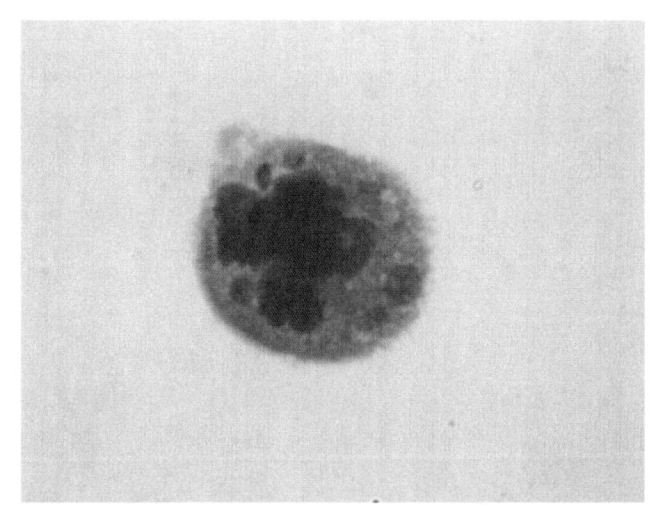

Fig. 5 Apoptotic cell

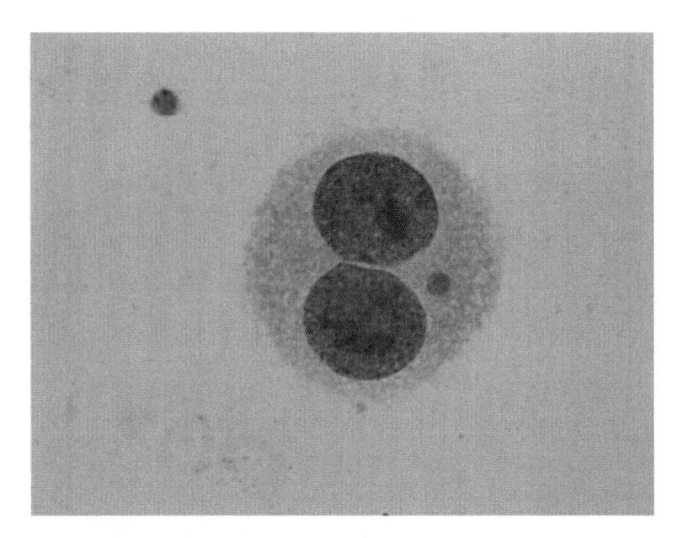

Fig. 6 Binucleate cell with one micronucleus

(a) Binucleated cells have two nuclei equal in size, staining pattern and intensity sometimes touching, but not overlapping, each other (Fig. 2).

(b) *MNi* have the same staining pattern and intensity as the main nuclei. The diameter of MNi ranges from 1/16 to 1/3 of one of the main nuclei. MNi are not linked or connected to the main nuclei and the micronuclear boundary should be distinguishable from the nuclear boundary (Fig. 6).

(c) *Buds* are very similar to MNi with the exception that they are connected to one of the main nuclei. Sometimes the buds appear as a small protrusion of the main nuclei (Fig. 7).

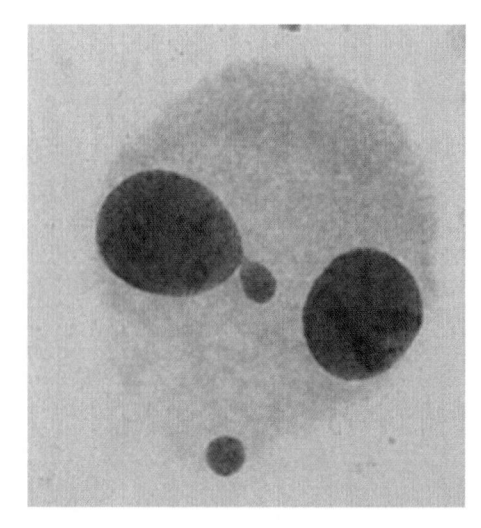

Fig. 7 Binucleated cell with one micronucleus and a bud

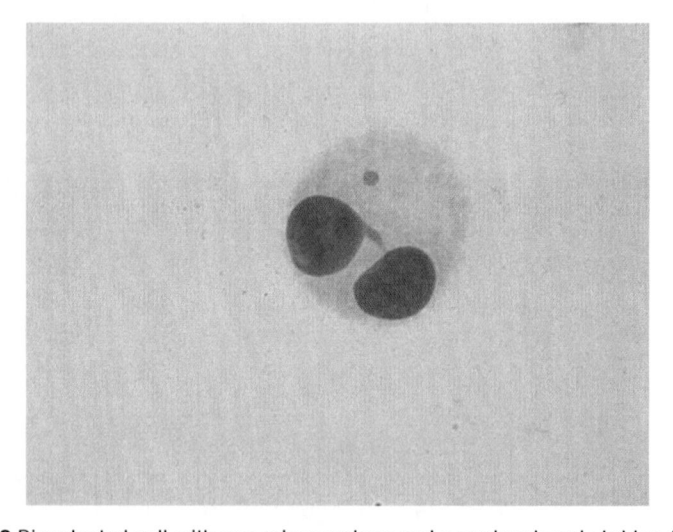

Fig. 8 Binucleated cell with one micronucleus and a nucleoplasmic bridge (NPB)

    (d) *NPBs* are continuous DNA-containing structures linking the two nuclei. The width of an NPB may vary but it should not exceed 1/4 of the diameter of the nuclei (Fig. 8).

3.2 BMNCyt Protocol

3.2.1 Buccal Cell Collection

1. Before starting the cell collection subjects are requested to rinse the mouth two times with 100 mL of water to remove cell debris.
2. Samples are collected by using two small-headed toothbrushes (one for each cheek), which are gently but firmly rotated ten times against the inside of the cheek wall in a circular motion to collect buccal cells.
3. Buccal cell samples are collected into two 30 mL containers labeled LC (left cheek) and RC (right cheek), each containing 20 mL of Saccomanno's fixative (*see* **Note 9**).

4. Buccal cell suspension fixed in Saccomanno's solution can be stored at 4 °C for months.

3.2.2 Buccal Cell Harvesting and Slide Preparation

1. Transfer the cell suspension from the containers into two centrifuge tubes and centrifuge for 10 min at $580 \times g$.

2. Aspirate the supernatant leaving approximately 2 mL of cell suspension. Briefly vortex the cells. Add 8 mL of buccal cell buffer at room temperature and resuspend the cells.

3. Centrifuge the cells for 10 min at $580 \times g$.

4. Aspirate the supernatant leaving approximately 1 mL of cell suspension and add 10 mL of buccal cell buffer.

5. Centrifuge the cell suspension for 10 min at $580 \times g$.

6. Aspirate off the supernatant leaving approximately 1 mL of cell suspension and add 5 mL of buccal cell buffer. Briefly vortex the cells (*see* **Note 10**).

7. Homogenize the cell suspension for 2–3 min using a handheld tissue homogenizer at medium intensity (*see* **Note 11**).

8. To increase the number of clearly separated cells, pass the cell suspension 5–6 times into a syringe using an 18 G needle.

9. Pool the cell suspension from the left and right cheek tubes into a 20 mL syringe, pass the cells through a 100 μm nylon filter held in a swinnex holder, and collect the filtered cells in a 15 mL centrifuge tubes (*see* **Note 12**).

10. Centrifuge the cell suspension for 10 min at $580 \times g$.

11. Remove the supernatant and resuspend the cells in 1 mL of buccal cell buffer.

12. Dilute 100 μL of cell suspension into 900 μL of buccal cell buffer and count the cells using a counting chamber.

13. To further aid in the cellular disaggregation, add 50 μL of DMSO per mL of cell suspension.

14. Fix the cells using the required volume of ethanol:glacial acetic acid 3:1 to give a concentration of 80,000 cells per mL.

15. Using a Pasteur pipette drop 100–150 μL of cell suspension onto a pre-cleaned/ready-to-use microscope slide.

3.2.3 Slide Staining

1. Put the fixed slides (including a spare control slide) for 1 min each in Coplin jars with 50 % and then 20 % ethanol sequentially, and then wash with Milli-Q water for 2 min in Coplin jars (*see* **Note 13**).

2. Immerse the slides in a Coplin jar of 5 M HCl, for 30 min, and then rinse in running tap water for 3 min (since the cells are now fixed they will not be lost while rinsing). Include a negative control with each batch to check for efficacy of 5 M HCl treatment by placing a sample slide in Milli-Q water for 30 min instead of in 5 M HCl (*see* **Note 14**).

3. Drain the slides and place them in Coplin jar containing Schiff's reagent for 60 min in the dark at room temperature. Rinse the slides in running water for 5 min and then rinse again in Milli-Q.

4. Counterstain the slides in 0.2 % light green for 20–30 s and rinse well in Milli-Q water.

5. Immediately place the slides face down on Whatman No. 1 filter paper to blot away any residual moisture.

6. Place the slides on a slide tray and allow the slides to dry for about 20 min.

7. Examine the cells at ×100 and ×400 magnification to assess the efficiency of staining and the density of the cells.

8. Place the slides to be coverslipped on tissue paper and follow **steps 6–9** of Subheading 3.1.3 of the CBMN Cyt protocol.

3.2.4 Slide Scoring for Buccal MN

1. Slides should be coded and scored blindly using transmitted light and under fluorescence microscopy at 1,000× magnification.

2. MN and other nuclear anomalies are identified under transmitted light and confirmed under fluorescence (*see* **Note 15**).

3. As a first step 1,000 cells are scored for viability. The photomicrographs of the different buccal cell types considered in the BMN Cyt are reported in Fig. 9. The frequency of basal cells, differentiated cells, and cells with anomalies associated with cell death is evaluated.

    (a) *Basal cells* are oval or roundish in shape with a large nucleus:cytoplasm ratio relative to differentiated cell. They are typically smaller in size than differentiated cells (Fig. 9a).

    (b) *Differentiated cells* are angular and flat in shape with a smaller nucleus:cytoplasm ratio compared with the basal cell. They are larger than basal cells (Fig. 9b).

    (c) *Buccal cells with condensed chromatin* are differentiated cells with areas of tightly packed chromatin in the nucleus usually in a striated parallel pattern (Fig. 9c).

    (d) *Karyorrhectic cells* are characterized by extensive nuclear chromatin aggregation with a loss of nuclear integrity indicative of nuclear fragmentation (Fig. 9d).

    (e) *Pyknotic cells* have a small shrunken nucleus (diameter ranges from 1/3 to 2/3 of that of a normal nucleus) that is intensely and uniformly stained (Fig. 9e).

    (f) *Karyolytic cells* have nuclei that are completely depleted of DNA which appear as Feulgen-negative ghostlike images (Fig. 9f).

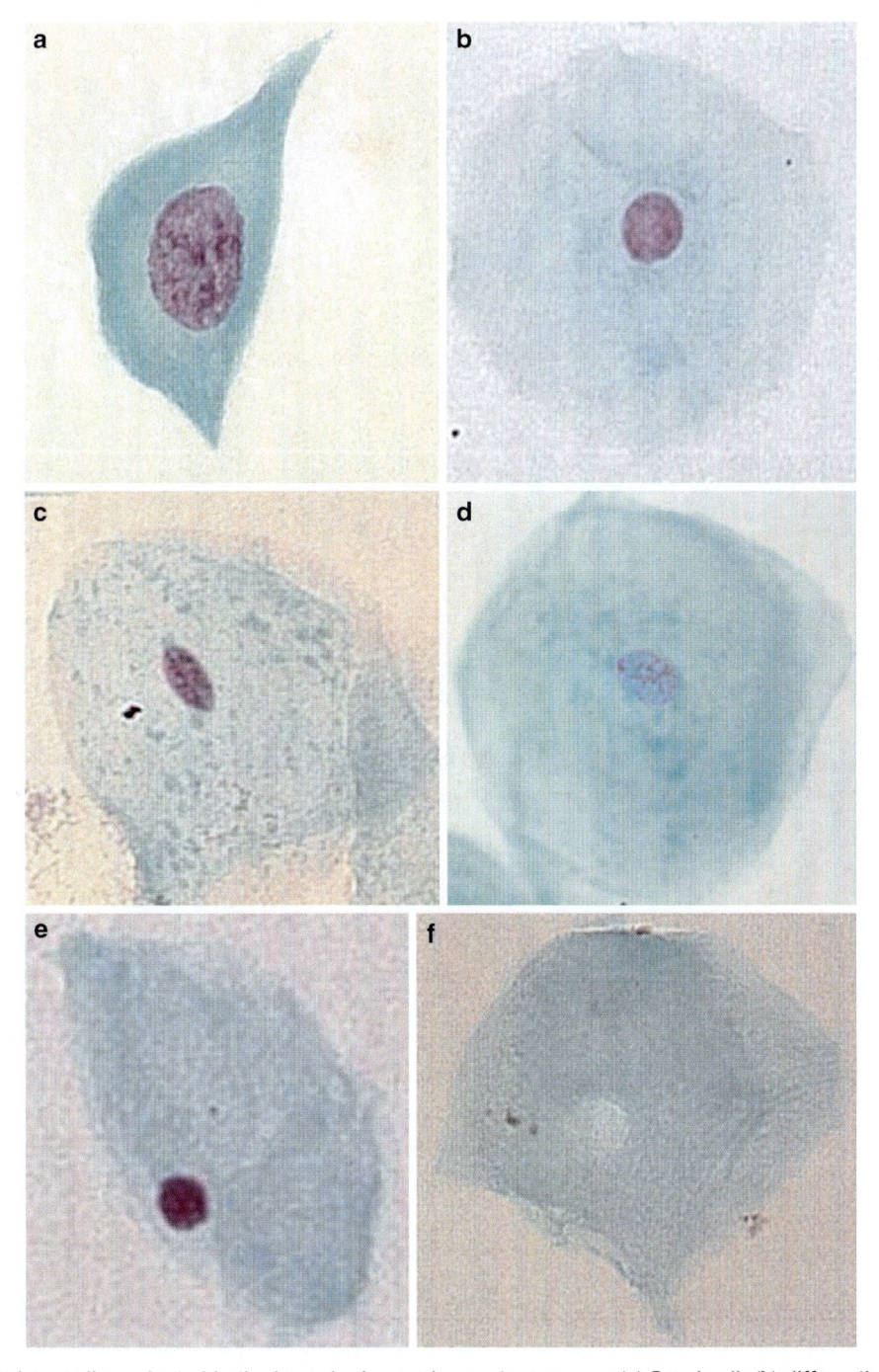

Fig. 9 Various cells evaluated in the buccal micronucleus cytome assay. (**a**) Basal cell; (**b**) differentiated cell; (**c**) buccal cell with condensed chromatin; (**d**) karyorrhectic cell; (**e**) pyknotic cell; (**f**) karyolytic cell

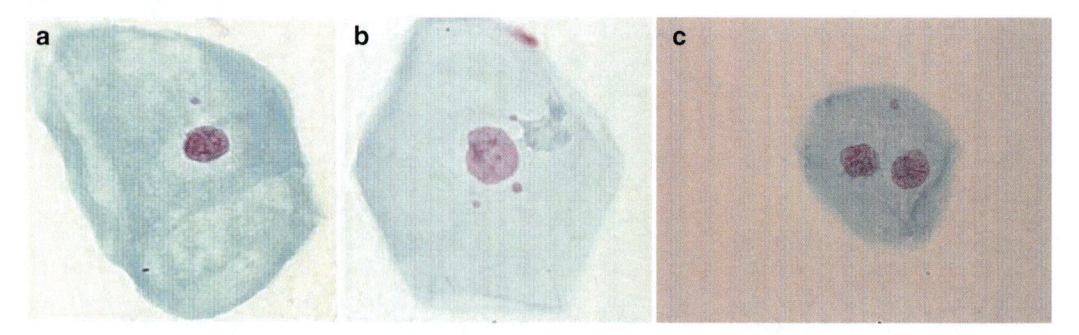

Fig. 10 Micronucleus as observed in the buccal micronucleus cytome assay. (**a**) and (**b**) Mononucleated buccal cell with MN; (**c**) binucleated buccal cell with MN

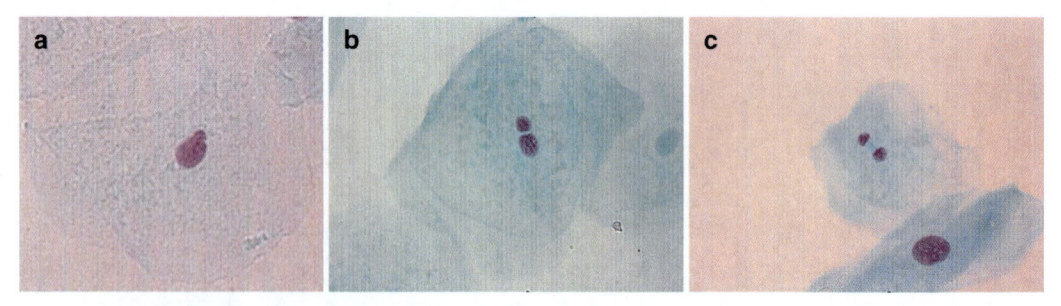

Fig. 11 Buds as observed in the buccal micronucleus cytome assay. (**a**) Mononucleated cell with nuclear bud; (**b**) and (**c**) cells with "broken egg" buds

4. Only differentiated cells are scored for nuclear alterations: MN and buds. A minimum of 2,000 differentiated cells for treatment are suggested to be scored.

    (a) *MNi* are round or oval Feulgen-positive bodies not linked or connected with the main nucleus with the same texture and stain intensity as the main nucleus but with a size that is 1/3–1/16 of the nucleus (Fig. 10a–c).

    (b) *Buds* appear as small nuclear Feulgen-positive bodies connected with the main nucleus or as small protrusions of the nuclei. Buds usually have 1/3–1/16 diameter of the main nucleus but in some cases could be more than 1/2 the main nucleus (broken eggs; Fig. 11a–c).

4 Notes

1. Avoid repeated refreezing and thawing of the FBS.
2. This material is toxic and a possible teratogen. It must always be purchased in sealed vials. The preparation of this reagent must be carried out in a cytoguard cabinet and the following personal protection must be used: Tyvek gown, P2 dust mask, double-nitrile gloves, and safety glasses. *Preparation*: Take the 50 mg vial of Cyt-B from −20 °C and allow it to reach room

temperature. Sterilize the top of the rubber seal with ethanol, but do not remove the seal. Vent the vial seal with a sterile needle and using a syringe add 2.5 mL of sterile DMSO and mix gently (Cyt-B should dissolve readily in DMSO). Remove the 2.5 mL from the vial and eject into a sterile 50 mL Falcon tube. Again add 2.5 mL of sterile DMSO into the vial and remove the contents into the 50 mL Falcon tube. Pipette 20 mL of sterile DMSO into the tube to reach a final volume of 25 mL Cyt-B solution (2 mg/mL) and mix gently. Transfer into a plastic sterile bottle, and add 141.7 mL of 0.9 % isotonic saline to reach a final volume of 166.7 mL. This gives a final concentration of 300 µg/mL Cyt-B solution. Mix and dispense adequate volumes into sterile 2 mL cryogenic capped vials to make multiple aliquots and store at −20 °C for up to 12 months.

3. Blood samples can be stored before culturing up to 24 h between 15 and 22 °C without any significant effect on cell survival, mitogen stimulation, and MN frequency [26]. The time and storage temperature are critical for obtaining a sufficient number of in vitro-dividing lymphocytes.

4. Hypotonic treatment is a critical step for slide preparation using whole blood cultures. Mild treatment is suggested to avoid the loss of necrotic and apoptotic cells and cells with NPBs.

5. The use of prefix provides a mild cell fixation avoiding the formation of cell clumps.

6. This step is introduced to stop or interrupt the process of cell preparation, when large number of samples needs to be processed simultaneously.

7. Detailed scoring criteria established by the HUMN project (http://www.humn.org) were described in previous papers [19, 24]. For a more comprehensive photo gallery refer to Fenech [19] and Fenech et al. [24]. Visual scoring, even though a time-consuming process requiring specific expertise, is still the preferred and only method of scoring slides for the complete set of biomarkers in micronucleus cytome assays. The automated scoring of MN is now possible allowing a high throughput for this specific biomarker and also can reduce the variability due to the subjective evaluation. Several automated image analysis systems were developed and they are at an advanced stage of validation [27–30].

8. In exceptional circumstances, such as folate deficiency, additional nuclear anomalies may be formed defined as "fused" nuclei (FUS), "circular" nuclei (CIR), and "horse-shoe" nuclei (HS) and could be included as additional CIN biomarkers in the lymphocyte CBMN-Cyt assay [31].

9. The method for buccal cell collection should be kept constant in order to obtain samples with a homogeneous distribution of the different cell types. It was observed that repeated vigorous

brushing on the same area can increase the percent of basal cells [25]. The use of Saccomanno's fixative allows preserving the cell suspensions at 4 °C for months before processing.

10. The washing procedure with buffer solution helps to remove bacteria and cell debris, which could confound the scoring.

11. The homogenization step allows single-cell suspension to be obtained.

12. The cell sampling is performed on the inside of both cheeks to maximize cell collection and to obtain a homogeneous cell suspension, avoiding unknown biases that may be caused by sampling one cheek only. After the first steps of cell processing the cell suspensions from both cheeks are pooled together and then filtered to discard cell clumps.

13. The sequential treatment with 50 and 20 % ethanol and then wash with Milli-Q water hydrate the cells before treatment with 5 M HCl.

14. Schiff's reagent is a solution that combines chemically with aldehydes to form a bright red product. The treatment of slides with 5 M HCl converts some of the deoxyribose in DNA to aldehydes which can be then detected by Schiff's reagent.

15. The staining technique is a critical step in BMCyt. The Feulgen–light green procedure is recommended to score the cell types in bright field and reconfirm the nuclear alterations using far-red fluorescence [25]. The use of staining procedures that are not specific for DNA are discouraged because they yield false-positive results [32]. Detailed scoring criteria were described in previous papers [25]. For a more comprehensive photo gallery refer to Thomas et al. (2009) [25].

References

1. Fenech M, Morley AA (1985) Measurement of micronuclei in lymphocytes. Mutat Res 147:29–36
2. Fenech M (2000) Mathematical model of the in vitro micronucleus assay predicts false negative results if micronuclei are not specifically scored in binucleated cells or in cells that have completed one nuclear division. Mutagenesis 15:329–336
3. Fenech M (2006) Cytokinesis-block micronucleus assay evolves into a "cytome" assay of chromosomal instability, mitotic dysfunction and cell death. Mutat Res 600:58–66
4. Fenech M (2001) The role of folic acid and vitamin B12 in genomic stability of human cells. Mutat Res 475:57–67
5. Fenech M, Aitken C, Rinaldi J (1998) Folate, vitamin B12, homocysteine status and DNA damage in young Australian adults. Carcinogenesis 19:1163–1171
6. Thomas P, Wu J, Dhillon V, Fenech M (2011) Effect of dietary intervention on human micronucleus frequency in lymphocytes and buccal cells. Mutagenesis 26(1):69–76
7. Dhillon VS, Thomas P, Iarmarcovai G et al (2011) Genetic polymorphisms of genes involved in DNA repair and metabolism influence micronucleus frequencies in human peripheral blood lymphocytes. Mutagenesis 26:33–42
8. El-Zein R, Vral A, Etzel CJ (2011) Cytokinesis-blocked micronucleus assay and cancer risk assessment. Mutagenesis 26(1):101–106
9. Bonassi S, Znaor A, Ceppi M et al (2007) An increased micronucleus frequency in peripheral blood lymphocytes predicts the risk of

cancer in humans. Carcinogenesis 28: 625–631
10. Bonassi S, El-Zein R, Bolognesi C et al (2011) Micronuclei frequency in peripheral blood lymphocytes and cancer risk: evidence from human studies. Mutagenesis 26(1):93–100
11. Murgia E, Ballardin M, Bonassi S et al (2008) Validation of micronuclei frequency in peripheral blood lymphocytes as early cancer risk biomarker in a nested case-control study. Mutat Res 639:27–34
12. Migliore L, Coppedè F, Fenech M et al (2011) Association of micronucleus frequency with neurodegenerative diseases. Mutagenesis 26(1):85–92
13. Andreassi MG, Barale R, Iozzo P et al (2011) The association of micronucleus frequency with obesity, diabetes and cardiovascular disease. Mutagenesis 26(1):77–83
14. Holland N, Bolognesi C, Kirsch-Volders M et al (2008) The micronucleus assay in human buccal cells as a tool for biomonitoring DNA damage: the HUMN project perspective on current status and knowledge gaps. Mutat Res 659:93–108
15. Kirsch-Volders M, Mateuca RA, Roelants M et al (2006) The effects of GSTM1 and GSTT1 polymorphisms on micronucleus frequencies in human lymphocytes in vivo. Cancer Epidemiol Biomarkers Prev 15:1038–1042
16. Bonassi S, Neri M, Lando C et al (2003) HUMN collaborative group effect of smoking habit on the frequency of micronuclei in human lymphocytes: results from the Human MicroNucleus project. Mutat Res 543:155–166
17. Fenech M, Bonassi S (2011) The effect of age, gender, diet and lifestyle on DNA damage measured using micronucleus frequency in human peripheral blood lymphocytes. Mutagenesis 26(1):43–49
18. Fenech M, Holland N, Chang WP et al (1999) The HUman MicroNucleus Project: an international collaborative study on the use of the micronucleus technique for measuring DNA damage in humans. Mutat Res 428:271–283
19. Fenech M, Chang WP, Kirsch-Volders M et al (2003) HUman MicronNucleus project. HUMN project: detailed description of the scoring criteria for the cytokinesis-block micronucleus assay using isolated human lymphocyte cultures. Mutat Res 534:65–75
20. Bonassi S, Fenech M, Lando C et al (2001) HUman MicroNucleus project: international database comparison for results with the cytokinesis-block micronucleus assay in human lymphocytes—I. Effect of laboratory protocol, scoring criteria, and host factors on the frequency of micronuclei. Environ Mol Mutagen 37(1):31–45
21. Fenech M, Bonassi S, Turner J et al (2003) HUman MicroNucleus project, intra- and inter-laboratory variation in the scoring of micronuclei and nucleoplasmic bridges in binucleated human lymphocytes. Results of an international slide–scoring exercise by the HUMN project. Mutat Res 534:45–64
22. Fenech M, Bolognesi C, Kirsch-Volders M et al (2007) Harmonisation of the micronucleus assay in human buccal cells: a human micronucleus (HUMN) project (www.humn.org) initiative commencing in 2007. Mutagenesis 22:3–4
23. Bonassi S, Biasotti B, Kirsch-Volders M et al (2009) State of the art survey of the buccal micronucleus assay: a first stage in the HUMN(XL) project initiative. Mutagenesis 24:295–302
24. Fenech M (2007) Cytokinesis-block micronucleus cytome assay. Nat Protoc 2(5):1084–1104
25. Thomas P, Holland N, Bolognesi C et al (2009) Buccal micronucleus cytome assay. Nat Protoc 4:825–837
26. Lee TK, O'Brien K, Eaves GS et al (1999) Effect of blood storage on radiation-induced micronuclei in human lymphocytes. Mutat Res 444:201–206
27. Decordier I, Papine A, Vande LK et al (2011) Automated image analysis of micronuclei by IMSTAR for biomonitoring. Mutagenesis 26:163–168
28. Darzynkiewicz Z, Smolewski P, Holden E et al (2011) Laser scanning cytometry for automation of the micronucleus assay. Mutagenesis 26:153–161
29. Varga D, Johannes T, Jainta S et al (2004) An automated scoring procedure for the micronucleus test by image analysis. Mutagenesis 19:391–397
30. Bolognesi C, Balia C, Roggieri P et al (2011) Micronucleus test for radiation biodosimetry in mass casualty events: evaluation of visual and automated scoring. Rad Measurements 46:169–175
31. Bull CF, Mayrhofer G, Zeegers D et al (2012) Folate deficiency is associated with the formation of complex nuclear anomalies in the cytokinesis-block micronucleus cytome assay. Environ Mol Mutagen 53:311–323
32. Nersesyan A, Kundi M, Atefie K et al (2006) Effect of staining procedures on the results of micronucleus assays with exfoliated oral mucosa cells. Cancer Epidemiol Biomarkers Prev 15:1835–1840

Chapter 11

Flow Cytometric Determination of Micronucleus Frequency

Azeddine Elhajouji and Magdalena Lukamowicz-Rajska

Abstract

During the last two decades the micronucleus (MN) test has been extensively used as a genotoxicity screening tool of chemicals and in a variety of exploratory and mechanistic investigations. The MN is a biomarker for chromosomal damage or mitotic abnormalities, since it can originate from chromosome fragments or whole chromosomes that fail to be incorporated into daughter nuclei during mitosis (Fenech et al., Mutagenesis 26:125–132, 2011; Kirsch-Volders et al., Arch Toxicol 85:873–899, 2011). The simplicity of scoring, accuracy, amenability to automation by image analysis or flow cytometry, and readiness to be applied to a variety of cell types either in vitro or in vivo have made it a versatile tool that has contributed to a large extent in our understanding of key toxicological issues related to genotoxins and their effects at the cellular and organism levels. Recently, the final acceptance of the in vitro MN test guideline 487 (OECD Guideline for Testing of Chemicals, In vitro mammalian cell micronucleus test 487. In vitro mammalian cell micronucleus test (MNVIT). Organization for Economic Cooperation and Development, Paris, 2010) together with the standard in vivo MN test OECD guideline 474 (OECD Guideline for The Testing of Chemicals, Mammalian erythrocyte micronucleus test no. 474. Organization for Economic Cooperation and Development, Paris, 1997) will further position the assay as a key driver in the determination of the genotoxicity potential in exploratory research as well as in the regulatory environment. This chapter covers to some extent the protocol designs and experimental steps necessary for a successful performance of the MN test and an accurate analysis of the MN by the flow cytometry technique.

Key words Micronucleus test, In vitro, In vivo, Flow cytometry, Genotoxicity, Human lymphocytes, Rodents

1 Introduction

In recent years a lot of attention has been paid to the micronucleus test (MNT) in terms of measurement of genotoxic potential of chemicals, radiation bio-dosimetry, and population biomonitoring. Having the potential to detect accurately both clastogenic and aneugenic chemicals, the assay is widely used in a variety of cell systems [1, 2]. Although the in vivo rodent MNT (OECD guideline 474) [3] is a part of the genoxtoxicity testing battery since years, the in vitro (OECD guideline No. 487) [4] was recently accepted by the ICH guideline as an alternate for chromosome aberration test [5] and was thoroughly validated by European Center for the Validation of Alternative Methods (ECVAM) workshop participants [6]. The MNT can be performed in vitro (using various cell systems, with and without metabolic activation incorporation) and in vivo (most commonly using rodent peripheral erythrocytes or bone marrow mainly using rats or mice; or in biomonitoring studies in humans). However micronuclei are expressed only in the population of cells that completed mitosis. Therefore, for the in vitro MNT test, demonstration of the cell division is mandatory. Already in 1985 Fenech and Morley [7] presented a cytochalasin B (Cyto B) micronucleus test that currently is a "gold standard" assay for in vitro primary human lymphocyte (HuLy) MNT. Cyto B is an inhibitor of actin polymerization that induces cytokinesis block [6–8]. Therefore, use of Cyto B enables proliferation monitoring. Cells that did not go through mitosis in culture will appear as mononucleated cells, the cells that divided once will be binucleated, and the cells that divided twice or more times will be polynucleated [7]. As a consequence, cytotoxicity assessment and the MN induction may be analyzed simultaneously. The methodology is, additionally, a very attractive tool for the determination of mitotic slippage [9], apoptosis [10], and cell proliferation [11]. Moreover, the Cyto B HuLy MNT may be regarded as a multiparametric "cytome" assay aiming at the determination of not only MN frequency and division status but also detection of nuclear buds, nucleoplasmic bridges, and apoptotic or necrotic cells [12]. Nucleoplasmic bridges may serve as direct evidence for DNA mis-repaired breaks or may help in identification of mechanism DNA repair or pathway defects.

1.1 In Vitro Micronucleus Test

In the option 1 of the International Conference on Harmonization (ICH) Guidance (S2)R1 on Genotoxicity Testing and Data Interpretation for Pharmaceuticals Intended for Human Use [5], the in vitro chromosomal aberration test or the in vitro micronucleus assay should be used. However, a high number of irrelevant positive results in comparison to in vivo rodent genotoxicity tests (referred to as false-positive results) have been reported. The reason for in vitro false-positive calls may be the application of an under/oversensitive cytotoxicity parameter or use of DNA repair or p53-deficient cell systems. Deficiency in the phase I and phase II metabolism may also find its contribution to the problem. Misleading in vitro results lead to compound misclassification and as a result to excessive in vivo testing. As agreed during the ECVAM workshop, the best hope for the reduction of false-positive results is applying the test systems, which are p53 and DNA repair proficient and have defined phase 1 and phase 2 metabolism, covering a broad set of enzyme forms. There is evidence that primary HuLy as well as other human cell-based systems (HepG2, TK6, MCL-5, and 3D skin models, primary cells, or co-culturing with metabolically competent cells) may show a reduced susceptibility to oversensitivity when compared to rodent cell line systems [13]. Ideally cells used for in vitro testing should fulfill all expectations described by ECVAM workshop participants. Moreover the cells should be of human origin and be karyotypically stable. The aim for the in vitro assay is to detect the majority of genetic endpoints with a relatively high specificity (without a decrease in the ability to detect the in vivo genotoxins and DNA-reactive, mutagenic carcinogens; [13]).

Conveniently, automation of MNT is feasible and already used routinely [14–34]. Presently, two types of automated MNTs are used to increase the throughput of tested compounds and decrease the time required for analysis. The advantages of both systems are the increase in analysis speed and elimination of scorer subjectivity present during nonautomated slide examination. The first automation approach includes image analysis [15, 16, 19–21, 30–32]. This technology is based on the computerized microscopic analysis of slides. A different MNT automation approach is flow cytometry, where nuclei and MN particles are analyzed as separate events following the cell lysis [14, 17, 18, 22–29, 33, 34]. Initially, Nüsse and Kramer [25] using ethidium bromide (EB) staining developed a two-step method to improve the separation of small MN from the main nuclei for flow cytometry-based MNT. Since the publication, a number of groups attempted to improve the staining or the analysis techniques. In order to remove artifacts and debris, a combination of Hoechst 33342 for DNA staining and Thiazole Orange, for RNA staining, have been used [26]. Another group using double-membrane staining significantly improved the separation of MN from nonspecific debris originating from membrane fragments [34]. Moreover, the analysis methodology introducing height and area values of the EB fluorescence enhanced the debris separation [27]. A later study approached the mathematical calculation of energy transfer between EB and Hoechst 33258 as an improvement of the data analysis [28]. Nevertheless, the main disadvantage of the flow cytometry-based MNT was the necessity to ascertain that the level of apoptosis in the culture is not excessive. A high rate of cell death may induce an increased rate of DNA-containing particles, originating from the, e.g., apoptotic bodies that show the same characteristics as MN and potentially may interfere with the MN gating area. In this context, promising results have been presented by Viaggi et al. [33] that improved the separation of apoptotic and necrotic HuLy by incorporation of magnetic separation. In this method, prior to staining procedures, the lymphocyte cell cultures were separated using antibody-conjugated (anti-CD2+) magnetic beads. Such a procedure enabled separation of lymphocytes from the artifacts and debris resulting in a significant reduction of MN-like events. Another approach was the staining suggested by Avlasevich et al. [14] and further developed and improved by Bryce et al. [17, 18].

The method successfully reduced the number of apoptotic bodies and necrotic cells from the MN scoring region by incorporation of membrane-impermeable dye—ethidium monoazide (EMA). The great advantage of the EMA incorporation is its covalent binding to the DNA upon photo-activation with a visible light source [35]. Due to this chemical characteristic DNA origination from the dead and dying cells remains stained with the dye and therefore may be excluded from the final MN gating area. The same dye was used to identify DNA originating from apoptotic/necrotic cells in primary HuLy micronucleus test described by our laboratory [22]. The assay incorporates the carboxy fluorescein succinimidyl ester (CFSE) staining as a division determination marker. Moreover, the technology provided a very high correlation with the "gold standard" Cyto B assay. In this approach authors apply additionally DAPI as a DNA dye. Recently, this technology was further validated using 30 compounds and miniaturized to meet the expectations of a high throughput and decrease the analysis time [22, 24].

1.2 In Vivo Micronucleus Test

The in vivo MNT is well established as a standard in vivo assay for genotoxicity assessment at the chromosomal level [36, 37], the mouse and rat being the most used species. Peripheral blood erythrocytes have been accepted as an appropriate target for micronucleus (MN) assessment for both acute and cumulative damage when animals are treated continuously for 4 weeks or more [4, 5]. When a bone marrow erythroblast develops into an immature erythrocyte, the main nucleus is extruded and the micronucleus that has been formed may remain behind in the cytoplasm. Detection of micronuclei is facilitated in these cells because they lack a main nucleus and an increase in the frequency of micronucleated immature erythrocytes in treated animals is an indication of induced chromosomal damage. Newly formed micronucleated immature erythrocytes can be assessed directly in the bone marrow or in the peripheral blood of rodents after a specific staining by either manual scoring using a microscope or automated analysis using image analysis or flow cytometry. The rodent in vivo MNT is especially relevant for the assessment of the genotoxic potential of the test compound and for mechanistic investigation because factors of in vivo metabolism, pharmacokinetics, and DNA repair processes are incorporated. The use of flow cytometry to analyze rat peripheral blood erythrocytes for the in vivo micronucleus assay is expected to increase the sensitivity of the test and allow assessment of the genotoxic effects at doses comparable to those that are relevant to human exposure [38–41]. The major advantage of incorporating the MN analysis into conventional chronic toxicology studies is the reduction of the number of animals used and the amount of compound required. This combination would also allow kinetic aspects and general toxicology observations to be used for the interpretation of genotoxicity results. Data from the Collaborative Study Group for the Micronucleus Test [42] suggest that rat peripheral blood is appropriate for the enumeration of MN, if scoring is limited to the youngest fraction of reticulocytes (types I and II) as classified by Vander et al. [43]. By restricting the analysis to the youngest reticulocytes, immature erythrocytes are scored prior to being captured by the spleen [44, 45]. This technique is based on the specific labeling of the young polychromatic cells with CD71 antigen on their surface, which is lost during the maturation process [39–41]. The purpose of this book chapter is to provide protocol guidances and reference for laboratories that plan to adopt the flow cytometry technology for either in vitro MN test (both human primary lymphocytes and cell lines) or in vivo in rodents.

2 Materials

2.1 TK6 Cell Line Flow Cytometry-Based Micronucleus Test Protocol

The protocol is based on the methodology using the In vitro MicroFlow kit provided by the Litron Laboratories.

1. Dimethylsulfoxide (DMSO).
2. Culture medium: RPMI 1640, GlutaMAX I, 25 mM HEPES (abbreviated in the text as RPMI 1640), supplemented with 2 % (v/v) penicillin–streptomycin (10,000 IU/mL to 10,000 μg/mL stock), with or without 10 % horse serum (heat inactivated) serum.
3. Staining agent: 12.5 mg/mL ethidium monoazide (EMA) in DMSO, stock solution (Litron Laboratories). Store until use in −20 °C.
4. Lysis solution 1 (Litron Laboratories) stored in 4 °C until use.
5. Lysis solution 2 (Litron Laboratories) stored in 4 °C until use.
6. Pipettes: 2, 5, 10, 25, 50 mL.
7. 25 cm^2 culture flasks.
8. Cell incubator able to provide 37 °C and humid atmosphere with 5 % CO_2.
9. Centrifuge able to provide the 600×g centrifuge conditions.
10. 15 mL Falcon centrifuge polystyrene tubes.
11. 96-multiwell U-bottom culture plates.
12. Flow cytometer equipped with argon (488 nm) laser. The cytometer should use the pulse analysis system, capable of providing a height, width, and area value for each parameter.

    The template preparation should follow the scheme presented in Fig. 1. If the 96-multiwell format protocol is performed, the flow cytometer should be equipped with a 96-multiwell sampler (high-throughput system—HTS) (*see* **Note 1**).
13. Visible light source (60 W light bulb).

2.2 Primary Human Lymphocyte Flow Cytometry-Based Micronucleus Test Protocol

1. Vacutainer CPT tubes (e.g., Becton Dickinson, Franklin Lakes, NJ, USA).
2. Wash buffer solution: Phosphate-buffered saline (PBS) with 2 or 5 % fetal bovine serum (FBS).
3. DMSO.

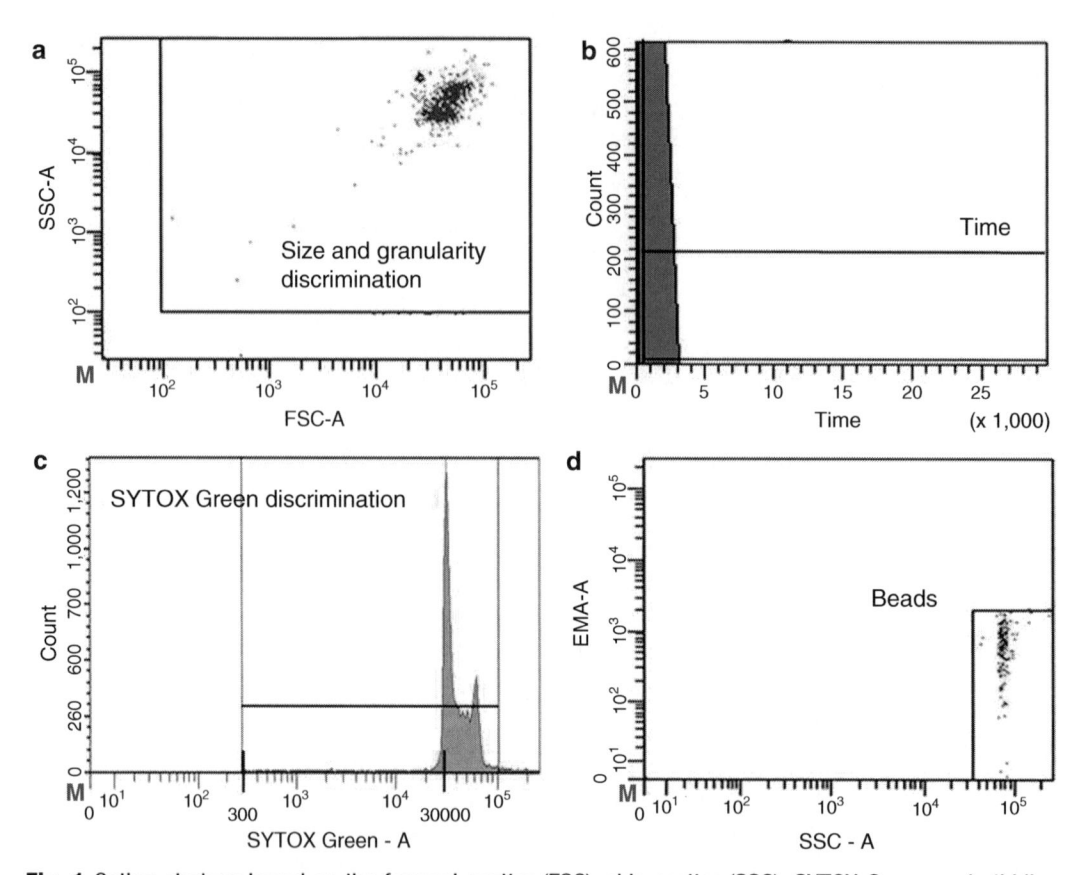

Fig. 1 Gating strategy based on the forward scatter (FSC), side scatter (SSC), SYTOX Green, and ethidium monoazide (EMA)-associated fluorescence. Plot (**a**)—Discrimination based on the size (FSC) and granularity (SSC) of analyzed events; accepted events should show 1 % of size and granularity if compared to the main nuclei population. Plot (**b**)—Based on time parameter events selected should be gated whenever the SYTOX Green signal is stable and may vary dependent on the flow cytometer, and therefore should be assessed individually for each machine. Plot (**c**)—Events that show 1/100 or higher size and SYTOX green fluorescence as the G1 population are accepted. Plot (**d**)—Based on EMA fluorescence and granularity characteristics the 6 μm green fluorescent beads are selected (required for the nuclei/bead cytotoxicity parameter). Plot (**e**)—Classified events should show 1/100 or higher size and SYTOX green fluorescence if compared to G1 population. Plot (**f**)—Classified events should show 1/100 or higher granularity and SYTOX green fluorescence if compared to G1 population. Plot (**g**)—Events expressing positive EMA staining (apoptotic/necrotic events) are excluded from further analysis. Plot (**h**)—Events that are accepted in the gating presented in the Plots (**a**)–(**g**) should be analyzed for the nuclei and MN (1/100th to 1/10th of G1 population SYTOX Green and size characteristics). Plot (**i**)—Based on the SYTOX Green fluorescence events classified as nuclei follow with the analysis in order to determine the number of cells in G1, S, or G2/M cell cycle phase. Adapted from [17, 23]

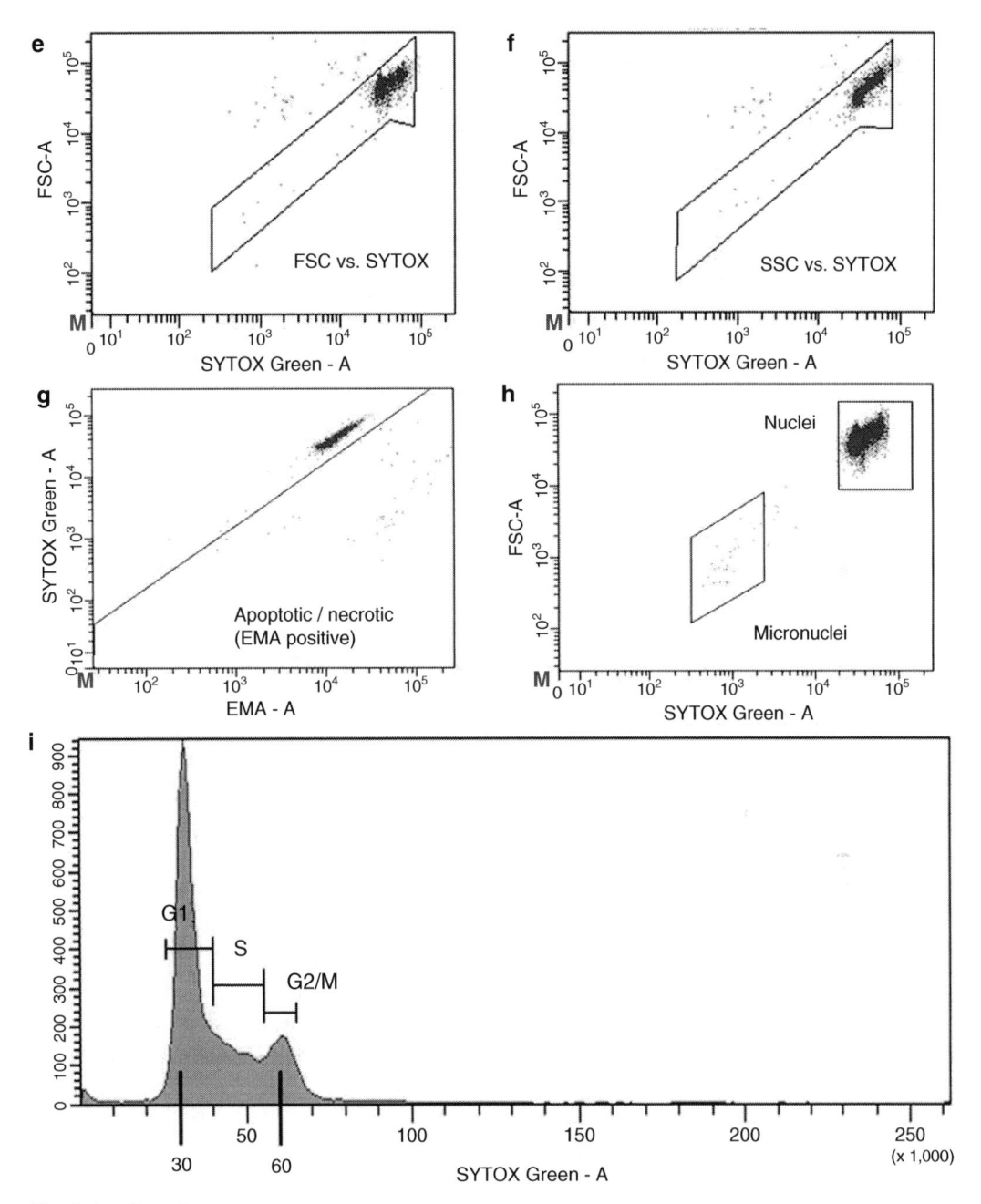

Fig. 1 (continued)

4. Carboxyfluorescein diacetate succinimidyl ester (CFDA-SE): 5 mM CFSE stock solution. Dissolve 25 mg of desiccated CFDA-SE powder (which equates to 44.8 µmol) in 8.96 mL of DMSO. Store in −20 °C.

5. Culture medium: RPMI 1640, GlutaMAX I, 25 mM HEPES, supplemented with 1 % (v/v) penicillin–streptomycin (10,000 IU/mL to 10,000 μg/mL), and 10 or 15 % FBS.

6. 3.75 % Phytohemagglutinin (PHA).

7. Pipettes: 2, 5, 10, 25, 50 mL.

8. 25 cm² culture flasks.

9. 96-multiwell U-bottom culture plates.

10. Cell incubator able to provide 37 °C and humid atmosphere and with 5 % CO_2.

11. Centrifuge able to provide the $1,800 \times g$ centrifuge conditions.

12. 15 mL Falcon centrifuge polystyrene tubes.

13. Staining agent: 12.5 mg/mL EMA in DMSO stock solution. Dilute 5 mg of EMA in 400 μL of DMSO. Store until use in −20 °C.

14. Lysis solution 1: 0.584 mg/mL NaCl, 1 mg/mL sodium citrate, 0.3 μg/mL IGEPAL-630, 250 μg/mLS RNase A, 1 μg/mL DAPI in Millipore water. Filter sterilize the solution through 0.22 μm filters and store in 4 °C until use.

15. Lysis solution 2: 85.6 mg/mL sucrose, 15 mg/mL citric acid, and 1 μg/mL of DAPI in Millipore water. Filter sterilize the solution through 0.22 μm filters and store in 4 °C until use.

16. Flow cytometer as described in **item 14** of Subheading 2.1.

2.3 In Vivo Peripheral Blood Flow Cytometry-Based Micronucleus Test Protocol

The protocol is based on the methodology using the In vivo MicroFlow kit provided by the Litron Laboratories.

1. MicroFlow Micronucleus Analysis Kits (Litron, NY): The kits contain methanol, heparin, washing buffer (Hanks' balanced salt solution; HBSS), RNase, anti-CD71-FITC antibody, anti-platelet antibody (phycoerythrin), propidium iodide (PI), malaria bio-standard blood samples, as well as positive and negative control samples.

2. PBS.

3. Anticoagulant: 500 USP units heparin/mL phosphate buffer solution. Dissolve 263.16 mg of heparin lithium salt (190 USP units) (Sigma Aldrich, Switzerland) in 100 mL of PBS (Dulbecco, Oxoid; to give 500 USP units heparin/mL PBS).

4. Methanol.

5. Eppendorf tubes.

6. Falcon tubes 15 mL.

7. Flow cytometer: As described in **item 14** of Subheading 2.1.

3 Methods

3.1 TK6 Cell Line Flow Cytometry-Based Micronucleus Test Protocol

This protocol can be carried out in TK 6 cells either cultured in 25 cm² flask format or in 96-well plate format as discussed below.

3.1.1 Cell Culturing and Treatment for 25 cm² Flask Cultures (See **Note 2**)

Culture of the cells should be performed 24 h before the scheduled treatment.

1. Estimate the cell number using the cell counter or other reliable assays.
2. Prepare the required volume of cell suspension of 0.2×10^6 cells/mL, in the acclimated culture medium (RPMI 1640 supplemented with 10 % (v/v) heat-inactivated horse serum and 2 % (v/v) penicillin–streptomycin).
3. Prepare at least duplicate cultures in 25 cm² culture flasks in 5 mL of cell suspension per flask.
4. Cultivate the cells for 24 h in 37 °C in humid atmosphere and with 5 % CO_2.
5. Prepare 100× concentrated solution, as compared to the requested test concentration in the appropriate solvent (e.g., DMSO; *see* **Note 3**).
6. 24 h since the culture initiation add 50 µL prepared solutions directly to the cell culture and mix gently by pipetting up and down with 5 mL pipette.

3.1.2 Cell Culture and Treatment for 96-Well Plate Format

1. Cells should be cultivated in the culture medium in a culture flask (e.g., 175 cm² culture flask) 24 h before the requested treatment time.
2. Approximately 1 h before the treatment initiation time, prepare 100× concentrated solution in the appropriate solvent (e.g., DMSO).
3. Using the 100× concentrated solution, prepare the 2× concentrated solution (20 µL of 100× solution + 980 µL culture medium) of the highest and second highest tested concentration in the culture medium (*see* **Note 4**).
4. Add 200 µL per well of the 2× highest and 2× second highest concentration (prepared in **step 2**) to the last and second last rows of wells, respectively.
5. Prepare required volume of the culture medium with 2 % of solvent (e.g., 1 mL of medium + 2 % solvent per concentration × number of tested concentrations).
6. Pipette 100 µL of the medium with 2 % content of the solvent to each well, but NOT in the two highest concentrations.

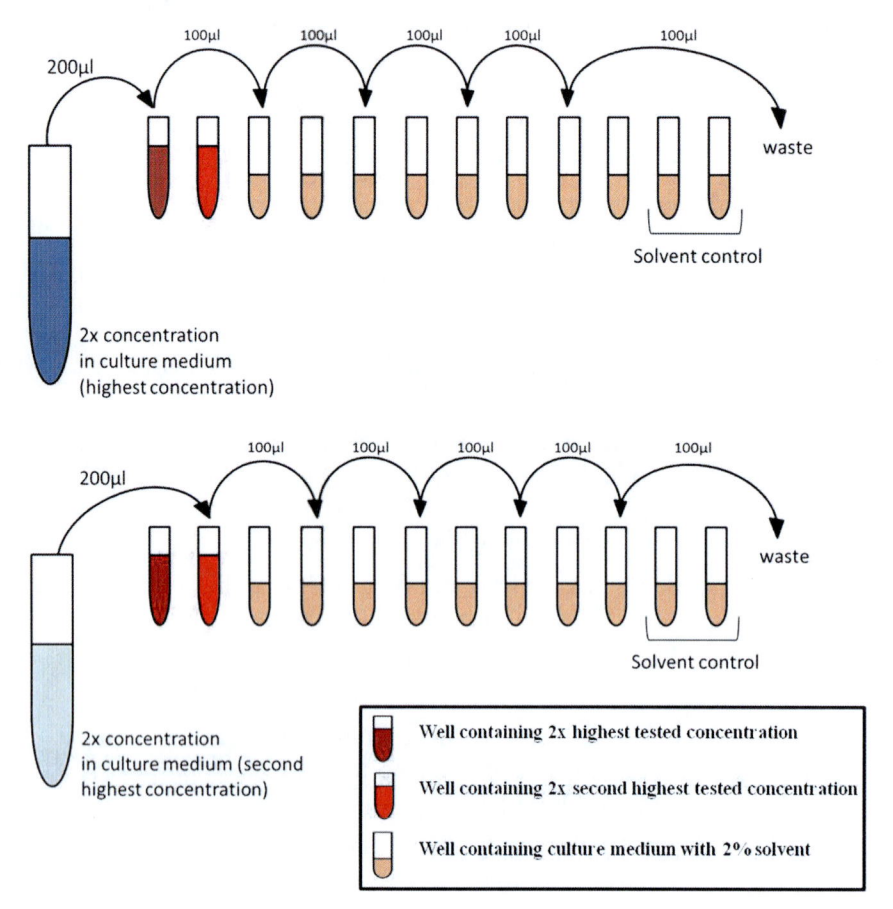

Fig. 2 The serial dilution scheme: Transfer 200 μL of the 2× solution (prepared in culture medium) of the highest and the second highest concentration. As follows transfer 100 μL to every second well mixing well every time. When either of the last or the second last well is mixed aspirate 100 μL and place in the waste container

7. Prepare the serial dilution (Fig. 2) transferring 100 μL of the highest concentration (placed in the last well row) to the third last well row and as follows to the fifth, seventh, etc.

8. Perform the same procedure for the second highest concentration (*see* Fig. 2) by transferring 100 μL of the second highest concentration (placed in the second last well row) to the fourth last well row and as follows to the sixth, eighth, etc.

9. Discard 100 μL from the lowest and second lowest concentration well row (*see* Fig. 2).

10. Prepare the required volume of cell suspension of 0.4×10^6 cells/mL, in the acclimatized culture medium (*see* **Note 5**).

11. Add one drop of green fluorescent latex beads to each 10 mL of cell culture.

12. For multi-well plate cell number assessment, sample approximately 0.5 mL of the volume to at least 3 FACS analysis tubes and analyze using the template as presented in Fig. 3 (*see* **Note 6**).

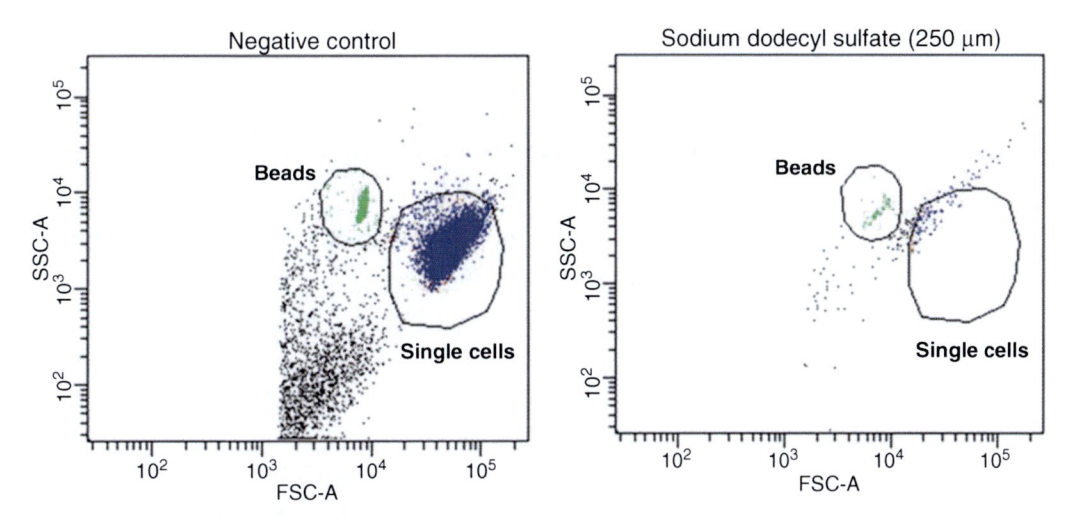

Fig. 3 The cell and bead assessment template: Beads should represent tight and unified population and should be distinctive from the cell population. In order to obtain the highest accuracy in the measurement the single cell population should be located as high on the forward scatter (FSC-A) axis as possible and between third and fourth log of the side scatter (SSC-A) axis

13. Pipette 100 μL of the cell suspension into each of the test wells.
14. Mix the cell suspension with the test item solution previously pipetted into the wells by gently pipetting up and down.
15. Cover the plates and cultivate cell cultures for 24 h in 37 °C in humid atmosphere and with 5 % CO_2 (*see* **Note 7**).

3.1.3 Cell Number Assessment

1. At harvest remove the cultures from the incubator.
2. Mix cells gently by pipetting up and down.
3. For the 25 cm² flask format transfer approximately 0.5 mL of the cell suspension to the flow cytometry tube. For the 96-multiwell format the cell number assessment should be performed directly from the plate. Proceed with the cell and bead number measurement using the template presented in Fig. 3 (*see* **Note 8**).
4. When the analysis is finished, immediately proceed to the steps in Subheading 3.1.5.

3.1.4 Harvest and Staining for 25 cm² Flasks

1. At harvest, transfer the cell cultures into 15 mL centrifuge tubes and centrifuge at $600 \times g$ for 5 min.
2. Discard the supernatant.
3. Resuspend cells by gentle tapping.
4. Place the cell cultures on ice.
5. Prepare EMA (nuclei acid dye A) working solution (*see* **Notes 9 and 10**).
6. Keeping the cells on ice, add 300 μL of EMA solution.

7. Make sure that all cells come into contact with the working solution.
8. Place the tubes on ice and approximately 15 cm under a visible light source and expose to light for 30 min.
9. Remove the light source.
10. Add 5 mL of cold buffer (PBS with 2 % FBS) to each tube.
11. Centrifuge at $600 \times g$ for 5 min.
12. Discard the supernatant.
13. Resuspend cell pellets in the remaining solution volume and maintain at room temperature until the next steps (try not to be later than 30 min).
14. Add 0.5 mL of lysis solution 1 to each tube.
15. Gently mix directly after adding the lysis solution 1.
16. Incubate in darkness at room temperature for 1 h.
17. Add 0.5 mL of lysis solution 2 to each tube.
18. Place the tubes in 4 °C until the analysis.

3.1.5 Harvest and Staining for 96-Well Plate Format

1. At harvest, centrifuge the plates at $300 \times g$ for 5 min.
2. Discard the supernatant.
3. Resuspend cells by gentle tapping.
4. Place the cell cultures on ice.
5. Prepare EMA (nuclei acid dye A) working solution (*see* **Notes 9** and **10**).
6. Keeping the cells on ice add 50 µL of EMA solution per well.
7. Make sure that all cells come into contact with the working solution.
8. Place the plates on ice and approximately 15 cm under a visible light source and expose to light for 30 min.
9. Remove the light source.
10. Add 200 µL of cold buffer (PBS with 2 % FBS) to each well.
11. Centrifuge at $300 \times g$ for 5 min.
12. Discard the supernatant.
13. Resuspend the cell pellets and maintain at room temperature until the next steps (try not to be longer than 30 min).
14. Add 100 µL of lysis solution 1 to each well.
15. Gently mix directly after adding the lysis solution 1.
16. Incubate in darkness at room temperature for 1 h.
17. Add 100 µL of lysis solution 2 to each well.
18. Place the plates in 4 °C until the analysis.

3.1.6 Flow Cytometry Analysis

Analyze the tubes and plates in a flow cytometer using the template in Fig. 1 (for the 96-well plates *see* **Note 11**).

3.1.7 Data Analysis and Interpretation

1. In order to define the genotoxic potential of the tested compound, the number of scored MN (Fig. 1, plot h) should be referred to the number of scored nuclei (events accepted in the gating strategy presented in Fig. 1, plot a–g) multiplied by 100 %. So obtained MN [%] frequency should be expressed as relative value of the negative control sample result.

2. EMA-positive events could serve as the apoptosis/necrosis control.

3. Cytotoxicity parameters: Cell/bead ratio (the ratio of single cells to beads), nuclei/bead ratio (number of scored nuclei, plot h, Fig. 1) referred to the number of scored beads (plot d, Fig. 1), or any assay performed additionally may be used. However, as recommended by OECD Guideline for testing of chemicals [4] highest tested concentration should aim 45 ± 5 % of cell survival. In order to determine the most reliable cytotoxicity parameter the validation within the laboratory using the standard positive and negative controls is recommended [17, 18, 23].

4. An example of the results obtained, where samples were treated with different compounds, is given below (Fig. 4). All samples were compared to negative control sample.

    (a) 125 µM sodium dodecyl sulfate (SDS), a non-genotoxic cell death inducer that results in increase in EMA-positive event frequency, however with no genotoxic effect observed.

    (b) 0.03 µg/mL mitomycin C (MMC) classified as a clastogen for which increase in EMA-positive events and MN frequency increase should be noted.

    (c) 0.029 µM colchicine—an aneugen. An induction of cell death (EMA-positive events) and genotoxic effect (MN frequency increase) should be observed.

3.2 Primary Human Lymphocyte Flow Cytometry-Based Micronucleus Test Protocol

The primary HuLy MNT requires HuLy culturing and therefore Biohazard Level 2 safety laboratory standards should be applied before the procedure may be performed. The methodology is based on the protocol developed in house and published [22, 24].

3.2.1 Blood Sampling

1. Sample the blood directly into the vacutainer CPT tubes.

2. Centrifuge the tubes at $1,800 \times g$ for 15 min or at $1,500 \times g$ for 20 min.

3. Gently remove the tubes from the centrifuge.

4. Aspirate the cloudy peripheral blood mononucleated cell (PBMC) layer and place it in the separate 15 mL polystyrene centrifuge tube with cap (*see* **Note 12**).

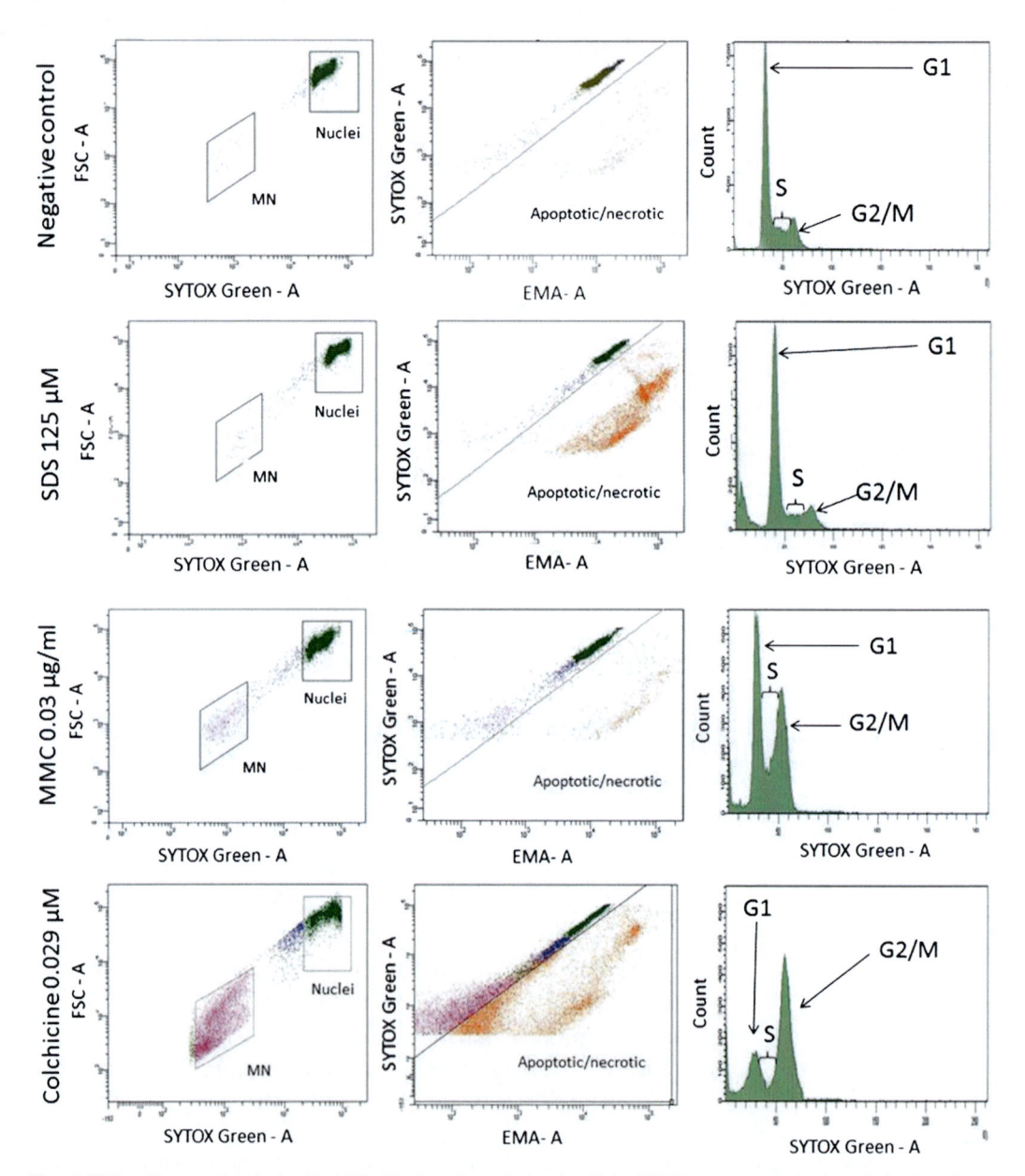

Fig. 4 TK6 cells were treated with 125 μM of sodium dodecyl sulfate (SDS), a non-genotoxic cell death inducer, 0.03 μg/mL of mitomycin C (MMC) classified as a clastogen, and an aneugen 0.029 μM of colchicine. FSC-forward scatter, SSC-side scatter, EMA-etidium monoazide, MN-micronuclei, G1, S, G2M-indicate the cell cycle. The letter "A" next to the parameter on the X- or Y-axis indicates the "area value" of the flow cytometry parameter (e.g., EMA-A means EMA-Area)

3.2.2 *Culture Initiation and CFSE Staining Protocol for 25 cm² Culture Flask Format (See **Note 13**)*

1. Wash PBMCs three times with PBS containing 5 % FBS, each time centrifuging at $300 \times g$ for 7 min.
2. Estimate the cell number using the Kova Glasstic slides with grids (Hycor) or other reliable assays.

3. Prepare the auto-fluorescence control cell cultures (0.7×10^6 cells/mL) using the culture medium consisting of RPMI 1640 supplemented with 15 % (v/v) of FBS, 1 % (v/v) penicillin–streptomycin, and 3.75 % PHA in two 25 cm² culture flasks with 5 mL of cell culture per flask.

4. Prepare the cell suspension of 100×10^6 cells/mL.

5. Prepare 2× concentrated working solution of CFSE by mixing 10 μL of 5 mM CFSE stock solution with 5 mL of PBS.

6. Add 2× concentrated CFSE working solution to the cell suspension in equal volume.

7. Incubate for 10 min at room temperature in darkness. Mix gently once or twice.

8. Block reaction by adding maximal possible volume of PBS with 5 % FBS at 4 °C.

9. Immediately centrifuge cells at $300 \times g$ for 5 min.

10. Discard the supernatant.

11. Repeat the washing step two more times.

12. Resuspend cells in the culture medium without PHA.

13. Estimate the cell number using the Kova Glasstic slides with grids (Hycor) or other reliable assays.

14. Prepare the nondividing control cell cultures (0.7×10^6 cells/mL) using the culture medium without PHA (consists of RPMI 1640 medium supplemented with 15 % (v/v) of FBS and 1 % (v/v) penicillin–streptomycin) in two 25 cm² culture flasks/5 mL of cell culture per flask.

15. Use the remaining PBMCs to prepare cultures of 0.7×10^6 cells/mL, for each of the two 25 cm² culture flasks per tested concentration (5 mL of cell culture per flask). At least one culture set should be reserved for the positive control compound.

16. Cultivate cells for 24 h at 37 °C in humid atmosphere and with 5 % CO_2.

3.2.3 Culture Initiation and CFSE Staining Protocol for 96-Multiwell U-Bottom Culture

1. Follow **steps 1** and **2** of Subheading 3.2.2.

2. Prepare the auto-fluorescence control cell cultures as given in **step 3** of Subheading 3.2.2 for four wells of 96-multiwell U-bottom plate with 300 μL of cell culture per well.

3. Follow through **steps 4–12** of Subheading 3.2.2.

4. Prepare the nondividing control cell cultures (0.7×10^6 cells/mL) using the culture medium without PHA consisting of RPMI 1640 supplemented with 15 % (v/v) of FBS and 1 % (v/v) penicillin–streptomycin, in four wells of 96-multiwell U-bottom plate in 300 μL of cell culture per well.

5. Use the remaining PBMCs to prepare cultures of 0.7×10^6 cells/mL for ≥4 wells of 96-multiwell U-bottom plate per tested concentration in 300 μL of cell culture per well. At least one culture set should be reserved for the positive control testing.

6. Cultivate cells for 24 h in 37 °C in humid atmosphere and with 5 % CO_2.

3.2.4 Treatment of Cells in 25 cm² Flask Cultures

1. Approximately 1 h before the treatment time, prepare the compound solutions.

2. Prepare 100× concentrated solution; the test compound should be dissolved in the appropriate solvent (e.g., DMSO; see **Note 3**).

3. At 24 h after culture initiation, add prepared solutions (for volume recommendation see **Note 3**) directly to the cell culture and mix gently by pipetting up and down with 5 mL pipette.

4. For a noncontinuous treatment (e.g., 3-h treatment with the recovery time) see **Note 14**.

3.2.5 Treatment of Cells in 96-Multiwell Format

1. Prepare 100× concentrated solution of requested treatment concentration in the appropriate solvent. As follows, dilute prepared solution 25 times (four times concentrated if compared to final requested concentration) using culture medium (960 μL of culture medium + 40 μL of 100× concentrated solution).

2. Place prepared solution in cell culture incubator, in order to acclimate to 37 °C.

3. At the treatment time, centrifuge the plate cell cultures for 5 min at $300 \times g$.

4. Gently remove the plates from the centrifuge and aspirate 75 μL of culture medium (see **Note 15**).

5. Add 75 μL of the prepared treatment solutions and acclimated to 37 °C to the cell cultures and mix gently by pipetting up and down.

6. Place the culture in the cell incubator and maintain until next steps.

7. For non-continuous treatment in 96-well format, e.g., 3-h treatment with the recovery time, see **Note 16**.

3.2.6 Harvest and Staining for Cultures in 25 cm² Flasks

1. Remove the EMA stock solution and DAPI stock solution from −20 °C to thaw.

2. At harvest, transfer the cell cultures into 15 mL centrifuge tubes and centrifuge at $600 \times g$ for 5 min.

3. Discard the supernatant.

4. Resuspend cells by gentle tube tapping.
5. Place the cell cultures on ice.
6. Prepare required, calculated amount of EMA working solution: Dilute EMA stock solution 100 times in PBS containing 2 % of FBS.
7. Keeping the cells on ice, add 300 μL of EMA solution.
8. Make sure that all cells come into contact with the working solution.
9. Place the tubes under a visible light source and expose to light for 25 min.
10. Remove the light source.
11. Add 5 mL of cold PBS buffer (PBS with 2 % FBS) to each tube.
12. Centrifuge at $600 \times g$ for 5 min.
13. Discard the supernatant.
14. Resuspend the cells in the remaining solution volume and maintain at room temperature until the next steps (try not to be longer than 30 min).
15. Add 0.5 mL of lysis solution 1 in each tube. Gently mix directly after adding the lysis solution 1.
16. Incubate in darkness, at room temperature for 1 h.
17. Mix gently and incubate for another 30 min.
18. 10 min before the end of incubation time, adjust the pH of the lysis solution 2 to reach pH 8–10.
19. Add 0.5 mL lysis solution 2 in each tube.
20. Analyze the specimens directly after the incubation time using the template in Fig. 5, *see* **Note 17**.

3.2.7 Harvest and Staining for Cultures in 96-Well Format

1. Remove the EMA stock solution and DAPI stock solution from −20 °C to thaw.
2. At harvest, centrifuge the plates at $300 \times g$ for 5 min.
3. Discard the supernatant.
4. Resuspend the cells by gentle plate tapping.
5. Place the cell cultures on ice.
6. Prepare EMA working solution: Dilute EMA stock solution 100 times in PBS with 2 % of FBS.
7. Keeping the cells on ice add 60 μL of EMA solution.
8. Make sure that all cells come into contact with the working solution.
9. Place the plates under a visible light source and expose to light for 25 min.

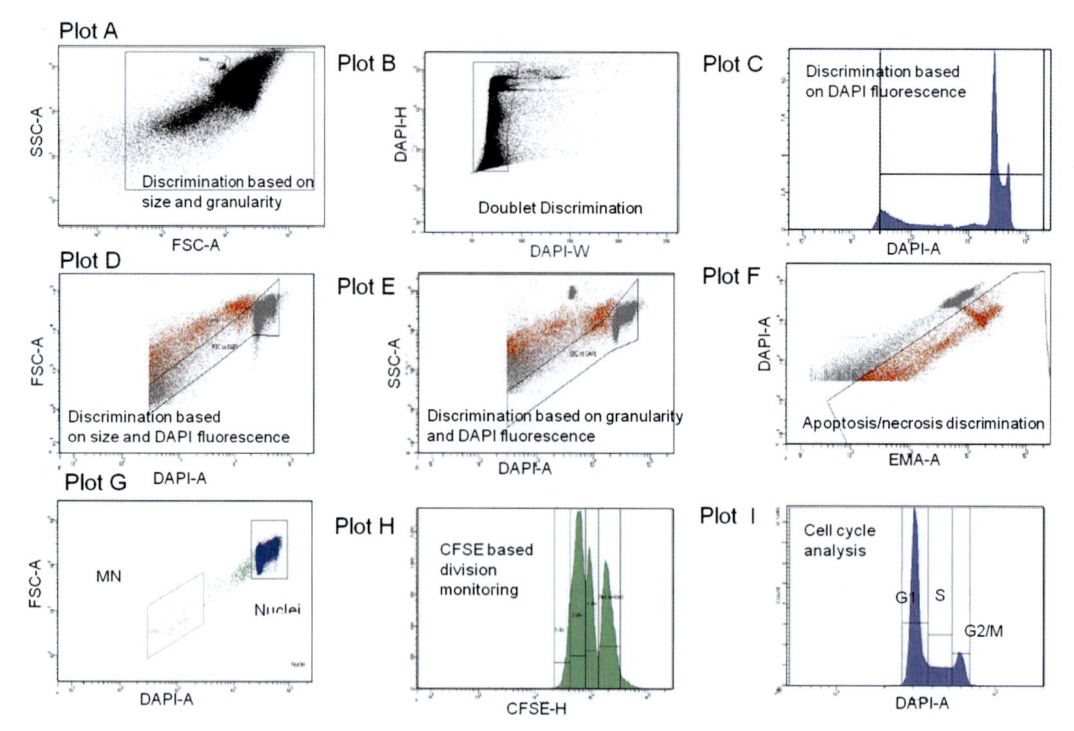

Fig. 5 The template preparation and gating strategy based on FSC, SSC, DAPI fluorescence, EMA fluorescence, CFSE fluorescence. Plot (**a**) (FSC-A vs. SSC-A): Discrimination based on size (FSC) and granularity (SSC). Plot (**b**) (DAPI-W vs. DAPI-A): Discrimination based on an increased value of the DAPI fluorescence. Plot (**c**) (histogram on DAPI-H): Excludes the events that show DAPI fluorescence lower than 1 % of the G1 population. Plot (**d**) (DAPI-A vs. FSC-A): The FSC vs. DAPI gate discrimination based on size and the fluorescence of 1 % of the G1 population. Plot (**e**) (DAPI-A vs. SSC-A): The SSC vs. DAPI gate includes all nuclei and all events that show the granularity and the fluorescence of 1 % of the G1 population. Plot (**f**) (EMA-A vs. DAPI-A): The EMA-positive and EMA-negative gating strategy exclude the apoptotic/necrotic events from the final population. Plot (**g**) (DAPI-A vs. FSC-A): Classified events should fulfill expectations described in Plots (**a**)–(**f**). MNs were defined as events between 1 and 10 % of DAPI-associated fluorescence of the main 2n nuclei. Plot (**h**) (Histogram CFSE-H): Nuclei accepted in "nuclei" gate were analyzed for proliferation status (reduced CFSE fluorescence). Plot (**i**) (Histogram DAPI-A): Additionally, nuclei accepted in final classification may be analyzed for DNA content using DAPI fluorescence. Adapted from Lukamowicz et al. (2011) [22]

10. Remove the light source.
11. Add 200 μL of cold buffer (PBS with 2 % FBS) to each well.
12. Centrifuge at $300 \times g$ for 5 min.
13. Discard the supernatant.
14. Resuspend the cells in the remaining solution volume and maintain at room temperature until the next steps (try not to be longer than 30 min).
15. Add 100 μL of lysis solution 1 into each well.
16. Gently mix directly after adding the lysis solution 1.

17. Incubate in darkness in room temperature for 1 h.
18. Mix gently and incubate for another 30 min.
19. 10 min before the end of incubation time adjust the pH of the lysis solution 2 to reach pH 8–10.
20. Add 100 µL of lysis solution 2 to each well.
21. Analyze the specimens directly after the incubation time using the HTS (*see* **Note 18**).

3.3 In Vivo Peripheral Blood Flow Cytometry-Based Micronucleus Test Protocol

The protocol is based on the methodology using the In vivo MicroFlow kit provided by the Litron Laboratories.

3.3.1 Preparation of the Sample Tubes for Blood Sampling

1. One day before blood sampling, label two Falcon tubes (15 mL, polypropylene) for each animal (at least with study number, animal number, and dose). Add 2 mL of methanol and store at −80 °C.
2. One day before blood sampling, label 2 mL Eppendorf tubes (at least with study number, animal number, and dose). Add 350 µL of 500 USP units heparin/mL or MicroFlow® anticoagulant/diluent from Litron Laboratories and store at 2–8 °C.

3.3.2 Blood Sampling from Rats/Mice

1. Following either acute or repeated animal treatment 50–120 µL of peripheral blood samples are collected for flow cytometric analysis.
2. Allow the blood to drip into a small dish.
3. Immediately afterwards, transfer 100 µL of blood by pipette into the Eppendorf tube containing anticoagulant solution, mix gently, and keep the vials in iced water.
4. Blood/anticoagulant mixture can be stored under refrigeration at 2–8 °C for up to 24 h.

3.3.3 Fixing the Blood Samples in Methanol

1. Using a suitable pipette, transfer 180 µL of the blood–anticoagulant mixture to each Falcon tube containing 2 mL ultracold methanol stored at approx. −80 °C.
2. After adding blood mixture, immediately store the tubes again at approx. −80 °C. This blood/anticoagulant/methanol mixture can be stored up to 2 years.

3.3.4 Sample Staining

1. Remove the tube containing the methanol/blood/anticoagulant mixture from −80 °C freezer.
2. Mix the contents briefly, and add 12 mL of MicroFlow® buffer solution (from Litron Laboratories) and store on ice.

3. Then centrifuge the blood/methanol/MicroFlow buffer solution mixture at 4 °C and at $300 \times g$ for 5 min.

4. Carefully pour off the supernatant (ensuring that the blood pellet is not touched) and pipette off the remaining MicroFlow buffer solution.

5. Then resuspend pellet using the vortex mixer and store at 2–8 °C.

6. Protect the sample rack from light with aluminum foil.

7. Initially, incubate at 2–8 °C for 30 min and then at room temperature for 30 min.

8. After incubating at room temperature, place the samples on ice until measured on the flow cytometer.

9. Immediately prior to analysis, 1.5–2 mL of ice-cold propidium iodide (PI) staining solution (MicroFlow® DNA Stain) is added to malaria parasite *Plasmodium berghei-infected* blood sample (biological standard, Litron, NY) for flow cytometer setup and calibration following the Litron kit calibration steps (*see* **Note 19**).

10. After the flow cytometer is calibrated, 1.5–2 mL of ice-cold PI is also added to one sample at a time and each is analyzed.

3.3.5 Flow Cytometry Analysis

1. MN analysis by flow cytometry is based on the use of an immunochemical reagent (anti-CD71-FITC conjugate) that differentially labels reticulocytes and (mature) erythrocytes in combination with propidium iodide staining, to detect MN in poly- and normochromatic erythrocytes [46].

2. The flow cytometry MNT is carried out on a BD flow cytometer tuned to 488-nm excitation.

3. The cytometer should use the pulse analysis system, capable of providing a height, width, and area value for each parameter. The template preparation should follow the scheme presented in Fig. 6.

4. A total of 10,000 MN-RETs are analyzed per sample.

5. The CD71-FITC, phycoerythrin, and PI fluorescence signals are detected in the respective fluorescence channels (stock filter sets).

6. Erythrocytes infected with the malaria parasite *Plasmodium berghei* (biological standard, from Litron, NY) are used to model MN-containing cells and to set up and calibrate the instrument (47, Fig. 6).

7. Excitation of cells is provided by using the 488-nm line of an argon laser.

8. Fluorescence is acquired through forward scatter (FSC) and side scatter (SSC) and is analyzed using the Diva package (Becton Dickinson, NJ).

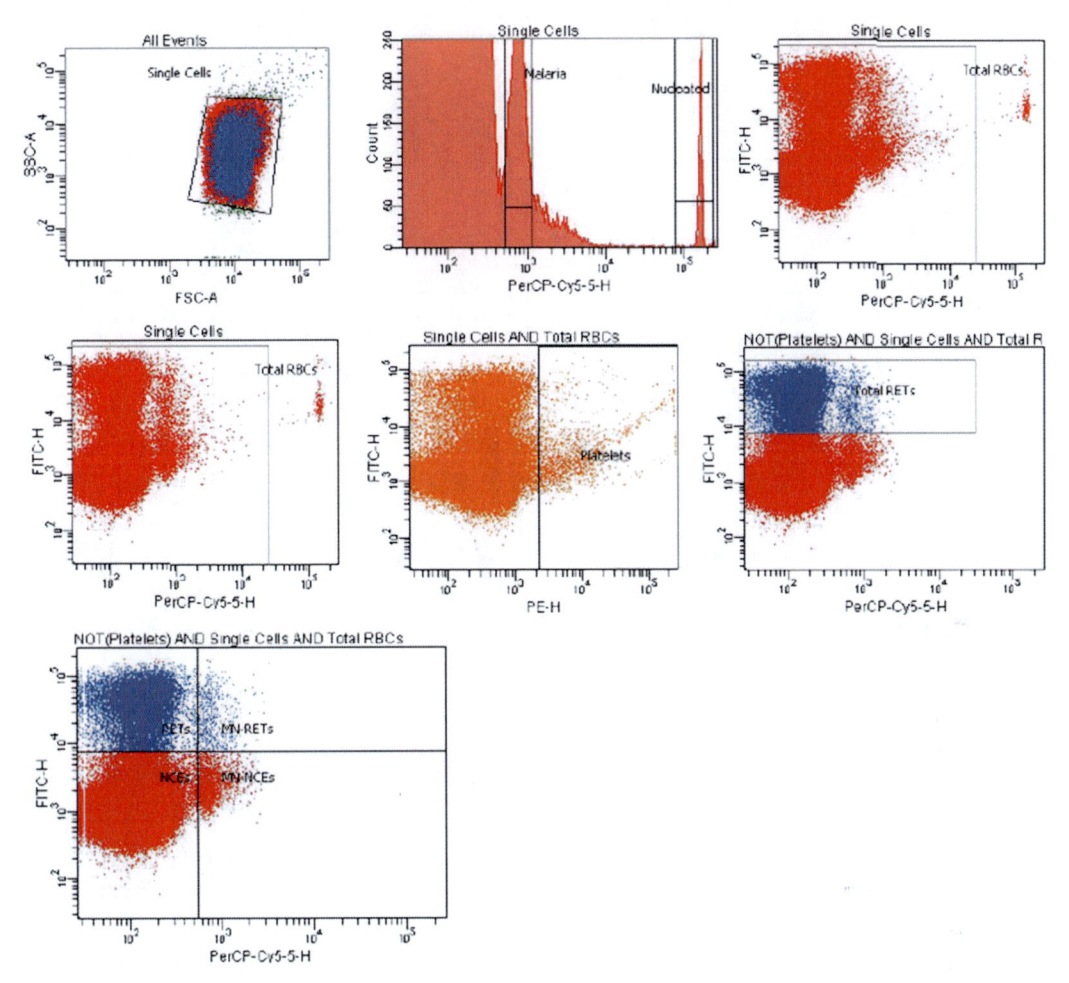

Fig. 6 The template preparation and gating strategy based on FSC, SSC, CD71-FITC, phycoerythrin, and PI fluorescence signals. *Plasmodium berghei* (biological standard purchased from Litron, NY) were used to model MN-containing cells and to set up and calibrate the instrument

9. The percentage of RET among all erythrocytes (RET plus normochromatic erythrocytes, NCE) and the percentage of RET with MN (MN-RET) are determined (Fig. 7).

4 Notes

1. It is important that the flow cytometer should be maintained clean and should be calibrated every time the measurement is performed. The low maintenance quality may affect the analysis.

2. Recommendation: TK6 human lymphoblastoid cells used for the study should be maintained between 3 and 10 passages.

3. In order to avoid the medium dilution, the compound solution should not exceed 1 % of the final culture volume (50 μL of treatment solution/5 mL cell cultures).

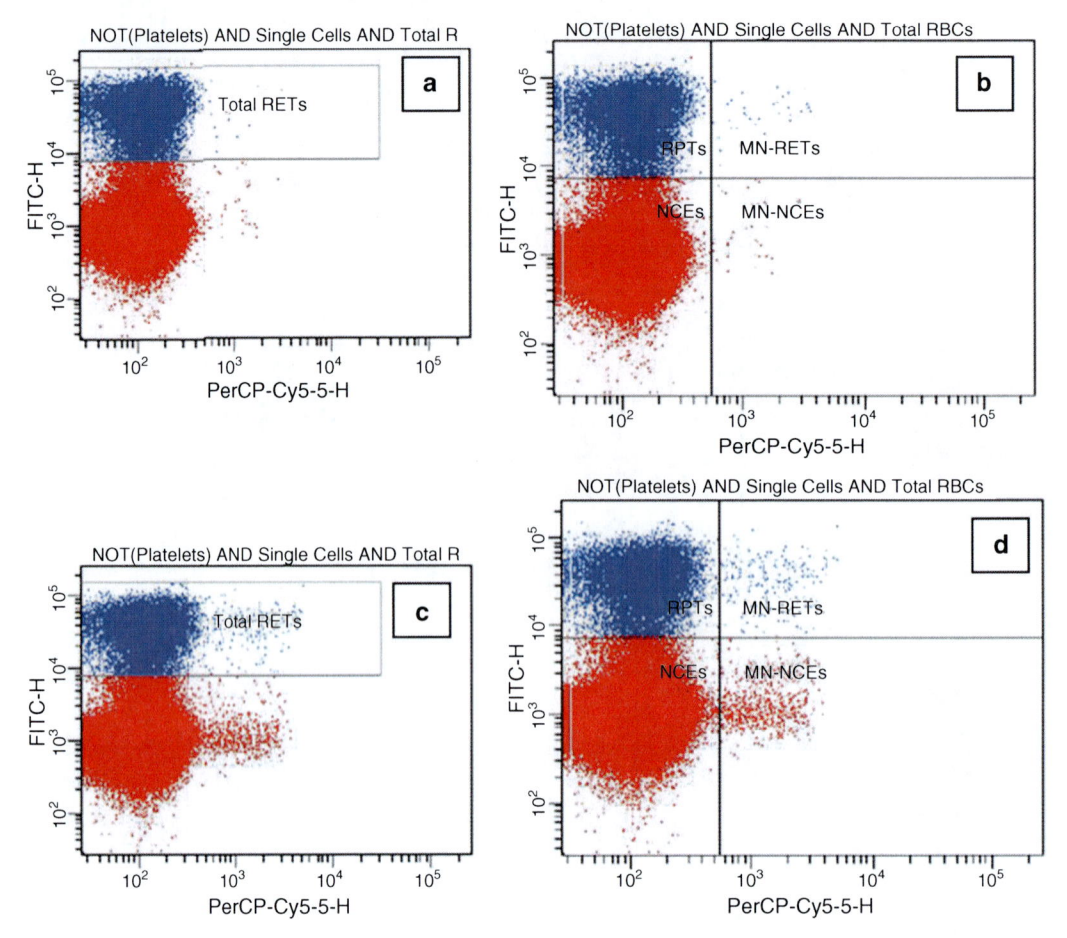

Fig. 7 Determination of the percentage of RET among all erythrocytes (RET plus normochromatic erythrocytes, NCE), the percentage of RET with MN (MN-RET), and the percentage of NCE with MN (MN-NCEs). The gating strategy is based on FSC, SSC, CD71-FITC, phycoerythrin, and PI fluorescence signals. Plots (**a**) and (**b**) are from a sample of a vehicle-treated animal. Plots (**c**) and (**d**) are from a sample of a genotoxic-treated animal

4. If four wells are tested per concentration prepare at least 1 mL of these solutions.

5. 100 μL is required per one well multiplied by the number of wells; prepare approximately 3 mL additionally to measure the for initial cells to beads ratio.

6. Until the measurement is performed, store the tubes at 4 °C for initial cells to beads ratio.

7. For short treatment, e.g., 3-h treatment with the recovery time (culture flask format and for 96-multiwell format only if the TK6 protocol is regarded):

    (a) For the treatment follow **steps 5** and **6** of Subheading 3.1.1 for flasks or **steps 2–15** of Subheading 3.1.2 for 96-well plates.

    (b) Remove the cell cultures in flasks or plate from the incubator after incubation for 3 h with test material.

(c) Transfer the cell cultures from culture flasks into the 15 mL tubes with caps (for culture flask format only).

(d) Centrifuge the cell cultures at 600×g for 5 min (culture flasks) or 300×g for 5 min (96-multiwell plates).

(e) Aspirate the supernatant from tubes/wells of plate.

(f) Resuspend the cells by gently tapping tubes/plates.

(g) Add fresh culture medium acclimated to 37 °C.

(h) Repeat **steps d–g**.

(i) Place cultures in the cell incubator and maintain until the next steps.

8. For the 96-multiwell plates place the plate in the HTS device and analyze the number of cells and beads using the template presented in Fig. 3 and using the HTS settings as presented below:

    (a) Sample flow rate 3 µL/s
    (b) Sample volume 3 µL
    (c) Mixing volume 100 µL
    (d) Mixing speed 100 µL/s
    (e) Number of mixes 2
    (f) Wash volume 400 µL

9. Before starting the harvesting process, remove the nucleic acid dye A and nucleic acid dye B from −20 °C to thaw.

10. Dilute EMA (nuclei acid dye A) stock solution 100 times in previously prepared buffer (PBS with an addition of 2 % FBS). The final amount of nucleic acid dye A (EMA working solution) for each experiment will be dependent on the number of samples. For culture flask format include 300 µL of nuclei acid dye A working solution per tube and for 96-multiwell format include 50 µL of nuclei acid dye A working solution per well.

11. For the 96-multiwell plates prior to analysis acclimate the plate to room temperature (approximately 0.5 h) and then analyze using the template in Fig. 1 and using the following HTS settings:

    (a) Sample flow rate 0.5 µL/s
    (b) Sample volume 75 µL
    (c) Mixing volume 100 µL
    (d) Mixing speed 100 µL/s
    (e) Number of mixes 2
    (f) Wash volume 400 µL

12. Avoid aspirating the serum and avoid placing the pipette too deep in the tube (to avoid touching the gel layer): it may block the pipette and as a result may lead to the loss of the material.

13. In order to avoid the fluorescence loss, during culture preparation, treatment, staining, and analysis, a light exposure should be avoided.

14. For the 3-h treatment with the recovery time in 25 cm² flasks:
    (a) For the treatment follow **steps 1–3** of Subheading 3.2.4 only for the HuLy MNT.
    (b) Remove the cell cultures from the incubator.
    (c) Transfer the cell cultures into the 15 mL tubes with caps.
    (d) Centrifuge the cell cultures at $600 \times g$ for 5 min.
    (e) Aspirate the supernatant.
    (f) Tap the tube to resuspend the cells.
    (g) Add fresh culture medium acclimated to 37 °C.
    (h) Centrifuge the cell cultures at $600 \times g$ for 5 min.
    (i) Tap the tube to resuspend the cells.
    (j) Add fresh culture medium acclimated to 37 °C.

15. Avoid placing the pipette too deep in the well and therefore disrupting and aspirating the cell pellet.

16. For the 3-h treatment with the recovery time in 96-well plates only for the HuLy MNT:
    (a) For the treatment follow **steps 1–6** of Subheading 3.2.5.
    (b) Remove the cell cultures from the incubator after the incubation.
    (c) Centrifuge the cell cultures in plates at $300 \times g$ for 5 min.
    (d) Aspirate the supernatant carefully.
    (e) Tap the plate to resuspend cells.
    (f) Add fresh culture medium acclimated to 37 °C to reach 300 μL final volume.
    (g) Centrifuge the cell cultures at $300 \times g$ for 5 min.
    (h) Tap the plate to resuspend the cells.
    (i) Add fresh culture medium acclimated to 37 °C to reach 300 μL final volume.
    (j) Place cultures in the cell incubator and maintain until next steps.

17. Do not analyze later than 4 h since the end of the staining procedure. If specimens need to be analyzed later than directly after the staining procedure, store in darkness on ice. Maintain the temperature in the measurement room low. High temperature may damage the specimen.

18. For 96-multiwell format analyze the specimens directly after the incubation time using the following HTS settings:
    (a) Sample flow rate—0.5 μL/s
    (b) Sample volume—75 μL

(c) Mixing volume—100 μL

(d) Mixing speed—100 μL/s

(e) Number of mixes—2

(f) Wash volume—400 μL

19. Adjust FL1 (FITC) PMT voltage so that the high CD71-positive RETs are located just above the lower green demarcation on the Y-axis of Plot F (see Fig. 6 in vivo flow). Adjust FL3 (PI/PerCP-Cy5) PMT voltage so that the NCEs are within the first decade of red fluorescence. Then proceed with different compensations as documented in the Litron kit manual.

References

1. Fenech M, Kirsch-Volders M, Natarajan AT et al (2011) Molecular mechanisms of micronucleus, nucleoplasmic bridge and nuclear bud formation in mammalian and human cells. Mutagenesis 26:125–132
2. Kirsch-Volders M, Plas G, Elhajouji A et al (2011) The in vitro MN assay in 2011: origin and fate, biological significance, protocols, high throughput methodologies and toxicological relevance. Arch Toxicol 85:873–899
3. OECD Guideline for The Testing of Chemicals (1997) Mammalian erythrocyte micronucleus test no. 474. Organization for Economic Cooperation and Development, Paris
4. OECD Guideline for Testing of Chemicals (2010) In vitro mammalian cell micronucleus test 487: In vitro mammalian cell micronucleus test (MNVIT). Organization for Economic Cooperation and Development, Paris
5. International Conference on Harmonization (2008) Guidance on genotoxicity testing and data interpretation for pharmaceuticals intended for human use S2(R1). Available at: http://www.ich.org/fileadmin/Public_Web_Site/ICH_Products/Guidelines/Safety/S2_R1/Step4/S2R1_Step4.pdf. Accessed 5 Mar 2012, Geneva
6. Corvi R, Albertini S, Hartung T et al (2008) ECVAM retrospective validation of in vitro micronucleus test (MNT). Mutagenesis 23:271–283
7. Fenech M, Morley AA (1985) Measurement of micronuclei in lymphocytes. Mutat Res 147:29–36
8. Carter SB (1967) Effects of cytochalasins on mammalian cells. Nature 213:261–264
9. Elhajouji A, Cunha M, Kirsch-Volders M (1998) Spindle poisons can induce polyploidy by mitotic slippage and micronucleate mononucleates in the cytokinesis-block assay. Mutagenesis 13:193–198
10. Decordier I, Cundari E, Kirsch-Volders M (2008) Mitotic checkpoints and the maintenance of the chromosome karyotype. Mutat Res 651:3–13
11. Kirsch-Volders M, Sofuni T, Aardema M et al (2003) Report from the in vitro micronucleus assay working group. Mutat Res 540:153–163
12. Fenech M (2006) Cytokinesis-block micronucleus assay evolves into a "cytome" assay of chromosomal instability, mitotic dysfunction and cell death. Mutat Res 600:58–66
13. Kirkland D, Pfuhler S, Tweats D et al (2007) How to reduce false positive results when undertaking in vitro genotoxicity testing and thus avoid unnecessary follow-up animal tests: report of an ECVAM Workshop. Mutat Res 628:31–55
14. Avlasevich SL, Bryce SM, Cairns SE et al (2006) In vitro micronucleus scoring by flow cytometry: differential staining of micronuclei versus apoptotic and necrotic chromatin enhances assay reliability. Environ Mol Mutagen 47:56–66
15. Bocker W, Muller WU, Streffer C (1995) Image processing algorithms for the automated micronucleus assay in binucleated human lymphocytes. Cytometry 19:283–294
16. Bocker W, Streffer C, Muller WU et al (1996) Automated scoring of micronuclei in binucleated human lymphocytes. Int J Radiat Biol 70:529–537
17. Bryce SM, Avlasevich SL, Bemis JC et al (2008) Interlaboratory evaluation of a flow cytometric, high content in vitro micronucleus assay. Mutat Res 650:181–195
18. Bryce SM, Bemis JC, Avlasevich SL et al (2007) In vitro micronucleus assay scored by flow cytometry provides a comprehensive evaluation of cytogenetic damage and cytotoxicity. Mutat Res 630:78–91
19. Castelain P, Van Hummelen P, Deleener A et al (1993) Automated detection of cytochalasin-B blocked binucleated lymphocytes for scoring micronuclei. Mutagenesis 8:285–293

20. Decordier I, Papine A, Plas G et al (2009) Automated image analysis of cytokinesis-blocked micronuclei: an adapted protocol and a validated scoring procedure for biomonitoring. Mutagenesis 24:85–93
21. Frieauff W, Potter-Locher F, Cordier A et al (1998) Automatic analysis of the in vitro micronucleus test on V79 cells. Mutat Res 413:57–68
22. Lukamowicz M, Kirsch-Volders M, Suter W et al (2011) In vitro primary human lymphocyte flow cytometry based micronucleus assay: simultaneous assessment of cell proliferation, apoptosis and MN frequency. Mutagenesis 26:763–770
23. Lukamowicz M, Woodward K, Kirsch-Volders M et al (2011) A flow cytometry based in vitro micronucleus assay in TK6 cells—validation using early stage pharmaceutical development compounds. Environ Mol Mutagen 52:363–372
24. Lukamowicz-Rajska M, Kirsch-Volders M, Suter W et al (2012) Miniaturized flow cytometry-based in vitro primary human lymphocyte micronucleus assay-validation study. Environ Mol Mutagen 53(4):260–270
25. Nusse M, Kramer J (1984) Flow cytometric analysis of micronuclei found in cells after irradiation. Cytometry 5:20–25
26. Nusse M, Marx K (1997) Flow cytometric analysis of micronuclei in cell cultures and human lymphocytes: advantages and disadvantages. Mutat Res 392:109–115
27. Roman D, Locher F, Suter W et al (1998) Evaluation of a new procedure for the flow cytometric analysis of in vitro, chemically induced micronuclei in V79 cells. Environ Mol Mutagen 32:387–396
28. Schreiber GA, Beisker W, Bauchinger M et al (1992) Multiparametric flow cytometric analysis of radiation-induced micronuclei in mammalian cell cultures. Cytometry 13:90–102
29. Slavotinek A, Miller E, Taylor GM et al (1995) Micronucleus frequencies in lymphoblastoid cell lines measured with the cytokinesis-block technique and flow cytometry. Mutagenesis 10:439–445
30. Tates AD, van Welie MT, Ploem JS (1990) The present state of the automated micronucleus test for lymphocytes. Int J Radiat Biol 58:813–825
31. Varga D, Johannes T, Jainta S et al (2004) An automated scoring procedure for the micronucleus test by image analysis. Mutagenesis 19:391–397
32. Verhaegen F, Vral A, Seuntjens J et al (1994) Scoring of radiation-induced micronuclei in cytokinesis-blocked human lymphocytes by automated image analysis. Cytometry 17:119–127
33. Viaggi S, Braselmann H, Nusse M (1995) Flow cytometric analysis of micronuclei in the $CD2^{+/-}$ subpopulation of human lymphocytes enriched by magnetic separation. Int J Radiat Biol 67:193–202
34. Wessels JM, Nusse M (1995) Flow cytometric detection of micronuclei by combined staining of DNA and membranes. Cytometry 19:201–208
35. Soejima T, Iida K, Qin T et al (2007) Photoactivated ethidium monoazide directly cleaves bacterial DNA and is applied to PCR for discrimination of live and dead bacteria. Microbiol Immunol 51:763–775
36. Heddle JA (1973) A rapid in vivo test for chromosomal damage. Mutat Res 18:187–190
37. Schmid W (1975) The micronucleus test. Mutat Res 31:9–15
38. Abramsson-Zetterberg L, Grawe J, Zetterberg G (1999) The micronucleus test in rat erythrocytes from bone marrow, spleen and peripheral blood: the response to low doses of ionizing radiation, cyclophosphamide and vincristine determined by flow cytometry. Mutat Res 423:113–124
39. Cammerer Z, Schumacher MM, Kirsch-Volders M et al (2010) Flow cytometry peripheral blood micronucleus test in vivo: determination of potential thresholds for aneuploidy induced by spindle poisons. Environ Mol Mutagen 51:278–284
40. Torous DK, Dertinger SD, Hall NE et al (2000) Enumeration of micronucleated reticulocytes in rat peripheral blood: a flow cytometric study. Mutat Res 465:91–99
41. Torous DK, Hall NE, Dertinger SD et al (2001) Flow cytometric enumeration of micronucleated reticulocytes: high transferability among 14 laboratories. Environ Mol Mutagen 38:59–68
42. Wakata A, Miyamae Y, Sato S et al (1998) Evaluation of the rat micronucleus test with bone marrow and peripheral blood: summary of the 9th collaborative study by CSGMT/JEMS. MMS. Collaborative Study Group for the Micronucleus Test. Environmental Mutagen Society of Japan. Mammalian Mutagenicity Study Group. Environ Mol Mutagen 32:84–100
43. Vander JB, Harris CA, Ellis SR (1963) Reticulocyte counts by means of fluorescence microscopy. J Lab Clin Med 62:132–140
44. Hayashi M, MacGregor JT, Gatehouse DG et al (2000) In vivo rodent erythrocyte micronucleus assay. II. Some aspects of protocol design including repeated treatments, integration with toxicity testing, and automated scoring. Environ Mol Mutagen 35:234–252
45. Hayashi M, Sofuni T, Ishidate M Jr (1984) Kinetics of micronucleus formation in relation to chromosomal aberrations in mouse bone marrow. Mutat Res 127:129–137

46. Cammerer Z, Elhajouji A, Suter W (2007) In vivo micronucleus test with flow cytometry after acute and chronic exposures of rats to chemicals. Mutat Res 626:26–33

47. Tometsko AM, Torous DK, Dertinger SD (1993) Analysis of micronucleated cells by flow cytometry. 1. Achieving high resolution with a malaria model. Mutat Res 292:129–135

Chapter 12

Fluorescence In Situ Hybridization (FISH) Technique for the Micronucleus Test

Ilse Decordier and Micheline Kirsch-Volders

Abstract

In recent years, cytogenetics in combination with molecular methods has made rapid progress, resulting in new molecular cytogenetic methodologies such as fluorescence in situ hybridization (FISH). FISH is a molecular cytogenetic technique used for the detection of specific chromosomal rearrangements and applicable to many different specimen types. It uses fluorescently labeled DNA probes complementary to regions of individual chromosomes. These labeled DNA segments hybridize with the cytological targets in the sample and can be visualized by fluorescence microscopy in interphase nuclei or on metaphase chromosomes. Here, we describe the FISH methodology with centromeric probes for human cells, which is used in combination with the cytokinesis-block micronucleus assay and which allows discrimination between mutagens inducing DNA breakage (clastogens) or chromosome loss (aneugens).

Key words DNA probe, Hybridization, Immunodetection, Micronucleus, Clastogens, Aneugens

1 Introduction

Fluorescence in situ hybridization (FISH) with DNA probes allows the visualization of defined nucleic acid sequences in particular cellular or chromosomal sites by hybridization of complementary fluorescently labeled probe sequences within intact metaphase or interphase cells. FISH is used for various purposes, including analysis of chromosomal damage, gene mapping, clinical diagnostics, molecular toxicology, and cross-species chromosome homology. FISH is usually applied to standard cytogenetic preparations on microscope slides, but it can be used on slides of formalin-fixed tissue, blood or bone marrow smears, and directly fixed cells or other nuclear isolates. The basic principle of the method is that single-stranded DNA will bind or anneal to its complementary DNA sequence. Thus, a DNA probe for a specific chromosomal region will recognize and hybridize to its complementary sequence on a metaphase chromosome or within an interphase nucleus. Both have to be in single-strand conformation; therefore the DNA

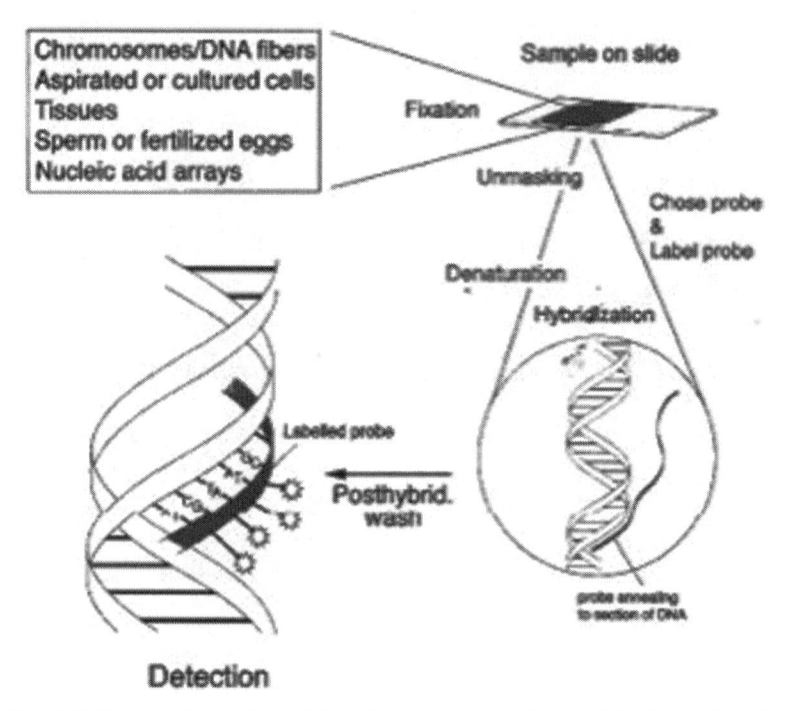

Fig. 1 Schematic illustration of the fluorescence in situ hybridization principle (adapted from Rautenstrauß and Liehr, 2002)

probe and the target DNA must be denatured (Fig. 1). Typical probe length is between 250 bp and 1 kb, depending on the application.

The standard FISH protocol includes the following steps, each of which is crucial for obtaining a successful FISH result: sample preparation and fixation, denaturation of probe and sample, hybridization of a fluorescently labeled probe to sample (annealing), post-hybridization washing, and detection. The probe signal can then be seen through a fluorescent microscope and the sample DNA can be scored for the presence or the absence of the signal. There are many types of probes which can be used for in situ hybridization. For molecular cytogenesis, three different types of probes are generally used: unique sequence probes which bind to single-copy DNA sequences in a specific chromosomal region or gene; probes for repetitive DNA sequences, centromeres (alpha satellite DNA), or telomeric sequences; and whole chromosome paints which are cocktails of unique sequence probes that recognize the unique sequences spanning the length of a particular chromosome. Previously, probes could be labeled by the researcher either directly with nucleotides coupled to a fluorochrome or indirectly with nucleotides coupled to a reporter molecule, which subsequently can be detected by conventional immunochemical methods. Nowadays, FISH is usually performed with pre-labeled commercially available probes.

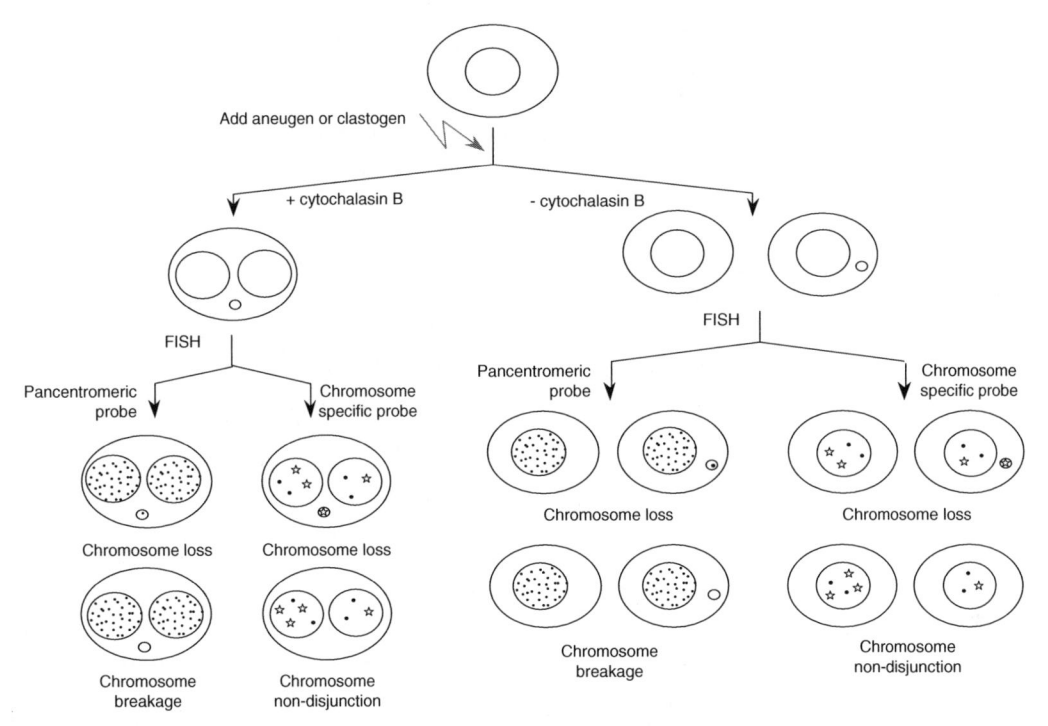

Fig. 2 Detection of chromosome loss and chromosome nondisjunction in the cytokinesis-block micronucleus assay combined with FISH

A widely applied application of the FISH in genetic toxicology is its combination with the in vitro cytochalasin-B block micronucleus (CBMN) methodology that allows discrimination between mutagens inducing DNA breakage (clastogens) or chromosome loss (aneugens) (Fig. 2). Combination of these methodologies allows the characterization of the genetic contents of MN, thereby distinguishing between MN originating from chromosome loss or breakage and determining the involvement of specific chromosomes and chromosome fragments in MN formation [1]. The simultaneous use of the CBMN assay and FISH allows achievement of a higher sensitivity for the adequate hazard assessment of mutagens and a better understanding of the biological mechanisms involved and the possibility to address thresholds for induction of aneuploidy versus clastogenicity. In addition, FISH with centromeric chromosome-specific probes also allows an accurate analysis of nondisjunction (unequal distribution of unique homologous chromosome pairs in the daughter nuclei Fig. 2). This is very helpful to perform risk assessment of compounds with threshold type of dose–responses. FISH with centromeric and chromosome-specific centromeric probes has been widely applied in genetic toxicology to elucidate the genotoxic potential of various mutagenic compounds and for the detection of chromosome damage induced in vitro and in vivo by chemical and physical agents (reviewed in refs. 2, 3).

Our laboratory used the MN assay in combination with FISH for the in vitro demonstration of thresholds for microtubule inhibitors, aneugenic compounds binding specifically to alpha-tubulin and inhibiting tubulin polymerization such as nocodazole, a chemotherapeutic drug, or carbendazim, a pesticide [4, 5]. The two endpoints reflecting aneuploidy were studied in vitro in human lymphocytes: chromosome loss and nondisjunction. To assess chromosome loss, the detection of centromere-positive versus centromere-negative MN by FISH with a general alphoidcentromeric probe was performed on cytochalasin-B-blocked binucleates, resulting from cultures exposed to the spindle poisons. For chromosome nondisjunction, the same compounds were investigated on cytokinesis-blocked binucleated lymphocytes in combination with FISH using chromosome-specific centromeric probes for chromosome 1 and chromosome 17. This allowed the accurate evaluation of nondisjunction, since artifacts were excluded from the analysis, as only binucleates with the correct number of hybridization signals were taken into account. We demonstrated dose dependency of the aneugenic effects and the existence of thresholds for the induction of chromosome nondisjunction and chromosome loss by these spindle inhibitors (lower for nondisjunction than for chromosome loss).

Recently, Lindberg et al. [6] used directly labeled pan-centromeric and pan-telomeric DNA probes to assess the content of MN in cultured binucleated lymphocytes of unexposed, healthy subjects. The authors suggested the combined centromeric and telomeric FISH methodology as a practical method to enhance the specificity of the MN assay allowing detection of MN harboring terminal/interstitial fragments, acentric/centric fragments, chromatid-type/chromosome-type fragments, and entire chromatids/chromosomes. The specificity of the assay should further be evaluated in vitro using model mutagens such as inducers of chromatid breaks and exchanges, chromosome breaks, and aneugens.

In conclusion, the MN assay combined with FISH allows to characterize the occurrence of different chromosomes in MN and to identify potential chromosomal targets of mutagenic substances. It permits to discriminate between aneugenic and clastogenic effects and has been applied in many studies elucidating mechanisms of genomic instability and mode of action of various genotoxic agents.

2 Materials

1. 20× Saline-sodium citrate buffer (SSC): 3 M NaCl, 0.3 M sodium citrate, pH 7, filter through Whatman 41 filter and store at room temperature.

2. RNase A: 100 μg/mL RNAse A in 2× SSC. Aliquot the stock RNAse A solution in volumes of 500 μL and store at −20 °C.

3. Pepsin: 40 units/mL pepsin prepared in 10 mM HCl, aliquot stock in volumes of 50 µL and store at −20 °C.
4. Ethanol: 100 % ethanol, also prepare 50 and 75 % dilution series with water.
5. 2 µg/mL DAPI in antifade mounting medium and/or 5 µg/mL ethidium bromide in antifade medium.
6. 0.075 M KCl.
7. Fixative: 3:1 methanol/acetic acid mixture. Freshly prepared just before use.
8. Formaldehyde.
9. Fluorescence microscope, filters, and optional triple-band-pass filter.
10. Glass slides.
11. Plastic coverslips for incubation and hybridization steps.
12. Heat block.
13. Coplin jars for washing steps.

3 Methods

3.1 Sample Preparation and Fixation

1. Transfer cells into 15 mL Falcon tubes (*see* **Note 1**).
2. Centrifuge cell suspension at $129.5 \times g$ for 8 min at room temperature.
3. Discard supernatant with water pump till only 500 µL remains.
4. Hypotonic shock: Add 3 mL of cold (4 °C) 0.075 M KCl to the cells on vortex (1000 rpm) or with dispenser.
5. Immediately centrifuge at $129.5 \times g$ for 8 min at room temperature.
6. Discard supernatant with water pump till 500 µL remain, and resuspend by patting.
7. Fixation: Add the volume of a full Pasteur pipette of 3:1 freshly prepared methanol/acetic acid mixture, drop by drop (slowly at the start, a little faster afterwards), on vortex (1,000 rpm) to the cell suspension, followed by four drops of formaldehyde (*see* **Note 2**).
8. Centrifuge at $129.5 \times g$ for 8 min at room temperature.
9. Discard supernatant with water pump till 500 µL of supernatant remains, and resuspend by patting.
10. Repeat fixation twice (slowly at the start, a little faster afterwards) without formaldehyde (if necessary for time management, tubes can rest overnight after second fixation).
11. Centrifuge the cells at $129.5 \times g$ for 8 min at room temperature.

12. Discard the supernatant till 100–200 µL of the supernatant (dependant on pellet size) remains.
13. Resuspend the pellet in 600 µL of methanol/acetic acid.
14. Drop the fixed cells onto slides using a Pasteur pipette, leave to dry (overnight), and store at −20 °C if not immediately used.

3.2 Slide Preparation for FISH (See Note 3)

1. Incubate the slides in a 100 mL coplin jar with 0.05 % RNase A solution in 2× SSC buffer for 60 min at 37 °C (*see* **Note 4**).
2. Wash the slides twice in 2× SSC for 5 min at room temperature in a coplin jar (100 mL).
3. Incubate the slides with 0.005 % pepsin prepared in 10 mM HCl for 10 min at 37 °C in a coplin jar (100 mL) (*see* **Note 5**).
4. Wash the slides in 1× PBS for 5 min at room temperature in a coplin jar.
5. Dehydrate the slides in an ethanol series: 50 %, then 75 %, and then finally in 100 % ethanol, in a coplin jar, for 5 min each at room temperature.
6. Air-dry the slides.

3.3 Hybridization (See Note 6)

1. Put the probe, at the volume cited by the supplier, onto the slide and cover with a coverslip.
2. Denature the probe and sample (cells on the slides) simultaneously for 4 min at 90 °C on a heat block.
3. Hybridize at 37 °C overnight in a humidity chamber (*see* **Note 7**).

3.4 Post-hybridization Wash and Counterstaining (See Notes 8–10)

1. Wash the slides in 2× SSC at room temperature in a coplin jar (100 mL) to remove the coverslip (*see* **Note 11**).
2. Wash the slides in 1× PBS for 5 min at room temperature in a coplin jar.
3. Dehydrate slides in an ethanol series: 50 %, 75 %, and then 100 % ethanol, in a coplin jar, for 5 min each at room temperature.
4. Counterstain the slides either with two to three drops of 5 µg/mL ethidium bromide in antifade mounting medium or with two to three drops of 2 µg/mL DAPI in antifade mounting medium and cover with a coverslip (*see* **Note 12**).
5. Leave the slides for at least 1 h at 4 °C before analyzing.

3.5 Analysis (See Notes 13 and 14)

1. Analyze the slides with a fluorescence microscope.
2. Per culture and per concentration at least 500 cytochalasin-B-blocked binucleated cells are scored with a maximum of 1,000 binucleate cells per culture.
3. The standard scoring criteria for MN (1/3 diameter, no overlap, shape) are used [7].

4. The MN in binucleated cells are examined for the presence of one or more spots and classified as centromere positive (MNCen+) or centromere negative (MNCen−), the latter showing no centromeres [5, 8].

5. For chromosomal nondisjunction, the scoring is restricted to binucleated cells having the diploid number for the analyzed chromosomes (two spots in for each of the two probes) to avoid artifacts in scoring of cells where the staining is not accurate. The distribution of the signals for both probes between the binucleated cells is scored as 2/2, 1/3, and 0/4. The events involving in both chromosomes analyzed are scored independently of each other but recorded in parallel per cell.

4 Notes

1. FISH technique for the micronucleus test can be performed in vitro in lymphocytes and cell lines. In this assay using human lymphocytes, the cells are cultured in the presence of phytohemagglutinin to stimulate mitosis. After 24 h of stimulation, the test compound is added for an appropriate time. At 44 h of culture, cytochalasin-B is added and at 72 h the cells are harvested. In case of cell lines cytochalasin-B should be added during the first cell cycle following the start of the treatment and the cells should be harvested prior to the second mitosis. For cell lines in suspension, cells are allowed to grow for 24 h. After 24 h, the medium is replaced with fresh medium and the test compound is added; cytochalasin-B should be added during the first cell cycle following the start of the treatment and the cells should be harvested prior to the second mitosis. For adherent cell lines, cells are seeded into 25 cm^2 flasks and are allowed to attach for 24 h. After 24 h, the medium is replaced with fresh medium and the test compound is added; cytochalasin-B should be added during the first cell cycle following the start of the treatment and the cells should be harvested prior to the second mitosis by trypsinization. A detailed protocol and a summary of the key steps in preparation of lymphocytes/cell lines for the cytokinesis-block assay can be found in [1].

2. Formaldehyde in the fixation is used for the preservation of the general organelle structure of the cell.

3. The protocol described above can be applied to human lymphocytes/cell lines. Depending on the type of cells used, the concentrations of RNase and pepsin may differ and should be tested to obtain an optimal penetration of the probe.

4. RNase treatment serves to remove endogenous RNA and may improve the signal-to-noise ratio in DNA–DNA hybridizations.

5. Pepsin can significantly improve probe penetration and serves to increase target accessibility by digesting the protein that surrounds the target nucleic acid.

6. From the hybridization step on, protect the slides and probes from light, as the probes are fluorescently labeled.

7. Prepare a humidity chamber by inserting a piece of Whatman paper on the bottom of a beaker and tissue papers at the sides, soak them with water, put a coplin jar with the slides in horizontal position into the beaker, cover with parafilm before putting into the incubator, and protect from the light.

8. The protocol described here is a standard protocol for FISH. Depending on the brand and supplier of the probe of interest used, the protocol can be modified as indicated by the supplier. This is especially the case for the post-hybridization washing.

9. Do not allow the slides to dry out at any time during the wash or the detection procedures.

10. The temperature and buffer concentrations (stringency) of hybridization and washing are important steps, as lower stringency may result in nonspecific bonding of the probe to other sequences, and higher stringency may result in a lack of signal.

11. Coverslips should fall off during this wash. If not, gently help them to fall off with a pincet.

12. Be careful when putting the coverslip and avoid air bubbles in order to have a clear image under the microscope.

13. The probes will display a single fluorescent spot at the location of the centromere of the chromosome.

14. If many aspecific signals are observed, include one or more post-hybridization washing steps.

References

1. Kirsch-Volders M, Plas G, Elhajouji A et al (2011) The in vitro MN assay in 2011: origin and fate, biological significance, protocols, high throughput methodologies and toxicological relevance. Arch Toxicol 85:873–899
2. Norppa H, Falck GC (2003) What do human micronuclei contain? Mutagenesis 18:221–233
3. Hovhannisyan GG (2010) Fluorescence in situ hybridization in combination with the comet assay and micronucleus test in genetic toxicology. Mol Cytogenet 3:1–11
4. Elhajouji A, Van Hummelen P, Kirsch-Volders M (1995) Indicationsfor a threshold of chemically-induced aneuploidy in vitro inhuman lymphocytes. Environ Mol Mutagen 26:292–304
5. Elhajouji A, Tibaldi F, Kirsch-Volders M (1997) Indication forthresholds of chromosome non-disjunction versus chromosomelagging induced by spindle inhibitors in vitro in humanlymphocytes. Mutagenesis 12:133–140
6. Lindberg HK, Falck GC, Jarventaus H et al (2008) Characterizationof chromosomes and chromosomal fragments in humanlymphocyte micronuclei by telomeric and centromeric FISH. Mutagenesis 23:371–376
7. Fenech M, Chang WP, Kirsch-Volders M et al (2003) Human MicronNucleus project. HUMN project: detailed description of the scoring criteria for thecytokinesis-block micronucleus assay using isolated human lymphocyte cultures. Mutat Res 534:65–75
8. Kirsch-Volders M, Elhajouji A, Cundari E et al (1997) The in vitro micronucleus test: a multi-endpoint assay to detect simultaneously mitotic delay, apoptosis, chromosome breakage, chromosome loss and non-disjunction. Mutat Res 392:19–30

Chapter 13

Comparative Genomic Hybridization (CGH) in Genotoxicology

Adolf Baumgartner

Abstract

In the past two decades comparative genomic hybridization (CGH) and array CGH have become crucial and indispensable tools in clinical diagnostics. Initially developed for the genome-wide screening of chromosomal imbalances in tumor cells, CGH as well as array CGH have also been employed in genotoxicology and most recently in toxicogenomics. The latter methodology allows a multi-endpoint analysis of how genes and proteins react to toxic agents revealing molecular mechanisms of toxicology. This chapter provides a background on the use of CGH and array CGH in the context of genotoxicology as well as a protocol for conventional CGH to understand the basic principles of CGH. Array CGH is still cost intensive and requires suitable analytical algorithms but might become the dominating assay in the future when more companies provide a large variety of different commercial DNA arrays/chips leading to lower costs for array CGH equipment as well as consumables such as DNA chips. As the amount of data generated with microarrays exponentially grows, the demand for powerful adaptive algorithms for analysis, competent databases, as well as a sound regulatory framework will also increase. Nevertheless, chromosomal and array CGH are being demonstrated to be effective tools for investigating copy number changes/variations in the whole genome, DNA expression patterns, as well as loss of heterozygosity after a genotoxic impact. This will lead to new insights into affected genes and the underlying structures of regulatory and signaling pathways in genotoxicology and could conclusively identify yet unknown harmful toxicants.

Key words Comparative genomic hybridization, CGH, Array CGH, Microarray, Genomic imbalances, Genotoxicology

1 Introduction

For cytogenetics and clinical diagnostics, the introduction of a highly versatile molecular cytogenetic technique named fluorescence in situ hybridization (FISH) in the mid-1980s was revolutionary for the evaluation of cytogenetic aberrations and abnormalities on a molecular basis [1]. Since then, this technique has been further improved and carried out countless times. The plethora of applications being used range from fiber FISH [2], multiplex FISH [3], and spectral karyotyping [4] to combined binary ratio labeling

(COBRA) FISH [5]. Also, 3D FISH applications have been more recently developed to study, for example, nuclear chromosome compartments in interphase [6, 7]. Depending on the assay in use, the resolution varies from 5 kb to 5 Mb when examining fluorescent signals at the level of chromatin strands, interphases, or metaphases [8]. All the above assays have in common that complementary probes or probe sets, differentially labeled with one or more fluorescent dyes, are hybridized to the target DNA that needs to be investigated, such as metaphase chromosomes or interphase nuclei. Changes within the target DNA can then be visualized and evaluated. However, it has proved extremely difficult and challenging to prepare metaphase spreads from certain cells or tissues such as solid tumors. Thus, it was just a matter of time until the very extensively used multicolor FISH technique was modified to serve as the basis for comparative genomic hybridization (CGH). Virtually at the same time, two groups, one in the USA at the University of California in San Francisco [9] and the other group in Heidelberg, Germany [10], recognized the principles of CGH.

1.1 Chromosomal Comparative Genomic Hybridization

The conventional chromosomal CGH method exclusively compares whole genomes for copy number changes, e.g., gain and loss of chromosomal DNA sequences. The major difference between CGH and the commonly employed FISH technique is the use of metaphase chromosomes serving only as a hybridization matrix for comparative genomic hybridization and are not representing target structures to be analyzed. Metaphase chromosome spreads on glass slides originate from cultures of stimulated lymphocytes from healthy individuals, either self-prepared or commercially bought. The quality of the chromosome spreads, the condensation degree of the chromosomes, and the density of the spreads on the slide are important criteria for the selection of these slides for CGH. Two sets of DNA probes are co-hybridized onto the chromosomal matrix (metaphase chromosomes). These DNA probes are generated from isolated genomic DNA from cells of a healthy individual or individuals (control DNA) on one hand and from specific target cells/tissue, e.g., a solid tumor, on the other hand [11]. As a prerequisite, the DNA probes have to be differently labeled with two kinds of fluorochromes, for instance the control DNA in green [e.g., fluorescein isothiocyanate (FITC)] and the test DNA in red (e.g., Cy3 or rhodamine). After hybridization, the ratio of fluorescence intensities along the metaphase chromosomes now exclusively displays the cytogenetic information regarding genomic changes in the test genome in relation to the control DNA (Fig. 1).

Thus, there is no need for culturing and preparing metaphase spreads from target cells such as tumor cells in order to analyze copy number changes or numerical abnormalities in the genome of the studied cells. As the two sets of DNA probes are applied evenly

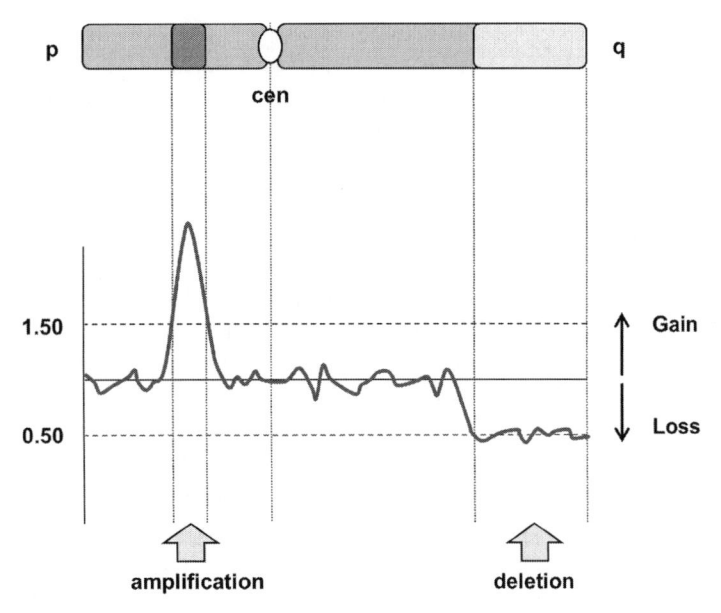

Fig. 1 Chromosomal CGH—representation of gain (on p-arm) and loss (terminal; on q-arm) of chromosomal material in the test genome in relation to the reference/control genome. After hybridization DNA probes made from genomic DNA of the test and the control cells, the fluorescence ratio of 1.0 = 2/2 indicates two copies of homologue sequence throughout the chromosome/genome—for both the control and test genomic DNA. The loss of one or even two homologues leads to ratios of 0.5 = 1/2 or 0 = 0/2, respectively. The latter then only shows the fluorescence of the fluorescently labeled control DNA

on the metaphase chromosomes, which serve as a hybridization matrix, the result is a balanced mix of the fluorescent dyes, e.g., appearing yellowish due to an even blending of green and red fluorescence intensities. Areas within the studied cells' chromosomal setup that are either lost or gained/amplified lead to an imbalanced mixture of the two probe sets at corresponding complementary sequences along the metaphase chromosomes. Subsequently, image analysis is employed to qualitatively and quantitatively evaluate information about copy number variations such as deletions, duplications and amplifications within the studied cells' genome. The lowest detection limit of over-expressed DNA in conventional CGH was found to be 0.25 Mb. In addition, for assessing minor amplifications the rule of thumb states that the smaller the duplication the higher the copy number (≥20 copies). The maximum resolution after losing one homologue is approximately 2 Mb, and for the loss of both homologues the resolution decreases to 1 Mb [12]. Subheading 3 of this chapter shows a standard protocol for chromosomal CGH. DNA probes employed in this protocol are directly labeled with fluorochromes by using nick translation.

Initially developed to investigate chromosomal changes in the genome within solid tumors, the chromosomal CGH method has also been employed to study the impact of genotoxins on the genome. Corso and Parry [13] developed the cell line MCL-5 for use with the CGH assay by transfecting the human B cell-derived lymphoblastoid cell line AHH-1 TK$^{+/-}$ with cDNAs from CYP1A2, CYP2A6, CYP2E1, CYP3A4, and EPHX1 in plasmids, thus expressing human cytochrome P450 enzymes as well as microsomal epoxide hydrolase. Being metabolically active makes these cells very useful as a screening tool for mutagenicity testing of chemicals [13]. In the following years toxicological studies have been undertaken determining copy number variations within the whole genome using various types of cell lines such as K562, MCF-7, and MCF-10A or by using animal tumor cells (e.g., rat gastric tumor cells). In particular, the resistance to various cytostatic drugs [14], the gastric tumor inducer and alkylating agent N-methyl-N'-nitrosoguanidine (MNNG) [15], xenoestrogens [16], and the soy isoflavone genistein [17] was evaluated. An interesting approach to evaluate environmental toxins such as mycotoxins and viruses in relation to carcinogenicity was carried out by Wong and colleagues. Using cells from human hepatocarcinomas from different geographic locations around the globe, they associated them with different risk factors such as aflatoxin intoxication or hepatitis B (HBV)/C (HCV) virus infection [18]. Subsequently, patterns of chromosomal gains and losses were successfully determined by chromosomal CGH. One finding indicated that HCV-related samples from Japan had a characteristically high incidence of an 11q13 gain in the tumor's genome [18]. With the completion of the Human Genome Project, sequence information became publicly available, revolutionizing biochemical research to carry out investigations on a genome-wide scale by using microarray technology [19].

1.2 Array CGH

In the late 1990s, the chromosomal CGH method was consequently further developed [20] using microarrays of genomic DNA fragments instead of whole metaphase chromosomes as hybridization targets, which significantly increased the resolution of the methodology. Early assays employed clones with large human DNA inserts in plasmids like bacterial artificial chromosomes (BAC), P1-derived artificial chromosomes (PAC), or yeast artificial chromosomes (YAC) but also other sources of well-characterized DNA sequences like cosmids or sub-cloned cDNA. The physical length of a single BAC spotted as an element on such an array can be up to 200 kb long. These so-called BAC arrays reached 3,000 to 30,000 spots and the resolution was found to be rather limited at around 3 Mb [21]. Such arrays are mostly being produced in-house using self-made or commercially bought arrayers.

Soon after or even in parallel to the development of BAC arrays, oligonucleotide-based microarrays (Fig. 2) were introduced with elements of 25–85mer high-density synthetic oligonucleotides or expressed sequence tags (EST) on up to 250,000 printed spots

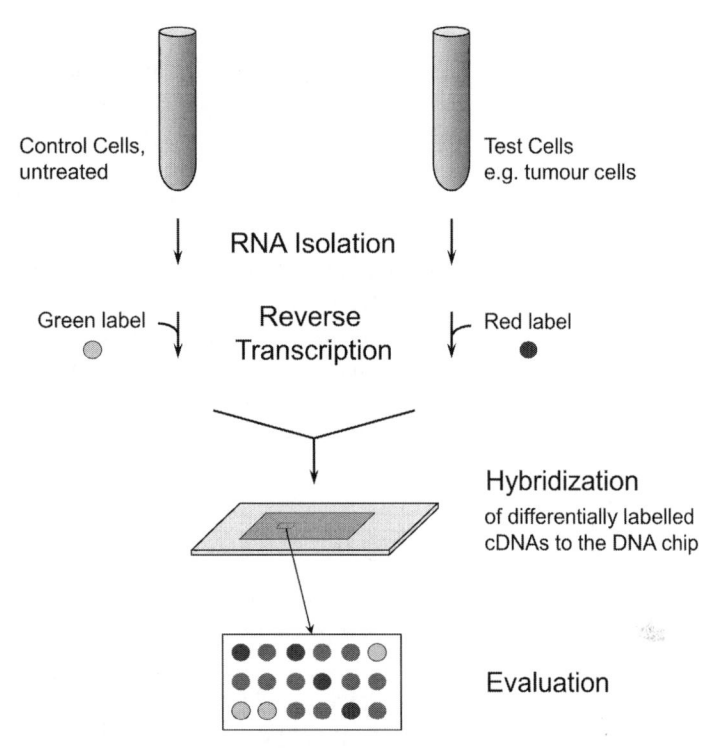

Fig. 2 Array CGH. Fluorescently labeled cDNAs generated from RNA of both control and test cells are hybridized onto a DNA chip (e.g., oligonucleotide microarray). Subsequent evaluation of the resulting dot matrix on the chip produces the gene expression profile of the studied cells

per slide/chip providing a higher resolution of 50–100 kb [22, 23]. On an array, the oligonucleotides are covalently bound to the surface of a glass slide or any other solid substrate and are conjugated with fluorochromes. After hybridization of the sample—fragmented genomic DNA also being labeled with fluorescent molecules—onto the oligonucleotides, the ratio of intensities of both fluorescent dyes can be visualized and subsequently evaluated. The amount of fluorescence conjugate bound to each microarray spot corresponds to the level of genes expressed in the examined cell [23, 24]. In the dawn of oligonucleotide arrays, a feasibility study showed that this technology can only provide a facile overview of gene expression responses relevant to drug metabolism and toxicology [25]. But with the increase in numbers of mapped oligonucleotides, resolution and efficiency increased leading even to variants of this array-based method. The exon array CGH assay for instance evaluates exogenic copy number variations, hence only targeting exons within genes [26, 27]. Although developed for preimplantation genetic diagnosis to assess numerical abnormalities on single blastomere array, CGH can also be employed to focus on the genotoxic impact at the single-cell level [28] using the single-cell array CGH assay where only one specific cell such as a blastomere is analyzed for its genomic copy number variations [29, 30].

A disadvantage of the above microarray CGH methods is unapparent at first as it is intrinsic to their setup. Due to the particular use of chromosomal DNA fragments as target structures, no information on zygosity can be obtained—this also applies to conventional chromosomal CGH. Thus, manufacturers of array-based chips are adding single-nucleotide polymorphism (SNP) nucleic acid sequences (25 bp long with a centered SNP) together with oligonucleotides onto microarrays. Moreover, straight SNP arrays are able to assess DNA sequence variations within chromosomes, individuals, or even species [31]. Besides distinguishing loss of heterozygosity events, these high-density SNP arrays also proved to be highly efficient for evaluating genome-wide copy number changes such as genomic amplifications or homozygous deletions [32, 33]. Modern, state-of-the-art high-resolution CGH arrays are nowadays capable of detecting and assessing segmental DNA copy number variations at kilobase pair resolution (spatial resolution at around 35 kb) with more than 900,000 SNPs and close to one million probes for the detection of copy number variation, e.g., the Genome-Wide Human SNP Array 6.0 from Affimetrix [34, 35]. Nevertheless, despite the growing number of very sophisticated microarray chips, common BAC arrays with a spatial resolution in the 150 kb range exhibit the highest signal-to-noise ratio when compared with oligonucleotide array platforms and might be better suited for the detection of single-copy aberrations [36].

For genotoxicity evaluation and toxicogenomics, both types of microarrays—BAC and oligonucleotide arrays—have been employed using mainly but not exclusively cell lines as target cells for toxicants. By using BAC array CGH containing approx. 6,500 BAC clones representing 0.5 Mb genomic resolution, Herzog et al. [37] evaluated copy number changes and chromosomal instability in mouse lung adenocarcinoma cells induced by the potent human lung carcinogen nicotine-derived nitrosamine ketone NNK (4-(methylnitrosamino)-1-(3-pyridyl)-1-butanone) present in tobacco smoke [37]. Significantly, more gross chromosomal changes were found in NNK-induced tumors compared with spontaneous tumors [37]. A 32K BAC re-array collection (CHORI) tiling path array CGH platform was used to characterize various cell lines with and without amplification of the EMSY gene. This gene plays a role as a potential oncogenic driver in the development of breast cancer. The ten cancer cell lines were treated with *cis*-platinum or olaparib to analyze for increased sensitivity to genotoxic therapies, i.e., for platinum salts or poly(ADP-ribose) polymerase (PARP) inhibitors in the presence of an amplified EMSY gene [38]. A cell line from normal human fetal colonic mucosa has been established by Soucek and colleagues and has been characterized for their mechanism of spontaneously acquired immortality [39].

Oligonucleotide arrays have been employed to evaluate genotoxins as well. Usually, cell lines but also human and animal tumor cells deriving from primary cells are the cells of choice to study the impact of genotoxins on the DNA, genes, and underlying regulatory and signaling pathways. At around the turn of this century, the NIEHS Microarray Center developed the so-called ToxChip with spotted human cDNAs containing clusters of up to 12,000 different cloned genes [40]. This array CGH chip technology allowed the screening and classification of toxicants by their gene expression pattern, i.e., mechanism of action. Accrued toxicant signatures then permitted the evaluation of unknown compounds by comparison with well-known toxicants. Heinloth et al. [41, 42] used these chips to assess the effect of 5 Gy γ-radiation, 7.5 J/m^2 UV-radiation, and oxidative stress (75 μM *tert*-butyl hydroperoxide) on dermal fibroblasts. The three treatments resulted in distinct patterns, indicating an involvement of ATM in regulation of transcription factors such as SP1, AP1, and MTF1 [41] as well as cyclin E-associated kinase activity reduction [42]. For a detailed study on multidrug resistance a custom-designed ABC-Tox microarray was also developed to focus exclusively on the ABC-transporter [43].

Genistein, a major soy isoflavone, has multiple properties and its impact on breast cancer is still controversial. Therefore, MCF-10A cells were treated with genistein for 3 months and then evaluated using conventional chromosomal CGH and also high-density oligonucleotide microarray CGH in order to detect small copy-number changes. A characteristic deletion on 9p21 was found. In general, long-term exposure might increase chromosomal imbalance [17]. Castagnola et al. [44] focused on oral, potentially malignant lesions of the mucosal epithelium with regard to cigarette smoke and alcohol consumption. A human oligonucleotide array showed significant copy-number aberrations in the genome of these cells, highlighting the potential transformation risk towards carcinoma [44]. Young Gprc5a-KO mice were treated with the tobacco carcinogen NNK and it was found that lung tumorigenesis was augmented by NNK in this mouse model. Microarray analysis revealed that gene expression changes induced by tobacco carcinogen(s) may be conserved between mouse and human lung epithelial cells [45]. In an extensive study, another group looked at the impact of hepatocarcinogens such as the mycotoxin aflatoxin as well as non-hepatocarcinogens on F344/N rat liver cells over a 2-year period. Gene expression patterns revealed that the formerly untested compounds myristicin and isosafrole may act in a hepatocarcinogenic manner [46]. Gene expression profiles can also be monitored in the yeast *Saccharomyces cerevisiae* strain S288C to assess the genotoxic impact of the mycotoxin citrinin suggesting that this compound found in food such as cereals and bread significantly induces not only oxidative stress response genes but also other genes associated with metabolism, cell response, defense, virulence, and energy [47]. Huang et al. [48] used a microarray expression and genotyping assay to assess gene expression and genotypes (heterozygous or homozygous) in the MCF-10F cell line after treating with the steroid hormone 17β-oestradiol (E_2), which is capable of inducing complete neoplastic transformation of the human breast MCF-10F epithelial cells. Functional profiling revealed progressive alterations in the integrin signaling pathway, apoptosis inhibition, and gain of tumorigenic cell surface markers. Estrogen exposure thus triggered phenotypic and genomic changes causing tumorigenesis, which confirms the role of E_2 in cancer initiation [48]. The development of microarray platforms for other species such as the very recent transcriptomic microarray platform for the Manila clam [49] will additionally allow the rapid evaluation of toxicants in our environment, more efficiently protecting wild life.

1.2.1 Array CGH Evaluation

A high-throughput method like array CGH evaluates in great detail copy number changes within the DNA across the whole genome; thus, the evaluation of a huge amount of data becomes increasingly challenging. Fluorescently labeled DNA probes consisting of equal rations of test and reference DNA co-hybridize to the mapped array DNA fragments on a slide or a chip. Array CGH intensity ratios, i.e., their transformation on the binary logarithmic ($\log_2$) scale, provide the most suitable information about genome-wide changes in copy number. In a perfect but theoretical situation without normalization or measurement errors, where for example all tumor cells have identical genomic alterations and are uncontaminated by cells from surrounding normal tissue, the normal copy-neutral ratio would correspond to $\log_2\left(\frac{2}{2}\right) = 0$ because reference and test DNA fragments both have two copies. It would indicate equal proportions of both DNAs and, thus, no copy number changes would be seen in the resulting profiles. The $\log_2$-transformed mean intensity ratios of single-copy losses and gains would exactly be $\log_2\left(\frac{1}{2}\right) = -1$ and $\log_2\left(\frac{3}{2}\right) = 0.6$, respectively. Multiple-copy gains or amplifications frequently found for oncogenes in tumors would relate to $\log_2\left(\frac{4}{2}\right) = 1$, $\log_2\left(\frac{5}{2}\right) = 1.3$, and so on. Loss of both copies on the other hand or deletions, which are often associated with tumor-suppressor mutations, would correspond to a negative infinite value ($-\infty$). In this hypothetical situation, the genomic alterations can be easily deduced from the data without statistical techniques. However, under real-world conditions, the $\log_2$ values may vary significantly from the calculated theoretical ratios. The main cause for this discrepancy can be found in the contamination of tumor samples with normal cells from the surrounding tissue as well as in the dependence between the fluorescence intensities of neighboring DNA fragments on the array [50].

Hence, efficient and robust statistical algorithms need to be developed in order to reliably characterize the CGH profiles. A recent popular approach to analyze and characterize array CGH data is the hidden Markov modelling of these data. This model favors a division of the signal into segments of constant copy number and a subsequent classification, which describes each segment as neutral, a loss, or a gain. A disadvantage of this technique is the sensitivity towards outliers triggering over-segmentation with the consequence that segments then incorrectly stretch across very short regions [51]. A modified hidden Markov model combines the necessity to account for the dependence between neighboring DNA fragments with an adopted Bayesian approach, which assumes informative priors for the model parameters. The strong point of this Bayesian, hidden Markov model is its reliability on essentially no tuning parameters; only the input of normalized $\log_2$ ratios is required, which is very convenient for the end user with little or no statistical training [50].

1.3 Toxicogenomics

In the wake of microarray technologies, new scientific fields have emerged, opening new possibilities for researchers. Toxicogenomics was one of them as it investigates how the genome reacts to hazardous substances and identifies genes that respond to groups or categories of chemicals by using high-throughput "omics" technologies such as genomic-scale mRNA expression (transcriptomics), cell- and tissue-wide protein expression (proteomics), as well as metabolite profiling (metabolomics) in combination with bioinformatics and conventional toxicology [52, 53]. In essence, toxicogenomics studies the relationship between the structure/activity of the genome and the detrimental biological effects of exogenous agents, enabling a multi-endpoint analysis of previously inaccessible information about the functional activity of biochemical pathways and differences among individuals and species [54].

Microarray technology in toxicological research allows the evaluation of toxin-modulated gene expression at the mRNA level and thus reveals molecular mechanisms of toxicology. With this technology the identification of genes and their products being involved in resistance or sensitivity to toxic compounds became possible [55]. A comparison of CGH technologies across microarray platforms such as BAC array, genotyping oligonucleotide array, and RNA expression microarray showed similar variations per probe. They all performed at an optimal level to detect known copy number variations when using for instance HL-60 cells. The evaluation of the performance of these platforms will become more important in the future as the array CGH technology continues to evolve [36]. Aardema and McGregor anticipated in 2002 that this new technology will spawn new families of biomarkers permitting the characterization and efficient monitoring of cellular perturbation. This will then lead to a better knowledge of the influence of genetic variation on toxicological outcomes and will allow the definition of environmental causes of genetic alterations and their relationship to human disease. Such an integrated approach will most likely amalgamate the fields of cell pathology, toxicology, molecular genetics, and genetic toxicology generating a comprehensive understanding of genetic control of cellular functions and of cellular responses to alterations in normal molecular structure and function [54]. This integrated approach has much to contribute to the early prediction of drug toxicity and adverse drug reactions for example [56].

Toxicogenomics bridges genotoxicity and carcinogenicity with methods being commercially available for high-throughput analyses. In the twenty-first century, computational toxicology for safety testing within a regulatory setting will significantly contribute to a substantial reduction of animal testing and human clinical trials [53]. Based on the analysis of gene expression using array-based toxicogenomics, it will become possible to screen for carcinogenicity and at the same time differentiate the possible modes of action of a detrimental compound [53].

1.3.1 Public Databases

The public Comparative Toxicogenomics Database (http://ctdbase.org/) provides a centralized resource based on literature linking the chemical to the gene and the disease, as well as the gene to the disease, to better comprehend the interaction of genes and gene products with environmental chemicals, and thus their effects on human health. This database integrates information about sequence, reference, species, microarray, and general toxicology and is also capable of visualizing cross-species comparisons of gene and protein sequences providing information for building complex interaction networks [57, 58]. To date the database holds close to 16 million toxicogenomic relationships. Other databases include Web resources such as the toxicological data network, TOXNET (http://toxnet.nlm.nih.gov/); the European chemical substances information system, ESIS (http://esis.jrc.ec.europa.eu/); or the carcinogenic potency database, CPDB (http://toxnet.nlm.nih.gov/cpdb/). An overview of these and other databases can be found in a reviewed and published list of toxicology databases [59] or at http://alttox.org/ttrc/resources/databases.html.

1.4 Regulatory Guidelines

In toxicology, it is crucial that the screening of toxicants as well as understanding of toxicity pathways is based on methodologies that are sound, proven, and embedded in regulatory guidelines. The field of "omics" has become an important tool for the evaluation of general and reproductive toxicology, the carcinogenicity potential of pharmaceuticals, and several other types of toxicity, and may eventually replace the use of animals. Much progress has been made in the last decade to standardize procedures; however, challenges remain for the assessment of pharmaceuticals for regulatory purposes in terms of off-target toxicological effects or issues of interpretation [60]. The use of microarray technology in toxicology marks the advent of toxicogenomics allowing monitoring of the expression level of thousands of genes on a genome-wide scale [61].

In 2007, the US National Research Council (NRC) released a report, "Toxicity Testing in the twenty-first century: A Vision and a Strategy," that proposes a paradigm shift for toxicity testing, involving a fundamental change in toxicity assessment towards the identification of serious perturbations of toxicity pathways [62, 63]. Perhaps with this proposed vision, toxicity testing will be significantly reduced or maybe the use of animals even completely eradicated in toxicity testing. It will focus much more closely on human significance and higher throughput, enabling a wider coverage of toxicants, life stages, and sensitive subpopulations [64]. Methodologies such as microarrays will help to transform current toxicology tests towards an in vitro toxicity pathway test approach as proposed by the NRC [64, 65]. Functional genomics as envisioned by the NRC then involves the assessment of gene alterations as well as protein and metabolite profiling upon chemical exposure. This will allow the identification of target pathways, establish dose relationships, and map underlying structures for pathway activations in a toxicological context [64, 66–68].

The microarray technology is a particularly powerful tool for identifying new genotoxic substances and their mechanisms of action. It also allows determination of no-effect levels and different susceptibility levels for tissues and cell types. However, besides the consideration of cost and effort, this microarray technology also requires careful planning and choice of analysis methodology to get optimal results [69]. Certain parameters such as suitable signal-to-noise ratios, low standard deviations (SD) of intensity ratios, as well as an optimized cot-1 DNA-to-probe DNA ratio are essential for a successful employment of this technology. It has to be taken into account that genomic array CGH uses a more complex probe mixture and shows lower copy number variations compared with pure expression arrays. Thus, it is crucial to indicate the percentage of spots that provided acceptable values of intensity ratios (typically >97 %, signal-to-noise ratio >2, and SD < 10 % for the duplicates) [70]. It is further important to define threshold values to avoid false-negative results, i.e., to eliminate false positives without removing true positives. Such a threshold for deletions and duplications can be described as the mean $\pm 4 \times$ SD [71]. This cutoff level results in one false positive for every four analyses (for an array with 3,500 loci) as 99.9936 % of the fragments fall within the normal range. For a valid diagnosis, this value of four times SD has to be below the detection limit of an autosomal duplication since the $\log_2$ of the ratio of duplication ($\log_2\left(\dfrac{3}{2}\right) = 0.6$) is closer to the normal ratio ($\log_2\left(\frac{2}{2}\right) = 0$) than to the ratio of a chromosomal loss ($\log_2\left(\frac{1}{2}\right) = -1$). Thus, this threshold can be defined as the difference between the $\log_2$-transformed mean intensity ratio of duplicated loci and twice the SD, i.e., $4 \times SD \leq \left(\left(\log_2\left(\frac{3}{2}\right)\right) - 2 \times SD\right)$ or $SD \leq 0.096$.

It is therefore imperative for quality control that data are obtained with criteria that address the quality and threshold values. In particular, it is crucial to report the number of clones on an array with successful hybridization, a minimum threshold for normal clones, and that the cutoff threshold for the SD of the $\log_2$-transformed intensity ratios did not exceed 0.096 [70].

Exponentially increasing amounts of data from highly adaptive and high-throughput microarray assays will entail a crucial role for bioinformatics with a demanding need for standardization. In 2001, the "Minimum Information About a Microarray Experiment" (MIAME) database [72] was set up to help structuring and channelling the enormous amounts of generated data for biostatistical analyses or even meta-analyses. Hence, bioinformatics is becoming more and more essential when using array chip technology to understand modes of action and the regulatory cellular networks involved in the toxicity pathway function [64]. Targeting gene expression patterns of murine lung cells from more than 250 microarrays, Taylor et al. [73] were able to show by using a set of network algorithms that nuclear factor erythroid 2-related factor (Nrf2) is a direct regulator of proteins involved in oxidative stress response after exposure to reactive oxygen species. Bioinformatics has also been utilized to predict novel transcriptional targets of Nrf2. Thus, network inference algorithms operating on high-throughput gene expression data have the potential to identify regulatory and signaling pathway relationships implicated in disease [73].

Standardization, quality control, and sound biostatistical algorithms are key to publish valid and comparable results, especially with regard to toxicological and toxicogenomic databases, which may hold results of a vast quantity of experiments from different laboratories around the world.

2 Materials

2.1 DNA Labeling

1. Nick translation buffer (10×): 500 mM Tris–HCl, pH 7.5, 100 mM MgSO$_4$, 1 mM dithiothreitol, 500 μg/mL bovine serum albumin (BSA). For 20 mL buffer, combine 10 mL of 1 M Tris–HCl (pH 7.5), 2 mL of 1 M MgSO$_4$, 0.2 mL of 0.1 M dithiothreitol, and 10 mg of BSA. Mix and adjust to final volume with pure water and freeze aliquots at −20 °C.

2. DNase I solution: 0.4 mU/μL DNase I, 40 mM Tris–HCl, pH 7.5, 6 mM MgCl$_2$, 2 mM CaCl$_2$. Prepare fresh always. For 15 mL of DNAse buffer (10×), combine 6 mL of 1 M Tris–HCl (pH 7.5), 0.9 mL of 1 M MgCl$_2$, and 0.3 mL of 1 M CaCl$_2$. Mix and adjust to final volume with pure water and freeze 1 mL aliquots at −20 °C. For use, dilute 10× DNAse buffer after thawing, with pure water to a 1× concentration. Then mix the DNAse I enzyme (Invitrogen) with 1× DNAse buffer to yield a final enzyme concentration of 0.4 mU/μL.

3. TAE buffer (50×): 2 M Tris base, 1 M acetic acid, 50 mM EDTA, pH 8.0.

4. 1.5 % Agarose gel.

5. 0.5 μg/μL Genomic DNA.

6. 40 mM dNTPs (dATP, dCTP, dGTP, dTTP).

7. 1 mM dNTPs (dATP, dCTP, dTTP).

8. 1 mM Cy3-dCTP.

9. 1 mM Cy3-dUTP.

10. 1 mM Fluorescein-12-dATP.

11. 1 mM Fluorescein-12-dCTP.

12. 0.5 U/μL DNA Polymerase I.

13. Absolute ethanol.

14. 1 kb DNA ladder.

15. Loading dye.

16. 0.5 μL/mL Ethidium bromide.

17. Glycogen: 20 mg/mL solution.

2.2 Conventional CGH

All aqueous solutions have to be prepared with pure water, e.g., milliQ water.

1. 70 % Formamide solution, pH 7.0: 70 % (v/v) deionized formamide in 2× SSC, pH 7.0. For 100 mL, mix 70 mL of deionized formamide, 10 mL of 20× SSC, and 10 mL of pure water, adjust pH to 7.0 with 1 M HCl, and then make up to the final volume with pure water.

2. 50 % Formamide solution, pH 7.0: 50 % (v/v) deionized formamide in 2× SSC. For 100 mL, mix 50 mL of deionized formamide, 10 mL of 20× SSC, and 30 mL of pure water. Adjust pH to 7.0 with 1 M HCl and then make up to final volume with pure water.

3. SCC buffer (20×), pH 7.0: 3 M NaCl, 0.3 M sodium citrate, adjust to pH 7.0. For 500 mL, dissolve 87.7 g of NaCl and 44.1 g of sodium citrate dihydrate in 400 mL of pure water; adjust to the final pH with 1 M NaOH and then to final volume with pure water.

4. Master mix: 71.4 % (v/v) deionized formamide, 14.3 % (w/v) dextran sulfate, 2.86× SSC. Mix 5.5 mL of deionized formamide, 1 mL of 20× SSC, and 1 g of dextran sulfate and heat for 2–3 h at 70 °C to completely dissolve the dextran sulfate. Adjust the pH to 7.0 with 1 M HCl and make up to final volume of 7 mL with pure water. Store 1 mL aliquots at −20 °C.

5. PN buffer, pH 8.0: 0.1 M NaH_2PO_4 (monobasic), 0.1 M Na_2HPO_4 (dibasic), Adjust to pH 8.0 with titration by adding monobasic solution (pH ~4.5) to the dibasic solution until a pH of 8.0 is reached. Then, after measuring the reached volume, add 0.1 % (v/v) IGEPAL® CA-630 (0.1 mL per 100 mL of phosphate buffer).

6. PNM buffer, pH 8.0: 5 % Nonfat dry milk powder, 0.02 % sodium azide, make a suspension with PN buffer, adjust pH with 1 M NaOH while stirring, stir overnight at 37 °C, centrifuge for 10 min at 3,000×g, sterile-filter the supernatant using 0.45 μm filter to avoid bacterial growth, aliquot, and store working solution at 4 °C and rest of aliquots at −20 °C.

7. DNA counterstaining solution (antifade): 0.05 μg/mL 4′,6-Diamidino-2-phenylindole dihydrochloride (DAPI), make a solution with 2× SCC buffer and store aliquots at −20 °C.

8. Antifade solution: 0.2 M Tris base solution, pH 8.0, 90 % (w/v) glycerol, 2.33 g of 1,4-diazabicyclo[2.2.2]octane (DABCO), store aliquots at −20 °C.

9. RNase A solution: 10 mM Tris–HCl, pH 7.5, 15 mM NaCl, 10 mg/mL RNase A, boil the stock solution for 15 min and let it slowly cool down to room temperature, store aliquots at −20 °C. Use a freshly prepared working solution, as a 1:100 dilution in 2× SSC, pH 7.0.

10. Pepsin stock solution: 10 % Pepsin. Dissolve 200 mg pepsin in 2 mL of pre-warmed water (37 °C); freeze 50 μL aliquots at −20 °C. As a freshly prepared working solution, use 10–50 μL of pepsin stock solution per 100 mL of 0.01 M HCl, pH 2.3 (pepsin end concentration of 10–50 μg/mL).

11. 1.5 % Agarose gel.

12. Loading dye.

13. Cot-1 DNA.

14. Salmon sperm DNA.

15. Sodium acetate.

16. Magnesium chloride.

17. Formaldehyde.

3 Methods

3.1 DNA Labeling

Occasionally, the amount of isolated DNA from test cells or single cells is not enough. Then, it is recommended to amplify the DNA before DNA labeling by using degenerate oligonucleotide-primed PCR (DOP-PCR; [74]) or primer extension pre-amplification (PEP; [75]).

3.1.1 Nick Translation

Nick translation is used to incorporate fluorescently labeled deoxynucleotides into the DNA. Two enzymes, DNase I randomly nicking the DNA and bacterial DNA polymerase I, are employed in this assay. The latter enzyme possesses a $5' \rightarrow 3'$ exonuclease activity to remove 1–10 nucleotides starting from a nick in the DNA as well as a $5' \rightarrow 3'$ polymerizing gap-filling activity, elongating the DNA from the 3' hydroxyl terminus mediating nick translation along the strand. By encountering a nick on the opposite strand, a double-strand break will be generated. For DNA polymerization, this template-dependent enzyme requires deoxynucleotides (dNTP) and bivalent Mg^{2+} ions. To enhance the efficiency of DNA labeling the employed DNA polymerase I lacks the $3' \rightarrow 5'$ proofreading activity, which is found in the native enzyme. This assay produces a range of differently sized probes resulting in smaller fragments over time.

1. Prerequisites: Set the water bath to 16 °C, set the heating blocks to 37 and 65 °C, cool down absolute ethanol at −20 °C, and prepare agarose gel and buffers during the procedure.

2. Mix reactions from Table 1 in 500 μL vials, e.g., Eppendorf vials, and always pipette enzymes on ice (*see* **Note 1**).

3. Incubate reactions at 16 °C for 90 min and then put them on ice.

4. Run 10 μL volume of each reaction on a 1.5 % agarose gel—the range of the length of the labeled DNA fragments on the gel should run as a smear from 0.3 to 2.3 kb (*see* Subheading 3.1.2).

5. Continue incubation at 16 °C for another 15–30 min if the fragments are still too large or stop reaction at 65 °C for 15 min accordingly.

6. Remove excess nucleotides, by using Nucleotide Removal Kit commercially available (e.g., Bio Spin 30 columns [Bio-Rad] or the QIAquick from QIAGEN).

7. Ethanol-precipitate the labeled DNA (*see* Subheading 3.1.3).

8. Keep the resuspended DNA in the fridge before use. For longer storage, store at −20 °C.

Table 1
Exemplary mixtures for labeling normal and test genomic DNA by using nick translation

Reagents	Reference DNA (µL)	Test DNA (µL)
Water (pure, autoclaved)	53.4	53.4
Genomic DNA (0.5 µg/µL)	12	12
Nick translation buffer (10×)	10	10
dATP (40 mM)	–	2
dCTP (40 mM)	–	–
dGTP (40 mM)	2	2
dTTP (40 mM)	2	–
dATP (1 mM)	0.8	–
dCTP (1 mM)	0.8	0.8
dTTP (1 mM)	–	0.8
Cy3-dCTP (1 mM) (PerkinElmer)	–	4
Cy3-dUTP (1 mM) (PerkinElmer)	–	4
Fluorescein-12-dATP (1 mM) (PerkinElmer)	4	–
Fluorescein-12-dCTP (1 mM) (PerkinElmer)	4	–
DNase I solution (5–10 µL depending on DNA)	5	5
DNA polymerase I (0.5 U/µL) (Life Technologies)	6	6
Total volume	100	100

The test DNA can be for instance from a tumor

3.1.2 Agarose Gel Electrophoresis

1. Prepare a 1.5 % normal melting point agarose gel in 1× TAE buffer.
2. Mix 10 µL of reaction volume with 1 µL of loading dye—as a marker use a 1 kb standard (e.g., GeneRuler, Fermentas).
3. Mix 4 µL of water with 1 µL of 1 kb ladder and 1 µL of loading dye (blue/orange).
4. Fill the wells of the gel with the standard and the reactions.
5. Use 1× TAE buffer as the electrophoresis running buffer.
6. Run the electrophoresis at 100 V, 300 mA, for approximately 20 min.
7. Stain the DNA, by using 0.5 µL/mL of ethidium bromide or SybrGreen I, either during preparation of the gel or by submersing the gel in an ethidium bromide solution.
8. Document the agarose gel—the length of the labeled DNA fragment should be in the range of 0.3–2.3 kb.

Table 2
Exemplary mixture of various probe DNA for CGH

Reagents	(μL)
Reference DNA, fluorescein labeled	20
Test DNA, rhodamine labeled	20
Human cot-1 DNA (1 mg/mL)	40
Salmon sperm DNA (10 mg/mL) (Sigma)	1
Sodium acetate (3 M, pH 5.0); 1/10 volume	8.1
Ethanol, absolute, −20 °C; 2.5× volume	202.5
Total volume	291.6

3.1.3 Ethanol Precipitation

1. Add to each reaction 1 μL of 20 mg/mL glycogen per 50 μL of volume and mix thoroughly. Glycogen will help to precipitate the DNA in an ethanol solution.
2. Add −20 °C cold ethanol (2.5 times the reaction volume).
3. Mix by inverting the vial three times.
4. Put vials for a minimum of 30 min in the freezer (better for 1 h or overnight).
5. Centrifuge at the highest speed at 4 °C for 30 min using a tabletop centrifuge.
6. Discard the supernatant, and place the vials upside down on paper tissue.
7. Dry the pellet for 5 min using a Speed-Vac centrifuge (no heating) or at 37 °C for 20–30 min in a drying cabinet.
8. Resuspend each DNA pellet in pure water (20–80 μL) and shake at 37 °C for 30 min.

3.2 Chromosomal CGH

Prerequisites: Set the water bath to 72 °C, warm up both water and Coplin jar with 70 % formamide at the same time, set heating blocks to 37 and 76 °C, cool down 70 % ethanol at −20 °C, and warm up washing solutions, (a) 2× SSC, (b) 50 % formamide, and (c) 0.2× SSC, to 42 °C. Avoid exposure to light when working with fluorochromes.

3.2.1 DNA Precipitation

1. Mix reaction (as given in Table 2) in a 500 μL vial, e.g., Eppendorf vial (*see* **Note 2**).
2. Precipitate at −20 °C overnight (alternatively at −80 °C for 30 min).
3. Centrifuge with the highest speed at 4 °C for 30 min using a tabletop centrifuge.
4. Remove the supernatant.

5. Wash the pellet with 250 μL of 70 % ethanol (−20 °C cold).
6. Centrifuge with the highest speed at 4 °C for 10 min using a tabletop centrifuge.
7. Remove the supernatant and place the vials upside down on tissue paper.
8. Dry the pellet for 5 min using a Speed-Vac centrifuge (no heating) or at 37 °C for 20–30 min in a drying cabinet.
9. Resuspend the pellet in 3 μL of pure water. This is the "probe DNA."

3.2.2 Preparation of the Hybridization Mix

1. Add 7 μL of master mix (*see* **Note 3**) to the resuspended probe DNA (from **step 9** of Subheading 3.2.1).
2. Mix thoroughly.
3. Incubate at 37 °C for 30 min.

3.2.3 Denaturation and Pre-annealing of the Probe DNA

1. Denature the probe DNA at 76 °C for 7 min.
2. Incubate at 37 °C for 45 min to allow for pre-annealing of cot-1 DNA to repetitive DNA sequences, e.g., *alu* sequences, in the reference and test DNA (*see* **Note 4**).
3. In the meantime warm up formamide solution and start denaturing metaphase slides (*see* Subheading 3.2.5) 15 min before the end of the pre-annealing step.

3.2.4 Pretreatment of Metaphase Chromosomes with RNase A and Pepsin

The following pretreatment steps are optional; however, they might be crucial for optimal hybridization of the DNA probes, especially when self-made slides are used.

1. Apply 150 μL of 100 μg/mL RNase A solution onto the slides.
2. Cover the slides with a plastic coverslip, e.g., a piece of Parafilm.
3. Incubate at 37 °C for 60 min in a wet box.
4. Shortly dip slides in 2× SSC.
5. Wash three times for 5 min in 2× SSC at room temperature.
6. Place slides in an empty Coplin jar and fill it up with pepsin working solution.
7. Incubate at 37 °C for 3–10 min (*see* **Note 5**).
8. Rinse shortly in PBS. Wash twice for 5 min each in PBS at room temperature.
9. Incubate for 5 min in PBS substituted with 50 mM $MgCl_2$.
10. Incubate for 4 min in 1 % formaldehyde to bind the chromosomes to the glass surface and to cross-link residual proteins on the slide; this also includes the deactivation of remaining pepsin (*the incubation time might vary from 1 to 15 min depending on the metaphase quality*).

11. Rinse shortly in PBS, and then wash for 5 min in PBS at room temperature.
12. Run slides through 70, 90, and 100 % ethanol series for 2 min each.
13. Air-dry slides.

3.2.5 Denaturation of the Target DNA

1. In a Coplin jar with 70 % formamide, denature metaphase slides (e.g., CGH Metaphase Target Slides from Abbott Molecular) at 72 °C for 5 min. (*Caution: Temperature decreases by 1°C per slide.*)
2. Transfer the slides into a Coplin jar with 70 % ethanol (−20 °C cold) and incubate for 2 min. Cold 70 % ethanol is favored in this step; dehydration and the drastic temperature change help to keep the strands of the denatured target DNA separated.
3. Incubate the slides for 2 min each in 90 % and absolute ethanol (at room temperature).
4. Allow the slides to air-dry.

3.2.6 In Situ Hybridization

1. Shortly pre-warm metaphase slides on a warming block at 37 °C.
2. Apply the hybridization mix onto the metaphase slides at an area with a high amount of metaphases.
3. Cover with a 22 × 22 mm² glass coverslip and seal it with rubber cement.
4. Incubate at 37 °C for 48–72 h in an incubator.

3.2.7 Washing Steps

Caution: Protect the slides from light exposure in all the subsequent steps. Do not ever let the targeted area(s) on the slides dry out.

1. Carefully remove the rubber cement.
2. Let the coverslip slide off in 2× SSC (this takes approx. 5 min).
3. Wash for 30 min in 50 % formamide solution at 42 °C.
4. Rinse shortly in 2× SSC.
5. Wash for 5 min in 2× SSC at 42 °C.
6. Wash twice for 5 min each in 0.2× SSC at 42 °C.
7. Place slides for 5 min in PN buffer at room temperature (*see* **Note 6**).

3.2.8 DNA Counterstaining

1. Incubate slides in DNA counterstaining solution for 30–60 s at room temperature.
2. Wash shortly in pure water.
3. Place slides for 5 min each in PN buffer at room temperature.
4. Drain the buffer, apply 40 μL antifade solution, and coverslip the slides.
5. Store slides in the dark at 4 °C until evaluation.

3.2.9 Evaluation

1. For the evaluation of the CGH slides a fluorescence microscope equipped with a set of suitable excitation and emission filters, CCD camera, and a computer is considered a prerequisite. In principle, it is possible to roughly analyze CGH slides manually; however, this is rather inaccurate. Analysis software is highly recommended to get an accurate quantitative and qualitative evaluation of the CGH. Companies like MetaSystems, Altlussheim, Germany, offer evaluation software (http://www.metasystems-international.com/isis/cgh) to analyze a set of metaphase for numerical abnormalities as well as copy number variations as seen in loss and gains of chromosomal material. Data interpretation then becomes more precise as the results from many metaphases are integrated for each chromosome (*see* **Note 7**).

2. A reliable evaluation requires the assessment of a number of metaphases (usually 5–15) with certain characteristics. The spreading has to be optimal with an even staining and only a few or none overlapping chromosomes.

3. Besides a low unspecific background, the hybridization pattern along the chromosomes has to be even with consistent intensities.

4. The correct settings of upper and lower thresholds are therefore very important.

5. For karyotyping, the inverted black-and-white image of the counterstained metaphase is used. The chromosomes are subsequently sorted for identification according to their number, size, and banding patterns by an experienced investigator. Some programs allow automatic sorting of the chromosomes but the resulting karyotype should always be additionally checked.

6. The fluorescence intensities along the axis of the chromosomes finally result in an intensity profile. Average profiles for all 24 different human chromosomes are then calculated by using the profiles from all analyzed metaphases—at least four copies of each chromosome should be analyzed.

7. Centromeres, the p-arms of acrocentric chromosomes (13, 14, 15, 21, and 22), and heterochromatin regions, i.e., 1q12, 9q12, 16q11.2, and Yq12, are usually excluded from interpretation [76–79].

4 Notes

1. Avoid exposure to light when working with fluorochrome-labeled nucleotides. Use drawers, covers, or boxes to protect the slides from direct sun light. Use yellow fluorescent lights in the lab if possible.

2. This DNA precipitation mix already contains all the DNAs needed for the hybridization process: both the labeled reference as well as the test DNA, human cot-1 DNA in excess (to block nonspecific hybridization of the probe to nontarget DNA, hence the pre-annealing step later on in the protocol), and salmon sperm DNA (which acts as a carrier DNA reducing the background by associating to nonbiological sites during hybridization). Sodium acetate and ethanol are used to precipitate the DNAs.

3. The master mix contains three crucial ingredients necessary for successful hybridization of DNA probes to target DNA. Formamide destabilizes hydrogen bonds within the DNA's double helix, as a consequence reducing the melting point of the DNA. Dextran sulfate is a polyanionic derivative of the glucan dextran acting as a crowder substance in the master mix. It helps to accelerate the hybridization process by reducing the access to water to nucleic acids. Monovalent cations like Na^+ from the SSC buffer interact with the negatively charged phosphate backbone of the DNA, thus increasing the affinity between the DNA probe and the target DNA.

4. This incubation is crucial as cot-1 DNA pre-anneals in this step with moderately repetitive sequences in the probe DNA reducing later on unspecific background signals on the target DNA throughout the genome. Cot-1 DNA consists of a fraction of DNA with a re-association coefficient of $c_0 \times t = 1$ mol s L^{-1}, combining time of incubation and DNA concentration in this term.

5. The incubation time depends in the quality and the age of the metaphase slides. It is recommended to use different time intervals to determine the optimal incubation time as well as different concentrations of pepsin as indicated in Subheading 2.2. Once established for a batch of slides, use the optimal concentration and time of exposure to digest superficial proteins for the rest of the slides. Avoid overexposure to pepsin as the chromosomes can detach from the glass surface of the slide.

6. If DNA probes have been indirectly labeled with reporter molecules like biotin or digoxigenin by using correspondingly conjugated nucleotides for the mixture in Table 1 then the detection procedure must be carried out after the above washing steps. For this example, no detection procedure is necessary as the DNA probes are directly labeled with fluorochromes.

7. The resolution of an experiment is in general defined by the number of bands that can be distinguished (high resolution ranges from 550 to 850 visible bands). Hence, it is directly proportional to the length of the metaphase chromosomes. The condensation degree which defines the length of the metaphase chromosomes can be influenced by the concentration

and the duration of exposure of colcemid to the cells during the culture of PHA-stimulated peripheral blood lymphocytes. Colcemid arrests the cells in metaphase by depolymerizing the microtubules of the spindle fiber. Commercially available metaphase spreads are usual showing long prometaphase chromosomes suitable for high-resolution CGH.

References

1. Pinkel D, Straume T, Gray JW (1986) Cytogenetic analysis using quantitative, high–sensitivity, fluorescence hybridization. Proc Natl Acad Sci U S A 83:2934–2938
2. Heng HH, Squire J, Tsui LC (1992) High-resolution mapping of mammalian genes by in situ hybridization to free chromatin. Proc Natl Acad Sci U S A 89:9509–9513
3. Speicher MR, Gwyn BS, Ward DC (1996) Karyotyping human chromosomes by combinatorial multi–fluor FISH. Nat Genet 12: 368–375
4. Schröck E, du Manoir S, Veldman T et al (1996) Multicolor spectral karyotyping of human chromosomes. Science 273:494–497
5. Tanke HJ, Wiegant J, van Gijlswijk RP et al (1999) New strategy for multi–colour fluorescence in situ hybridisation: COBRA: COmbined Binary RAtio labelling. Eur J Hum Genet 7:2–11
6. Cremer T, Cremer C (2001) Chromosome territories, nuclear architecture and gene regulation in mammalian cells. Nat Rev Genet 2:292–301
7. Bolzer A, Kreth G, Solovei I et al (2005) Three–dimensional maps of all chromosomes in human male fibroblast nuclei and prometaphase rosettes. PLoS Biol 3:e157
8. Speicher MR, Carter NP (2005) The new cytogenetics: blurring the boundaries with molecular biology. Nat Rev Genet 6:782–792
9. Kallioniemi A, Kallioniemi OP, Sudar D et al (1992) Comparative genomic hybridization for molecular cytogenetic analysis of solid tumors. Science 258:818–821
10. du Manoir S, Speicher MR, Joos S et al (1993) Detection of complete and partial chromosome gains and losses by comparative genomic in situ hybridization. Hum Genet 90:590–610
11. Kallioniemi OP, Kallioniemi A, Sudar D et al (1993) Comparative genomic hybridization: a rapid new method for detecting and mapping DNA amplification in tumors. Semin Cancer Biol 4:41–46
12. Zitzelsberger H, Lehmann L, Werner M et al (1997) Comparative genomic hybridisation for the analysis of chromosomal imbalances in solid tumours and haematological malignancies. Histochem Cell Biol 108:403–417
13. Corso C, Parry EM (1999) The application of comparative genomic hybridization and fluorescence in situ hybridization to the characterization of genotoxicity screening tester strains AHH-1 and MCL-5. Mutagenesis 14:417–426
14. Carlson KM, Gruber A, Liliemark E et al (1999) Characterization of drug-resistant cell lines by comparative genomic hybridization. Cancer Genet Cytogenet 111:32–36
15. Corso C, Parry JM (2004) Comparative genomic hybridization analysis of N-methyl-N′-nitrosoguanidine-induced rat gastrointestinal tumors discloses a cytogenetic fingerprint. Environ Mol Mutagen 43:20–27
16. Payne J, Jones C, Lakhani S et al (2000) Improving the reproducibility of the MCF-7 cell proliferation assay for the detection of xenoestrogens. Sci Total Environ 248:51–62
17. Kim YM, Yang S, Xu W et al (2008) Continuous in vitro exposure to low-dose genistein induces genomic instability in breast epithelial cells. Cancer Genet Cytogenet 186:78–84
18. Wong N, Lai P, Pang E et al (2000) Genomic aberrations in human hepatocellular carcinomas of differing etiologies. Clin Cancer Res 6:4000–4009
19. Clarke PA, te Poele R, Wooster R et al (2001) Gene expression microarray analysis in cancer biology, pharmacology, and drug development: progress and potential. Biochem Pharmacol 62:1311–1336
20. Solinas-Toldo S, Lampel S, Stilgenbauer S et al (1997) Matrix-based comparative genomic hybridization: biochips to screen for genomic imbalances. Genes Chromosomes Cancer 20: 399–407
21. Ylstra B, van den Ijssel P, Carvalho B et al (2006) BAC to the future! or oligonucleotides: a perspective for micro array comparative genomic hybridization (array CGH). Nucleic Acids Res 34:445–450
22. Brennan C, Zhang Y, Leo C et al (2004) High-resolution global profiling of genomic alterations with long oligonucleotide microarray. Cancer Res 64:4744–4748
23. Chan VSW, Theilade MD (2005) The use of toxicogenomic data in risk assessment: a regulatory perspective. Clin Toxicol (Phila) 43: 121–126

24. Amin RP, Hamadeh HK, Bushel PR et al (2002) Genomic interrogation of mechanism(s) underlying cellular responses to toxicants. Toxicology 181–182:555–563
25. Gerhold D, Lu M, Xu J et al (2001) Monitoring expression of genes involved in drug metabolism and toxicology using DNA microarrays. Physiol Genomics 5:161–170
26. Aradhya S, Lewis R, Bonaga T et al (2012) Exon-level array CGH in a large clinical cohort demonstrates increased sensitivity of diagnostic testing for Mendelian disorders. Genet Med 14:594–603
27. Wang J, Zhan H, Li F-Y et al (2012) Targeted array CGH as a valuable molecular diagnostic approach: experience in the diagnosis of mitochondrial and metabolic disorders. Mol Genet Metab 106:221–230
28. Hu DG, Webb G, Hussey N (2004) Aneuploidy detection in single cells using DNA array-based comparative genomic hybridization. Mol Hum Reprod 10:283–289
29. Fiegler H, Geigl JB, Langer S et al (2007) High resolution array-CGH analysis of single cells. Nucleic Acids Res 35:e15
30. Cheng J, Vanneste E, Konings P et al (2011) Single-cell copy number variation detection. Genome Biol 12:R80
31. Crotwell PL, Hoyme HE (2012) Advances in whole-genome genetic testing: from chromosomes to microarrays. Curr Probl Pediatr Adolesc Health Care 42:47–73
32. Zhao X, Li C, Paez JG et al (2004) An integrated view of copy number and allelic alterations in the cancer genome using single nucleotide polymorphism arrays. Cancer Res 64:3060–3071
33. Auer H, Newsom DL, Nowak NJ et al (2007) Gene-resolution analysis of DNA copy number variation using oligonucleotide expression microarrays. BMC Genomics 8:111
34. Le Scouarnec S, Gribble SM (2012) Characterising chromosome rearrangements: recent technical advances in molecular cytogenetics. Heredity (Edinb) 108:75–85
35. Schillert A, Ziegler A (2012) Genotype calling for the Affymetrix platform. Methods Mol Biol 850:513–523
36. Hester SD, Reid L, Nowak N et al (2009) Comparison of comparative genomic hybridization technologies across microarray platforms. J Biomol Tech 20:135–151
37. Herzog CR, Desai D, Amin S (2006) Array CGH analysis reveals chromosomal aberrations in mouse lung adenocarcinomas induced by the human lung carcinogen 4-(methylnitrosamino)-1-(3-pyridyl)-1-butanone. Biochem Biophys Res Commun 341:856–863
38. Wilkerson PM, Dedes KJ, Wetterskog D et al (2011) Functional characterization of EMSY gene amplification in human cancers. J Pathol 225:29–42
39. Soucek K, Gajdusková P, Brázdová M et al (2010) Fetal colon cell line FHC exhibits tumorigenic phenotype, complex karyotype, and TP53 gene mutation. Cancer Genet Cytogenet 197:107–116
40. Medlin JF (1999) Timely toxicology. Environ Health Perspect 107:A256–A258
41. Heinloth AN, Shackelford RE, Innes CL et al (2003) ATM-dependent and -independent gene expression changes in response to oxidative stress, gamma irradiation, and UV irradiation. Radiat Res 160:273–290
42. Heinloth AN, Shackelford RE, Innes CL et al (2003) Identification of distinct and common gene expression changes after oxidative stress and gamma and ultraviolet radiation. Mol Carcinog 37:65–82
43. Annereau JP, Szakács G, Tucker CJ et al (2004) Analysis of ATP-binding cassette transporter expression in drug-selected cell lines by a microarray dedicated to multidrug resistance. Mol Pharmacol 66:1397–1405
44. Castagnola P, Malacarne D, Scaruffi P et al (2011) Chromosomal aberrations and aneuploidy in oral potentially malignant lesions: distinctive features for tongue. BMC Cancer 11:445
45. Fujimoto J, Kadara H, Men T et al (2010) Comparative functional genomics analysis of NNK tobacco-carcinogen induced lung adenocarcinoma development in Gprc5a-knockout mice. PLoS One 5:e11847
46. Auerbach SS, Shah RR, Mav D et al (2010) Predicting the hepatocarcinogenic potential of alkenylbenzene flavoring agents using toxicogenomics and machine learning. Toxicol Appl Pharmacol 243:300–314
47. Iwahashi H, Kitagawa E, Suzuki Y et al (2007) Evaluation of toxicity of the mycotoxin citrinin using yeast ORF DNA microarray and Oligo DNA microarray. BMC Genomics 8:95
48. Huang Y, Fernandez SV, Goodwin S et al (2007) Epithelial to mesenchymal transition in human breast epithelial cells transformed by 17beta-estradiol. Cancer Res 67:11147–11157
49. Milan M, Coppe A, Reinhardt R et al (2011) Transcriptome sequencing and microarray development for the Manila clam, Ruditapes philippinarum: genomic tools for environmental monitoring. BMC Genomics 12:234
50. Guha S, Li Y, Neuberg D (2008) Bayesian hidden markov modeling of array CGH data. J Am Stat Assoc 103:485–497
51. Shah SP, Xuan X, DeLeeuw RJ et al (2006) Integrating copy number polymorphisms into array CGH analysis using a robust HMM. Bioinformatics 22:e431–e439
52. OECD (2009) Series on testing and assessment no. 100 – Report of the second survey

on available omics tools (ENV/JM/MONO (2008)35). Available at http://search.oecd.org/officialdocuments/displaydocumentpdf/?cote=env/jm/mono%282008%2935&doclanguage=en
53. Mahadevan B, Snyder RD, Waters MD et al (2011) Genetic toxicology in the 21st century: reflections and future directions. Environ Mol Mutagen 52:339–354
54. Aardema MJ, MacGregor JT (2002) Toxicology and genetic toxicology in the new era of 'toxicogenomics': impact of '-omics' technologies. Mutat Res 499:13–25
55. Vrana KE, Freeman WM, Aschner M (2003) Use of microarray technologies in toxicology research. Neurotoxicology 24:321–332
56. Ge F, He QY (2009) Genomic and proteomic approaches for predicting toxicity and adverse drug reactions. Expert Opin Drug Metab Toxicol 5:29–37
57. Mattingly CJ, Rosenstein MC, Davis AP et al (2006) The comparative toxicogenomics database: a cross-species resource for building chemical–gene interaction networks. Toxicol Sci 92:587–595
58. Davis AP, King BL, Mockus S et al (2011) The comparative toxicogenomics database: update 2011. Nucleic Acids Res 39:D1067–D1072
59. Young RR (2002) Genetic toxicology: web resources. Toxicology 173:103–121
60. Jacobs A (2009) An FDA perspective on the nonclinical use of the X-Omics technologies and the safety of new drugs. Toxicol Lett 186:32–35
61. Nuwaysir EF, Bittner M, Trent J et al (1999) Microarrays and toxicology: the advent of toxicogenomics. Mol Carcinog 24:153–159
62. Committee on Toxicity Testing and Assessment of Environmental Agents, National Research Council (2007) Toxicity testing in the 21st century: a vision and a strategy. The National Academies Press, Washington, DC
63. Andersen ME, Al-Zoughool M, Croteau M et al (2010) The future of toxicity testing. J Toxicol Environ Health B Crit Rev 13:163–196
64. Krewski D, Westphal M, Al-Zoughool M et al (2011) New directions in toxicity testing. Annu Rev Public Health 32:161–178
65. Bhattacharya S, Zhang Q, Carmichael PL et al (2011) Toxicity testing in the 21 century: defining new risk assessment approaches based on perturbation of intracellular toxicity pathways. PLoS One 6:e20887
66. Gatzidou ET, Zira AN, Theocharis SE (2007) Toxicogenomics: a pivotal piece in the puzzle of toxicological research. J Appl Toxicol 27:302–309
67. Harrill AH, Rusyn I (2008) Systems biology and functional genomics approaches for the identification of cellular responses to drug toxicity. Expert Opin Drug Metab Toxicol 4:1379–1389
68. Zhang Q, Bhattacharya S, Andersen ME et al (2010) Computational systems biology and dose–response modeling in relation to new directions in toxicity testing. J Toxicol Environ Health B Crit Rev 13:253–276
69. Grant GR, Manduchi E, Stoeckert CJ Jr (2007) Analysis and management of microarray gene expression data. Curr Protoc Mol Biol 19:Unit 19.6
70. Vermeesch JR, Melotte C, Froyen G et al (2005) Molecular karyotyping: array CGH quality criteria for constitutional genetic diagnosis. J Histochem Cytochem 53:413–422
71. Shaw-Smith C, Redon R, Rickman L et al (2004) Microarray based comparative genomic hybridisation (array-CGH) detects submicroscopic chromosomal deletions and duplications in patients with learning disability/mental retardation and dysmorphic features. J Med Genet 41:241–248
72. Brazma A (2009) Minimum Information About a Microarray Experiment (MIAME)-successes, failures, challenges. Scientific World Journal 9:420–423
73. Taylor RC, Acquaah-Mensah G, Singhal M et al (2008) Network inference algorithms elucidate Nrf2 regulation of mouse lung oxidative stress. PLoS Comput Biol 4:e1000166
74. Telenius H, Carter NP, Bebb CE et al (1992) Degenerate oligonucleotide-primed PCR: general amplification of target DNA by a single degenerate primer. Genomics 13:718–725
75. Zhang L, Cui X, Schmitt K et al (1992) Whole genome amplification from a single cell: implications for genetic analysis. Proc Natl Acad Sci U S A 89:5847–5851
76. du Manoir S, Kallioniemi OP, Lichter P et al (1995) Hardware and software requirements for quantitative analysis of comparative genomic hybridization. Cytometry 19:4–9
77. du Manoir S, Schröck E, Bentz M et al (1995) Quantitative analysis of comparative genomic hybridization. Cytometry 19:27–41
78. Piper J, Rutovitz D, Sudar D et al (1995) Computer image analysis of comparative genomic hybridization. Cytometry 19:10–26
79. Lundsteen C, Maahr J, Christensen B et al (1995) Image analysis in comparative genomic hybridization. Cytometry 19:42–50

Chapter 14

The In Vitro Micronucleus Assay and Kinetochore Staining: Methodology and Criteria for the Accurate Assessment of Genotoxicity and Cytotoxicity

Bella B. Manshian, Neenu Singh, and Shareen H. Doak

Abstract

The in vitro micronucleus assay is currently one of the most commonly used test systems for the study of genotoxic effects of chemicals. It is considered the preferred method for measuring chromosome damage as it allows the determination of both chromosomal loss and breakage. The type of chromosomal damage induced can be distinguished by using the kinetochore or pan-centromeric staining using molecular probes that label the centromeric regions of chromosomes allowing the determination of aneugenic (chromosome loss) or clastogenic (chromosome breakage) agents.

In this chapter, we provide a description of the basic principles and methods of the in vitro micronucleus assay with detailed explanations of the scoring criteria for the genotoxicity and cytotoxicity endpoints by manual or automated analysis.

Key words Kinetochore, Micronucleus, In vitro, Cytokinesis-block, Mononuclear, Automated

1 Introduction

1.1 Micronucleus Assay

Micronuclei (MNi) are miniature nuclei formed by the condensation of acentric chromosomal fragments or total lagging chromosomes that fail to be incorporated into the daughter nuclei during cell division [1], and instead are left behind during the anaphase stage of mitosis or meiosis (Fig. 1). They therefore represent a DNA damage event within a cell and the micronucleus frequency following exposure to an exogenous agent provides a measure of genotoxicity.

The micronucleus (MN) assay was first accepted in 1997 by the Organization for Economic Cooperation and Development (OECD) for the detection of damage to chromosomes or the mitotic apparatus in mammalian erythrocytes (TG 474) following in vivo testing of chemicals [2]. Following this, in 2010, the guidelines were approved for the in vitro micronucleus test (TG 487) [3].

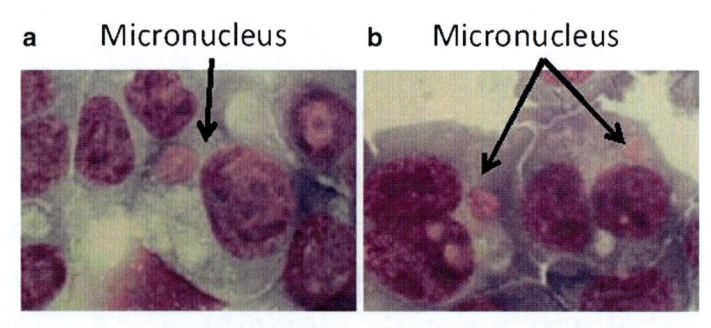

Fig. 1 Mononucleated (a) and binucleated (b) cells with micronuclei present. Cells stained with Giemsa

The MN assay is a sensitive quantitative measure of the degree of gross chromosomal damage induced by a test agent, which can subsequently be classified as a clastogenic event (chromosomal fragmentation), or aneugenic damage (changes in chromosome copy number). The conventional MN method relies on scoring MNi in mononuclear cells which is a simple and rapid assay. One of its main advantages is the ability to score MNi under a light or epifluorescence microscope relatively easily. However, this methodology has become less favored for in vitro studies due to its inability to discriminate between MN events occurring in actively dividing versus nondividing cells. For MNi to be correctly interpreted as damage induced by a test agent it is a prerequisite for the cell to have undergone division in the presence of this exogenous substance. Therefore, it is essential to understand the kinetics of the experimental cells in order to correctly interpret the observed MN events and this is difficult to do when using the mononuclear version of the MN assay. Over the years, various attempts have been made to improve the method by differentiating actively dividing from nondividing cells. These attempts involved using different chemicals [4, 5]. However, most of these chemicals proved to induce MNi themselves [4, 6]. Eventually, Fenech and Morley applied cytochalasin-B to the MN assay [6]. Cytochalasin-B is a mycotoxin compound isolated from the mold *Helminthosporium dematioideum* (discovered by Carter and colleagues [7]) that inhibits cytokinesis without interfering with nuclear division [8] by disrupting microfilament structure [7]. Thus, incorporation of this compound into the MN assay results in the formation of binucleate cells when they are actively dividing. Consequently, detection of MNi in such binucleate cells indicates that those cells have undergone cell division in the presence of the test agent and as a result have accrued chromosomal damage. This methodology has since been widely applied in studies focused on quantifying MNi induced by various agents and in different cell types [9].

The inclusion of cytochalasin-B into the MN assay has led to the test system being named the cytokinesis-blocked micronucleus assay (CBMN) and is often the preferred in vitro method as it limits the scoring of cells to only those that have undergone mitosis in the presence of the test compound. However, one of the disadvantages of using the CBMN method is the possible toxic effect of cytochalasin-B on some cell types and at certain concentrations [10, 11]. However, optimization studies have demonstrated that 3–4 micogram/mL cytochalasin-B induces the maximum number of binucleated cells without any additional genotoxic effect, with a higher (but less than two-fold) frequency of MNi in binucleated cells compared to those seen in mononucleated cells. One consideration to keep in mind when using cytochalasin-B is the amount of time allowed after cells are incubated with cytochalasin-B [12]. The most suitable time is one and a half cell cycle or prior to the start of the next mitosis. Clearly, cell cycle time will be different for different cell lines; therefore, it is important to obtain this information prior to the start of experiments. Cytochalasin-B can be added simultaneously during exposure of the cells to the test substance. If there are concerns that cytochalasin-B might interfere with the uptake of the test substance into the cells, 4–6 h are allowed for the cells to be exposed to the test substance before the addition of cytochalasin-B. Alternatively, in experiments where the chemical needs to be washed away after a specific period of treatment, cytochalasin-B can be added after the washing step.

It is worth noting that the use of cytochalasin-B is mandatory when using blood lymphocytes due to variability of cell cycle and cell division times within the same culture and between separate cultures [13].

1.2 Measures of Toxicity

It is imperative that sensitive and reliable methods are employed for assessing toxicity of a given test compound as a cytotoxicity of $\geq 55 \pm 5$ % is likely to induce chromosomal damage as a secondary effect of cell death mechanisms, thereby confounding the true genotoxicity interpretation [14, 15]. There are a number of different ways in which cytotoxicity can be assessed (as will be discussed below) and it is vital that it is accurately determined to avoid misleading positive results. Therefore, this ensures the credibility of in vitro genotoxicity data determined from an MN assay.

1.3 Kinetochore Staining

The characterization of the contents of MNi induced by test agents is essential for understanding the type of damage induced and to infer mechanisms of genotoxicity [14]. MNi may comprise either (1) acentric chromosome fragments (i.e., lacking a centromere) or (2) whole chromosome(s) that fails to migrate to its respective spindle poles during the anaphase stage of cell division. Kinetochore staining thus involves identification of MNi containing a centromere (kinetochore positive, K+) or lacking a centromere (kinetochore negative, K−). MNi that are K+ indicate the presence of a whole chromosome, which represents an aneugenic effect.

In contrast, those MNi lacking a centromere (K−) contain acentric chromosome fragments, which are the consequence of a clastogenic effect. The ratio of K− to K+ MNi therefore determines the mode of action behind the observed genotoxicity and can thus help to classify a given test substance as a clastogen or an aneugen [15].

The detection of clastogenic or aneugenic effects leading to structural or numerical aberrations, respectively, can also be determined with fluorescent pan-centromeric and/or chromosome-specific probes [16]. This approach of identifying different origins of MNi via probes has proven to be very useful and is being widely applied to aid the understanding of mechanisms of action.

2 Materials

2.1 Micronucleus Assay

1. Mitomycin-C: An alkylating, DNA cross-linking agent that has been shown to induce significant levels of MNi and thus is a useful positive control [17, 18].

2. Cytochalasin-B: 2 micogram/mL Cytochalasin-B. Prepare stock solution by mixing cytochalasin-B purchased from the manufacturers with dimethyl sulfoxide (DMSO) and aliquots are stored at −70 °C (*see* **Note 1**).

3. Phosphate-buffered saline (PBS).

4. Hypotonic solution: 0.56 % KCl.

5. Fixative 1: 5:1:6 parts of methanol:acetic acid:0.9 % NaCl, respectively.

6. Fixative 2: 5:1 parts of methanol:acetic acid, respectively.

7. VECTASHIELD with 4′,6-diamidino-2-phenylindole (DAPI).

8. Giemsa stain: 20 % Giemsa in PBS, pH 6.8.

9. Xylene.

10. DPX mounting medium.

11. Cells: Various established cell lines, or primary cultures, can be used for the in vitro MN assay. Two important considerations for cell selection are (a) p53 competency [19] and (b) their background level of MNi, as only cells with lower than 2.5 % (in the absence of cytochalasin-B) and 3 % (in the presence of cytochalasin-B) MN frequency should be utilized [20]. This rule was later changed to state that cells with 0.5–2.5 % MN frequency with stable background levels are acceptable. Considering that cells from different sources would differ in their background levels, the adequacy of the cell for the MN assay would have to be determined by establishing a historical background level for the negative, solvent, and positive controls in the specific cell line [3].

The cells that have so far been used with the in vitro MN assay include human peripheral blood lymphocytes, and established cell lines such as the human lymphoblastoid AHH-1 [21], MCL5 [22], and TK6 [19] cells. It is important to note that when donors are selected for the harvest of human peripheral lymphocytes age and sex should be taken into account as these may have an important effect on the observed background MN frequencies. Furthermore, the use of samples from older women should be avoided due to the increased MNi formation by the X-chromosome that tends to increase with age [23]. Therefore, it is recommended to obtain peripheral blood lymphocytes from young men approximately 18–35 years of age.

The cell type selected is not a limiting factor and others that have been successfully applied with this methodology include the HepG2 (liver hepatocellular carcinoma); BEAS-2B (human bronchial epithelial), and HFF-1 (human foreskin fibroblast) cell lines [24–26]. However, current recommendations have indicated that to minimize misleading positive results, human lymphocytes and TK6 cells are the most appropriate for genotoxicity testing [19].

12. Metabolic activation: Some chemicals are only toxic when metabolized into their active form. However, not all cell lines are metabolically competent. Therefore, attention should be paid either to use of cell lines that have metabolically activating enzymes such as the cytochrome P450 family or that an exogenous source is used for the metabolic activation. MCL-5 cells are one such example. These cells are human B-lymphoblastoid cells derived from AHH-1 cells which have high levels of CYP1A1 enzyme [27]. MCL-5 cells are engineered to express human cytochrome P-450 CYP1A1 and five transfected human cDNAs encoding CYP1A2, CYP2A6, and microsomal epoxide hydrolase (mEH) metabolizing enzymes [28].

    (a) In the absence of metabolically competent cells the S9 fraction, which is a cofactor-supplemented postmitochondrial fraction taken from rodent liver treated with enzyme-inducing agents, is the most commonly used exogenous metabolizing system [29]. S9 can be prepared from male Sprague Dawley rats induced with Aroclor 1254. A 10 % solution of S9 mix is prepared by adding the following components:

500 mM sodium phosphate buffer (pH 7.4)	20 mL
180 mg/mL D-glucose 6-phosphate	0.85 mL
25 mg/mL nicotinamide adenine dinucleotide phosphate (NADP)	12.6 mL
250 mM magnesium chloride ($MgCl_2$)	3.2 mL

(continued)

(continued)

150 mM potassium chloride (KCl)	22 mL
1 mg/mL L-histidine (in 250 mM MgCl$_2$)	4 mL
1 mg/mL D-biotin	4.88 mL
S9	10 mL
Purified water	22.48 mL

A final concentration between 1 and 10 % (v/v) S9 fraction is used [30]. However, the concentration might need to be adjusted to the individual chemicals used in the study.

13. Negative and positive controls: All experiments should be performed in the presence of concurrent negative (solvent or vehicle) and positive controls. Treatments and preparation of the control groups should be identical to those of the treatment groups.

    Positive controls should be selected based on their ability to produce significant levels of MNi. If possible, it is recommended to select a positive control that belongs to the same chemical class as the substance under study, but this is not always necessary, as the positive control is to demonstrate that the assay is functioning appropriately. Another consideration is the choice of the dose of the positive control. This should be a concentration at which a clear significant increase in MNi frequency is observed in the absence of excessive cytotoxicity. Some examples of positive controls recommended include ethyl methanesulfonate, ethyl nitrosourea, mitomycin-C, cyclophosphamide (monohydrate), and triethylenemelamine.

2.2 Kinetochore Assay

1. Slides in methanol (to ensure cleanliness).
2. 90 % methanol.
3. Pepsin: 25–50 μg/mL pepsin prepared in 0.01 M HCl (pH 2.7–3). Store this solution at −20 °C.
4. PBS.
5. 50 mM MgCl$_2$ in PBS: Mix 50 mL of 1 M MgCl$_2$ with 950 mL of PBS.
6. 70, 80, and 95 % ethanol series.
7. 20× SSC solution: Dissolve 87.6 g of NaCl and 44.1 g of sodium citrate in 500 mL of double-distilled water. Adjust the pH to 7.4 using concentrated HCl and autoclave.
8. 2× SSC solution: A 10× dilution of 20× SSC.
9. RTU human chromosome pan-centromeric paints (various commercial suppliers are available).
10. 70 % and 50 % formamide diluted to appropriate level in 2× SSC.

11. Primulin: 0.5 µg/mL primulin solution. Dilute primulin powder stock in PBS or deionized water. The solution is kept at 4 °C in the dark.
12. Bovine serum albumin (BSA): 1 % BSA. Add 1 g of BSA (Fraction V) to 100 mL of PBS. Allow to dissolve by stirring on a liquid stirrer. If desired, add 0.2 mL of 10 % sodium azide. Store at 4 °C.
13. CREST antibody: Diluted 1:1 with PBS (Antibodies Incorporated, Davis, USA).
14. FITC-conjugated goat anti-human IgG, stored at −20 °C.
15. VECTASHIELD® Mounting Medium with DAPI: DAPI is a nucleic acid stain that preferentially labels double-stranded DNA. Using Vectashield with DAPI protects the latter from photobleaching and preserves its fluorescence.

3 Methods

3.1 Micronucleus Assay

The MN assay can be performed using one of the three variations of the MN methodology as illustrated in Fig. 2. Here we describe the sequential method as followed routinely in most laboratories. For the variations *see* **Notes 2–4**.

1. Test concentrations (dose): Prepare concentrations up to 100 µl total volume in the appropriate diluent. Acceptable chemical concentrations to be used in the in vitro MN methodology should be determined before the initiation of the experiments. This information might be already available from previous work, but if there is no data available then concentrations should be selected carefully to avoid misleading positive responses and those producing high levels of cytotoxicity [3].

2. Cells are maintained and routinely passaged when they are 80 % confluent.

3. To initiate the MN assay, seed cells at $1-1.5 \times 10^5$ cells/mL in 25 cm^2 flasks in 10 mL of medium (or in 75 cm^2 flasks in case of adherent cells) and incubate for 1 cell cycle at 37 °C in an atmosphere of 5 % CO_2 to allow the cells enough time to reach their growth phase. If using cell types, such as TK6 lymphoblastoid cells that reach exponential phase after 1–3 h of seeding (*see* **Note 2**), cells are seeded at $2-3 \times 10^5$ cells/mL.

4. After 1 cell cycle examine cells under a light microscope. Cells need to be approximately 80 % confluent (this is essential for adherent cells). TK6 cells can be checked either after 1 cell cycle (if $1-1.5 \times 10^5$ cells/mL are used) or after 1–3 h (if $2-3 \times 10^5$ cells/mL are used) (*see* **Note 5**).

5. Each experiment is accompanied with a negative diluent control and a positive control in parallel experiments. 0.01 µg/mL

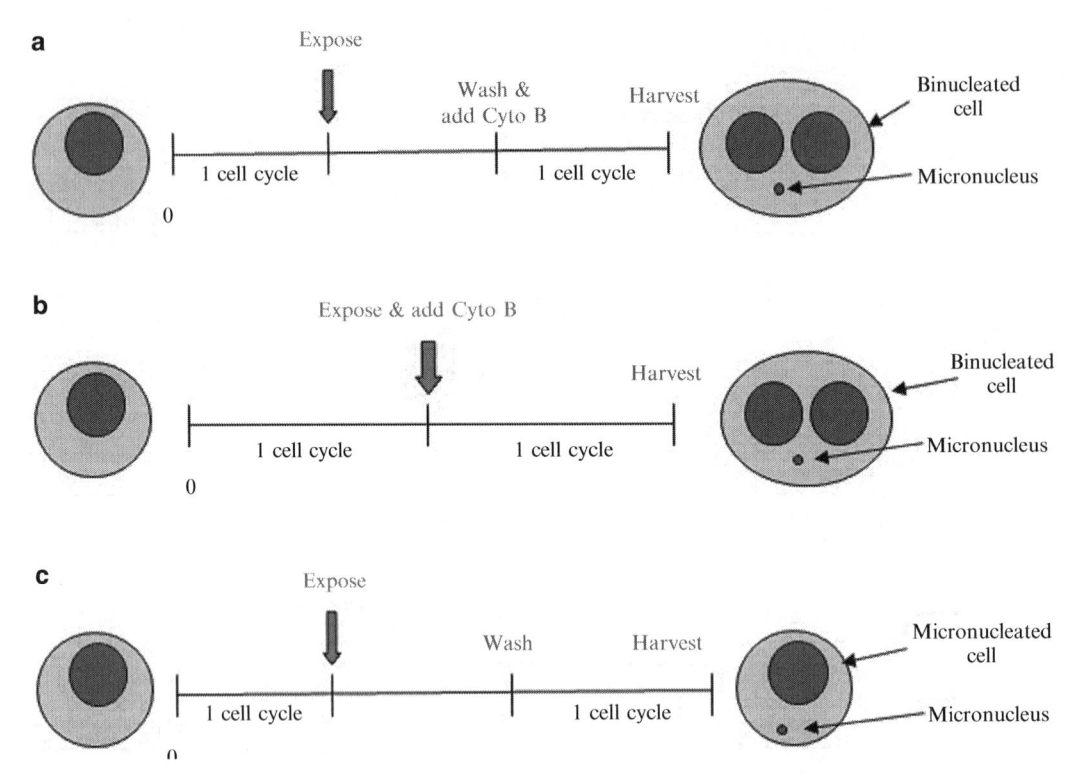

Fig. 2 Schematic illustrating the various ways of performing the MN assay. (**a**) Sequential CBMN assay, (**b**) simultaneous CBMN, (**c**) mononucleated MN assay

mitomycin-C is a commonly used and OECD-TG 474-recommended positive control for the MN assay [2]. All treatments are performed in duplicates or triplicates.

6. Expose the cells to the test substance and incubate under normal growth conditions for the required period which in this example is 1 cell cycle (Fig. 2a). Exposure time for chemicals usually does not exceed 24 h due to their ability to diffuse into the cells within this time period and due to considerations of their half-life.

7. Remove the test agents by washing twice with sterile PBS. In case of suspension cells, aliquot the cells into centrifuge tubes, and centrifuge at 124 rcf (g-force) for 10 min. Discard the supernatant and the cells are then resuspended in 10 mL of sterile PBS. This is repeated twice. Cells are then reseeded in complete culture media (this will differ according to the cells used).

8. Add cytochalasin-B at a concentration of 3–6 µg/mL to each culture. Cells are then incubated in the presence of cytochalasin-B for 1.5–2.0 cell cycles.

9. Each sample is harvested and processed separately. Adherent cells are detached using trypsin–EDTA. Suspension or detached

adherent cells are poured into centrifuge tubes and harvested by centrifugation. The settings of the centrifuge will differ according to the cell type. Most adherent cells can be spun down at 279 rcf for 8 min or 124 rcf for 10 min for suspension cells, as some cells, for example TK6 cells, are sensitive and will be harvested at 124 rcf (g-force) for 10 min.

10. Different techniques may be used in slide preparation according to the scoring method to be used (*see* **Note 6**). The cell cytoplasm should be retained to allow the detection of MNi and (in the cytokinesis-block method) enable reliable identification of binucleated cells.

11. After harvesting, resuspend the cells in approximately 2 mL serum-free media.

12. Glass slides that were pre-placed into 90 % ethanol must be polished with a tissue and coded.

13. One hundred microliters of the cell suspension is cytospun onto polished and labeled glass slides for 5 min at 279 rcf. For sensitive cells such as TK6 it is preferable to cytospin at 124 rcf. Alternatively, if a cytospin is not available, then the cells must be re-suspended in a small volume of Carnoy's fixative (acetic acid:methanol, 1:3) and spread on pre-cleaned, coded slides.

14. Remove the slides and examine under a light microscope to ensure that a suitable cytodot is present where cells are not too close, overlaid, or too sparse. If the density of the cells is too low or high, adjust the quantity of the remaining cell suspension.

15. Fix the cytodots on the slides in 95 % ice-cold methanol for 10 min at −20 °C and allow to air-dry.

16. Once dry, stain the slides in 20 % Giemsa for 12 min at room temperature to visualize the nuclei, MNi, and cytoplasmic membrane. (The cells can also be stained with acridine orange dye, *see* **Note 7**.)

17. Rinse slides in staining buffer (PBS) twice and then allow to air-dry.

18. Dehydrate slides from excess water by dipping into xylene for approximately 10 s.

19. Apply a drop of DPX to the cytodot and then press a 22 × 22 mm coverslip on top.

20. Leave the slides for several hours/overnight to allow the glue to set.

21. Image slides under a light microscope at ×100 magnification using immersion oil.

22. *Scoring and analysis*: For each slide the number of MNi in a minimum of 1,000 binucleated or mononucleated cells per

replicate should be scored and blindly analyzed to calculate the MN frequency. In the CBMN manual scoring system all cell types should be scored until 1,000 binucleate cells have been identified. Thus, the final score-sheet should indicate the proportion of mononucleated, binucleated, tri-nucleated, tetra-nucleated, and multinucleated cells, in addition to metaphase, apoptotic, and necrotic cells per 1,000 binucleate cells observed. From this information, the Proliferation Index (explained in Subheading 3.2) can be derived which is used for the analysis of cytotoxic effects.

The criteria for scoring MNi are detailed in **Notes 8** and **9**. The number of MNi in a single cell can be 0–3 or more, but this would only be scored as a single micronucleated cell, i.e., the individual number of MNi per cell is not considered [31]. Automated scoring can also be performed on the slides (*see* **Note 10**).

3.2 Measures of Toxicity

1. Initiating cell culture: For suspension cells: Seed the cells at $1.0–1.5 \times 10^5$ cells/mL in 10 mL culture and count cells in each culture flask (initial cell count) under sterile conditions, in separate cultures from the MN assay flasks. For adherent cells: cells are detached by trypsinization, harvested by centrifugation, and split into each flask at $1.0–1.5 \times 10^5$ cells/mL in 10 mL culture media.

2. Treat cells with the chemical agent at the exact same concentrations and conditions to the MN assay cultures.

3. Subsequent to treatment with a test substance for the required time-points, wash the cells, suspend the cells in fresh medium, and incubate the cells for 1 cell cycle.

4. Count the cells (final or post-treatment cell count).

5. Use the formulas in **steps 6** and **7** below to obtain the cytotoxicity values.

6. *In the absence of cytochalasin-B*: In studies without cytochalasin-B, it is important to assess cytotoxicity in cell count-based methods.

    Cytotoxicity based on relative increase in cell counts (RICC) or on relative population doubling (RPD) is recommended [32], as both take into account the proportion of the cell population which has divided. Most reports indicate that relative cell count (RCC) is not the most appropriate method as it can underestimate cytotoxicity [33]):

$$\text{RCC} = \frac{\left(\text{Final count treated cultures}\right)}{\left(\text{Final count control cultures}\right)} \times 100$$

$$\text{RICC} = \frac{\begin{pmatrix} \text{Increase in number of cells} \\ \text{in treated cultures}(\text{final} - \text{initial}) \end{pmatrix}}{\begin{pmatrix} \text{Increase in number of cells} \\ \text{in control cultures}(\text{final} - \text{initial}) \end{pmatrix}} \times 100$$

$$\text{RPD} = \frac{(\text{No. of Population doublings in treated cultures})}{(\text{No. of Population doublings in control cultures})} \times 100$$

where

Population doubling =

$$\left[\log\left(\frac{\text{Posttreatment cell number}}{\text{Initial cell number}} \right) \right] / \log 2$$

Thus, an RICC or an RPD of 55 % indicates 45 % cytotoxicity.

7. *In the presence of cytochalasin-B*: The method described in **step 1** can be applied here to determine cytotoxicity; however this will require satellite cultures for each dose and each replicate. Alternatively, cytotoxicity can be measured in the cytochalasin-blocked MN assay by taking into consideration the relative frequencies of mononucleate, binucleate, and multinucleate cells that form following treatment.

The cytokinesis-block proliferation index (CBPI) is a measure of the average number of cell cycles that each cell passes through, in the presence of cytochalasin-B [32, 34, 35]. The CBPI can be used to calculate the cytoxicity from the following formula:

%Cytotoxicity = $100 - 100\{(\text{CBPI}_T - 1)/(\text{CBPI}_C - 1)\}$

where T = test substance treatment and C = untreated control.
CBPI is calculated according to the formula

$$\text{CBPI} = \frac{((M1 + 2M2 + 3(M3 + M4 + M5))}{N}$$

where M1–M5 represents the number of cells which are mononucleate, binucleate, trinucleate, tetranucleate, and multinucleate, respectively, and N is the total number of scored cells [36].

Thus, a CBPI of 1 (all cells are mononucleate) is equivalent to 100 % cytotoxicity.

In the presence of cytoB, the evaluation of cytotoxicity can also be based on replicative index (RI), which indicates the relative number of nuclei in treated cultures compared to control cultures and can be used to calculate the % cytotoxicity.

$$\text{Cytotoxicity} = 100 - \text{RI}:$$

$$\text{RI} = \frac{\dfrac{\left(\begin{array}{c}(\text{No. of binucleated cells}) + \\ (2 \times \text{No. of multinucleate cells})\end{array}\right)}{(\text{Total number of cells})_T}}{\dfrac{\left(\begin{array}{c}(\text{No. of binucleated cells}) + \\ (2 \times \text{No. of multinucleate cells})\end{array}\right)}{(\text{Total number of cells})_C}} \times 100$$

where T = test substance treatment and C = untreated control.

Thus, an RI of 55 % indicates 45 % cytotoxicity.

3.3 Centromeric Labelling

Centromeric labelling can be performed by staining either the kinetochore or the pan-centromeric region. The slide preparation description provided below is for both methods of staining.

3.3.1 Pan-Centromeric Chromosome Paint, Human (Cambio 1695-F-01)

This protocol has been validated on the human B lymphoblastoid TK6 and human gastric AGS cell lines.

1. *Pre-paration of slides*: Following the MN assay, cytocentrifuge the cells (2×10^5 cells/mL) at 279 rcf for 5 min onto slides rinsed in methanol and polished with a tissue. Allow slides to air-dry. Fix the slides in 90 % methanol for 10 min at −20 °C. The fixed slides can be stored at −20 °C indefinitely.

2. *Pre-treatment of interphase cell preparation*: Pre-warm both the sample slides and 25–50 μg/mL 0.01 M HCl pepsin (pH 2.7–3) to 37 °C in an incubator for 10 min (*see* **Note 11**). Apply several drops of pepsin to the cytodot area on each slide and leave at 37 °C for 30 s (e.g., for TK6 cells). Pepsin is a protease digestion enzyme that is used to permeabilize the cell membrane to enhance the penetration of the labelling probe. Place the slides in PBS for 5 min to arrest the pepsin treatment followed by a further 5 min in 50 mM $MgCl_2$/PBS solution (this solution is used to protect cells and chromosomes from further erosion by the pepsin). Dip slides for 1 min in 0.05 μg/mL primulin solution (a red cytoplasmic stain). Dehydrate the slides in 70 %, 80 %, and 95 % ethanol for 2 min each. This will prepare them for the denaturation step.

3. *Denaturation and hybridization*: Warm the probes (ready-to-use human chromosome pan-centromeric paints from Cambio) to 37 °C for 5 min and mix them well. Take 5 μl of probe per slide and add to a microcentrifuge tube. Denature chromosomes on slide in 70 % formamide in 2× SSC for 2 min at 70 °C. Immerse in ice-cold 70 % ethanol and dehydrate through a series of alcohol washes: 70 %, 90 %, and then 100 %. Dry the slides at room temperature. Denature probe for 10 min

at 85 °C. Immediately chill on ice. Apply 5 µl of probe to slide, and hybridize for approximately 16 h at 37 °C in a humidified chamber.

4. *Post-hybridization wash*: Remove the coverslip, and wash for 5 min at 37 °C in 2× SSC. Wash the slides twice in 50 % formamide/2× SSC at 37 °C, for 5 min each time. Wash slides in 2× SSC, twice for 5 min each time. Counterstain with 10 µl VECTASHIELD® Mounting Medium with 1.5 µg/mL of DAPI. Following this step, the slides are subsequently scored as described below in Subheading 3.3.3.

3.3.2 Kinetochore Staining Protocol

1. Place a humidified chamber in an incubator at 37 °C to pre-warm. A humidified chamber can be made using a box with a lid, by placing tissue saturated with water into the bottom, with tubes to lay the slides on and to raise them above the wet surface.

2. Remove slides previously fixed following the micronucleus assay from storage at −20 °C and rehydrate in PBS for 1 min at room temperature.

3. Dip slides for 1 min in 0.5 µg/mL primulin solution to stain the cytoplasm of the cell.

4. Wash the slides by dipping in PBS and replace in fresh PBS.

5. Dilute CREST antibody 1:1 with PBS (Antibodies Incorporated, Davis, USA).

6. Drain excess PBS from the slide without disrupting the cytodot.

7. Apply 25 µl of the diluted antibody gently on the cytodot and put a plastic coverslip (which allows the retention of more fluid on the slide than glass coverslips; these can easily be cut from light plastic packaging material) on the top.

8. Place the slides in humidified chamber at 37 °C for 45 min for the antibody to hybridize to the DNA. Incubating samples in a humidified chamber prevents the formation of a temperature gradient and thus the evaporation of the antibody.

9. Remove the coverslips and wash the slides three times in 1 % BSA/PBS solution at room temperature, for 3 min each time, to block nonspecific binding of antibodies.

10. Wash the slides in PBS for 3 min at room temperature.

11. Apply 25 µl of FITC-conjugated goat anti-human IgG (diluted 1:100 with PBS) to each cytodot on the slide and put a plastic coverslip on the top.

12. Place the slides in humidified chamber at 37 °C for 45 min.

13. Wash slides three times in 1 % BSA/PBS at room temperature, for 3 min each, then once in PBS at room temperature for 3 min, and finally once in sterile water at room temperature for 3 min. These washes will remove any nonspecifically bound

antibodies. They will also prevent the formation of crystals if the slides are stored for later analysis. Shorter washes will result in high background fluorescence which will disrupt scoring while much longer washes will result in reduced signal.

14. Leave the slides to air-dry in the dark.

15. Counterstain with 10 μl of VECTASHIELD® Mounting Medium containing 1.5 μg/mL DAPI. The slides are subsequently scored as described below in Subheading 3.3.3.

3.3.3 Imaging/Scoring

Slides are mounted onto a fluorescence microscope and visualized under UV (DAPI, excitation: 358 nm; emission: 463 nm; for nuclei), red (primulin, excitation: 410 nm; emission: 550 nm; for cytoplasm), and green (FITC, excitation: 495 nm; emission: 517 nm; for centromeric/kinetochore regions) filters. For each dose, MNi from 100 binucleated cells need to be scored for the presence or the absence of centromeric or kinetochore signals (from pan-centromeric or kinetochore staining, respectively)—this should be achieved by scoring 50 cells with MNi per sample replicate.

4 Notes

1. To assess cytotoxicity when including cytochalasin-B in the MN assay, a separate satellite culture needs to be set up, in which the total culture volume can range from 1 to 10 mL for suspension cells. For adherent cells, the cells should be grown in 75 cm^2 flasks. Previous work has shown that cytochalasin-B concentration is important for the survival of cells and for the prevention of false-positive results. According to the OECD guidelines (TG 487) [13], 3–6 μg/mL is recommended; however, it is best to investigate the ideal concentration according to the cell line used [37]. The aim of this test is to achieve 50 % or more binucleated cells in the negative control cultures. Cytotoxicity based on cell counts can also be performed if cytochalasin-B has been included in the assay.

2. The sequential version of the CBMN assay is an extended method (Fig. 2a). This methodology is used when it is suspected that cytochalasin-B might interfere with the uptake or the function of the test substance (this is mostly true for nanomaterials).

3. *CBMN assay—simultaneous method*: Simultaneous exposure is the shortened version of the CBMN assay (Fig. 2b). However, this methodology only applies for experiments where exposure period is for 1.5–2.0 cell cycles. In this method, cells are seeded as described in **steps 2–4** of Subheading 3.1. Cells are exposed to the test material simultaneously in the presence of 3–6 μg/mL of cytochalasin-B for up to 2.0 cell cycles or before the

initiation of the next mitosis. Cells are harvested, and slides prepared as described in **steps 9–22** of Subheading 3.1, and MN scored as described in **step 23** of Subheading 3.1.

The simultaneous version of the CBMN assay is a more commonly used methodology as it saves an additional day in the experiment time. Moreover, it is a methodology that has been well accepted by the chemical toxicology committees such as the OECD based on previous research that showed no difference in sequential or simultaneous methods of the CBMN assay. However, this method has proven to be much more problematic in the case of particle toxicology studies since the uptake of nanomaterials has been shown to be hindered in the presence of cytochalasin-B [38]. Cytochalasin-B has the capacity to interfere with the formation of the actin-filaments required for endocytosis; hence it prevents the cellular uptake of nanomaterials [39]. Thus, it is important to avoid co-exposure of cells to nanomaterials simultaneously with cytochalasin-B when using the CBMN assay. Consequently, the sequential CBMN assay is recommended for nanomaterial studies. This provides the option for post-treatment with cytochalasin-B (i.e. cells are exposed to the nanomaterials for the required period of time, then the nanomaterials are removed by washing several times, and the culture medium is supplemented with cytochalasin-B). Alternatively, it is also possible to perform the CBMN assay with nanomaterials using a delayed co-treatment method where cells are treated with the nanomaterial for 3–24 h, followed by the addition of cytochalasin-B for the remainder of the treatment time. To date, there are no studies that have been performed to determine which of these procedures is favorable and thus further investigation is required [32, 39].

4. The mononucleated MN assay is performed in the absence of cytochalasin-B. In this method, cells are seeded as described in **steps 2–4** of Subheading 3.1. Treat the cells when they are at their exponential phase (1 cell cycle or 1 h to several hours with some cell lines such as the TK6 cells). Cells are harvested, and slides prepared as in **steps 9–22** of Subheading 3.1. Scoring MNi in this assay is similar to scoring MNi in binucleated cells, with the exception that only mononucleated cells containing MNi are scored here. Only 1,000 mononucleated cells are scored along with any tri-nucleated, tetra-nucleated, and multi-nucleated cells, plus metaphase, apoptotic, and necrotic cells.

It is important to note that in the CBMN assay, MNi are scored in cells that are prevented from completing cell division. Hence, it follows that when one MN is scored in one binucleated cell, if that cell were allowed to continue to divide the MNi would only be located in one of the subsequent daughter cells. Thus, only one MN would be scored out of two mononuclear cells. This might lead to reduced frequency

Table 1
Metafer software settings for binucleated MicroNuclei setup found on the MSearch submenu of the Metafer system optimized for the TK6 human lymphoblastoid suspension cell line

	Nuclei	Micronuclei
Object threshold	30 %	8 %
Minimum area	20 μm^2	1.5 μm^2
Maximum area	400 μm^2	55 μm^2
Maximum relative concavity of depth	0.9	0.9
Maximum aspect ratio	1.5	4.0
Maximum distance between	30 μm^2	25 μm^2
Maximum area asymmetry	70 %	–

of MNi discovered in data from the mononuclear MN assay as compared to the CBMN assay for the same substance. Another disadvantage of this methodology is that it is unable to discriminate between MN events from actively dividing versus nondividing cells. This means that the cell scored might not have yet had the chance to undergo cell division, and hence cellular repair.

5. An hour is sufficient for certain cell types to return to exponential growth.

6. *Automated micronucleus assay*: The Metafer system was first introduced by MetaSystems in 1987 [40] and can be used to automate the scoring of binucleated or mononucleated cells containing MNi. These events are identified according to a set of parameters stored in the software, known as the classifier (Table 1). These parameters specify the morphology of the nuclei based on size and shape, and the distance between the nuclei and their size ratio, and a region of interest is denoted which limits the MNi identified in that area only and rejects other objects detected outside this region.

   (a) The protocol used to prepare the slides has been optimized for use with this system. The software settings are also altered according to the cell line used and whether mono- or binucleated cells are being scored. In general, the parameters used for detecting MNi in lymphoblastoid cells include the following:

   (b) Cells are harvested and fixed in hypotonic treatment 0.56 % KCl and centrifuged immediately for 10 min.

   (c) Then the cell pellet is resuspended in fixative 1: methanol:acetic acid:0.9 % NaCl (5:1:6 parts), incubated for 10 min at room temperature, and then centrifuged at 4 °C.

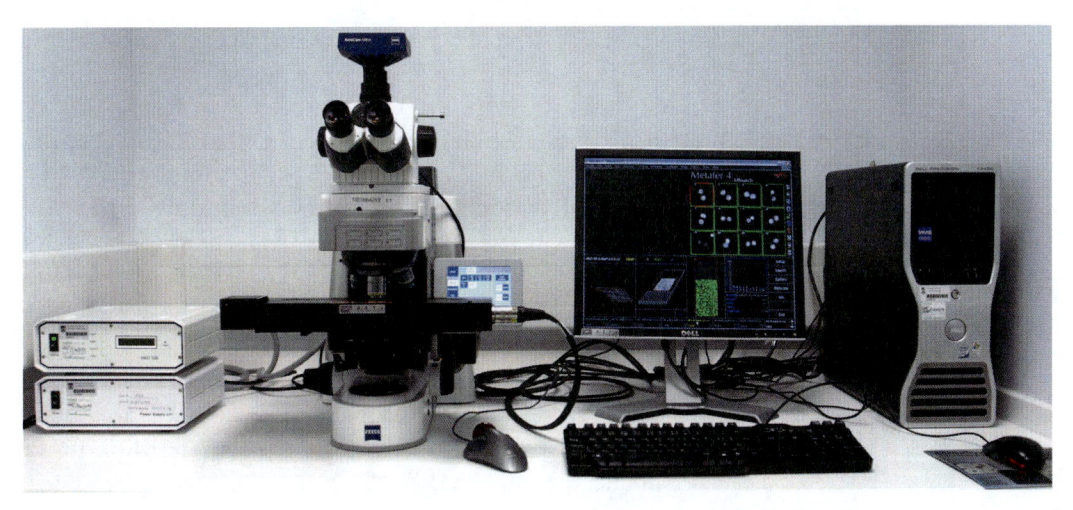

Fig. 3 A Carl Zeiss microscope connected to a computer with the Metafer software (picture courtesy of Mr. John Wills)

(d) Finally, the cell pellet is fixed with fixative 2: methanol:acetic acid (5:1 parts), incubated for 10 min at room temperature, and centrifuged at 4 °C. This step (without the incubation) is repeated four times.

(e) Cells are left in the second fixative at least overnight. The next day, cells are dropped onto slides that have been pre-soaked with fixative 2.

(f) These slides are allowed to dry and are stained with DAPI for 10 min.

(g) Slides are mounted onto the Metafer stage and scored for MNi in either the binucleated or the mononucleated cells. Up to eight slides can be loaded on the stage.

(h) After selecting the appropriate classifier settings (according to the cell type being analyzed) the microscope is focused on each slide and the analysis is initiated. The instrument performs the search scan and presents results as images in the gallery and as a histogram where the number of cells with 0, 1, 2, 3, and more MNi is presented (Fig. 3).

(i) The results can be visually inspected from the gallery images in order to confirm the data presented by the software. In this system, it is not possible to score tri-, tetra-, and multinucleated cells. However, it is possible to readily score more than 1,000 and up to 10,000 cells per slide per replicate per dose adding to the statistical robustness of the results.

7. Slides can be either stained with Giemsa and scored under a light microscope or stained with acridine orange for scoring under an epifluorescence microscope. 0.25 mg/mL Acridine

orange in PBS: Prepare a stock solution of 25 mg/mL by diluting a 0.25 g/mL concentrated solution in 10 mL PBS. Aliquot this stock into 1 mL volumes and store at −20 °C. Dilute 1 mL of this acridine orange stock in 100 mL (non-sterile) PBS and use for staining. Acridine orange staining: Dip slides in 0.25 mg/mL acridine orange for 10 s. Rinse twice in staining buffer (PBS at pH 7.8). Leave in fresh staining buffer for 1 h in the dark. Leave slides to air-dry in the dark. Mount 22×22 mm coverslip with PBS over the cytodot and view immediately under an epifluorescent microscope.

8. *Criteria for scoring nuclei*: The scoring criteria listed in the HUman MicroNucleus (HUMN) project are the following (for more details and illustrations of the parameters listed here refer to Fenech et al. [33]):

9. *Criteria for scoring MNi*: MNi are morphologically (size and shape) similar to but smaller than the nuclei of the cell being investigated. Fenech and colleagues have introduced a set of criteria by which it is easier to identify MNi and distinguish them from false positives [33]. The HUMN project led by Prof. Michael Fenech involving 30 laboratories worldwide aimed to identify variables in the methods used across different laboratories and to standardize these methodologies to generate more reliable and comparable results [33]. The HUMN project provides detailed instructions for scoring MNi in the CBMN system in lymphocytes. These criteria might change slightly with other cell lines and therefore need to be taken into account. These criteria are as follows:

    (a) The cells should be binucleated.

    (b) The cytoplasmic membrane of the cell should be clearly distinguishable from that of adjacent cells and should be intact.

    (c) Both nuclei of a binucleated cell should have relatively similar sizes, and similar staining and intensity of color.

    (d) The two nuclei of a binucleated cell should co-occur within the boundaries of the same cytoplasmic membrane and should be intact with a clear view of the nuclear membrane.

    (e) The nuclei should be separate from one another. They may touch but not overlap. However, if the nuclei are overlapping but the boundaries of each nucleus are clearly visible then this cell can be scored. Sometimes nuclei are connected by very thin nucleoplasmic bridges. This is still considered acceptable if the bridge is no wider than 1/4th of the diameter of the nucleus.

    (f) MNi are only scored in binucleated cells. Additional scoring of MNi in mononucleated cells could help in understanding levels of toxicity. Tri-, tetra-, and multinucleated and necrotic and apoptotic cells are not scored for the occurrence of MNi.

(g) MNi are smaller than nuclei. According to the HUMN project the size of MNi in human lymphocytes can be determined in proportion to the mean diameter of the main nucleus of a cell (1/16th to 1/3rd of the diameter) or the total area of the nuclei in a binucleated cell (1/256th and 1/9th of the area of one of the main nuclei).

For example; diameter of main nucleus is 5 µm².

Area of one of the nuclei in BN cells is 12 µm².

Size of a MN is 5 µm/3 = 1.6 µm or 12 µm²/9 = 1.3 µm².

These values are approximations. No need to perform actual measurements to determine the size of MNi. These approximations are only presented to make it easy to distinguish MNi from other events:

(h) MNi are non-retractable, that is, they do not deflect light from a straight path by refraction. This should allow the examiner to distinguish MNi from other stained objects within the cells such as particles.

(i) MNi are not attached or linked to the nuclei. MNi-like bodies could be present that are attached to one of the nuclei with a thin bridge. These are known as "budding bodies" and are not scored as MNi. MNi can be in very close proximity to the nucleus and appear as though they are touching. In this scenario, if the boundaries of both the nucleus and the MNi are clearly distinguishable then these MNi are scored.

(j) Most MNi will have the same staining pattern and intensity as the main nuclei. However, occasionally MNi might stain more intensely.

Illustrated examples for these criteria can be found in the HUMN project report [33].

10. Automated MN assay scoring: Unlike manual scoring, the automated Metafer system is only able to identify mononucleated or binucleated cells separately and classify the presence or the absence of MNi in each cell analyzed. When scanning of a predetermined cell number is complete, a gallery of images will appear. For simplicity, we will only refer to analysis of binucleated cells from the CBMN assay below, but the principle is exactly the same for the analysis of mononuclear cells. Binucleated cells with MNi will appear at the top of the gallery and must be examined to ensure that those MNi identified are appropriate. Once all the identified MNi have been accepted or

rejected, it is necessary to go through the remaining binucleate gallery as the software may occasionally miss some MNi (if gallery images are not clear then confirmation can be achieved by examining the cells in question under the microscope). It is best to do this with a ×100 objective lens. Due to its automation many thousands of cells can be scored using the Metafer system providing statistically robust data sets.

11. Treatment times with pepsin can vary for different cell lines and therefore, need to be optimized according to the cell line under study.

Acknowledgment

We would like to express our thanks to the EPSRC for the funding award granted to S.H.D. that supported B.M. and N.S.

References

1. Fenech M (2007) Cytokinesis-block micronucleus cytome assay. Nat Protoc 2(5): 1084–1104
2. OECD (1997) Test No. 474: mammalian erythrocyte micronucleus test. OECD Publishing, Paris
3. OECD (2010) Test No. 487: in vitro mammalian cell micronucleus test. OECD Publishing, Paris
4. Pincu M, Bass D, Norman A (1984) An improved micronuclear assay in lymphocytes. Mutat Res 139(2):61–65
5. Fenech M, Morley AA (1985) Measurement of micronuclei in lymphocytes. Mutat Res 147(1–2):29–36
6. Fenech M, Morley AA (1986) Cytokinesis-block micronucleus method in human lymphocytes: effect of in vivo ageing and low dose X-irradiation. Mutat Res 161(2):193–198
7. Carter SB (1967) Effects of cytochalasins on mammalian cells. Nature 213(5073):261–264
8. Kolber MA, Broschat KO, Landa-Gonzalez B (1990) Cytochalasin B induces cellular DNA fragmentation. FASEB J 4(12):3021–3027
9. Fenech M, Holland N, Chang WP et al (1999) The HUman MicroNucleus Project–An international collaborative study on the use of the micronucleus technique for measuring DNA damage in humans. Mutat Res 428(1–2): 271–283
10. Bousquet PF, Paulsen LA, Fondy C et al (1990) Effects of cytochalasin B in culture and in vivo on murine Madison 109 lung carcinoma and on B16 melanoma. Cancer Res 50(5):1431–1439
11. Lunov O, Syrovets T, Loos C et al (2011) Differential uptake of functionalized polystyrene nanoparticles by human macrophages and a monocytic cell line. ACS Nano 5(3): 1657–1669
12. Kirsch-Volders M, Sofuni T, Aardema M et al (2000) Report from the In Vitro Micronucleus Assay Working Group. Environ Mol Mutagen 35(3):167–172
13. OECD (2006) OECD guideline for the testing of chemicals draft proposal for a new guideline 487: in vitro micronucleus test. OECD, Paris
14. Hovhannisyan GG (2010) Fluorescence in situ hybridization in combination with the comet assay and micronucleus test in genetic toxicology. Mol Cytogenet 3:17
15. Singh N, Jenkins GJ, Nelson BC et al (2012) The role of iron redox state in the genotoxicity of ultrafine superparamagnetic iron oxide nanoparticles. Biomaterials 33(1):163–170
16. Gonzalez NV, Nikoloff N, Soloneski S et al (2011) A combination of the cytokinesis-block micronucleus cytome assay and centromeric identification for evaluation of the genotoxicity of dicamba. Toxicol Lett 207(3):204–212
17. Renzi L, Pacchierotti F, Russo A (1996) The centromere as a target for the induction of chromosome damage in resting and proliferating mammalian cells: assessment of mitomycin C-induced genetic damage at kinetochores and centromeres by a micronucleus test in mouse splenocytes. Mutagenesis 11(2):133–138
18. Sgura A, Antoccia A, Ramirez MJ et al (1997) Micronuclei, centromere-positive micronuclei

and chromosome nondisjunction in cytokinesis blocked human lymphocytes following mitomycin C or vincristine treatment. Mutat Res 392(1–2):97–107
19. Fowler P, Smith K, Young J et al (2012) Reduction of misleading ("false") positive results in mammalian cell genotoxicity assays. I. Choice of cell type. Mutat Res 742(1–2):11–25
20. OECD (2004) OECD guideline for the testing of chemicals, draft proposal for a new guideline 487: in vitro micronucleus test. OECD, Paris
21. Parry JM, Parry EM, Bourner R et al (1996) The detection and evaluation of aneugenic chemicals. Mutat Res 353(1–2):11–46
22. Doherty AT, Ellard S, Parry EM et al (1996) An investigation into the activation and deactivation of chlorinated hydrocarbons to genotoxins in metabolically competent human cells. Mutagenesis 11(3):247–274
23. Catalan J, Autio K, Kuosma E et al (1998) Age-dependent inclusion of sex chromosomes in lymphocyte micronuclei of man. Am J Hum Genet 63(5):1464–1472
24. Ehrlich V, Darroudi F, Uhl M et al (2002) Fumonisin B(1) is genotoxic in human derived hepatoma (HepG2) cells. Mutagenesis 17(3):257–260
25. Manshian BB, Jenkins GJ, Williams PM et al (2013) Single-walled carbon nanotubes: differential genotoxic potential associated with physico-chemical properties. Nanotoxicology 7:144–156
26. Mersch-Sundermann V, Knasmuller S, Wu XJ et al (2004) Use of a human-derived liver cell line for the detection of cytoprotective, antigenotoxic and cogenotoxic agents. Toxicology 198(1–3):329–340
27. Crespi CL, Thilly WG (1984) Assay for gene mutation in a human lymphoblast line, AHH-1, competent for xenobiotic metabolism. Mutat Res 128(2):221–230
28. Crespi CL, Gonzalez FJ, Steimel DT et al (1991) A metabolically competent human cell line expressing five cDNAs encoding procarcinogen-activating enzymes: application to mutagenicity testing. Chem Res Toxicol 4(5):566–572
29. Maron DM, Ames BN (1983) Revised methods for the Salmonella mutagenicity test. Mutat Res 113(3–4):173–215
30. OECD (2007) In Vitro Mammalian Cell Micronucleus Test (MNvit), in OECD guideline for the testing of chemicals draft proposal for a new guideline 487. OECD, Paris
31. Fenech M (2010) The lymphocyte cytokinesis-block micronucleus cytome assay and its application in radiation biodosimetry. Health Phys 98(2):234–243
32. Lorge E, Hayashi M, Albertini S et al (2008) Comparison of different methods for an accurate assessment of cytotoxicity in the in vitro micronucleus test. I. Theoretical aspects. Mutat Res 655(1–2):1–3
33. Fellows MD, O'Donovan MR, Lorge E et al (2008) Comparison of different methods for an accurate assessment of cytotoxicity in the in vitro micronucleus test. II: practical aspects with toxic agents. Mutat Res 655(1–2):4–21
34. Kirsch-Volders M, Sofuni T, Aardema M et al (2003) Report from the in vitro micronucleus assay working group. Mutat Res 540(2):153–163
35. Surralles J, Xamena N, Creus A et al (1995) Induction of micronuclei by five pyrethroid insecticides in whole-blood and isolated human lymphocyte cultures. Mutat Res 341(3):169–184
36. International Atomic Energy Agency (2001) Cytogenetic analysis for radiation dose assessment - a manual. Technical reports series No. 405, 13, 122
37. Ellard S, Parry EM (1993) A modified protocol for the cytochalasin B in vitro micronucleus assay using whole human blood or separated lymphocyte cultures. Mutagenesis 8(4):317–320
38. Doak SH, Griffiths SM, Manshian B et al (2009) Confounding experimental considerations in nanogenotoxicology. Mutagenesis 24(4):285–293
39. Gonzalez L, Sanderson BJ, Kirsch-Volders M (2011) Adaptations of the in vitro MN assay for the genotoxicity assessment of nanomaterials. Mutagenesis 26(1):185–191
40. Schunck C, Johannes T, Varga D et al (2004) New developments in automated cytogenetic imaging: unattended scoring of dicentric chromosomes, micronuclei, single cell gel electrophoresis, and fluorescence signals. Cytogenet Genome Res 104(1–4):383–389

Part III

Tests for Primary DNA Damage

Chapter 15

γ-H2AX Detection in Somatic and Germ Cells of Mice

Eugenia Cordelli and Lorena Paris

Abstract

The phosphorylation of histone H2AX at serine 139 (γ-H2AX) is one of the first steps of DNA damage response and its detection is widely used as a sensitive marker for DNA double-strand breaks induced by ionizing radiation or other genotoxic agents. Immuno-stained phosphorylated histone can be measured in single cells by flow cytometry or single γ-H2AX foci can be visualized and counted microscopically in histological or cytological preparations. Animal studies are well recognized as important tools to study mechanisms of in vivo response to genotoxic stress. Tissues are composed by many cell types differing for function, differentiation, and proliferative capacity. In particular, due to the complexity of spermatogenesis and the heterogeneity of testicular cell subpopulations, an accurate characterization of damage in this tissue is difficult and requires an approach which allows the identification of damage in the different cellular compartments. This chapter presents techniques for γ-H2AX detection in mouse bone marrow and testicular cells. Furthermore, advantages and weaknesses of flow cytometric and microscopic methods are described.

Key words γ-H2AX, DNA damage, Bone marrow, Germ cells, Mouse, Immunocytochemistry, Flow cytometry

1 Introduction

Double-strand breaks (DSB) are among the most deleterious type of DNA damage. They cause toxic effects if not repaired or mutations if repaired incorrectly. One of the first steps in the cellular response to DSB is phosphorylation of serine 139 on histone H2AX (γ-H2AX) in the chromatin flanking the DNA double-stranded ends which form nuclear foci at DSB sites [1, 2]. A direct correlation has been found between H2AX phosphorylation and the number of DSBs induced by ionizing radiation [3]. Therefore, counting of γ-H2AX foci has been introduced and is now widely used for the quantitative evaluation of the induction and repair of DSBs by various cytotoxic agents, including ionizing radiation [4–7].

Wide variations among different cell types have been observed both in untreated cells and after genotoxic stress [4, 8].

Furthermore, H2AX histone can also become phosphorylated independently from DNA breaks. An example is the phosphorylation of the histone associated with sex chromosomes in XY body of pachytene spermatocytes or the presence of numerous large foci, not associated with DNA strand breaks, in pluripotent mouse embryonic stem cells [9–11]. Moreover, the removal of γ-H2AX foci does not always coincide with the actual time of DSB repair [4]. In particular, γ-H2AX foci have been observed at the sites of chromosomal aberrations (CA) even when the formation of these aberrations has been completed [12, 13] or long times after irradiation when strand breaks are no more evident with other tests [14]. Together with the reduction in number, changes in the morphology of foci have been observed with time upon genotoxic insult [4, 14].

Assessment of γ-H2AX relies on the use of antibodies directed against the phosphorylated histone which are currently commercially available. γ-H2AX foci can be detected microscopically by immunohistochemical or immunocytochemical assays or the level of phosphorylated histone can be measured by immunoblotting or flow cytometry. Microscopic and flow cytometric methods have the advantage to detect immunostaining in individual cells, thus allowing the identification of cell subpopulations with different levels of DNA damage in heterogeneous cell populations. This property is most significant in in vivo studies in which tissues are composed by many cell types differing for function, differentiation, and proliferative capacity.

With immunohistochemical and immunocytochemical approaches individual foci are scored and results are given as percentage of positive nuclei, or number of foci/nucleus. Alternatively, nuclei can be classified in different classes depending on the number of foci. Immunohistochemistry preserves the architecture of the tissue and different cell types are easily recognized, but overlapping among neighboring cells may cloud the scoring of foci.

1.1 Comparison Between Microscopic and Flow Cytometric Detection of H2AX Phosphorylation: Advantages and Drawbacks of the Two Methods

Microscopic quantification of foci by wide field fluorescence microscopy is the most frequently used method to assess H2AX phosphorylation. Nevertheless, factors such as foci overlapping, background staining, out-of-focus fluorescence, and autofluorescence may reduce the ability to clearly distinguish individual foci. Also, differences in the dimension of foci can lead to discrepancies among laboratories in what is considered a focus. Nevertheless, the method is extremely sensitive in the detection of ionizing radiation-induced DNA damage as γ-H2AX foci are induced in in vitro cell culture at doses as low as 1.2 mGy [15]. In the same study, a linear relationship between the number of foci per cell and the dose delivered was observed between 1.2 mGy and 2 Gy. A linear dose–response curve has been also observed from 0.1 to 1 Gy in cells from different mouse tissues irradiated in vivo [6].

At higher doses, the overlapping of foci can lead to an underestimation of the actual number of foci [16]. The minimum reported dose detection limit for flow cytometry-based γ-H2AX analysis is higher, approximately 200 mGy for irradiated cell cultures [17], in comparison to microscopic detection, but a linear dose–response curve is obtainable for a larger interval of doses. The main advantage of flow cytometry is the speed of the analysis which allows the simple detection of γ-H2AX in a large number of cells; the other side is the larger sample needed for flow cytometric than for microscopic analysis. Another advantage of flow cytometry is the possibility of multi-parametric analysis which allows correlating, within the same individual cells, the extent of H2AX phosphorylation with other parameters such as DNA content, apoptosis, and membrane markers [18]. An important consideration when comparing the two methods is that although both are based on immunofluorescence, flow cytometry detects total fluorescence intensity in each cell while microscopy allows the scoring of individual foci. This can lead to different results considering, for example, that a high number of small foci can give the same fluorescence intensity of a small number of large foci. This consideration highlights the complementarity between flow cytometry- and microscopy-based methods, especially to describe phenomena that vary in heterogeneous cell populations and are characterized by quantitative as well as topological variations of histone phosphorylation.

2 Materials

2.1 Isolation of Cell Suspensions

1. Phosphate-buffered saline (PBS), pH 7.4.
2. 0.02 % Na$_2$EDTA prepared in PBS.
3. 70 % Ethanol.
4. 1 mL Syringe.
5. Pasteur pipettes.
6. Petri dishes (35 mm).
7. Scissors.
8. Forceps.
9. Centrifuge.
10. 70 μm Nylon mesh.
11. 15 mL Conical centrifuge tubes.
12. Vortex.

2.2 Immunocytochemistry on Slide-Seeded Cell Suspensions

1. Microscope slides.
2. 2 % Paraformaldehyde.
3. PBS + 0.05 % Triton X-100.

4. Blocking buffer + Triton X-100 (BBT): PBS, 0.05 % Triton X-100, 5 % no-fat dry milk (*see* **Note 1**).

5. Blocking buffer (BB): PBS + 5 % no-fat dry milk (*see* **Note 2**).

6. Primary antibody: Mouse monoclonal antibody anti-phospho-histone H2AX (Upstate Biotechnology, Lake Placid, NY) diluted 1:1,000 in BBT solution (*see* **Note 3**).

7. Secondary antibody: Goat anti-mouse IgG conjugated with Alexa Fluor 488 (Molecular Probes, Eugene OR, USA) diluted 1:2,000 in BB solution (*see* **Note 3**).

8. PBS.

9. Antifade solution: Vectashield containing 1.5 μg/mL 4′,6-diamidino-2-phenylindole (DAPI) (Vector Laboratories, Inc. Burlingame, CA, USA).

10. Coverslips.

11. Cytocentrifuge (Cytospin 2, Shandon).

12. Parafilm (*see* **Note 4**).

13. Coplin jar.

14. Incubation box (*see* **Note 5**).

15. Micropipette.

16. Nail polish.

17. Fluorescence microscope equipped with 20 and 100× objectives.

2.3 Flow Cytometry

1. Tris-buffered saline (TBS), pH 7.4: 0.1 M Tris, 0.1 M NaCl, adjust pH to 7.4 with 1 N HCl.

2. TBS containing serum and Triton X-100 (TST): TBS + 4 % fetal bovine serum (FBS) + 0.1 % Triton X-100.

3. Primary antibody: Mouse monoclonal antibody anti-phospho-histone H2AX (Upstate Biotechnology, Lake Placid, NY) diluted 1:250 in TST solution (*see* **Note 3**).

4. 2 % TBS: TBS containing 2 % FBS.

5. Secondary antibody: Goat anti-mouse IgG conjugated with Alexa Fluor 488 (Molecular Probes, Eugene OR, USA) diluted 1:250 in TST solution (*see* **Note 3**).

6. Stock solution of 1 mg/mL propidium iodide (PI) prepared in PBS.

7. Stock solution of 1 mg/mL DNAse-free RNAse prepared in PBS.

8. Ice.

9. Centrifuge.

10. Rotating mixer.

11. 15 mL Conical centrifuge tubes.
12. Micropipette.
13. 5 mL BD Falcon round-bottom tubes.
14. Cytometer. FACSCalibur (Becton Dickinson, San Jose, CA, USA).

2.4 Immunohistochemistry on Testis Sections

1. Coplin jar.
2. Phosphate-buffered formaldehyde: 4 %. Dilute formalin 1:10 with phosphate buffer.
3. Xylene.
4. Absolute ethanol.
5. Ethanol series: 96, 90, 80, 70, 50 % Ethanol.
6. PBS.
7. Peroxidase quenching solution: 3 % hydrogen peroxide solution in methanol.
8. Blocking solution A (Kit Zymed).
9. Blocking solution B (Kit Zymed).
10. Primary antibody: Mouse monoclonal antibody anti-phospho-histone H2AX (Upstate Biotechnology, Lake Placid, NY) dilution 1:200 in PBS.
11. Secondary antibody: Reagent 2 (Kit Zymed).
12. Amino-ethyl-carbazole (AEC) solution: Reagent 3 (Kit Zymed).
13. Glycerol Vinyl Alcohol (GVA) mounting solution (reagent 5, Kit Zymed).
14. Distilled water.
15. Tap water.
16. Hematoxylin.
17. Nail polish.
18. Paraffin.
19. Incubation box (*see* **Note 5**).
20. Micropipette.
21. Microtome.
22. Microscopic slides.
23. Coverslips.
24. Oven.
25. A phase contrast research microscope with 20, 63, and 100× objectives.

3 Methods

3.1 Immunocytochemical Detection of γ-H2AX in Bone Marrow and Testicular Cells

3.1.1 Bone Marrow Cells: Isolation and Fixation

1. Sacrifice the animals according to regulatory guidelines applied in your country and remove both femurs.
2. Remove muscles and fat from the bones and cut the tips of the femurs.
3. With a 1 mL syringe containing 1 mL of PBS with 0.02 % EDTA at room temperature, flush the bone marrow from each femur into a conical centrifuge tube inserting the needle in the opening at both ends (repeat the flushing until the bone is white and empty).
4. Pass the cell suspension gently through the needle to remove any clumps (pay attention not to make air bubbles in the cell suspension).
5. Add 9 mL of PBS with 0.02 % EDTA.
6. Centrifuge the cell suspension at $180 \times g$ for 10 min and discard the supernatant.
7. Resuspend the pellet in 5 mL of 70 % ethanol (to avoid clumps, add ethanol drop by drop while mixing on the vortex).
8. Store the sample at −20 °C (fixed samples can be stored for several months). The fixed cell suspensions can be used both for immunocytochemical (*see* Subheading 3.1.3) and flow cytometric analysis (*see* Subheading 3.2).

3.1.2 Testis Cells: Isolation and Fixation

1. Sacrifice the animals according to regulatory guidelines applied in your country.
2. Dissect and remove the testis.
3. Place the testis in a Petri dish.
4. Cut one side of tunica albuginea and, with the aid of forceps, squeeze the seminiferous tubules out.
5. Add a few drops of PBS and mince the tubules with scissors.
6. Add 2 mL of PBS and pipette with Pasteur pipette.
7. Filter through a 70 μm nylon mesh into a conical centrifuge tube.
8. Centrifuge the cell suspension at $180 \times g$ for 10 min and discard the supernatant.
9. Resuspend the pellet in 5 mL of 70 % ethanol (to avoid clumps, add ethanol drop by drop while mixing on the vortex).
10. Store at −20 °C (fixed samples can be stored for several months). The fixed cell suspension can be used both for immunocytochemical (*see* Subheading 3.1.3) and flow cytometric analysis (*see* Subheading 3.2).

3.1.3 Immunocytochemistry

1. Cytospin an aliquot containing 10^5 cells (bone marrow cells: 500 rpm for 5 min; testis cells: 600 rpm for 5 min) onto a microscope slide (control the quality of the slide in terms of cell concentration).
2. Place slides into a Coplin jar containing 2 % paraformaldehyde to fix the cells and incubate for 30 min at room temperature.
3. Wash the slides in PBS+0.05 % Triton X-100 for 20 min to permeabilize the cells.
4. Drain the slides.
5. Put 100 μL of BBT solution on the slide, cover with a precut parafilm, and incubate for 20 min at 37 °C in a humid incubation box, to block nonspecific interactions and further permeabilize cells (*see* **Note 5**).
6. Remove the parafilm and drain the slides.
7. Put 100 μL of anti-γ-H2AX antibody on the slide, cover with precut parafilm, and incubate for 20 min at 37 °C and overnight at 4 °C in an incubation box.
8. Remove the parafilm and drain the slides.
9. Wash the slides in PBS+0.05 % Triton X-100 twice for 15 min each.
10. Put 100 μL of BB solution on the slide to prevent nonspecific interactions, cover with a precut parafilm, and incubate for 30 min at 37 °C in an incubation box.
11. Remove the parafilm and drain the slides.
12. Put 100 μL of fluorochrome-bound secondary antibody on the slide, cover with a precut parafilm, and incubate for 2 h at 37 °C in an incubation box.
13. Remove the parafilm and drain the slides.
14. Wash the slides in PBS twice for 15 min each.
15. Drain the slides.
16. Put on the slides 100 μL of antifade solution containing DAPI. Cover with coverslips.
17. Seal the edges of the coverslip with nail polish.
18. Proceed to the microscopic analysis.

3.1.4 Microscopy

1. View the slides using a fluorescence microscope at 1,000× magnification (*see* **Note 6**). In our experience, without the aid of a confocal microscope, it is difficult to count the number of foci per nucleus, when the number of foci exceeds 8–10 foci/nucleus. For this reason we generally use as a parameter the percentage of positive nuclei (considering that positive nuclei are those with one or more foci) or classify nuclei in different classes depending on the number of foci (i.e., 0–1 foci; 2–5 foci; 6–10 foci; >10 foci).

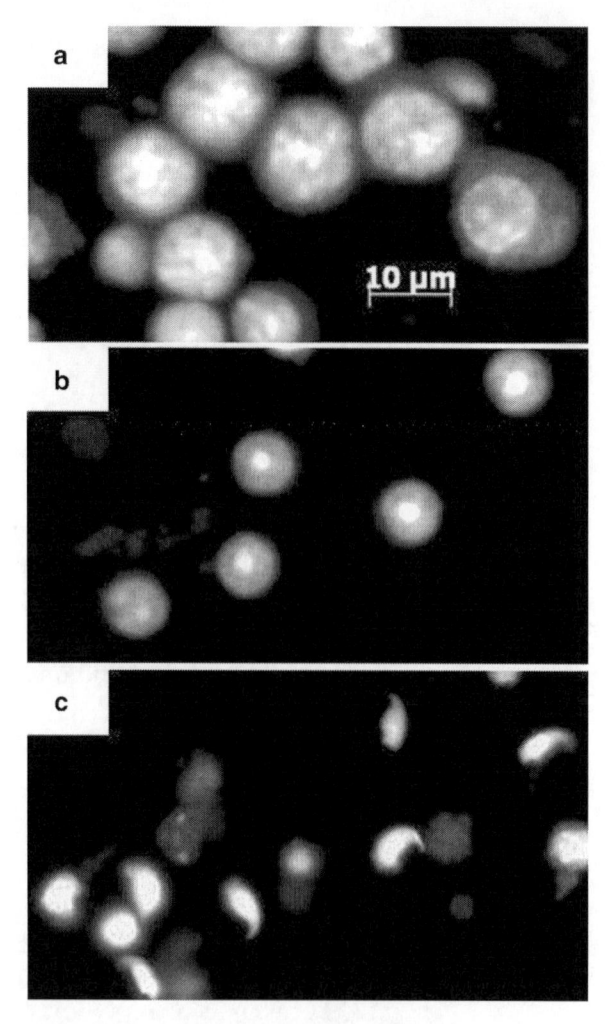

Fig. 1 Photomicrographs of DAPI-stained testicular cells spreads. (**a**) Spermatocytes; (**b**) round spermatids; (**c**) elongated spermatids

2. For bone marrow, count at least 100 nuclei. In addition to nuclei with foci, after irradiation it is possible to observe bone marrow nuclei with a bright diffuse γ-H2AX staining, corresponding to cells in early apoptotic stage. These cells are separately recorded from cells with foci.

3. In testis preparations, different types of cells can be observed and classified according to their shape and/or nuclear morphology [14, 19], e.g., spermatocytes, round spermatids, and elongated spermatids (Fig. 1). Among spermatocytes, those at pachytene stage are easily recognized by the γ-H2AX-positive XY body. The number of foci or the percentage of positive nuclei can be measured for each of these cell subpopulations (*see* **Note 7**). The percentage of spermatogonia and Sertoli cells is extremely small using the suggested protocol for

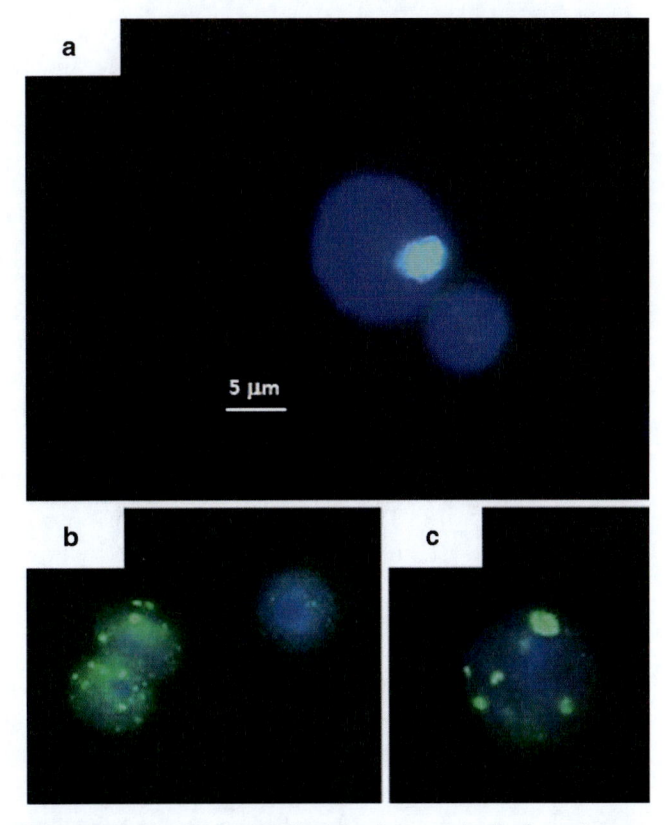

Fig. 2 Photomicrographs of γ-H2AX-stained mouse testicular cell suspensions. (a) Unirradiated cells: One pachytene spermatocyte showing a bright stained XY body and one round spermatid; (b) three round spermatids 1 h after 1 Gy; (c) one pachytene spermatocyte 1 h after 1 Gy. *Green*: γ-H2AX; *blue*: DNA stained with DAPI

disaggregation and finding these cells in the cytological preparation to count foci is arduous. In Fig. 2 representative images of γ-H2AX staining in isolated mouse testicular cells are reported.

3.2 Detection of γ-H2AX in Bone Marrow and Testicular Cells by Flow Cytometry

3.2.1 Sample Preparation

1. Transfer an aliquot of fixed cell suspension containing approximately 2×10^6 cells in a conical centrifuge tube (*see* **Note 8**).
2. Add 10 mL of TBS solution, to wash the cells. Centrifuge at $300 \times g$ for 5 min and discard the supernatant.
3. Resuspend the pellet in 1 mL of cold TST solution to block nonspecific interactions and further permeabilize cells and incubate for 10 min on ice.
4. Centrifuge at $300 \times g$ for 5 min and remove the supernatant.
5. Resuspend the pellet in 200 μL of anti-γ-H2AX antibody.
6. Keep for 2 h at room temperature in a rotating mixer.

7. Add 5 mL of 2 % TBS to wash the cells, centrifuge at 300×g for 5 min, and remove the supernatant.

8. Repeat **step 7**.

9. Resuspend the pellet in 200 μL of fluorochrome-bound secondary antibody. Keep for 1 h at room temperature in a rotating mixer in the dark.

10. Add 5 mL of TBS, centrifuge at 300×g for 5 min, and remove the supernatant.

11. Repeat **step 10**.

12. Resuspend the cells in 500 μL of PBS containing 5 μg/mL propidium iodide and 40 μg/mL DNAse-free RNAse (see **Note 9**). The stained cell suspension can be measured after 10 min or maintained at 4 °C in refrigerator overnight.

13. Transfer the cell suspension to a 5 mL BD Falcon round-bottom tube and analyze by flow cytometry.

3.2.2 Flow Cytometry

1. A negative control, in which the primary antibody is omitted, is processed in each measuring session, to provide a measure of the nonspecific signal. Its value is subtracted from the γ-H2AX mean values measured in the samples of the same measurement session, and used as threshold for γ-H2AX-positive cells.

2. Analyze cells by a FACSCalibur flow cytometer equipped with an air-cooled argon ion laser. Green (Alexa Fluor 488) fluorescence is collected after a 530±30 nm band-pass filter and red (propidium iodide) fluorescence after a LP620 long-pass filter.

3. Red fluorescence width is plotted versus red fluorescence area to discriminate and eliminate doublets from the measurements (Fig. 3). Results can be calculated on FL1 histogram as mean of FL1 peak or as percentage of positive cells.

4. In Fig. 4 representative histograms of a negative control, unirradiated cells, or cells 1 h after in vivo irradiation with 4 Gy are reported. Furthermore, different cell subpopulations can be distinguished based on their DNA content and the mean value of γ-H2AX fluorescence and the percentage of positive cells can be separately calculated for each cell subpopulation by the computer-interactive "gating" analysis. In particular, γ-H2AX fluorescence in G_1/G_0, S, or G_2/M phases of cell cycle can be assessed in bone marrow cells (Fig. 5).

5. In testis, different cellular subpopulations can be distinguished based on differences in their DNA content. Typical DNA content fluorescence intensity distribution histograms from adult mouse testicular cells are characterized by three main peaks representing cells with 1C, 2C, and 4C DNA content. The 1C region includes spermatids. Diploid G1-phase spermatogonia

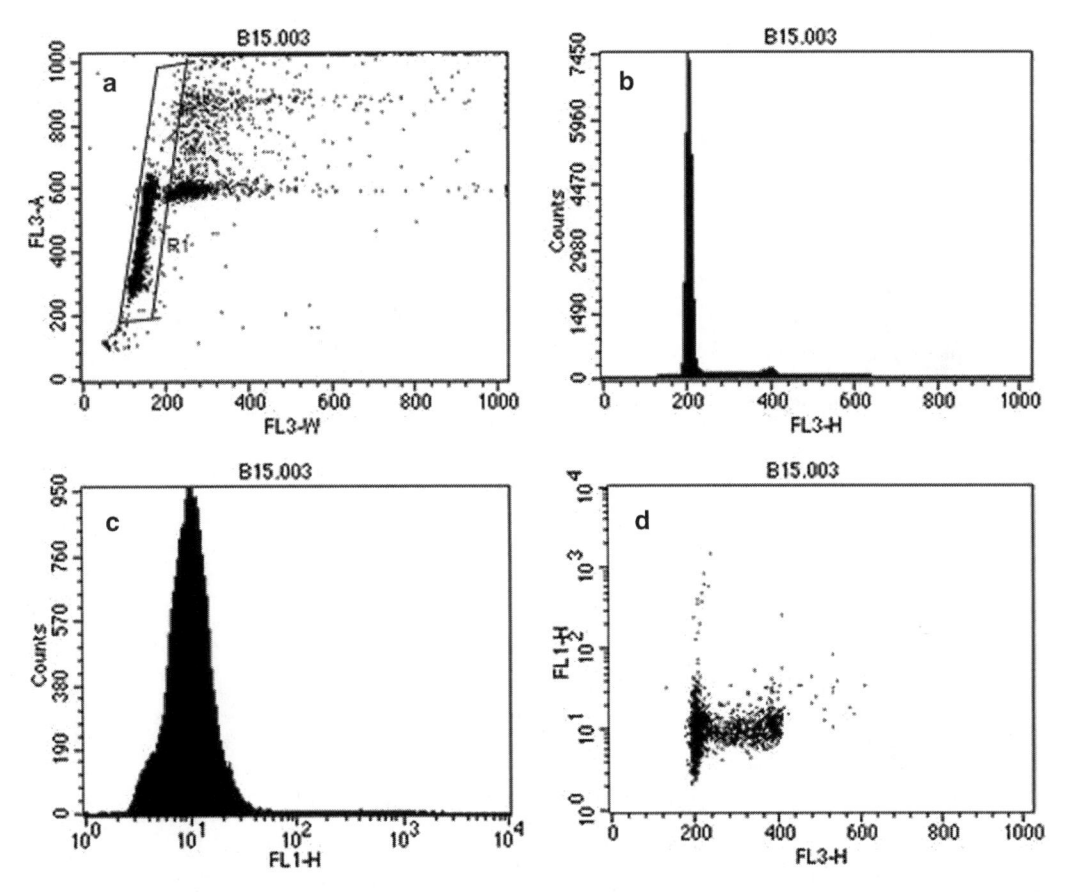

Fig. 3 Flow cytometric analysis—representative histograms and cytograms of unirradiated bone marrow cells. (**a**) Red fluorescence width (Fl3-W) plotted versus red fluorescence area (Fl3-A). The gate region R1 excludes cellular debris and doublets and captures single nuclei. Only gated events are included in histograms (**b**, **c**) and cytogram (**d**). (**b**) Univariate DNA content distribution histogram. On this histogram cells can be discriminated on the basis of DNA content and their relative position in the cell cycle (G1, S, G2/M). (**c**) Univariate γ-H2AX content distribution histogram. (**d**) Biparametric cytogram of DNA content (Fl3-H) versus γ-H2AX content (Fl1-H). Each cell is concurrently characterized by DNA and γ-H2AX content

and secondary spermatocytes are recorded in the second peak (2C), together with testis somatic cells. The third peak refers to cells with 4C DNA content (4C) which include some G2/M spermatogonia but, mostly, primary spermatocytes. Actively DNA-synthesizing S-phase cells (S) are located in the region between 2C and 4C peaks. The mean value of γ-H2AX fluorescence and the percentage of positive cells can be separately calculated for each of these cell subpopulations (Fig. 6). In testicular cells the evaluation of γ-H2AX fluorescence in different cell subpopulations is particularly important because of the heterogeneity of γ-H2AX content evidenced by the broad FL1 distribution histogram (Fig. 6c).

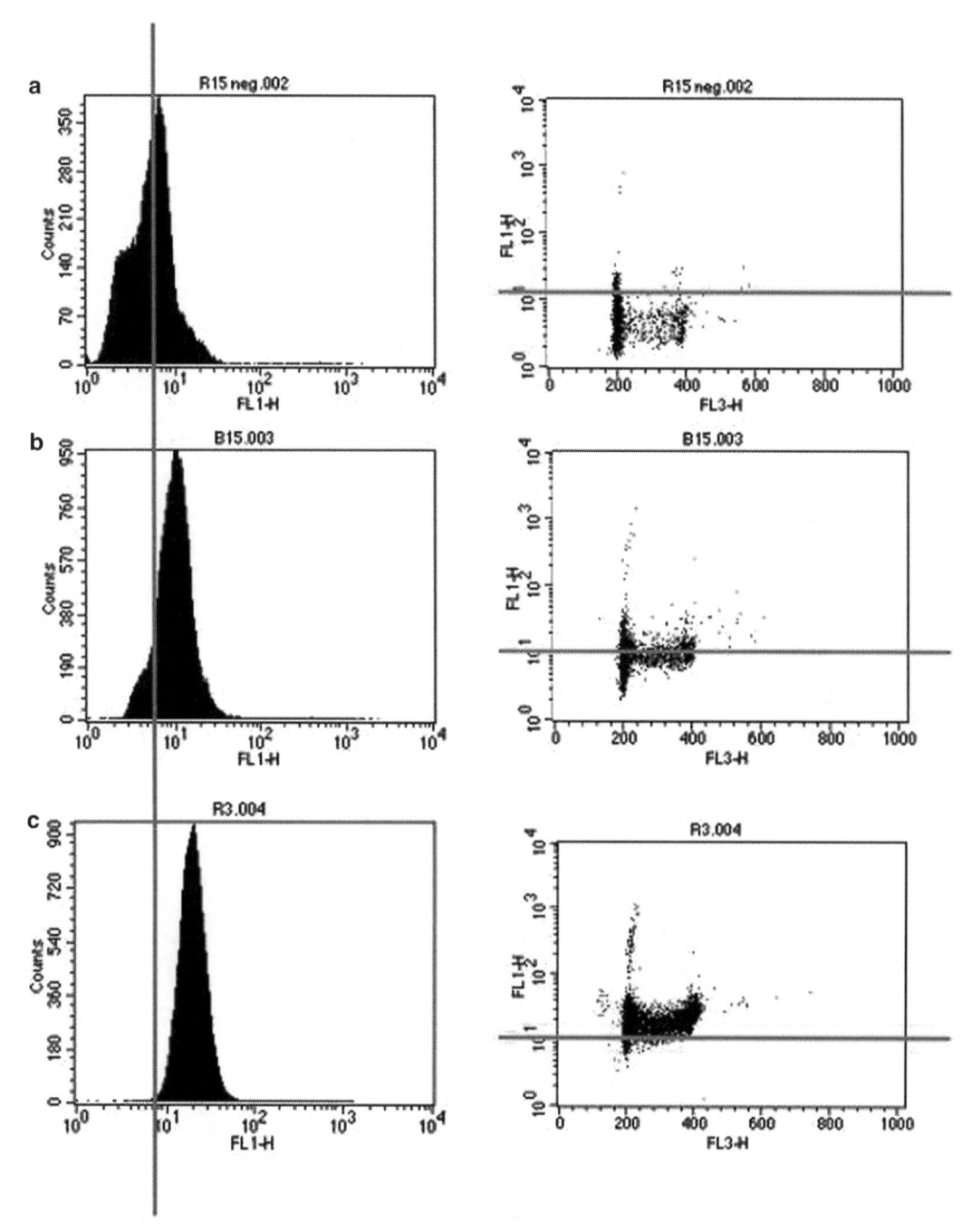

Fig. 4 Flow cytometric analysis—representative histograms and cytograms of bone marrow cells. *Left*: FL1 histograms; *right*: FL3 (DNA) versus FL1 (γ-H2AX) cytograms; **(a)** negative control, **(b)** unirradiated cells, **(c)** cells 1 h after in vivo irradiation with 4 Gy. The *vertical bar* on histograms marks the channel of mean FL1 fluorescence of negative control. The *horizontal bar* on cytograms divides the regions of positive and negative γ-H2AX cells

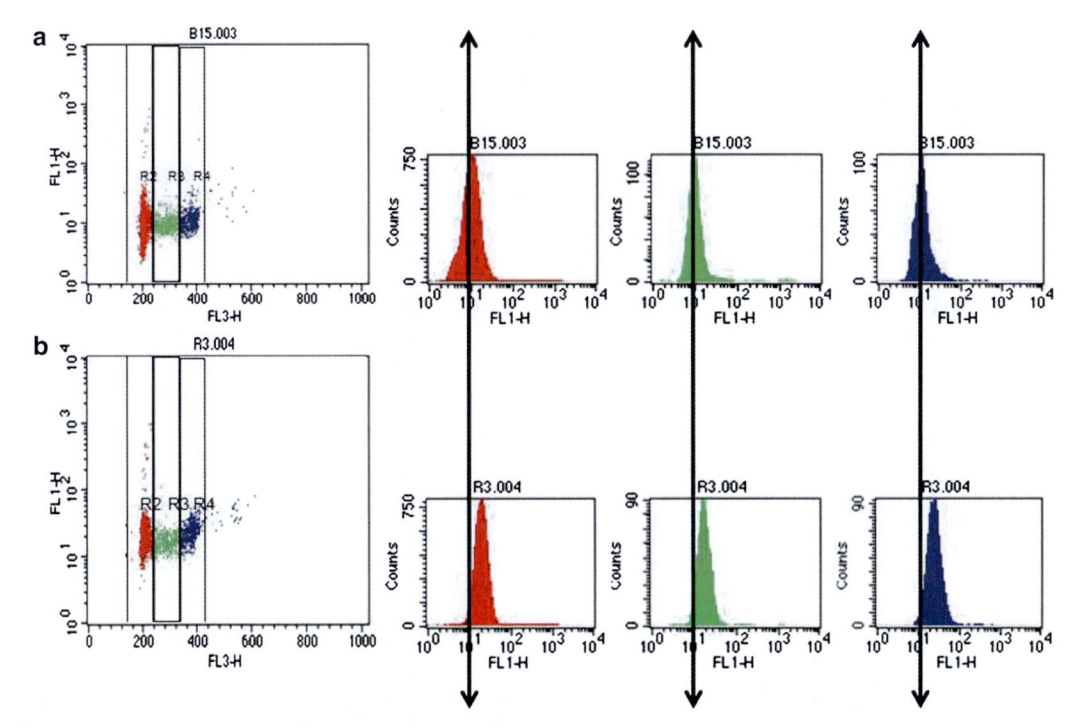

Fig. 5 Flow cytometric analysis of (**a**) unirradiated bone marrow cells or (**b**) cells 1 h after in vivo irradiation with 4 Gy. On the cytograms (*left*), three regions can be discriminated on the basis of DNA content: (R2) G_1/G_0 cells, (R3) S-phase cells, and (R4) G_2/M cells. γ-H2AX fluorescence can be separately evaluated in these subpopulations (*right*). The *vertical bar* on histograms marks, for each subpopulation, the channel of mean FL1 fluorescence of unirradiated cells

3.3 Immunohistochemical Detection of γ-H2AX in Testis Sections

3.3.1 Fixation of Sections

1. Fix the testes in 4 % phosphate-buffered formaldehyde for 24 h at room temperature (*see* **Note 10**).
2. Place the fixed testes in 50 % ethanol for 24 h.
3. Then place the testes in 70 % ethanol for 24 h (testes can be kept in ethanol up to 20–25 days).
4. Embed the testes sample in paraffin.
5. Cut 5 μm thick sections and mount on glass slides.
6. Dry the slides at 37 °C.

3.3.2 Sample Deparaffinization and Rehydration

1. Place the slides in an oven at 60 °C for 30 min.
2. Immerse the slides in xylene for 10 min × 3 times.
3. Immerse the slides in absolute ethanol for 3 min × 2 times.
4. Immerse the slides in 96 % ethanol for 3 min.
5. Immerse the slides in 90 % ethanol for 3 min.
6. Immerse the slides in 80 % ethanol for 3 min.

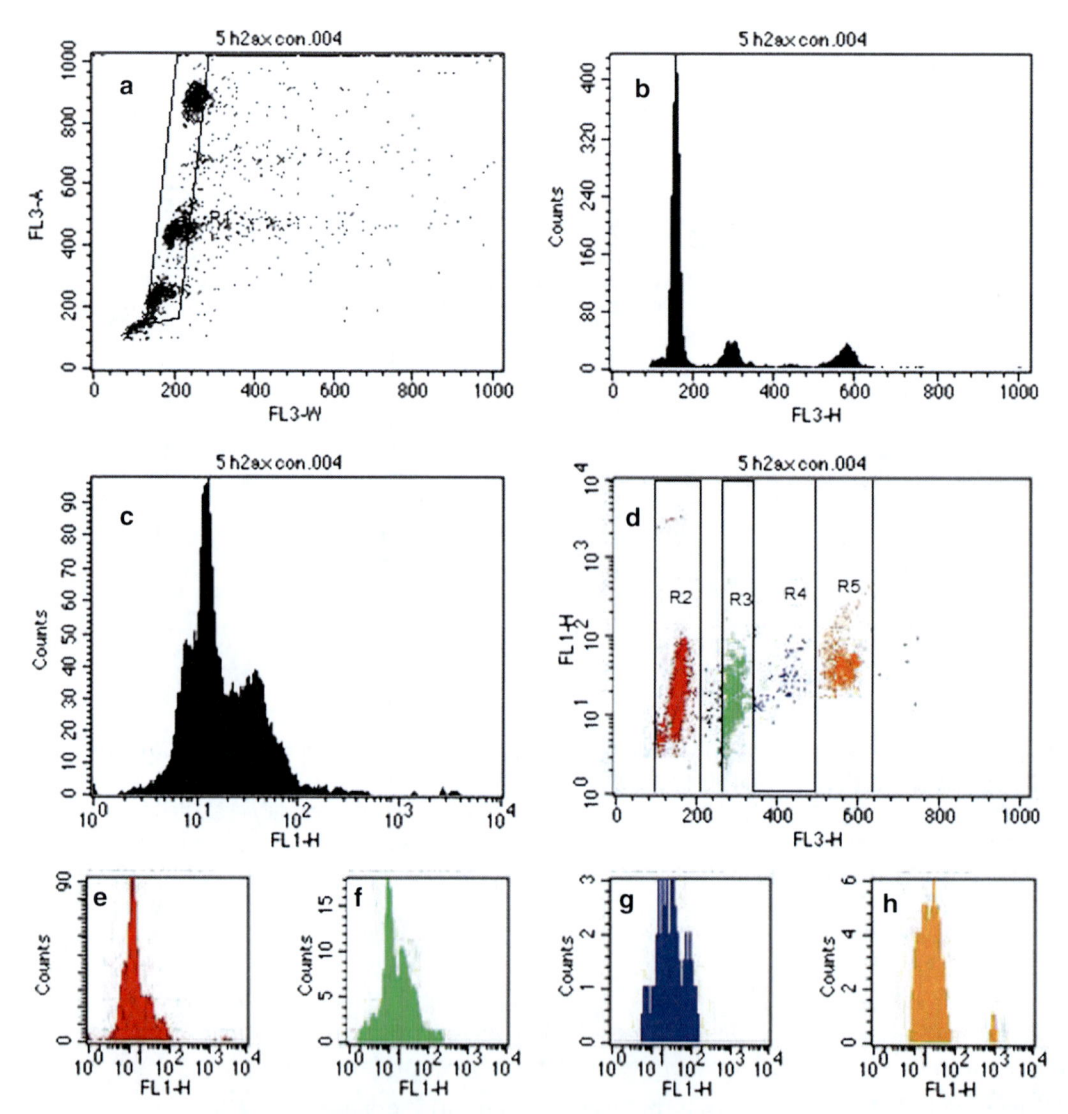

Fig. 6 Flow cytometric analysis—representative histograms and cytograms of unirradiated testicular cells. (**a**) Red fluorescence width plotted versus red fluorescence area to discriminate and eliminate doublets from the measurements. (**b**) FL3 (DNA) histogram. (**c**) FL1 (γ-H2AX) histogram. (**d**) FL3 versus FL1 cytogram in which four regions can be discriminated on the basis of DNA content: (R2) 1C cells, (R3) 2C cells, (R4) S-phase cells, and (R5) 4C cells. γ-H2AX fluorescence can be separately evaluated in these subpopulations (**e–h**)

7. Immerse the slides in 70 % ethanol for 3 min.
8. Wash the slides in PBS for 2 min × 3 times (do not allow testis section to dry from this point on).

3.3.3 Immunohistochemistry

Process rehydrated sections according to instructions provided by the supplier (we use HistoMouse™-Kit Zymed; Invitrogen, CA, USA).

1. Quench endogenous peroxidase by 10-min incubation in peroxidase quenching solution (3 % hydrogen peroxide in methanol).
2. Wash the slides in PBS for 2 min × 3 times.
3. Add two drops of blocking solution A, to block nonspecific interactions, cover with a coverslip, and incubate for 30 min at room temperature in an incubation box.
4. Wash the slides in distilled water.
5. Add two drops of blocking solution B, to block nonspecific interactions, cover with a coverslip, and incubate for 10 min at room temperature in an incubation box.
6. Wash the slides in distilled water.
7. Wash the slides in PBS for 2 min × 3 times.
8. Apply on the slide 100 µL of anti-γ-H2AX antibody, cover with a coverslip, and incubate for 60 min at room temperature in an incubation box.
9. Wash the slides in PBS for 2 min × 3 times.
10. Apply on the slide 100 µL of secondary antibody–peroxidase conjugate (reagent 2), cover with a coverslip, and incubate for 15 min at room temperature in an incubation box.
11. Wash the slides in PBS for 2 min × 3 times.
12. Apply 100 µL of cold chromogen AEC single solution (reagent 3) and incubate for 13 min at room temperature in an incubation box.
13. Rinse well with distilled water.
14. Counterstain the slides by immerging them in a Coplin jar containing hematoxylin for 20 s in order to have a good distinction of nuclei and cytoplasm.
15. Wash the slides in tap water and then rinse with distilled water.
16. Apply two drops of GVA mounting solution (reagent 5) to the slide and mount with a coverslip.
17. Air-dry the slides overnight and seal the edges of the coverslip with nail polish.

3.3.4 Microscopy

1. The organization of germ cells within the mouse seminiferous tubule has been widely studied [20] and the exact description of the Oakberg's 12-stage classification of seminiferous tubules is beyond the scope of this paragraph. However, in the section of tubules, cells in different stages of differentiation can be discriminated (Fig. 7).
2. In testis, H2AX becomes phosphorylated in different phases of germ cell differentiation. In particular, γ-H2AX intense staining is evident in the XY body in pachytene spermatocytes; cells in

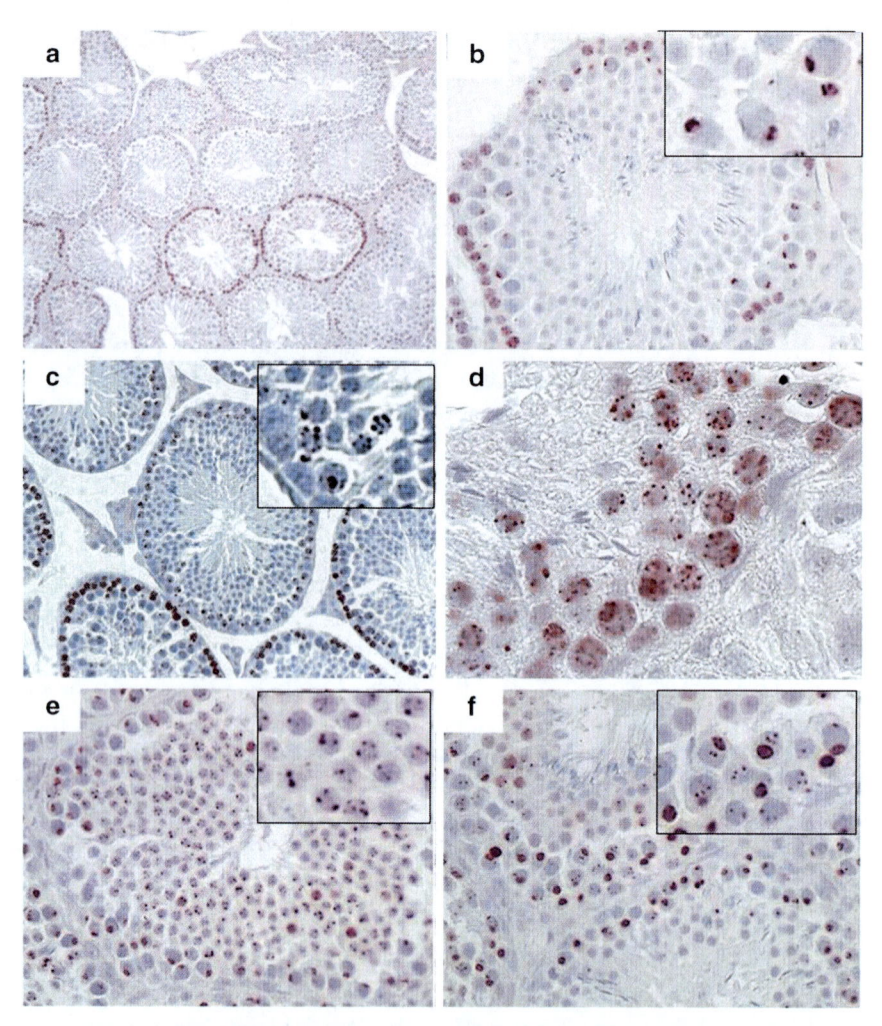

Fig. 7 Photomicrographs of γ-H2AX-stained mouse testicular sections. (**a**) 100× magnification of a section from an unirradiated testis showing several tubules, some of which have a marked nuclear staining in peripheral cells. (**b**) Higher magnification (400×) of a tubule showing the γ-H2AX intense staining of XY body in pachytene spermatocytes. (**c**) 200× magnification of a testis section 1 h after 1 Gy X-ray in which foci are visible. (**d**) 1,000× magnification of part of a tubule 1 h after 4 Gy X-ray. Numerous foci are visible in round spermatids and spermatocytes but not in elongated spermatids. Foci are still present 24 h (**e**) and 7 days (**f**) after irradiation (4 Gy X-ray)

the periphery of some tubules (including spermatogonia and preleptotene/leptotene spermatocytes) also show a marked nuclear staining; some elongating spermatids also show a diffuse staining probably associated with DNA breaks occurring during chromatin remodelling. These high base levels of phosphorylation of γ-H2AX make the measurement of induced phosphorylation difficult in some types of testicular cells; however, induced foci have been clearly detected in round spermatids and spermatocytes in pachytene (with stained XY body) or in other stages after X-ray irradiation.

3. Operatively, we score the slide at 200× magnification and measure the percentage of tubules showing a marked staining of their peripheral layer.

4. Then, γ-H2AX staining is analyzed at 630× magnification in primary spermatocytes, with or without XY body, and in round spermatids.

5. Depending on the level of damage, the number of foci can be counted or nuclei can be classified as positive or negative or divided in different classes on the basis of the number of foci.

4 Notes

1. Add 5 % (w/v) no-fat dry milk powder and 0.05 % Triton X-100 to PBS, mix, and centrifuge for 10 min at 10,625×g. Use the supernatant and discard the pellet.

2. Add 5 % (w/v) no-fat dry milk powder to PBS, mix, and centrifuge for 10 min at 10,625×g. Use the supernatant and discard the pellet.

3. Other options are commercially available. We only have experience with this antibody which demonstrated to work well in our experimental conditions.

4. Cut the parafilm according to the shape and the size of the microscopic slide. We prefer to use parafilm instead of coverslip because it is easier to remove after incubations.

5. You can use normal 50 or 100 place microscope slide boxes. Put on the base of the box a napkin embedded in water to maintain humidity. Place slides horizontally leaned on the lateral racks.

6. Operatively, we use the filter for DAPI to select a field that has multiple but not overlapping cells and we count the number of cells. On the same field, we change the filter to green fluorescence and count the number of cells with foci or the number of foci per cell. Then we put the filter for DAPI again and select another field and so on, scanning the slide until 100 cells are analyzed. In this way the selection of cells to be scored is not based on the presence or the absence of γ-H2AX signals.

7. Round spermatids are the more represented cell subpopulation and we count at least 100 of these cells. Spermatocytes are less frequent and we suggest counting a total of 100 spermatocytes assigning them to the class of "with XY body" or "without XY body" (keep in mind that they correspond to different developmental stages). Elongated spermatids can also be counted, even if in our experience, they do not show foci but only diffuse fluorescence and no foci are induced at short time after irradiation.

8. Even if for flow cytometric measurements the 5 mL Falcon round-bottom tubes are required, along with the immunostaining procedure we prefer to use the conical centrifuge tubes to limit the loss of cells during the several washes and centrifugations.
9. Prepare fresh staining solution starting from propidium iodide and RNAse stock solutions.
10. Bouin's fixation, although preserving testis tissue organization better than formaldehyde, partially impairs immunostaining and makes the foci less distinct.

References

1. Rogakou EP, Pilch DR, Orr AH et al (1998) DNA double-stranded breaks induce histone H2AX phosphorylation on serine 139. J Biol Chem 273:5858–5868
2. Fernandez-Capetillo O, Lee A, Nussenzweig M et al (2004) H2AX: the histone guardian of the genome. DNA Repair (Amst) 3:959–967
3. Sedelnikova OA, Rogakou EP, Panyutin IG et al (2002) Quantitative detection of 125IdU-induced DNA double-strand breaks with γ-H2AX antibody. Radiat Res 158:486–492
4. Belyaev IY (2010) Radiation-induced DNA repair foci: Spatio-temporal aspects of formation, application for assessment of radiosensitivity and biological dosimetry. Mutat Res 704:132–141
5. Horn S, Barnard S, Rothkamm K (2011) Gamma-H2AX-based dose estimation for whole and partial body radiation exposure. PLoS One 6:e25113
6. Rübe CE, Grudzenski S, Kühne M et al (2008) DNA double-strand break repair of blood lymphocytes and normal tissues analysed in a pre-clinical mouse model: implications for radiosensitivity testing. Clin Cancer Res 14:6546–6555
7. Smart DJ, Ahmed KP, Harvey JS et al (2011) Genotoxicity screening via the γ-H2AX by flow assay. Mutat Res 715:25–31
8. Hudson D, Kovalchuk I, Koturbash I et al (2011) Induction and persistence of radiation-induced DNA damage is more pronounced in young animals than in old animals. Aging 3:609–620
9. Banath JP, Bañuelos CA, Klokov D et al (2009) Explanation for excessive DNA single-strand breaks and endogenous repair foci in pluripotent mouse embryonic stem cells. Exp Cell Res 315:1505–1520
10. Blanco-Rodriguez J (2009) γ-H2AX marks the main events of the spermatogenic process. Microsc Res Tech 72:823–832
11. Fernandez-Capetillo O, Mahadevaiah SK, Celeste A et al (2003) H2AX is required for chromatin remodelling and inactivation of sex chromosomes in male mouse meiosis. Dev Cell 4:497–508
12. Forand A, Dutrillaux B, Bernardino-Sgherri J (2004) Gamma-H2AX expression pattern in non-irradiated neonatal mouse germ cells and after low-dose gamma-radiation: relationships between chromatid breaks and DNA double-strand breaks. Biol Reprod 71:643–649
13. Suzuki M, Suzuki K, Kodama S et al (2006) Phosphorylated histone H2AX foci persist on rejoined mitotic chromosomes in normal human diploid cells exposed to ionizing radiation. Radiat Res 165:269–276
14. Paris L, Cordelli E, Eleuteri P et al (2011) Kinetics of gamma-H2AX induction and removal in bone marrow and testicular cells of mice after X-ray irradiation. Mutagenesis 26:563–572
15. Rothkamm K, Löbrich M (2003) Evidence for a lack of DNA double-strand break repair in human cells exposed to very low X-ray doses. Proc Natl Acad Sci USA 100:5057–5062
16. Sak A, Stuschke M (2010) Use of γ-H2AX and other biomarkers of double-strand breaks during radiotherapy. Semin Radiat Oncol 20:223–231
17. MacPhail SH, Banáth JP, Yu TY et al (2003) Expression of phosphorylated histone H2AX in cultured cell lines following exposure to X-rays. Int J Radiat Biol 79:351–358
18. Tanaka T, Huang X, Halicka HD et al (2007) Cytometry of ATM activation and histone H2AX phosphorylation to estimate extent of DNA damage induced by exogenous agents. Cytometry A 71:648–661
19. Mahadevaiah SK, Turner JM, Baudat F et al (2001) Recombinational DNA double-strand breaks in mice precede synapsis. Nat Genet 27:271–276
20. Ahmed EA, de Rooij DG (2009) Staging of mouse seminiferous tubule cross-sections. Methods Mol Biol 558:263–277

Chapter 16

Multicolor Laser Scanning Confocal Immunofluorescence Microscopy of DNA Damage Response Biomarkers

Julian Laubenthal, Michal R. Gdula, Alok Dhawan, and Diana Anderson

Abstract

DNA damage through endogenous and environmental toxicants is a constant threat to both a human's ability to pass on intact genetic information to its offspring as well as somatic cells for their own survival. To counter these threats posed by DNA damage, cells have evolved a series of highly choreographed mechanisms—collectively defined as the DNA damage response (DDR)—to sense DNA lesions, signal their presence, and mediate their repair. Thus, regular DDR signalling cascades are vital to prevent the initiation and progression of many human diseases including cancer. Consequently, quantitative assessment of DNA damage and response became an important biomarker for assessment of human health and disease risk in biomonitoring studies. However, most quantitative DNA damage biomarker techniques require dissolution of the nuclear architecture and hence loss of spatial information. Laser scanning confocal immunofluorescence microscopy (LSCIM) of three-dimensionally preserved nuclei can be quantitative and maintain the spatial information. Here we describe the experimental protocols to quantify individual key events of the DDR cascade in three-dimensionally preserved nuclei by LSCIM with high resolution, using the simultaneous detection of Rad50 as well as phosphorylated H2AX and ATM and in somatic and germ cells as an example.

Key words DNA damage response, LSCIM, Biomonitoring, Immunostaining

1 Introduction

Efficient signalling and stable repair of DNA lesions, particularly DNA double-strand breaks (DSBs), is pivotal for genome stability, cellular homeostasis, and, hence, survival of the organism [1, 2]. An integrated signalling and genome-maintenance network, broadly defined as the DNA damage response (DDR) occurs in every eukaryote live cell upon the induction of DNA damage. DDR is initiated by the rapid recruitment and posttranslational modifications of DNA repair proteins at the vicinity of DNA lesions, visualized by immunostaining as DDR foci [3–5]. Activation of DDR signalling cascades has four key aims: (a) signalling the location of the DNA lesion, recruiting DNA repair factors, and finally initiating the repair machinery; (b) increasing the accessibility of the local chromatin structure around the lesion to the DNA repair machinery by chromatin remodelling; (c) arresting the cell cycle and thus preventing the transmission of DNA lesions to progeny cells until the lesions are repaired; and (d) in case of non-repairable DNA damage, initiation of apoptosis or cellular senescence programs to avert formation of permanent aberrations and their transmission to the offspring [3, 4, 6]. Failure in one of these pivotal DDR functions can lead to increased levels of genomic instability and DNA damage. The most harmful form of such DNA damage is a DSB and if left unrepaired or imperfectly repaired a single DSB can be lethal or provide an opportunity to initiate genetic aberrations, such as mutation, translocation, or deletion [1, 2, 7]. Because both genomic instability and DNA damage can transform a normal cell into an unstable, tumorigenic condition, they are considered as hallmarks of cancer development, progression, and spread [8, 9]. Accordingly, the quantification of DNA damage and DDR became reliable biomarkers in human health, such as in monitoring human populations for occupational, lifestyle, and environmental exposures to toxicants as well as in predicting the outcome and response to therapeutic agents in disease management [10–13]. Unfortunately, most quantitative biomarker techniques detecting DDR parameters, such as epifluorescence microscopy, flow cytometry, Comet assays, or Western blots require dissolution of the nuclear architecture and hence loss of spatial information [1, 2, 14, 15]. Laser scanning confocal immunofluorescence microscopy (LSCIM) on three-dimensionally conserved nuclei can both be quantitative and maintain the spatial information and, with the appropriate imaging software, be performed at a similar high-throughput level [16, 17]. LSCIM works well with any cell or tissue [15]. However, as for any human biomonitoring study, the selection of the most appropriate tissue should be based on the expected dynamics of the exposed population characteristics, exposure type, endpoint mechanisms, persistence of DNA damage, rates of cell turnover, sample timing, study aims, tissue accessibility, and possible ethical and social restrictions [18–20]. Thus, for biomonitoring of the DDR in somatic cells, lymphocytes are widely used because their cellular, metabolic, nuclear, and chromatin state reflects the overall exposure of the organism. Furthermore, they can also be easily obtained in a socially as well as ethically acceptable, minimally invasive route [21]. While there is currently no cell type available to monitor DDR in reproductive cells of the female by a noninvasive route, spermatozoa can be used to monitor the male reproductive cells. Spermatozoa are thought to reflect the overall extent of toxicant exposure to the male germ line and are available noninvasively as well as in large quantity for a lifetime [17, 22].

In a typical immunofluorescence experiment, separated lymphocytes or spermatozoa are first spread on a microscope slide at an appropriate cell concentration, followed by a partial removal of the lymphocyte membrane with lysis buffer or relaxation of the tightly packed spermatozoon chromatin with dithiothrietol solution. Then, the respective primary and secondary antibodies can be applied and after counterstaining of the DNA with 4′,6-diamidine-2-phenylindole (DAPI) the cells are ready for analysis by laser scanning confocal microscopy. Using the appropriate combinations of distinctive antibody species and their subclasses coupled to fluorochromes with different emission maxima, multiple antigens of the DDR can be visualized side by side using LSCIM [15]. Based on this, we describe here a method to simultaneously visualize three of the DDR's main regulators with high spatial resolution: phosphorylated ataxia-telangiectasia-mutated (pATM) protein kinase, phosphorylated histone H2A variant H2A.X (pH2A.X), and Rad50, a member of the structural maintenance of chromosome (SMC) protein family [4]. ATM- and RAD3-related (ATR) protein kinases are master regulators of genome stability maintenance and DDR by signaling the locations of DNA lesions, initiating cell cycle check points as well as DNA repair and apoptosis pathways [23, 24]. ATM is phosphorylated by a DNA strand break itself as well as by a cooperation of the Tip60 acetyltransferase and the Mre11–Rad50–Nbs1 (MRN11) complex [25, 26]. Formation of the MRN complex is also initiated by DNA lesions, and was shown to be essential for activating and recruiting un-phosphorylated ATM to DNA lesions [15, 27]. Activated ATM, along with activated ATR, can then phosphorylate serine 139 of H2A.X, on chromatin flanking DNA strand break sites [2]. This posttranslational modification of H2A.X, also termed γH2AX, is considered a hallmark indicator of the occurrence of DNA lesions and today's most reliable biomarker for DSBs [2, 17, 28].

2 Materials

2.1 Separation, Freezing, and Thawing of Lymphocyte

1. Anticoagulant sodium heparin-coated tubes.
2. Saline: 0.9 % NaCl (w/v) in double-distilled H_2O.
3. Density gradient solution for lymphocytes (*see* **Note 1**).
4. Centrifuge ($1,425 \times g$ is required).
5. Freezing medium: 90 % fetal calf serum, 10 % dimethylsulfoxide (v/v).
6. Cryo-vials.
7. Freezers (−20 and −80 °C are required).
8. Liquid nitrogen tank.
9. Cryogenic gloves.

2.2 Separation, Freezing, and Thawing of Spermatozoa

1. 50 mL sterile collection container.
2. 80 % density gradient solution for spermatozoa: Add 2 mL density gradient buffer to 8 mL density gradient solution (*see* **Note 2**).
3. 40 % density gradient solution for spermatozoa: Add 6 mL of density gradient buffer to 4 mL density gradient solution (*see* **Note 2**).
4. Centrifuge ($300 \times g$, $500 \times g$ are required).
5. Fetal bovine serum (FBS).

2.3 Preparation and Fixation of Lymphocytes

1. Paraformaldehyde (PFA): 4 % PFA in 1× PBS (w/v). Prepare fresh by heating PFA powder in 1× PBS up to 65 °C. Adjust the pH to 7.4 with 10 N NaOH and filter through a 0.45 μm filter. Store the solution in 10 mL aliquots at −20 °C. (*Caution*: PFA fumes are highly toxic by inhalation or skin contact. Always work in a ventilated fume hood and wear goggles.)
2. Centrifuge ($450 \times g$ is required).
3. Cytospin ($200 \times g$ is required) and cytospin slides.
4. Hydrophobic pen (PapPen©).

2.4 Preparation and Fixation of Spermatozoa

1. Phosphate buffer saline (PBS): 137 mM NaCl, 2.75 mM KCl, 1.45 mM KH_2PO_4, 15.25 mM Na_2HPO_4, pH 7.2.
2. Dithiothreitol (DTT): 10 mM DTT prepared in 0.1 M Tris–HCl, pH 8.0.
3. 2× saline–sodium citrate (SSC) buffer: 300 mM NaCl, 30 mM Na citrate, pH 7.1.
4. Inverted light microscope equipped with 40 and 100× phase contrast positive objectives.
5. Diamond pen.
6. Coplin jar.

2.5 Immunostaining

1. Blocking buffer: 4 % Bovine serum albumin (BSA) in 1× PBS (v/v).
2. Lysis buffer: 0.15 % Triton X-100 (v/v), 4 % BSA (v/v) in 1× PBS. This solution can be kept at 4 °C for up to 2 days.
3. Primary and secondary antibodies (*see* **Notes 3** and **4**).
4. Antifade solution: 0.223 g of 1,4-diazabicyclo[2.2.2]octane, 7.2 mL of glycerol, 0.8 mL of ddH_2O, and 2 mL of 1 M Trizma base, pH 8.0. Store in 1 mL aliquots at −80 °C and protect from light.
5. DAPI stock solution: 0.5 g/mL of DAPI (in antifade solution.
6. Coverslips, 22mm × 22mm.
7. Cotton swabs.
8. Nail polish.

2.6 Microscopy and Image Analysis

1. Immersion oil.

2. Epifluorescence microscope equipped with 40, 63, and 100× objectives; a triple-band filter for the simultaneous observation of three fluorochromes such as Cy3 (excitation: 554; emission: 568), fluorescein isothiocyantate (490; 525), DAPI (350, 470), or Cy5 (649; 666); an infrared fluorochrome visible only via a charge-coupled device (CCD) camera; and 100 W mercury illumination (HBO 100) bulb.

3. Laser-scanning confocal microscope (LSM510 META, Carl Zeiss) equipped with UV laser (Enterprise, emitting light of 364 nm wavelengths) and VIS lasers: Argon (488 nm), HeNe1 (543 nm), and HeNe2 (633 nm) and Plan Apochromat 63×/1.4 NA DIC oil objective lens. For 3D image analysis, the nuclei were scanned with a Z-axial distance of 200 nm yielding separate stacks of 8-bit grayscale images for each fluorescent channel with a pixel size of 100 nm. For each optical section, images were collected sequentially for all fluorophores.

4. Commercially available image analysis software or open-source software, such as ImageJ (http://rsbweb.nih.gov/ij/).

3 Methods

3.1 Separation, Freezing, and Thawing of Lymphocytes

1. Collect whole blood by venepuncture into anticoagulant sodium heparin-coated tubes.

2. Dilute whole blood 1:1 with saline in 50 mL tubes.

3. Slowly pipette 6 mL of the diluted blood on top of 3 mL of density gradient solution. Avoid mixing the two phases.

4. Spin down at $1,425 \times g$ for 20 min at room temperature.

5. Thoroughly retrieve lymphocytes from just above the boundary between the density gradient solution at the tube's bottom without mixing with the upper serum layer. Avoid disturbance of the density gradient solution's layer.

6. Mix the isolated lymphocytes with 10 mL saline.

7. Spin down at $1,200 \times g$ for 15 min at room temperature.

8. Remove carefully all supernatant so that 100 μL of isolated lymphocyte solution remains at the bottom of the tube.

9. Pipette this 100 μL of isolated lymphocyte solution into a cryo-vial and mix with 900 μL of freezing medium.

10. Freeze the cryo-vial for 6 h at −20 °C and then at −80 °C for up to 1 month. Use liquid nitrogen when storage for more than 1 month is required. *Caution.* Always wear cryogenic gloves to protect your hands and arms when working in hazardous, ultracold environments.

11. When using frozen lymphocytes for an immunostaining experiment, place cryo-vials in a 37 °C water bath for speedy thawing to avoid artificial cellular and DNA damage by ice crystal formation.

3.2 Separation, Freezing, and Thawing of Spermatozoa

1. Collect semen sample in a 50 mL sample sterile collection container (*see* **Note 5** for general requirements for the method).
2. Let the semen liquefy for 1 h at room temperature.
3. Pipette 2 mL of 80 % density gradient solution in a 15 mL conical tube and carefully layer with 2 mL of 40 % density gradient solution. Avoid mixing the layers.
4. Thoroughly layer 0.5 mL of liquefied semen sample onto the density gradient mixture.
5. Spin down at $300 \times g$ for 15 min. Do not use the centrifuge's brake function.
6. Carefully aspirate from the surface everything except the pellet and 2 mm liquid on top of the pellet (*see* **Note 6**).
7. Transfer the spermatozoa pellet to new 15 mL conical tube and wash with 5 mL of 1× PBS
8. Spin down at $500 \times g$ for 5 min. Do not use the centrifuge's brake function.
9. Remove the supernatant and resuspend the pellet in 1 mL of FCS in cryo-vials (*see* **Note 6**).
10. Freeze cryo-vials for 4 h at −10 °C, followed by 16 h at −20 °C and then at −80 °C. *Caution*. Always wear cryogenic gloves to protect your hands and arms when working in hazardous, ultracold environments.
11. When using frozen spermatozoa for an immunostaining experiment, place cryo-vials in water bath at 32 °C for rapid thawing.

3.3 Preparation and Fixation of Lymphocytes

1. Centrifuge lymphocytes at $500 \times g$ for 5 min, discard the supernatant, and resuspend the pellet in a suitable volume of 1× PBS, depending on the required lymphocyte concentration.
2. Fix lymphocytes with 1 mL of 4 % PFA for 10 min at room temperature to preserve antigen epitopes and the nuclear architecture (*see* **Notes 7** and **8**).
3. Wash the cells with 1 mL of 1× PBS, centrifuge at $450 \times g$ for 5 min at 4 °C, and remove the supernatant. Repeat this step twice.
4. Load cells onto a cytospin and centrifuge at $200 \times g$ for 2 min onto a cytospin slide.
5. Check concentration of lymphocytes using an inverted light microscope.
6. Draw a circle around the lymphocyte area on the slide with a hydrophobic pen to hold the antibody solutions within the target area.

3.4 Preparation and Fixation of Spermatozoa

1. Centrifuge spermatozoa at $500 \times g$ for 5 min, discard the supernatant, and resuspend the pellet in a suitable volume of 1× PBS, depending on the required spermatozoa concentration.
2. Pipette 7 µL of spermatozoa solution onto the far left side of a microscope slide.
3. Touch with a second microscope slide the spermatozoa drop in a 45° angle and allow the solution to spread along the edge of the upper slide.
4. Move the upper slide slowly along the entire length of the lower microscope slide to spread the solution in a thin layer and let air-dry at room temperature.
5. Check concentration of spermatozoa using an inverted light microscope.
6. Draw a circle around the spermatozoa area on the microscope slide with a diamond pen.
7. Incubate microscope slide in ice-cold 10 mM DTT for 30 min at 4 °C for decondensation (*see* **Note 9**).
8. Wash slides thoroughly in 2× SSC buffer for 5 min and air-dry at room temperature under constant humidity.
9. Evaluate spermatozoa decondensation degree with an inverted light microscope and eventually adjust DTT incubation times.
10. Fix spermatozoa slides with 4 % PFA for 10 min at room temperature (*see* **Notes 7** and **8**).
11. Remove 4 % PFA leftovers with three washes in 1× PBS for 5 min each and air-dry.
12. Draw a waterproof circle around the spermatozoa area on the slide with a hydrophobic pen to hold the antibody solutions within the target area.

3.5 Immunostaining

The slides prepared for lymphocytes in **step 6** of Subheading 3.3 and sperm cells in **step 12** of Subheading 3.4 are used for immunostaining.

1. Add 1× PBS into the circle without wetting its hydrophobic barrier and remove liquid by tilting the slide on a paper towel. Repeat this step twice.
2. Pipette a sufficient amount of lysis buffer to fill the circle and incubate for 10 min at room temperature (*see* **Note 10**).
3. To remove all detergent leftovers from the lysis step, incubate the slide in a coplin jar filled with 1× PBS for 5 min at room temperature. Change 1× PBS and wash again for 5 min.
4. Gently remove all liquid by tilting the slide on a paper towel and carefully dry the hydrophobic barrier using cotton swabs.
5. Add blocking buffer into the circle without wetting its hydrophobic barrier and incubate for 15 min at room temperature in

a humid chamber. Then remove blocking buffer by tilting the slide on a paper towel.

6. In the meantime, dilute primary antibodies in blocking solution, mix thoroughly, and spin down at 14,000×g for 2 min (*see* **Note 11**).

7. Pipette diluted antibodies into the circle and incubate for 16 h at 4 °C. If several haptens are used, apply all antibodies mixed in one solution (*see* **Notes 12** and **13**).

8. Remove unbound primary antibodies with three washes in 1× PBS for 5 min at room temperature each.

9. Again, get rid of all remaining liquid by tilting the slide on a paper towel and carefully dry the hydrophobic barrier using cotton swabs.

10. In the meantime, dilute secondary antibodies in blocking solution, mix thoroughly, and spin down at 14,000×g for 2 min (*see* **Note 11**).

11. Apply diluted secondary antibodies into the circle and incubate for 6 h at 4 °C. If several haptens are used, apply all antibodies mixed in one solution. At this step and the next ones, protect the cells from light with foil (*see* **Notes 14** and **15**).

12. Eliminate unbound secondary antibodies with three washes in 1× PBS for 5 min each at room temperature.

13. Remove excess liquid carefully with a Kimewipe© and apply 20 μL of 1:1,000 diluted DAPI (from the stock) in antifade solution to stain the DNA (*see* **Notes 16** and **17**).

14. Again, remove excess antifade solution around the circle with a Kimwipe© and cover the slide with a coverslip. Do not squash the cells.

15. Seal the edges of the coverslip with nail polish and let nail polish air-dry in the dark (*see* **Note 18**).

16. Before performing LSCIM, evaluate immunostaining results with a conventional epifluorescence microscope for specific binding and signal intensities (Figs. 1 and 2).

3.6 Microscopy and Image Analysis

1. Set up laser scanning confocal microscope at room temperature and apply immersion oil if needed.

2. Take Z-stacks in different places of the slide for each sample in 25–40 slices per stack to image the whole nucleus. Collect images for each optical section sequentially for all fluorochromes and differential interference contrast (DIC).

3. Analyze data on a "per nucleus basis" with ImageJ© software by applying "3D object counter" plug-in. Results of the analysis are presented as a total foci number per image; individual foci positions are not analyzed (*see* **Note 19**).

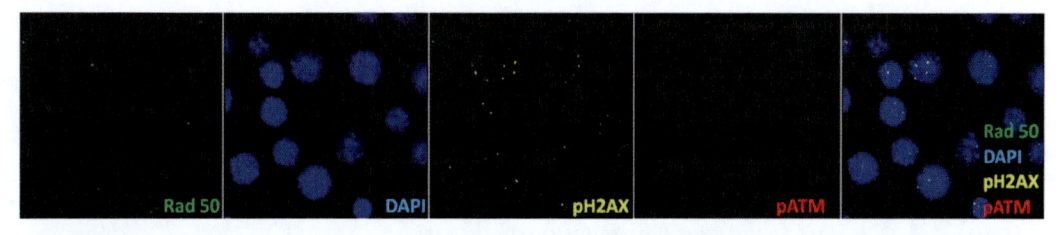

Fig. 1 Maximal projection of a Z-stack showing DAPI, RAD50, pATM, and pH2AX foci in human lymphocytes

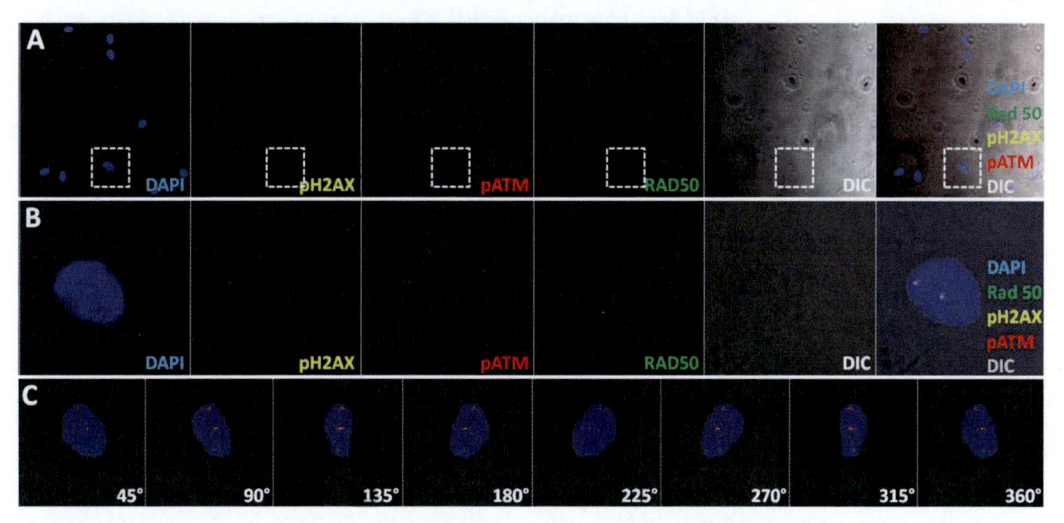

Fig. 2 Maximal projections of Z-stacks showing differential interference contrast (DIC) as well as DAPI, RAD50, pATM, and pH2AX foci in human spermatozoa. (**a**) Overview, showing eight spermatozoa without DDR signalling and one spermatozoa with DDR signalling (*white frame*). (**b**) Magnification of spermatozoa with activated DDR foci from (**c**) Magnification of spermatozoa with activated DDR foci from (**a**) shown from different angles of 45°, 90°, 135°, 180°, 225°, 270°, 315°, and 360°

4 Notes

1. A density (discontinuous) gradient solution is a solution containing sodium diatrizoate and polysaccharide, which retains only lymphocytes while all other leukocytes and erythrocytes pass through it. We use here the commercially available Lymphoprep© solution from Axis Shield (Oslo, Norway).

2. A density gradient solution for spermatozoa allows motile spermatozoa to be separated from seminal plasma as well as immature sperm, leukocytes, bacteria, epithelial cells, and cell debris. The gradient is achieved by layering solutions of two different densities and on top of them the semen sample. Only motile spermatozoa will proceed through both solutions, seen as a pellet after centrifugation. Although density gradients on spermatozoa are intended for usage in assisted reproduction

technology procedures, our laboratory and others use them in experiments where pure spermatozoa are desired. We use here the commercially available Pure-sperm© solution and buffer from Nidacon (Mölndal, Sweden).

3. Primary antibodies can be aliquoted and stored at −80 °C for up to 2 years in 50 % glycerol or according to the manufacturer's instructions.

4. We use here the following labelling strategy: a rabbit monoclonal to ATM (phospho S1981), a mouse monoclonal to γH2A.X (phospho S139), and a sheep monoclonal to Rad50 antibody.

5. Routine semen analysis must be performed within 2 h of sample collection according to the WHO criteria [29] to provide details on color, volume, viscosity, sperm concentration, pH, motility, morphology, and liquefaction in order to distinguish aspermia, azoospermia, hypospermia, oligozoospermia, asthenozoospermia, teratozoospermia, and normospermia. Sperm count measures the concentration of spermatozoa in the ejaculate. Over 16 million spermatozoa per mL are considered as normal. A lower sperm count is termed oligozoospermia. Total sperm count describes the total number of spermatozoa in the whole ejaculate. The lower reference limit is set to 40 million spermatozoa per ejaculate. Motility of spermatozoa is divided in four grades: Grade 1, no movement of any spermatozoa; grade 2, spermatozoa move their tails, but spermatozoa themselves cannot move; grade 3: spermatozoa move, but the direction of movement is curved or nonlinear; grade 4: spermatozoa move linearly and fast. A 50 % motility within 60 min of sample collection is considered as normal. The morphology defines the shape of the nucleus and tail of the spermatozoa. The total volume of sample is considered with 1.5 mL and up to 6.5 mL as normal. Lower volumes are an indicator of infertility. The pH level should be in the range of 7.2–8.0 to be considered as normal. Liquefaction describes the process in which the proteins of the ejaculate dissolve and the ejaculate becomes normally liquid within 20 min.

6. If no pellet is visible, remove all liquid except the lowest 0.5 mL.

7. PFA-based fixatives stabilize cells by reacting with basic amino acids to form methylene cross-links. The amount of methylene bridges to be formed depends mainly on four variables: concentration, pH, temperature, and fixation time of PFA. Hence, PFA over-fixation can lead to extensive cross-linking, which can mask epitopes needed to react with the antibodies. To avoid this, one can alter all four variables of PFA to optimize antibody penetration. Importantly, PFA-fixation is usually reversible.

8. PFA-fixed cells can be stored for several days at 4 °C before immunostaining experiments.

9. DTT is a reducing agent, which is used to cleave disulfide bonds in the DNA in order to leave a more relaxed chromatin structure of the spermatozoa during immunostaining with nuclear antibodies.

10. Detergents in the lysis solution ideally form tiny holes in the cell membrane, through which primary and secondary antibodies can subsequently pass into the cell. Even small increases of lysis time can easily result in total lysis of the cellular membranes and subsequently sticking and merging of multiple cells, rendering high-quality in situ analysis impossible. On the other, small decreases in lysis time may prevent successful penetration of the antibodies through the cell membrane.

11. With appropriate combinations of fluorochromes and filters, multiple antigens can be selectively immunostained in the same cell. It is important to define if the antibodies of one immunoreaction interact with antibodies used in the other, simultaneous immunoreactions. When such cross-reactions can be excluded (e.g., when combining anti-mouse, anti-rabbit, and anti-sheep antibodies in a three-color experiment), antibodies of identical layers (i.e., primary antibodies or secondary antibodies) can be mixed and simultaneously applied. However, if unwanted cross-reaction cannot be excluded (e.g., when combing two anti-rabbit antibodies of the same subclass), the immunostaining reaction for each antigen has to be completed successively, with the addition of appropriate blocking experiments to avoid cross-reaction. Generally, antibody cross talk is not expected when primary antibodies from different species such as mouse, rabbit, sheep, or monkey are used and the secondary antibodies are species specific and purified from cross-reacting elements. Similar applies for combining antibodies of different immunoglobulin isotypes (e.g., IgG and IgM), or subclasses (e.g., IgG1 and IgG2), which can be visualized with isotype- or subclass-restricted secondary antibodies.

12. As a general rule, primary and secondary antibody concentrations should always be tested in single immunostaining experiments before performing any multicolor experiments.

13. If primary antibodies do not stain optimally at 4 °C overnight, incubation should be performed for 3 h at 37 °C (lymphocytes) or 32 °C (spermatozoa).

14. To control the specificity of the primary antibody, lymphocytes or spermatozoa should be stained with the secondary antibody, but not with the primary. In that way, one can differentiate between background staining caused by an unspecific binding of the secondary antibody and staining of the targeted primary

antibody caused by a specific binding of the secondary antibody to the primary only.

15. Fluorescence immune-stained slides must be always kept out of sunlight and can be stored at −20 °C for years. Usually, re-staining with the secondary (fluorescent) antibody is required after thawing the slide.

16. An appropriate counterstaining is always necessary to test for preservation of normal nuclear morphology. As a certain proportion of nuclei always get damaged due to freezing/thawing and denaturation, counterstaining is also necessary to select well-preserved nuclei for the further analysis. The choice of the counterstaining dye is determined by the fluorochromes used, as well as by the wavelengths available for microscopy. DAPI (producing a blue fluorescence) is highly specific to DNA and is therefore the most popular counterstain. For red fluorescence, one can use 5 μg/mL propidium iodide (working concentration); for far-red fluorescence 1 μM TO-PRO-3 (working concentration); and for green fluorescence 10 μM SYTO 16 (working concentration).

17. Fading of the fluorescent signals of the antibody-labelled proteins as well as the DNA staining after a few minutes of UV light exposure are caused by oxidation of the antifade solution. The oxidized antifade solution can be identified by the color change of the solution from clear to brownish and is mostly caused by inappropriate storage of antifade solution at RT for too long. Hence, always store this solution at −20 °C and make it fresh frequently to obtain optimal signal intensities.

18. Sealing of coverslips with nail polish prevents the cells from moving and mixing with immersion oil during microscopy. If cell re-staining becomes necessary, simply peel off the nail polish in 1× PBS.

19. The software macro operates by evaluating the maximum intensity projection of a three-dimensional stack of images to produce a single two-dimensional intensity image, which is then processed. The images are normalized and the background noise corrected. The noise suppression is performed with minimum reduction in focus detection sensitivity. Nuclear boundaries and foci are automatically identified in images by a threshold algorithm. For identifying nuclear boundaries a DAPI staining is used. Minimum foci size of 5×5 (0.25 μm^2) pixels and maximum foci size of 200 pixels should be selected. Adjunct or overlapping nuclei have to be adequately segmented using watershed transformation. Cells that are partially on the edge of the image or deformed cells are excluded by the software and not counted. Cells with pan-nuclear staining or band-like staining are also not analyzed [16, 17].

References

1. Redon CE, Weyemi U, Parekh PR et al (2012) Gamma-H2AX and other histone post-translational modifications in the clinic. Biochim Biophys Acta 1819:743–756
2. Bonner WM, Redon CE, Dickey JS et al (2008) GammaH2AX and cancer. Nat Rev Cancer 8:957–967
3. Polo SE, Jackson SP (2011) Dynamics of DNA damage response proteins at DNA breaks: a focus on protein modifications. Genes Dev 25:409–433
4. Jackson SP, Bartek J (2009) The DNA-damage response in human biology and disease. Nature 461:1071–1078
5. Bekker-Jensen S, Mailand N (2010) Assembly and function of DNA double-strand break repair foci in mammalian cells. DNA Repair (Amst) 9:1219–1228
6. Ciccia A, Elledge SJ (2010) The DNA damage response: making it safe to play with knives. Mol Cell 40:179–204
7. Richards RI (2001) Fragile and unstable chromosomes in cancer: causes and consequences. Trends Genet 17:339–345
8. Hanahan D, Weinberg RA (2011) Hallmarks of cancer: the next generation. Cell 144:646–674
9. Hanahan D, Weinberg RA (2000) The hallmarks of cancer. Cell 100:57–70
10. Redon CE, Nakamura AJ, Zhang YW et al (2010) Histone gammaH2AX and poly(ADP-ribose) as clinical pharmacodynamic biomarkers. Clin Cancer Res 16:4532–4542
11. Mah LJ, El-Osta A, Karagiannis TC (2010) GammaH2AX as a molecular marker of aging and disease. Epigenetics 5:129–136
12. Gasser S, Raulet D (2006) The DNA damage response, immunity and cancer. Semin Cancer Biol 16:344–347
13. Gallo V, Khan A, Gonzales C et al (2008) Validation of biomarkers for the study of environmental carcinogens: a review. Biomarkers 13:505–534
14. Solovei I, Cremer M (2010) 3D-FISH on cultured cells combined with immunostaining. Methods Mol Biol 659:117–126
15. Harvath L (1999) Overview of fluorescence analysis with the confocal microscope. Methods Mol Biol 115:149–158
16. Zlobinskaya O, Dollinger G, Michalski D et al (2012) Induction and repair of DNA double-strand breaks assessed by gamma-H2AX foci after irradiation with pulsed or continuous proton beams. Radiat Environ Biophys 51:23–32
17. Laubenthal J, Zlobinskaya O, Poterlowicz K et al (2012) Cigarette smoke-induced transgenerational alterations in genome stability in cord blood of human F1 offspring. FASEB J 26(10):3946–3956
18. Eastmond DA, Hartwig A, Anderson D et al (2009) Mutagenicity testing for chemical risk assessment: update of the WHO/IPCS Harmonized Scheme. Mutagenesis 24:341–349
19. Anderson D, Zeiger E (1997) Human monitoring. Environ Mol Mutagen 30:95–96
20. Albertini RJ, Anderson D, Douglas GR et al (2000) IPCS guidelines for the monitoring of genotoxic effects of carcinogens in humans. International Programme on Chemical Safety. Mutat Res 463:111–172
21. Dusinska M, Collins AR (2008) The comet assay in human biomonitoring: gene-environment interactions. Mutagenesis 23:191–205
22. Sipinen V, Laubenthal J, Baumgartner A et al (2010) In vitro evaluation of baseline and induced DNA damage in human sperm exposed to benzo[a]pyrene or its metabolite benzo[a]pyrene-7,8-diol-9,10-epoxide, using the comet assay. Mutagenesis 25:417–425
23. Derheimer FA, Kastan MB (2010) Multiple roles of ATM in monitoring and maintaining DNA integrity. FEBS Lett 584:3675–3681
24. Lee JH, Paull TT (2007) Activation and regulation of ATM kinase activity in response to DNA double-strand breaks. Oncogene 26:7741–7748
25. Sun Y, Jiang X, Price BD (2010) Tip60: connecting chromatin to DNA damage signaling. Cell Cycle 9:930–936
26. Stracker TH, Petrini JH (2011) The MRE11 complex: starting from the ends. Nat Rev Mol Cell Biol 12:90–103
27. Rupnik A, Lowndes NF, Grenon M (2010) MRN and the race to the break. Chromosoma 119:115–135
28. Dickey JS, Redon CE, Nakamura AJ et al (2009) H2AX: functional roles and potential applications. Chromosoma 118:683–692
29. WHO (1999) WHO laboratory manual for the examination of human semen and sperm-cervical mucus interaction. Published on behalf of the World Health Organization by Cambridge University Press, Cambridge, UK

Chapter 17

The Comet Assay: Assessment of In Vitro and In Vivo DNA Damage

Mahima Bajpayee, Ashutosh Kumar, and Alok Dhawan

Abstract

Rapid industrialization and pursuance of a better life have led to an increase in the amount of chemicals in the environment, which are deleterious to human health. Pesticides, automobile exhausts, and new chemical entities all add to air pollution and have an adverse effect on all living organisms including humans. Sensitive test systems are thus required for accurate hazard identification and risk assessment. The Comet assay has been used widely as a simple, rapid, and sensitive tool for assessment of DNA damage in single cells from both in vitro and in vivo sources as well as in humans. Already, the in vivo Comet assay has gained importance as the preferred test for assessing DNA damage in animals for some international regulatory guidelines. The advantages of the in vivo Comet assay are its ability to detect DNA damage in any tissue, despite having non-proliferating cells, and its sensitivity to detect genotoxicity. The recommendations from the international workshops held for the Comet assay have resulted in establishment of guidelines. The in vitro Comet assay conducted in cultured cells and cell lines can be used for screening large number of compounds and at very low concentrations. The in vitro assay has also been automated to provide a high-throughput screening method for new chemical entities, as well as environmental samples. This chapter details the in vitro Comet assay using the 96-well plate and in vivo Comet assay in multiple organs of the mouse.

Key words DNA damage, Genotoxicity, Alkaline comet assay, In vitro, Chinese hamster ovary cells, In vivo, Mouse multiple organs

1 Introduction

Industrialization has resulted in an added burden of toxicants already existing in the environment. Industrial effluents, automobile exhaust, and new chemical entities being added each day into the environment adversely affect not only the humans but also the environment on the whole. It is quite likely that some chemicals will interact with the genetic blueprint "DNA" and induce damage in the form of mutations, leading to carcinogenesis. Therefore, risk assessment and hazard identification become imperative for monitoring human and environmental health.

The single-cell gel electrophoresis assay or the Comet assay is a versatile, simple, sensitive, and rapid tool for detecting primary

DNA damage at an early stage. It has become a popular test since its inception by Ostling and Johanson [1] and modification to the alkaline version by Singh et al. [2]. It has found wide application in in vitro, in vivo, as well as human biomonitoring studies, as it requires a small amount of sample and can be conducted in any proliferating or non-proliferating cell population [3]. The standardized and validated Comet assay can help in hazard identification and risk assessment [4, 5]. The assay has been adapted to detect DNA single- and double-strand breaks and alkaline labile sites (ALS), oxidative DNA damage, DNA cross-links, DNA adducts, apoptosis, and necrosis. Variations in the components of the assay have allowed it to be used in bacteria, cells from different organs of animals, germ cells (sperms), etc. The International Workgroup on Genetic Toxicology (IWGT) for in vitro genotoxicity tests recommended that the Comet assay be evaluated for inclusion in the in vitro test battery since it is not dependent on proliferating cells and also provides information on a wide range of DNA damage [6]. The critical point of the assay is to obtain viable cells for the assay so that there is no ambiguity in the explanation of the amount of observed damage in the DNA.

The duplex DNA is organized in the nucleus around the histone proteins, which help in its supercoiling and packaging. Interaction of the DNA with genotoxicants may cause DNA strand breaks. The Comet assay is based on the principle that strand breaks reduce the size of the large duplex DNA molecule, and under high pH the relaxed strands are pulled out during electrophoresis. The high salt concentration in the lysis step of the assay (pH 10) removes the cell membranes, histone proteins, cytoplasm, and nucleoplasm, and disrupts the nucleosomes, leaving behind the nucleoid, consisting of the negatively supercoiled DNA. The breaks present in the DNA cause local relaxation of supercoils and loops of DNA are then free [7]. A high pH (>13) at unwinding step facilitates denaturation of DNA (with disruption of hydrogen bonds between the double-stranded DNA) and expression of alkali-labile sites as frank breaks. During electrophoresis, this damaged DNA is pulled out toward the anode, thus forming the distinct "comet," with a head (intact DNA) and a tail (damaged DNA), visualized after fluorescent staining.

An overview of the method: A small number of viable cells obtained from a desired source (in vitro, in vivo, humans) are suspended in low-melting agarose gel and spread onto an agarose-pre-coated microscope slide. The cells are then sandwiched with another thin layer of agarose to prevent loss (Fig. 1). These cells are subjected to alkaline lysis to obtain nucleoids, which then undergo alkaline electrophoresis. After electrophoresis, the neutralization step allows some renaturation of the DNA, and the DNA is stained with a fluorescent dye (e.g., ethidium bromide). Cells with higher DNA damage display increased migration of chromosomal DNA from the nucleus toward the anode, which

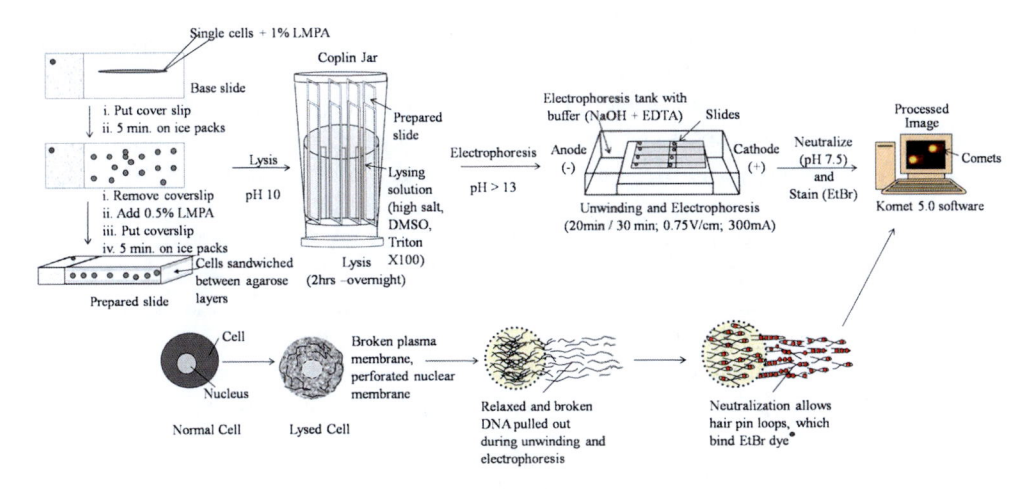

Fig. 1 A schematic representation of the Comet procedure

1.1 The In Vitro Comet Assay

resembles the shape of a comet, when viewed under a fluorescent microscope and hence the name "Comet assay." The amount of DNA migration indicates the amount of DNA damage in the cell.

The Comet assay has been performed in vitro in various cell lines of animal and human origin and utilized to understand the genotoxic activity of chemicals and compounds. The most frequently used adherent cell lines are Chinese hamster ovary/lung cells (CHO/CHL, V79), and human breast cancer cell line (MCF7). The suspension cell lines on the other hand include mouse lymphoma cells (L5178Y) and human lymphoblast (TK6). Cultured human peripheral lymphocytes are also widely used for in vitro Comet experiments. The animal cells have an impaired p53 function and can be used to test the genotoxicity of compounds that cannot be further metabolized. However, the addition of the S9 fraction makes these cells metabolically competent and in vitro tests should be conducted with as well as without S9. The cells derived from human origin (lymphocytes, TK6, HepG2) are metabolically competent and can be utilized for tests, circumventing any external metabolic activation. The Comet assay has been adapted in the multiwell plates for in vitro studies in our laboratory using the CHO cells for studying the genotoxicity of the individual metabolites of a pesticide [8] and benzene [9], as well as of herbicides [10]. The Comet assay adapted in multiwell plate method allows the use of small amounts of the test substance, saves on the disposals, and has also been automated providing a high-throughput screening [11–13]. Human peripheral lymphocytes have been used to reveal the DNA damaging potential in vitro of pesticide [8], benzene and its metabolites [14], and leachates of municipal sludge [15] in the Comet assay. Recently, genotoxicity of nanomaterials like oxide of zinc and titanium have also been tested in HepG2 [16–18], and human epidermal cells [19–21].

1.2 The In Vivo Comet Assay

The Comet assay conducted in vivo has been utilized as part of the current in vivo regulatory testing strategy, and has been recommended as a follow-up test for in vitro-positive results [22–24]. The international workgroup on genotoxicity testing considered integrating the Comet assay (especially liver Comet) along with the MN assay for repeated dose testing into standard toxicity testing to provide an alternative to independent assays [25]. Since the assay can be performed in cell from any tissue, it becomes a valuable tool for genotoxicity assessment of a compound in a specific organ as well as at the site of contact. Various intensive guidelines and recommendations are available for the assay as a result of international workgroups [26–29] and an OECD guideline is also being formulated.

The mouse multiple organ genotoxicity using the Comet assay for detection of 208 compounds was first conducted by Sasaki et al. [30]. Meanwhile, our studies have shown DNA damage in blood and bone marrow as well as in different organs of mice by pesticide [31], municipal sludge [32], argemone oil [33–36], and also nanomaterials [37] using the Comet assay.

In this chapter we describe the protocols followed for the in vitro Comet assay using CHO cells as well as the in vivo assay conducted on cells from different organs of mouse.

2 Materials

Prepare all the solutions in deionized water unless otherwise stated. All chemicals used should be of analytical reagent (AR) grade.

2.1 Cell Culture Components for In Vitro Comet Assay

1. Incomplete Ham's F12 medium: Dissolve Dulbecco's Ham's F12 nutrient mixture (1 L packet) 10 mL of antibiotic/antimycotic solution and 300 mg of L-glutamine in 900 mL of autoclaved water. Add 2 g of sodium bicarbonate and adjust pH to 7.4, using 1 N HCl. Make up the volume to 1,000 mL, filter sterilize the solution using 0.22 μm filter, and store at 4 °C (*see* **Note 1**).

2. Fetal bovine serum (FBS): Ready-to-use solution (*see* **Note 2**).

3. Complete Ham's F12 medium: Add 10 % FBS to incomplete Ham's F12 medium, filter sterilize using a 0.22 μm syringe filter, and store at 4 °C (*see* **Note 1**).

4. Trypsin–EDTA solution: 0.25 % solution. The ready-to-use solution is thawed, aliquoted into 15 mL tubes, and stored at −20 °C. This is done to avoid the repeated freezing and thawing which helps to preserve the enzyme activity of trypsin. Working solution: 0.005 % trypsin in PBS, freshly prepared before each use.

5. Trypan blue dye: 0.4 % ready-to-use solution.

6. Laminar hood.

7. CO_2 incubator.

8. Serological pipettes and motorized aspirator/pipettor.
9. Tissue culture flasks (25cm^2).
10. 96-well plates.

2.2 In Vivo Components

1. Anticoagulant: 1,000 U/mL heparin sodium salt. 10 μL of heparin used to coat a microfuge tube for collection of 50–100 μL blood.
2. Hank's balanced salt solution (HBSS): Add and mix contents (10.4 g) of 1 L packet of Dulbecco's HBSS in 990 mL of water in a conical flask, adjust pH to 7.4 using 0.1 N HCl, quantity sufficient (q.s.) to 1,000 mL. Filter sterilize using 0.22 μm syringe filter and store at 4 °C.
3. Phosphate-buffered saline (PBS; Ca^{2+}, Mg^{2+} free): Mix and dissolve 1 L packet (10 g) of Dulbecco's PBS in 990 mL of water, adjust pH to 7.4 using 0.1 N HCl, and make up the volume (q.s) to 1,000 mL (*see* **Note 3**).
4. Mincing solution: HBSS, 20 mM EDTA, and 10 % DMSO, make fresh and chill (4 °C) before use.
5. Lymphocyte separating medium: For example Histopaque-1077/Lymphoprep/Ficoll for isolation of lymphocytes from whole blood. Ready-to-use solution. Store at 4–8 °C.
6. 5,6-Carboxyflourescein dye: Ready-to-use solution.
7. Dissection instruments: Fine scissors, forceps, scalpel.
8. Petri dishes, 50 mL tubes with screw-caps.
9. 1 mL syringes with 21-gauge needles.
10. Ice buckets.
11. RPMI- medium: Mix and dissolve 10.4g (1 L packet) of Roswell Park Memorial Institute (RPMI)-1640 medium, and 300 mg of L glutamine in 990 mL of autoclaved water. Add 10 mL of antbiotic/antimycotic solution and 2g of sodium bicarbonate, adjust pH to 7.2 and q.s to 1L, filter sterilize the solution using 0.22 μm syringe filter and store at 4°C (*see* **Note 1**).

2.3 Comet Assay Components

1. Normal melting agarose (NMA): 1 % NMA. Add 1 g NMA in 90 mL of water. Shake occasionally while heating and bring to boil to dissolve the agarose. Finally q.s. to 100 mL with water (*see* **Note 4**). Make fresh for each use. Stabilize temperature at 60 °C in dry bath before making base slides.
2. Low-melting-point agarose (LMPA): 1 and 0.5 % LMPA. For 1 % LMPA, add 500 mg in 50 mL of PBS. Shake occasionally while heating and bring to boil to dissolve the agarose. Avoid vigorous boiling. Microwave at low power for short intervals (*see* **Note 5**). Aliquot and keep at 4 °C for a week. For 0.5 % LMPA, dilute 1 % LMPA with equal volume of PBS (prepare just before making slides).

3. Lysing solution: 2.5 M NaCl, 100 mM disodium ethylenediaminetetraacetic acid (EDTA disodium salt), 10 mM Trizma base. Weigh 146.1 g of NaCl, 37.2 g of EDTA, and 1.2 g of trizma, add to about 700 mL of dH$_2$O, and begin stirring the mixture. Add ~8 g of NaOH and allow the mixture to dissolve. Adjust the pH to 10 using HCl or NaOH. q.s. to 890 mL with water (*see* **Note 6**). Store the solution at room temperature.

4. Final lysing solution: Lysing solution, pH 10, 1 % Triton X-100, 10 % dimethylsulfoxide (DMSO). Mix and refrigerate for at least 30 min at 4 °C before use (*see* **Note 7**).

5. Electrophoresis buffer: 300 mM NaOH, 1 mM EDTA in water.

   Prepare stock solutions of 10 N NaOH: 200 g dissolved in 500 mL of distilled water (*see* **Note 8**) and 200 mM EDTA: 14.89 g dissolved in 200 mL of water and adjust pH to 10 (*see* **Note 9**).

   For use, prepare 1× working buffer: 300 mM NaOH/1 mM EDTA from the stock solutions. Add 30 mL of 10 N NaOH and 5 mL of 200 mM EDTA stock solutions, mix well, and q.s. to 1 L with chilled water (*see* **Note 10**).

6. Neutralization buffer: 0.4 M Tris base, pH 7.5. Weigh 48.5 g Tris base and add to 800 mL of water with constant stirring, adjust pH to 7.5 with concentrated >10 N HCl (*see* **Note 11**), q.s. to 1 L with water, and store the solution at room temperature. Refrigerate for 30 min before use (*see* **Note 12**).

7. Staining solution: 2 μg/mL ethidium bromide (EtBr). Prepare 10× stock solution of 20 μg/mL. Dissolve 10 mg in 50 mL of water; store at room temperature.
   For use prepare 1× solution—mix 1 mL of stock with 9 mL of water (*see* **Note 13**).

8. End-frosted microscope slides (24×72 mm), coverslips (24×60 mm), slide boxes, slide trays, dry bath, coplin jars, electrophoresis tank, Pasteur pipettes, pipette bulbs, ice packs.

3 Methods

3.1 Cell Culture and Obtaining Cells for In Vitro Comet Assay

Here we describe the method followed for CHO cells (*see* **Note 14**).

1. All steps should be performed under aseptic conditions in a laminar hood.

2. CHO cells are obtained and kept frozen in −80 °C. The cells are thawed and cultured in 5 mL of complete Ham's F12 medium (CF12) at 37 °C and 5 % CO$_2$ till a monolayer of cells is obtained in the flask (*see* **Note 15**).

3. Exponentially growing cells about 70–80 % confluent are used for the assay (*see* **Note 16**).

4. Remove the exhausted CF12 medium from the culture flask. Wash the cells with incomplete F12 medium (IF12). Add 1.5 mL of 0.25 % trypsin to detach the cells from the surface of the flask (*see* **Note 17**).

5. Add 2 mL of fresh CF12 medium to the flask. Flush well to form a homogenous suspension of cells and transfer to a round-bottomed centrifuge tube.

6. Centrifuge at $250 \times g$ for 10 min.

7. Discard the supernatant and resuspend the pellet in CF12 to obtain 1×10^6 cells per mL. Add 100 μL of this suspension (to contain 10,000 cells) to each well in a 96-well plate. For each concentration, duplicate wells are plated. Maintain the plate in a humidified incubator with 5 % CO_2 and 37 °C for 20–24 h (*see* **Note 18**).

8. Next day, prepare the various concentrations of the test compound in IF12 (*see* **Note 19**). If the compound is soluble in water then use the medium directly for making the test concentrations, or dissolve the compound in an appropriate vehicle (e.g., DMSO) and then dissolve in IF12. Use IF12 as a control. A positive control group, e.g., 2 mM ethylmethane sulfonate (EMS), should always be included.

9. Take out the plate from incubator and aspirate the medium. Wash the cells twice with 100 μL of IF12 (*see* **Note 20**). Then add to duplicate wells, different concentrations of the test substance in 100 μL IF12 medium. Incubate the plate in an incubator at 37 °C for 3–6 h depending on the experiment.

10. After incubation of the cells with the test substance, aspirate the medium from the wells. Wash the cells with 100 μL of PBS.

11. Aspirate the PBS. Add 50 μL of 0.005 % trypsin to each well. Incubate at 37 °C for 3–5 min. Add 100 μL of CF12 to each well. Flush 2–3 times with a pipette and transfer into centrifuge tubes (*see* **Note 21**).

12. Centrifuge the cells at $500 \times g$ for 5 min.

13. Remove the supernatant. Resuspend the pellet in 50 μL PBS and use 10 μL for assessing cell viability (*see* **step 14**) and for the Comet assay (proceed to Subheading 3.3). Prepare one slide from each sample.

14. Trypan blue dye exclusion test is carried out to assess cell viability. Add and mix 10 μL of trypan blue dye to 10 μL of the cell suspension from **step 13** above. Let stand for 5 min. Count the number of dead (blue) and live (shiny) cells on a hemocytometer to calculate the percent viability and number of cells left after treatment (*see* **Note 22**).

15. The above experiment is conducted in triplicate to arrive at conclusive results.

3.2 Obtaining Cells from Multiple Organs of Mouse for In Vivo Comet Assay

1. Experiments are planned according to the Comet assay guidelines [26–29].

2. Animal care and dosing: Mice or rats can be used for the assay or other animal may be used if justified. Either gender of the animals may be used appropriately, depending on the study. Follow the animal caring policy laid down by the Animal Ethics Committee of your institute. For the present example, male Swiss albino mice (6–8 weeks old, 22 ± 2 g) are used and are raised on a commercial pellet diet and water ad libitum. 4–5 Animals are used for each treatment group and caged separately. Acclimatize the animals for at least five days before the start of the study. They can be treated once in 24 h with multiple treatment schedules using appropriate route of exposure. It is recommended that multiple dose levels should be assessed [28]. If a single acute treatment is given, sampling may be done at 3–6 h or 23–26 h, and with multiple treatments sampling of the animals is done 3–6 h after the last dose [27]. If historical data is available only two animals should be used for the negative and positive controls.

3. Dissolve the test substances and administer in an appropriate vehicle/solvent which themselves are not toxic to the animal. Prepare fresh test material samples for every dosing. Negative and positive controls must be included for each experiment. Route of exposure should be either by intraperitoneal injection or oral. However, any other route may be used if justified. The maximum volume of the administered dose should be based on the body weight of the animal, and should not exceed 20 mL/kg body weight. Also the volume of liquid administered should be equal at all doses.

4. Collection of blood: First, 20–50 µL of blood is drawn from the tail vein, and collected into heparinized microcentrifuge tubes.

5. Animals are sacrificed by cervical dislocation. Lay the animal on the dissection table and wipe the ventral surface with ethanol.

6. Collection of organs: The brain dissected out first and quickly transferred into the cold HBSS solution kept in 50 mL tubes in ice. Then, dissect the animals by making an incision at the ventral side, anterior to the urethral opening and with a scissors, cut along the ventral midline to expose the viscera. Remove and collect the organs (e.g., liver, kidney, and spleen or other organs as required) immediately in 10 mL of cold HBSS kept in 50 mL tubes in ice (*see* **Note 23**).

7. Dissect out the femur, and clean thoroughly to remove tissue around the bone. Using a syringe with a 21-gauge needle, flush the bone marrow in 1 mL of FBS already aliquoted into microcentrifuge tubes (*see* **Note 24**).

8. Lymphocyte preparation: Lymphocytes are isolated from whole blood using Histopaque-1077. Take 20 µL of blood and mix with 1 mL RPMI-1640 medium. Layer this mixture over 100 µL of Histopaque1066, in a microcentrifuge tube, and centrifuge at 500×g for 3 min. Take the interface of the medium/Histopaque containing the lymphocytes and add to 1 mL of RPMI-1640 medium. Centrifuge again at 500×g for 3 min to pellet the lymphocytes, and resuspend in 100 µL of PBS for the Comet assay.

9. Preparation of a single-cell suspension from organs: Place 0.2 g of each organ in 1 mL of freshly prepared chilled mincing solution in a Petri dish and chop into pieces with a scalpel or a scissors. (DMSO in the mincing solution prevents lipid peroxidation and chelates the free radicals preventing additional DNA damage.) Allow the pieces to settle and transfer the supernatant into a microcentrifuge tube. Take the cells from the middle of the tube containing single cells for the experiment. (Other ways of preparing single-cell suspension may be used, e.g., isolated nuclei [30] or cells separated by enzyme action [38].)

10. Count the cells using a hemocytometer/cell counter. Approximately 20,000 cells are required for duplicate slides for the Comet assay. Take the required amount (5–10 µL from bone marrow sample, 40–50 µL from liver, kidney, and spleen) of the respective cell suspension, mix with 1 mL of PBS, and centrifuge to settle the cells. Remove the supernatant and add 110 µL of PBS to form a cell suspension. Proceed for the viability test (10 µL; **step 11**) and Comet assay (100 µL, Subheading 3.3).

11. Cytotoxicity should be reported with Comet data as it can influence the result of the assay [28]. Check the cytotoxicity in the cells from different organs using 5,6-carboxyfluorescein and in lymphocytes or bone marrow cells using trypan blue dye. Place 10 µL of cell suspension in a microcentrifuge tube, and add 10 µL of carboxyfluorescein or trypan blue dye. Mix and let stand for at least 2 min. Load 10 µL onto a hemocytometer and count the number of cells under a fluorescent microscope. (Alternatively, load 10 µL on a glass slide, put a coverslip, and count 100 cells under a fluorescent microscope.) Record the number of viable cells (green) and dead cells (red) and calculate the % viability. For using trypan blue follow **step 14** from Subheading 3.1.

3.3 Comet Assay

1. Prepare base slides to be used for the assay several days in advance or one day before the experiment.

2. Hold the frosted end of the slide with forceps, dip in methanol, and pass it over a blue flame. The alcohol will burn away the

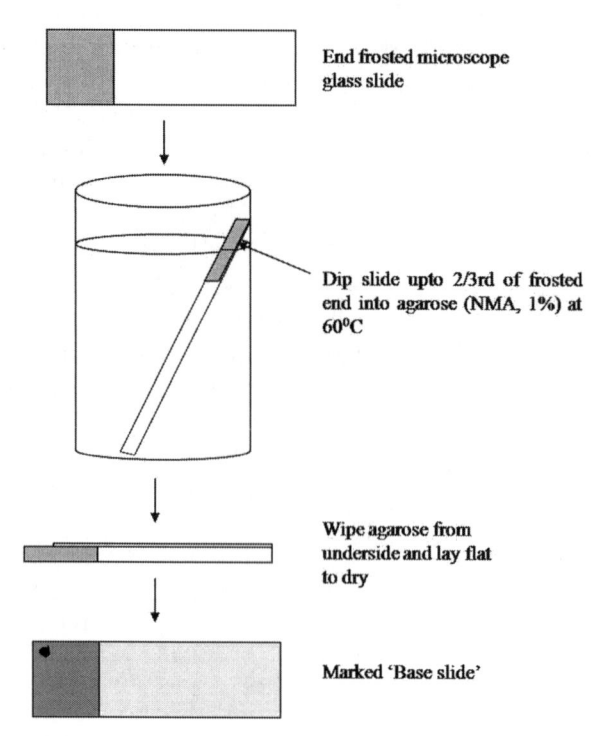

Fig. 2 Preparation of the base slide

machine oil and dust. Keep the cleaned slide separately for further use. Pre-cleaned slides can also be used to avoid this step.

3. Melt 1 % NMA in the microwave or on a hot plate. Transfer to a 100 mL beaker and place it in a 60 °C dry/water bath to cool and stabilize the temperature (*see* **Note 25**).

   Dip the slides up to one-third of the frosted area and gently remove (*see* Fig. 2). Wipe the underside of slide to remove agarose and lay the slide in a tray or on a flat surface to dry. This is the "base" slide. The slides may be air dried or warmed at 50 °C for quicker drying. Store the slides at room temperature until needed; avoid humidity conditions (*see* **Note 26**).

4. Melt 1 and 0.5 % LMPA and maintain at 37 °C dry bath. Code the base slides in pencil/diamond marker for the samples.

5. Mix the cell suspension of the cells of interest in PBS with 1 % LMPA as follows:

   (a) In vitro—40 µL of treated or untreated CHO cells from one well of the 96-well plate (from **step 13** of Subheading 3.1 above) + 40 µL of 1 % LMPA.

   (b) In vivo—100 µL of cell suspension (from **step 11** of Subheading 3.2 above) + 100 µL of 1 % LMPA (for making slides from blood *see* **Note 27**).

80 μL of the mixture is added onto each pre-coated base slide. Place a coverslip (24 × 60 mm) and put the slide on a slide tray resting on ice packs (~5 min) until the agarose layer hardens (*see* **Note 28**). Usually, slides were prepared in duplicate; however, it is now recommended to prepare three slides from each in vivo sample for better statistical power [27].

6. Gently slide off the coverslip and add a third layer of 80 μL of 0.5 % LMPA to the slide (*see* **Note 29**). Replace the coverslip and return the slide to the slide tray resting on ice packs until the agarose layer hardens (~5 min).

7. Lysing: Remove the coverslip from the gel by gentle sliding and slowly lower the slides into coplin jars containing cold, freshly prepared final lysing solution. Protect from light and refrigerate overnight or for a minimum of 2 h. This is the first pause in the assay, and slides may be stored for at least 1 week in cold lysing solution without affecting the results (*see* **Note 30**).

8. Unwinding: Remove the slides gently from the lysing solution. Place the slides side by side in the horizontal gel box near one end, sliding them as close together as possible. Fill the buffer reservoirs with freshly made and cold 1× electrophoresis buffer (pH > 13) until it completely covers the slides (avoid bubbles over the agarose). Slides are left in the alkaline buffer for 20 min to allow for unwinding of the DNA and expression of the alkali-labile damage (*see* **Note 31**).

9. Electrophoresis: Turn on the power supply to 24 V (~0.74 V/cm) and adjust the current to 300 mA by raising or lowering the buffer level. Depending on the purpose of the study and on the extent of migration in control samples, electrophoresis is carried out on the slides for 10–40 min (*see* **Note 32**).

10. Neutralization: Turn off the power. Gently drain the buffer, take out the slides, and place on a drain tray. Dropwise coat the slides with neutralization buffer (pH 7.4), and leave for at least 5 min. Drain the slides and repeat the procedure two more times (*see* **Note 33**).

11. Staining: Stain the slides with 80 μL of 1× EtBr, leave for 5 min, and then dip in chilled distilled water to remove excess stain. Place a coverslip over it and keep in a slide box and at 4 °C till scoring (*see* **Note 34**).

12. The slides may also be stored and scored later either before staining or for archival purposes after staining. This is the second pause in the assay. For this, follow through neutralizing step (**step 10** above). Then, drain the slides of the neutralization buffer, briefly dip in water, and then dip them for 20 min in cold 100 % ethanol or 100 % methanol for dehydration. Remove and dry in air. Then place the slides in an

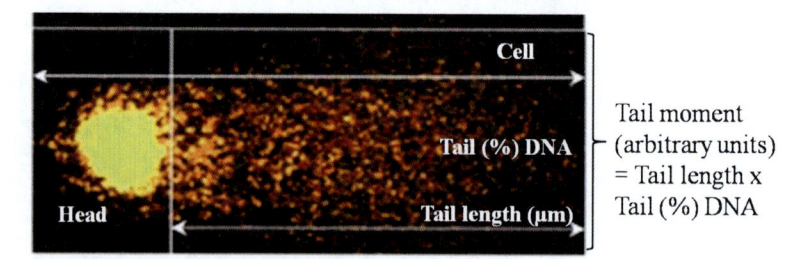

Fig. 3 Parameters evaluated for the Comet assay (http://www.cometassayindia.org/definitions.htm)

oven at 50 °C for 30 min to ensure complete drying of the slides. Store in a slide box in a dry area. When required, rehydrate the gel by dipping slides in chilled distilled water for 30 min and stain with EtBr as in **step 11** above and cover with a fresh coverslip. After scoring, remove the coverslip again, rinse in 100 % alcohol to remove stain, let dry, and store for archival purposes.

13. Visualization of DNA damage: Gently wipe the slide and coverslip with tissue paper, to blot the excess moisture before viewing under the microscope. Use the 40× objective on a fluorescent microscope to give a final magnification of 400×. Score 100 random cells from each slide per sample (see **Note 35**). The microscope is linked to a charge-coupled device (CCD) camera to acquire images using software (e.g., Komet 5) for the analysis of the quantitative and qualitative extent of DNA damage. Generally, the Comet parameters recorded are Olive tail moment (OTM, arbitrary units), tail DNA (%), and tail length (distance of migration of the DNA from the nucleus, μm; TL) (Fig. 3). The comets may also be scored manually as also discussed in Chapter 18.

14. Compare the amount of migration per cell, the number of cells with increased migration, and the extent of migration among damaged cells.

3.4 Statistics

Details of experimental design and statistical analysis used in Comet assay have been well described [39]. The experimental unit for in vitro studies is the culture and for in vivo studies is the animal [40]. Recently recommendations for statistics for the Comet assay have also been published by a special interest group of the statisticians in the pharmaceutical industry [41]. Parametric tests such as Student's t-tests and ANOVA and nonparametric equivalents are appropriate for analyzing experiments [8–10, 14, 15]. Linear trends can be used for dose-related effects and will have higher statistical power [31].

4 Notes

1. The medium should be prepared in a laminar hood, using autoclaved syringe filters or sterile automated filter components and kept at 37 °C in CO_2 incubator to rule out the possibility of any contamination. In case of contamination, bacterial and or yeast growth can be seen as a white powdery deposit at the bottom of the bottle, or the medium may turn translucent. Such medium should be properly discarded after autoclaving the contents. To prevent the loss of the entire medium due to contamination, store incomplete medium in two 500 mL bottles, instead of one 1 L bottle. Also make smaller amounts of the complete medium at one time.

2. FBS should be kept at −20 °C till use. Thaw the serum by keeping the bottle at 37 °C in a water bath. If heat inactivation is required, heat the contents of the bottle at 56 °C for 30 min in a water bath. Care should be taken to start the incubation time only when the contents have reached the required temperature. Aliquot FBS aseptically into 50 mL tubes and then in 10 mL tubes for single-time use to avoid repeated freezing and thawing of the serum.

3. Store PBS solution at room temperature. To prevent contamination of the whole solution make aliquots and store. If ready-made PBS is not available, it can be prepared using sodium and potassium salts. Weigh 80 g of sodium chloride (NaCl), 11.5 g anhydrous sodium phosphate dibasic (Na_2HPO_4) OR 29 g of sodium phosphate dibasic heptahydrate ($Na_2HPO_4 \cdot 7H_2O$), 2 g of potassium chloride (KCl), and 2 g of potassium dihydrogen phosphate (KH_2PO_4). Prepare a 10× stock solution by dissolving all the salts in 900 mL dH_2O and adjust the pH to 7.4 using 0.1 N HCl. This solution may be autoclaved after adjusting volume to 1,000 mL and dispensed in 50 mL aliquots. To get a 1× solution for the assay, add 10 mL aliquot to 90 mL deionized water.

4. Take care to completely dissolve NMA and do not overheat the solution by vigorous boiling or allow the solution to boil over, as it can cause a change in the concentration. This may impact the base layer formation.

5. Take care to completely dissolve LMPA and do not overheat the solution or allow the solution to boil over as it can cause a change in the concentration, which will in turn affect the movement of DNA and hence the results. 1 and 0.5 % LMPA can be prepared separately or 1 % LMPA can be diluted with equal volume of PBS. Also make aliquots of 5–10 mL in glass vials with stoppers and refrigerate at 4–8 °C. For use, bring the aliquots to room temperature before melting on a heating

pad/water bath/microwave. Do not store the leftover LMPA after the experiment, as it may result in change of concentration of agarose on reheating.

6. The volume is kept at 890 mL as the DMSO and Triton X-100 will make up the final working solution to 1 L. If not adding DMSO, make up that volume with lysing solution.

7. The purpose of adding DMSO in the lysing solution is to scavenge the free radicals generated by the iron released from hemoglobin when blood or animal tissues are used. It is not needed for other situations, e.g., where cell line is being used, or where the slides will be kept in lysing for a brief time only. For one coplin jar, ~40 mL of the final lysing solution is required; thus calculate the number of coplin jars required for the test and accordingly make the final lysing solution.

8. Take care while making the 10 N NaOH solution. It is an exothermic reaction and glass bottles tend to become hot. Make the solution in a bottle surrounded by crushed ice. Take 400 mL water, add a few pellets of NaOH at a time, and completely dissolve before proceeding. Once all the pellets have dissolved q.s to 500 mL and store at room temperature.

9. EDTA has poor solubility in water and it dissolves only at pH 8. Thus add ~2 g of NaOH to the solution while dissolving the EDTA. Once dissolved, adjust the pH to 10 with 1 N NaOH.

10. Make fresh 1× working buffer before each electrophoresis run. The total volume of the final buffer depends on the gel box capacity. Before use, ensure that the pH of the buffer is >13.

11. Adjust the pH using 10 N concentrated HCl first to reduce the gap between starting and required pH and then use lower concentration 1 or 0.1 N HCl to avoid a sudden fall in the pH below the required pH.

12. The neutralizing solution should be chilled to keep the integrity of the agarose intact.

13. Adequate precautions and care should be taken while handling EtBr as it is a proven carcinogen. Always wear gloves while preparing and handling the EtBr. Proper disposal should also be followed.

14. CHO is an adherent cell line. The steps given in this protocol can be followed for adherent cells, e.g., CHL cells (or V79), human hepatoma cells (HepG2), and other epithelial cell lines, with respective media being used for the different cells. The use of cell type and cell line should be justified for the study. Concentration of trypsin used may differ for the cell lines used; however care should be taken to use low concentration of trypsin so that it does not cause toxicity itself. Non-adherent or suspension cells, e.g., human lymphoblast cell line—JM1 or mouse lymphoma cells L5178Y, and primary cells such as

isolated lymphocytes may also be used. In these cells there will be no step for trypsinization. The cells are centrifuged to pellet the cells. The medium/test material can then be aspirated and cells resuspended in PBS for the subsequent steps.

15. Cells are obtained from an authorized cell culture center as a monolayer, which are then frozen in FBS with 10 % DMSO (freezing medium) and kept at −80 °C or liquid nitrogen for lengthy storage. Care should be taken to thaw the frozen vial (at −80 °C) very quickly. Place the vial in water maintained at 37 °C, immediately transfer the contents into a round-bottomed tube with 5 mL of CF12, and centrifuge at $250 \times g$. The pellet is resuspended in 5 mL of CF12 and transferred into a 25cm² tissue culture flask. The culture flasks are maintained at 37 °C and 5 % CO_2. At 70 % confluency, the cells are dislodged using 0.25 % trypsin, and seeded into fresh flasks with 5 mL of CF12; this is called "subculture or passage." Cells are used for experiments after at least two passages from start of culture. Also, one set of experiments should be carried out on the same passage numbers.

16. 70–80 % confluence ensures that cells are healthy while 100 % confluence will have cells, which might be stressed for nutrition.

17. Avoid over-trypsinization of the cells. Add trypsin, rotate the flask to cover the surface of the cells, and then aspirate the excess trypsin. Incubate the flask at 37 °C for 2–5 min. Assess the cells under a microscope. When all the cells become round, add CF12 to stop the reaction of trypsin and flush the cells to detach them from the flask. The FBS in the medium acts as an inhibitor for trypsin.

18. The 96-well plates allow the use of small amounts of the test substances as well as save on the disposals. 24-well or even 48-well plates can also be used. The volume of the test material, medium, and trypsin used will then change accordingly.

19. The following recommendations are made for the in vitro testing using the Comet assay [26]:

    (a) *Test substances.* Test substances should be freshly prepared for each experiment. Solid test substances should be dissolved in an appropriate solvent, which is stable, does not react with the test substance or medium, e.g., DMSO, and diluted in medium/PBS. Wherever possible aqueous solvent should be preferred. Liquid substances should be added directly or diluted in test medium.

    (b) *Exposure.* The cells (monolayer or suspension) should be treated with various concentrations of test material in the presence and absence of external metabolic activation (S9) for 3–6 h. The highest concentration tested should not induce cytotoxicity and with each experiment, cytotoxicity should be evaluated. No single method for cytotoxicity determination has been decided on, but dose selection and assay results based on cytotoxicity should be justified.

For the Comet assay, duplicate cultures should be analyzed at each concentration along with negative and positive controls. At least three analyzable concentrations should be tested. For non-cytotoxic compounds, maximum tested concentration should be 5 μL/mL, 5 mg/mL, or 0.01 M whichever is lowest. For insoluble compounds the highest dose should be a concentration above the limit of solubility in culture medium at the end of exposure, if there is no toxicity at the lowest insoluble concentration. However, if toxicity is observed at higher than the lowest insoluble concentration then more than one concentration with precipitation may be tested. The precipitate should not interfere with the experiment completion.

(c) *Metabolic activation.* Both the absence and presence of external source of metabolic activation should be included in the study. The postmitochondrial fraction (S9; obtained after centrifugation at 9,000×g) prepared from liver homogenate of male rats treated with enzyme-inducing chemicals such as Arochlor 1254, or a mixture of phenobarbitone and β-naphthoflavone, is used. The S9 is mixed with cofactors (such as sodium phosphate buffer, NADP, glucose-6-phosphate, and $MgCl_2$) and used between concentrations of 1–10 % (v/v) of the final test medium. Add 10 μL of S9 mix in each well and 90 μL of the test concentrations prepared in IF12 medium for the assay.

(d) *Controls.* Both negative and positive controls should be included in the study, with and without metabolic activation. Negative control will be solvent in treatment medium.

Positive controls not requiring metabolic activation: methyl methanesulfonate (MMS); ethyl methanesulfonate (EMS); and ethyl-nitroso-urea. Positive controls requiring metabolic activation: benzo-a-pyrene (BaP) and cyclophosphamide (CP).

(e) *Analysis and results.* Slides should always be coded and scored blind (without the knowledge of code). At least 25 cells from each duplicate slide and at least 50 cells per culture must be scored. The study unit is the culture and all statistical analysis should be made on the response of the culture. The results of each experiment should be verified in independent experiments. For determining a positive result, a concentration-dependent increase in DNA damage at one or more concentrations tested should be observed. For a positive response the corresponding cytotoxicity data should also be considered to rule out the possibility of increase in DNA damage not associated with genotoxicity. A reproducible result in independent experiments is the clear call for positive or negative results. A positive result indicates that the test substance induces DNA damage in cultured mammalian cells under those test conditions, while a negative result shows that the test substance does not induce DNA damage in cultured mammalian cells under the test conditions employed.

20. While adding the test concentrations, aspirate the medium/PBS of only the concerned wells. Do not aspirate the medium/PBS of all the wells at once, as it will cause drying of the cells in the latter wells.

21. Very low concentration of trypsin (0.005 %) is used because higher concentrations increase DNA damage. Take care to aspirate almost all the medium/PBS before adding trypsin because the trypsin is already dilute. Observe the cells under a microscope. When the cells have become rounded, flush 2–3 times and add complete medium to stop the trypsin action. Do not wait for all the cells to come in solution as then they will be over-trypsinized. Try and use 200 μL micro-tips with cut ends (for a larger orifice) while flushing and transferring cells to maintain the integrity of the cells.

22. Alternatively, load 10 μL on a glass slide and put a coverslip. Now count a total of 100 cells and calculate the number of dead (blue) and live (shiny) cells. The dead cells have a compromised cell membrane which enables the cells to take up the dye and appear blue, while the live cells with an intact cell membrane appear shiny under the light microscope.

23. The HBSS should be cold and kept in ice to prevent any further damage to the cells.

24. Take care while cleaning the femur, as it can break. Cut the head/trochanter as well as the other end. Fix the 21-guage needle of the syringe in the head, hold between the index finger and thumb, and flush the FBS through the femur. If some marrow is still left, aspirate the same FBS into the syringe and repeat the above procedure till the femur becomes clear.

25. NMA at 60 °C is in molten form and maintaining this temperature is necessary while making the slides. Below this temperature the agarose starts becoming thicker, forming lumps, and will not stay on the slide.

26. Slides should be labeled at the frosted end consistently with graphite pencil before storage to avoid confusion of the side with/without gel. Do not use permanent marker for labeling, as the ink gets washed off during electrophoresis. If the gels are not sticking to the slides properly, it may be due to unclean

slides, humidity, not dipping the slide properly up to the frosted end in NMA at the time of base slide preparation, or low NMA concentration. Cleaning the slides with methanol, avoiding humid conditions, taking care while dipping slides in NMA, and/or increasing the concentration of NMA to 1.5 % should eliminate the problem.

27. If using whole blood for the slides, either (a) or (b) can be followed.

    (a) Mix ~5 µL whole blood with 75 µL LMPA (0.5 %) and layer onto one base slide.

    (b) Alternatively, take 20 µL blood and dilute with 80 µL PBS, add 100 µL LMPA (1 %), and layer 80 µL each onto two base slides.

    (c) A small volume of blood can be added to 1 mL RPMI medium and stored cold (refrigerated or on ice) for an extended time until processed. The cells must then be centrifuged and care taken to remove as much supernatant as possible before adding 75 µL of 0.5 % LMPA/5 µL blood to the pellet. If sample storage is necessary, this sample can be flash-frozen in liquid nitrogen and stored at −70 °C until processed. Flash-freezing will not affect DNA integrity; however, cell survival is compromised.

28. The amounts indicated are based on using No.1, 24×60 mm coverslips. Proportional volumes can be used for coverslips differing in size. One slide per well is prepared for in vitro, while two slides per sample are prepared for in vivo.

29. Make sure that the slide trays resting on ice packs are cold enough for the gel to set; else while removing coverslips by sliding, the gel might break. The final concentration of LMPA in the second and third layers should be same to prevent uneven migration of DNA in the two layers.

30. The lysing solution should be cold to maintain the integrity of the gel. The slide preparation should be performed under dim yellow light to prevent additional DNA damage.

31. The longer the exposure to alkali, the greater is the expression of alkali-labile damage.

32. Different gel boxes will require different voltage settings to correct for the distance between the anode and the cathode. The goal is to obtain migration among the control cells without it being excessive. The optimal electrophoresis duration differs for different cell types. A lower voltage, amperage and a longer electrophoresis time may allow for increased sensitivity. Electrophoresis time, current, and voltage are variables that affect the assay performance [42].

33. Use a glass pipette with a rubber bulb for this step; add the buffer slowly in drops, as a stream of buffer may break the gel.

34. Take adequate precautions while working with EtBr as it is a known carcinogen. Spread a layer of tissue paper before keeping the slides in the slide box to ensure that the excess water is absorbed and slides are moist till scoring.

35. Although any image analysis system may be suitable for the quantitation of SCGE data, we use a Komet 5/Komet 6 image analysis system (developed by Kinetic Imaging, Ltd. (Liverpool, UK, and now marketed by ANDOR technology), attached to a fluorescence microscope (Leica DM LB 100T, Leica, Germany) equipped with appropriate filters (N2.1, excitation wavelength of 515–560 nm and emission wavelength of 590 nm).

Acknowledgements

The authors gratefully acknowledge the funding from CSIR, New Delhi, under its network projects (NWP34, NWP35) and OLP 009. The funding from UK India Education and Research Initiative (UKIERI) standard award to Institute of Life Sciences, Ahmedabad University, Ahmedabad, India (IND/CONT/E/11-12/217), and the University of Bradford is gratefully acknowledged. Funding from the European Union Seventh Framework Programme (FP7/2007-2013) under grant agreement no. 263147 (NanoValid—Development of reference methods for hazard identification, risk assessment and LCA of engineered nanomaterials) is also acknowledged.

References

1. Ostling O, Johanson KJ (1984) Microelectrophoretic study of radiation-induced DNA damage in individual mammalian cells. Biochem Biophys Res Commun 123:291–298
2. Singh NP, McCoy MT, Tice RR et al (1988) A simple technique for quantitation of low levels of DNA damage in individual cells. Exp Cell Res 175:184–191
3. Dhawan A, Bajpayee M, Parmar D (2009) Comet assay: a reliable tool for the assessment of DNA damage in different models. Cell Biol Toxicol 25:5–32
4. Dusinska M, Collins AR (2008) The comet assay in human biomonitoring: gene–environment interactions. Mutagenesis 23:191–205
5. Brendler-Schwaab S, Hartmann A, Pfuhler S et al (2005) The in vivo comet assay: use and status in genotoxicity testing. Mutagenesis 20:245–254
6. Pfuhler S, Fellows M, van Benthem J et al (2011) In vitro genotoxicity test approaches with better predictivity: summary of an IWGT workshop. Mutat Res 723(2):101–107
7. Collins AR, Oscoz AA, Brunborg G (2008) The comet assay: topical issues. Mutagenesis 23(3):143–151
8. Bajpayee M, Pandey AK, Zaidi S et al (2005) DNA damage and mutagenicity induced by endosulfan and its metabolites. Environ Mol Mutagen 47(9):682–692
9. Pandey AK, Gurbani D, Bajpayee M et al (2009) *In silico* studies with human DNA topoisomerase-II alpha to unravel the mechanism of in vitro genotoxicity of benzene and its metabolites. Mutat Res 661(1–2):57–70
10. Patel S, Bajpayee M, Pandey AK et al (2007) In vitro induction of cytotoxicity and DNA strand breaks in CHO cells exposed to cypermethrin, pendimethalin and dichlorvos. Toxicol In Vitro 21(8):1409–1418
11. Kiskinis E, Suter W, Hartmann A (2002) High throughput Comet assay using 96-well plates. Mutagenesis 17(1):37–43

12. Stang A, Brendamour M, Schunck C et al (2010) Automated analysis of DNA damage in the high-throughput version of the comet assay. Mutat Res 698(1–2):1–5
13. Stang A, Witte I (2010) The ability of the high-throughput comet assay to measure the sensitivity of five cell lines toward methyl methanesulfonate, hydrogen peroxide, and pentachlorophenol. Mutat Res 701(2):103–106
14. Pandey AK, Bajpayee M, Parmar D et al (2008) Multipronged evaluation of genotoxicity in Indian petrol-pump workers. Environ Mol Mutagen 49(9):695–707
15. Bakare AA, Pandey AK, Bajpayee M et al (2007) DNA damage induced in human peripheral blood lymphocytes by industrial solid waste and municipal sludge leachates. Environ Mol Mutagen 48(1):30–37
16. Sharma V, Anderson D, Dhawan A (2011) Zinc oxide nanoparticles induce oxidative stress and genotoxicity in human liver cells (HepG2). J Biomed Nanotechnol 7(1):98–99
17. Shukla RK, Kumar A, Gurbani D et al (2013) TiO(2) nanoparticles induce oxidative DNA damage and apoptosis in human liver cells. Nanotoxicology 7(1):48–60
18. Sharma V, Anderson D, Dhawan A (2012) Zinc oxide nanoparticles induce oxidative DNA damage and ROS-triggered mitochondria mediated apoptosis in human liver cells (HepG2). Apoptosis 17(8):852–870
19. Sharma V, Shukla RK, Saxena N et al (2009) DNA damaging potential of zinc oxide nanoparticles in human epidermal cells. Toxicol Lett 185:211–218
20. Sharma V, Singh SK, Anderson D et al (2011) Zinc oxide nanoparticle induced genotoxicity in primary human epidermal keratinocytes. J Nanosci Nanotechnol 11(5):3782–3788
21. Shukla RK, Sharma V, Pandey AK et al (2011) ROS-mediated genotoxicity induced by titanium dioxide nanoparticles in human epidermal cells. Toxicol In Vitro 25(1):231–241
22. COM (United Kingdom Committee on Mutagenicity of Chemicals in Food, Consumer Products and the Environment) (2000) Guidance on a strategy for testing of chemicals for mutagenicity.
23. U. S. Food and Drug Administration (2012) Guidance for industry. www.fda.gov/downloads/Drugs/Guidances/ucm074931.pdf
24. International Committee on Harmonization (ICH) guideline S2(R1) (2011) Guidance on genotoxicity testing and data interpretation for pharmaceuticals intended for human use. http://www.ich.org/fileadmin/Public_Web_Site/ICH_Products/Guidelines/Safety/S2_R1/Step4/S2R1_Step4.pdf
25. Rothfuss AA, Honma MM, Czich AA et al (2011) Improvement of in vivo genotoxicity assessment: combination of acute tests and integration into standard toxicity testing. Mutat Res 723:108–120
26. Tice RR, Agurell E, Anderson D et al (2000) Single cell gel/Comet assay: guidelines for in vitro and in vivo genetic toxicology testing. Environ Mol Mutagen 35:206–221
27. Hartmann A, Agurell E, Beevers C et al (2003) Recommendations for conducting the in vivo alkaline Comet assay. Mutagenesis 18:45–51
28. Burlinson B, Tice RR, Speit G et al (2007) In vivo comet assay workgroup, part of the fourth international workgroup on genotoxicity testing. Fourth International Workgroup on Genotoxicity testing: results of the in vivo Comet assay workgroup. Mutat Res 627: 31–35
29. Vasquez MZ (2012) Recommendations for safety testing with the in vivo comet assay. Mutat Res 747:142–156
30. Sasaki YF, Sekihashi K, Izumiyama F et al (2000) The comet assay with multiple mouse organs: comparison of comet assay results and carcinogenicity with 208 chemicals selected from the IARC monographs and U.S. NTP Carcinogenicity Database. Crit Rev Toxicol 30(6):629–799
31. Patel S, Pandey AK, Bajpayee M et al (2006) Cypermethrin-induced DNA damage in organs and tissues of the mouse: evidence from the comet assay. Mutat Res 607(2):176–183
32. Bakare AA, Patel S, Pandey AK et al (2012) DNA and oxidative damage induced in somatic organs and tissues of mouse by municipal sludge leachate. Toxicol Ind Health 28(7):614–623
33. Ansari KM, Chauhan LK, Dhawan A (2005) Unequivocal evidence of genotoxic potential of argemone oil in mice. Int J Cancer 112(5): 890–895
34. Ansari KM, Dhawan A, Khanna SK et al (2005) In vivo DNA damaging potential of sanguinarine alkaloid, isolated from argemone oil, using alkaline Comet assay in mice. Food Chem Toxicol 43(1):147–153
35. Das M, Ansari KM, Dhawan A et al (2005) Correlation of DNA damage in epidemic dropsy patients to carcinogenic potential of argemone oil and isolated sanguinarine alkaloid in mice. Int J Cancer 117(5):709–717
36. Ansari KM, Dhawan A, Khanna SK et al (2006) Protective effect of bioantioxidants on argemone oil/sanguinarine alkaloid induced genotoxicity in mice. Cancer Lett 244(1):109–118
37. Sharma V, Singh P, Pandey AK et al (2012) Induction of oxidative stress, DNA damage and apoptosis in mouse liver after sub-acute

oral exposure to zinc oxide nanoparticles. Mutat Res 745(1–2):84–91
38. Burlinson B (2012) The in vitro and in vivo Comet assays. In: Parry JM, Parry EM (eds) Genetic toxicology: principal and methods. Methods in molecular biology, vol 817. Humana, New York, pp 143–163
39. Lovell DP, Thomas G, Dubow GR (1999) Issues related to the experimental design and subsequent statistical analysis of in vivo and in vitro comet studies. Teratog Carcinog Mutagen 19(2):109–119
40. Lovell DP, Omori T (2008) Statistical issues in the use of the comet assay. Mutagenesis 23:171–182
41. Bright J, Aylott M, Bate S et al (2011) Recommendations on the statistical analysis of the Comet assay. Pharm Stat 10(6):485–493
42. Azqueta A, Gutzkow KB, Brunborg G et al (2011) Towards a more reliable Comet assay: optimizing agarose concentrations, alkaline unwinding time and electrophoresis conditions. Mutat Res 724:41–45

Chapter 18

The Comet Assay in Human Biomonitoring

Diana Anderson, Alok Dhawan, and Julian Laubenthal

Abstract

Human biomonitoring studies aim to identify potential exposures to environmental, occupational, or lifestyle toxicants in human populations and are commonly used by public health decision makers to predict disease risk. The Comet assay measures changes in genomic stability and is one of the most reliable biomarkers to indicate early biological effects, and therefore accepted by various governmental regulatory agencies. The appeal of the Comet assay lies in its relative simplicity, rapidity, sensitivity, and economic efficiency. Furthermore, the assay is known for its broad versatility, as it can be applied to virtually any human cell and easily adapted in order to detect particular biomarkers of interest, such as DNA repair capacity or single- and double-strand breaks. In a standard experiment, isolated single cells are first embedded in agarose, and then lysed in high-salt solutions in order to remove all cellular contents except the DNA attached to a nuclear scaffold. Subsequent electrophoresis results in accumulation of undamaged DNA sequences at the proximity of the nuclear scaffold, while damaged sequences migrate towards the anode. When visualized with fluorochromes, these migrated DNA fragments resemble a comet tail and can be quantified for their intensity and shape according to internationally drafted guidelines.

Key words Comet assay, Single-cell electrophoresis, DNA damage, Double-strand breaks, Single-strand breaks, Alkali-labile sites

1 Introduction

Biomarkers can be broadly defined as indicators and predictors of mechanisms, events, or changes in a biological system. Theoretically, anything that can be measured in the human body may represent a biomarker [1]. In human biomonitoring studies, however, biomarkers are typically more restricted and defined as indicators of exposure, effect, and susceptibility with the ultimate goal of identifying causes of inherited or environmentally induced adverse outcomes in humans [2, 3]. As most organs cannot be reached by a socially and ethically acceptable minimally invasive route, responses are usually monitored in surrogate tissue, such as peripheral lymphocytes or buccal cells [4]. A frequently used biomarker of exposure, effect, or susceptibility is DNA damage, which is a critical risk factor for aging, heritable diseases, or the development of cancer. DNA damage can be measured with chemical-specific biomarkers such as aflatoxin–DNA adducts, which gives a highly sensitive indication of aflatoxin exposure through diet but is not environmental exposure specific [5, 6]. However, most DNA damage biomarkers are environmental exposure specific while being less chemical specific, hence providing a more general measure of DNA damage. A biomarker represented by benzo(a)pyrene DNA adducts can detect benzo(a)pyrene from a variety of sources to which the human is exposed daily, including active and passive cigarette smoke, grilled food, or air pollution [7]. For such complex mixtures, these exposure-specific biomarkers provide an index of overall genotoxic insult and their lack of chemical specificity is seen as an advantage [8].

The Comet assay is considered a biomarker of universal DNA damage that may reflect a category of genotoxic agent and is frequently used today in human biomonitoring studies as a very early indicator for a biological effect [9–11]. The Comet assay was adapted from the nucleotide sedimentation technique and introduced in 1984 as the single-cell electrophoresis assay by Ostling and Johanson [12] to detect DNA double-strand breaks (DSBs) under neutral pH conditions. Today, the 1988 method developed by Singh et al. [13], and later modified by Olive et al. [14], the alkaline version (pH ≥ 13) of the assay is most frequently used, allowing the simultaneous detection of single- and double-strand breaks as well as single-strand breaks (SSBs) associated with incomplete excision repair sites and alkali-labile sites (ALSs). Moreover, other types of DNA damage such as DNA cross-links (e.g., thymine dimers) and oxidative DNA damage can also be studied using antibodies or specific DNA repair enzymes with the alkaline version. Finally, in 2000 an international consortium set up the first guidelines for a standardized protocol which included crucial parameters such as slide preparation, electrophoresis, lysis conditions, or quantification of DNA damage [15]. A method for human monitoring in the Comet assay was also described in the WHO/IPCS document in 2000 [16].

In a standard experiment (Fig. 1), nucleated cells are embedded in low-melting-point agarose on a microscope slide already covered with a layer of agarose. The slides containing the cells are then immersed in a high-salt solution containing detergents to remove the membrane (cytoplasmic and nuclear). Because the DNA is wrapped around an octamer of histone proteins which compress about 2 m of DNA to fit within the small nucleus, the lysis step also degrades the histone proteins to relax the DNA structure. Following the lysis, the slides are placed on a horizontal agarose gel tank and electrophoresed. After cell lysis, DNA still remains supercoiled due to different levels of DNA compaction, and is likely to still be attached to the nuclear matrix, but at the sites of DNA strand breaks, supercoils are more relaxed. During electrophoresis,

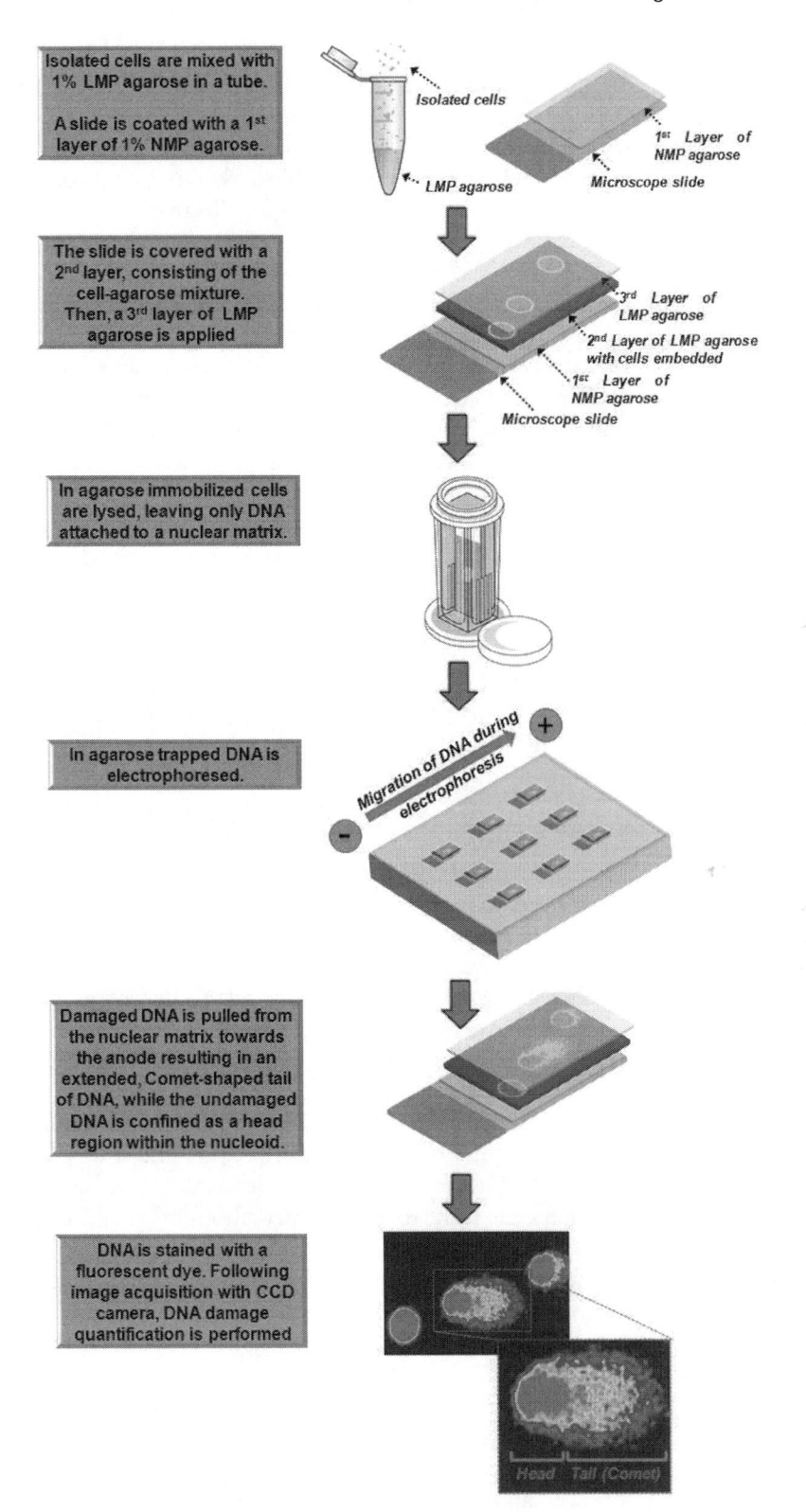

Fig. 1 Schematic illustration of the standard Comet assay experiment

these relaxed loops of damaged DNA are pulled from the nuclear matrix towards the anode resulting in an extended, comet-shaped tail of DNA, while the undamaged DNA is confined as a head region within the nucleoid. Variation of electrophoresis pH conditions and incubation times establish the DNA damage type and sensitivity level which can be detected. Electrophoresis under alkaline conditions detects double- and single-strand breaks, as well as ALSs (at $pH \geq 13$) [17, 18]. At neutral electrophoresis conditions (pH 8–9), primarily DSBs quantify, and only a small amount of SSBs, in turn, due to the relaxation of supercoiled loops containing the breaks, contribute to the formation of the comet tail [14, 19]. Since toxicants induce primarily SSBs and ALSs, the alkaline Comet assay is more commonly used. Important variations of this protocol include further decondensation approaches prior to lysis, unwinding, and electrophoresis when assessing damage in condensed spermatozoal DNA [7, 20] and the addition of DNA repair enzymes such as endonuclease III (Endo III) and formamidopyrimidine DNA glycosylases (FPG) to convert oxidized pyrimidines or purines into SSBs, in order to detect oxidative DNA damage [21, 22]. Furthermore, using image analysis software, the Comet assay even detects apoptotic cells, which are termed in the Comet assay as "ghosts" or "hedgehogs" due to their faint and small head and large dispersed tails [23]. However, these cells are so damaged that the image analysis software is unable to assign a degree of damage. Since they may constitute a bias in the interpretation of the results, it was recommended by Tice et al. [15] to exclude apoptotic cells from genotoxicity analysis.

In human biomonitoring studies, the Comet assay can provide crucial information on risk assessment of environmental, occupational, and lifestyle exposures, which may form the basis for new legislative laws initiated by public health decision makers [4, 16]. Given this pivotal role, it is imperative that the Comet assay is standardized and validated and that ethical guidelines are followed so that misinterpretation, misapplication, or inappropriate release of data are avoided. Participating in a cohort study must always be voluntary, not create any health risks, and proceed only with informed consent from each individual. Thus, each participant does have the right to be informed about his/her individual results [4, 16]. However, since control values can overlap exposed values and vice versa, results are usually used only on a group basis as opposed to an individual basis. General and local ethical guidelines on human experimentation have to be strictly followed and prior to commencing any study, all relevant applications should have been made to the local committees for ethics in research. Also, as with any other biomarker used in human biomonitoring, the principles of Good Laboratory Practice and Good Clinical Practice should be followed and studies only commenced when the assay is fully standardized and validated [9, 16, 24].

Theoretically, the Comet assay can be performed on any mammalian tissue, given that a single-cell suspension can be produced [16]. However, in human biomonitoring, selection of the most appropriate tissue should be based on the expected dynamics of the exposed population characteristics, exposure type, endpoint mechanisms, persistence of DNA damage, rate of cell turnover, sample timing, study aims, tissue accessibility, as well as possible ethical and social restrictions [4, 9, 16]. Interestingly, it was found that more than 88 % of all Comet assay studies were performed on human blood cells, while other tissues, such as exfoliated nasal, buccal, or bladder cells, accounted for just about 10 % [25]. That is mainly because blood (a) circulates through the whole organism, representing an estimate of the overall extent of toxicant exposures, and (b) can be acquired by a socially and ethically acceptable, minimally invasive route. Particularly for human biomonitoring studies, which usually involve many samples to be collected and analyzed at the same time or within a specific time frame, isolated lymphocytes from whole blood are considered the (somatic) material of choice [11, 26]. Isolated lymphocytes can be frozen for many months as numerous studies found no increases in endogenous levels of DNA strand breaks between fresh and frozen lymphocytes, while only few reported very slight but consistent differences in DNA damage [9, 27, 28]. However, as a general rule, blood samples should be collected in the morning and, in case of studies lasting longer than 1 year, during the same season in each year to avoid seasonal variations, mixed with lithium heparin or sodium heparin to avoid clotting, transported under cool conditions but avoiding direct ice contact and lymphocytes isolated at the same day of venepuncture [9, 16]. Most importantly, collecting samples from all exposed individuals and after that from all control individuals or vice versa must be avoided; also samples from exposed and control individuals must be handled in the same way [28]. If blood samples are transported by air, they must under no circumstances be subjected to X-irradiation at the airport and, thus, all required shipping permits must be obtained well in advance from the respective airline and usually not the airport itself. However, long-term freezing of whole blood is thought to be more sensitive to DNA damage compared with lymphocytes and therefore primarily used in biomarker studies involving DNA extraction [9, 16]. In recent years, spermatozoa have become increasingly used for human monitoring of the reproductive function, as they are thought to reflect the overall extent of toxicant exposure to the male germ line [7]. However, the collection of spermatozoa samples in large cohort studies remains difficult for ethically and/or social reasons [7, 10, 29]. So far, no comparable tissue is available to measure the reproductive function of the female in human biomonitoring studies.

Following all these guidelines and recommendations can minimize the effect of confounding variables (i.e., exposures which are not investigated) or technical artifacts (e.g., induced by handling differences) which may result in biased estimation of exposure effects. The extreme case of a biased estimation is the detection of a causal effect where none exists or a true effect is unseen [10]. Typical technical variability examples include lack of uniform sampling, shipping, processing, and freezing of samples as well as improper use of reagents/solutions, preparation of comet slides, lysis and electrophoresis conditions, as well as scoring methods and designs [16, 19, 24].

Oxidative DNA damage can have a pivotal role in the development and progression of many diseases and therefore has become a valuable biomarker in human biomonitoring with the Comet assay. Intrinsic DNA-repair mechanisms, namely, nucleotide excision repair and base excision repair, involve the enzymatic generation of DNA nicks to replace oxidized nucleotides, such as 8-oxo-2-deoxyguanosine [19, 22]. Also bacterial enzymes, such as Endo III and FPG, can recognize and remove, respectively, oxidized pyrimidines and altered purines and subsequently create DNA nicks. Hence, addition of these repair enzymes after cell lysis during a basic Comet assay experiment can form DNA breaks at the sites of such oxidative DNA damage and subsequently increase the amount of damaged DNA, to be quantified in the comet's tail [30, 31].

In conclusion, the Comet assay combines the advantages of an in situ assay with the simplicity of high-throughput technique for the detection of DNA damage, on a low cost, simple, fast, and reproducible basis. That is why the Comet assay is currently the most widely applied method for detecting DNA damage at a single-cell level and it is one of the reference methods to evaluate DNA damage and repair in human biomonitoring studies [16, 17]. In this chapter we describe the Comet assay as applied to human biomonitoring studies.

2 Materials

2.1 Lymphocyte Separation, Freezing, and Thawing

1. Anticoagulant sodium heparin-coated tubes.
2. Saline: 0.9 % NaCl (w/v) in double-distilled H_2O.
3. Density gradient solution (*see* **Note 1**).
4. Centrifuge ($1,425 \times g$ is required).
5. Freezing medium: 90 % fetal calf serum and 10 % dimethyl sulfoxide (v/v).
6. Cryo-vials.

7. Freezers (−20 and −80 °C are required).
8. Liquid nitrogen tank.
9. Cryogenic gloves.

2.2 Slide Preparation

1. Microscope slides with frosted end.
2. Metal plate and ice-box to hold multiple microscope slides.
3. Water bath or heat block.
4. Agarose: 1 % (w/v) normal-melting-point (NMP) agarose prepared in double-distilled water (ddH$_2$O).

2.3 Agarose Embedding and Lysis of Cells

1. Phosphate buffer saline (PBS): 137 mM NaCl, 2.75 mM KCl, 1.45 mM KH$_2$PO$_4$, 15.25 mM Na$_2$HPO$_4$, pH 7.2.
2. 0.5 and 1 % (w/v) low-melting-point (LMP) agarose in PBS.
3. Alkaline lysis solution: 2.5 M NaCl, 100 mM Na$_2$EDTA, 10 mM Tris base in ddH$_2$O. This solution can be kept at room temperature for up to 2 weeks. Prior to use, add 10 % DMSO (v/v) and 1 % Triton X-100 (v/v) and cool for 1 h at 4 °C (*see* **Notes 2** and **3**).
4. Centrifuge (200 × *g* is required).
5. Water bath or heat block.
6. Coverslips (22 × 50 mm).
7. Coplin jar.

2.4 Electrophoresis

1. Alkaline electrophoresis buffer: 300 mM NaOH, 1 mM Na$_2$EDTA, prepared in ddH$_2$O. Adjust pH to 13.2 with 10 M NaOH.
2. Neutralization buffer: 0.4 M Tris base prepared in ddH$_2$O. Adjust pH to 7.5 with 10 M HCl.
3. Horizontal electrophoresis tank. The tank's size determines the number of slides and consequently experiments performed in one electrophoresis run, as well as the respective power supply parameters required.
4. 20 V power supply (0.8 V/cm on platform, 300 mA).
5. 10–50 mL Plastic syringe.

2.5 Staining

1. Tris–EDTA buffer: 1 mM Na$_2$EDTA, 10 mM Tris–HCl, pH 8.
2. Staining solution: 20 μg/mL ethidium bromide in Tris–EDTA buffer. Stock solutions (1:100 in ddH$_2$O) can be stored at −20 °C for up to 6 weeks (*see* **Note 4**). *Caution*. Ethidium bromide is classified as mutagenic and possibly carcinogenic.

2.6 Quantification

1. Epifluorescence microscope equipped with two objectives (20× and 40×); fluorescence filter (510 nm absorption; 595 nm emission); 100 W mercury illumination bulb.
2. Charge-coupled device (CCD) camera.
3. Computer workstation.
4. Image analysis software is available as freeware, such as CometScore (http://autocomet.com/products_cometscore.php), or can be purchased, such as Comet 6.0 (Andor, UK). Alternatively homemade comet scoring software can be easily programmed using open-source imaging software, such as ImageJ (http://rsbweb.nih.gov/ij/).

2.7 The Neutral Comet Assay (Rest of the Requirements Are Similar to Subheadings 2.1–2.6 Above)

1. Neutral lysis solution: 30 mM EDTA, 0.5 % SDS (w/v) in ddH$_2$O. Adjust pH to 8.0 with 1 M NaOH (instead of lysis solution in **step 3** of Subheading 2.3).
2. Neutral electrophoresis buffer: 90 mM Tris base, 90 mM boric acid, 2 mM EDTA prepared in ddH$_2$O. Adjust pH to 8.5 with 1 M NaOH (instead of alkaline electrophoresis buffer in **step 1** of Subheading 2.4).

2.8 The Sperm Comet Assay (Other Requirements Are Similar to Subheadings 2.1–2.6 Above)

1. Agarose: 2 % (w/v) and 1 % (w/v) LMP agarose in PBS instead of agarose in **step 2** of Subheading 2.2.
2. Lysis solution: Add 10 mM dithiothreitol to the alkaline lysis solution in **step 7** of Subheading 2.3.
3. Lysis solution with 0.05 g/mL proteinase K.

2.9 The FPG Comet Assay

1. Repair enzyme buffer: 40 mM HEPES, 0.5 mM EDTA, 0.1 M KCl, 0.2 mg/mL bovine serum albumin, add small volumes of KOH until pH 8.0 is reached. (One can make a 20× stock solution and freeze at −20 °C.)
2. Endo III and/or FPG.
3. Moist chamber: Use a sealable plastic box and line with moist paper towels.

3 Methods

See **Notes 5–7** for guidelines and general requirements for the method.

3.1 Lymphocyte Separation, Freezing, and Thawing

1. Collect whole blood by venepuncture into anticoagulant sodium heparin-coated tubes. Blood may also be collected by finger prick method into heparinized tubes (*see* **Note 8**).
2. Dilute whole blood 1:1 with saline in 50 mL tubes.

3. Slowly pipette 6 mL of the diluted blood on top of 3 mL of density gradient solution. Avoid mixing the two phases.
4. Spin down at 1,425 ×g for 20 min at room temperature.
5. Thoroughly retrieve lymphocytes from just above the boundary between the density gradient solution at the tube's bottom without mixing with the upper serum layer. Avoid disturbance of the density gradient solution's layer.
6. Mix the isolated lymphocytes with 10 mL of saline.
7. Spin down at 1,200 ×g for 15 min at room temperature.
8. Remove carefully all supernatant so that 100 μL of isolated lymphocyte solution remains at the bottom of the tube.
9. Pipette this 100 μL of isolated lymphocyte solution into a cryo-vial and mix with 900 μL of freezing medium.
10. Freeze the cryo-vial for 6 h at −20 °C and then at −80 °C for up to 1 month. Use liquid nitrogen when storage for more than 1 month is required. *Caution.* Always wear cryogenic gloves to protect your hands and arms when working in hazardous, ultracold environments.
11. When using frozen samples for a Comet assay experiment, place cryo-vials in a 37 °C water bath for speedy thawing to avoid artificial cellular and DNA damage by ice crystal formation.

3.2 Slide Preparation

1. Rinse the microscope slides with 100 % ethanol and flame for a short time.
2. Melt the 1 % (w/v) NMP agarose solution in a microwave. Prevent liquid overflow or excessive evaporation.
3. Place the melted agarose in a water bath maintained at 60 °C.
4. Dip slides briefly into the melted agarose solution, and remove excess agarose by wiping the back of the slide with a tissue.
5. Let the slides dry overnight at room temperature. Agarose-pre-coated slides can be stored for up to 7 months in an airtight box at room temperature (*see* **Note 9**).

3.3 Agarose Embedding and Lysis of Cells

1. Place a metal plate on top of ice in an appropriate ice storage box.
2. Melt LMP agarose in PBS to make 0.5 and 1 % agarose solutions. Keep agarose melted in a water bath at 37 °C.
3. Wash lymphocytes in PBS by gently vortexing and spin down at 200 ×g for 5 min.
4. Discard all supernatant and resuspend the cells in 1 mL of fresh PBS. Repeat **step 3**.
5. Discard all the supernatant and finally resuspend the pellet in 100 μL of PBS.

6. Mix the cell suspension with an equal amount of 1 % LMP agarose solution by carefully pipetting the mixture up and down. *Caution*. Avoid any air bubbles during mixing as this can result in artificial DNA damage. (For making slides with whole blood *see* **Note 10**.)

7. Drop 100 μL of the agarose/cell suspension to the center of the pre-coated agarose slide and cover the agarose layer with a coverslip.

8. Leave the coverslipped slide on the ice-cold metal plate to let the agarose set for 10 min.

9. Once the agarose is set, remove the coverslip and apply the last (third) layer of 100 μL of 0.5 % LMP agarose solution and cover with a coverslip.

10. Repeat **step 8** (*see* **Note 11**).

11. Once the agarose is set on all slides, remove coverslips and incubate slides at 4 °C for 1 h in a vertical position in a Coplin jar with ice-cold alkaline lysis solution (*see* **Note 12**).

12. Remove the alkaline lysis solution carefully from the Coplin jar and wash the slides three times for 15 min in Coplin jars filled with ddH$_2$O at 4 °C. Do not shake the slides during the washing steps in order to prevent any damage to the thin agarose layers (*see* **Note 13**).

13. Transfer the slides to the electrophoresis tank and place them close to the anode to ensure lack of variability between the slides (*see* **Note 14** and Fig. 1).

3.4 Electrophoresis and Staining

1. Fill the tank with ice-cold electrophoresis buffer so that all slides are covered with ~4 mm layer of buffer and close the tank. Avoid bubbles on top of the slides. Place the tank into a cold room or closed refrigerator set at 4 °C to avoid melting of the agarose layers during electrophoresis (*see* **Note 15**).

2. Pre-incubate slides in the alkaline electrophoresis buffer for 30 min (*see* **Note 16**).

3. Electrophorese the DNA at a constant voltage of 0.75 V/cm with the current reaching 300 mA, for 30 min. If the current is too high or too low, remove or add alkaline electrophoresis buffer using a 10–50 mL syringe. A current between 290 and 310 mA is acceptable (*see* **Note 17**).

4. Remove all slides from the tank and drain the slides' edges with a paper towel.

5. Drop-wise add neutralization buffer on all slides, leave for 5 min, and then drain. Repeat this wash three times.

6. Rinse the slides carefully with ddH$_2$O to avoid loss of the agarose layers, and remove excess liquid from the slides.

7. Staining: Stain the electrophoresed DNA with 60 μL of ethidium bromide solution and cover with a coverslip (*see* **Note 18**).

3.5 Quantification

Usually 50 cells in three repeats are scored for one comet experiment. There are four different ways for quantification:

1. Comet tail formation assessment by eye: Here, cells are graded in five categories according to the approximate amount of DNA in the tail and each cell is assigned a score on the scale from 0 to 4: 0 = <5 % of the DNA is in the tail (undamaged cell with a faint comet tail); 1 = 5–20 % (low-level damage); 2 = 20–40 % (medium-level damage); 3 = 40–95 % (high-level damage); 4 = >95 % (total damage). For 50 scored cells the overall score (i.e., the sum of scores of the 50 cells) should be between 0 and 200 arbitrary units.

2. Semi-automated image analysis: Here cells are selected by the scorer manually, while the CCD camera captures the selected comets and the image analysis software quantifies the signal intensity as well as the diameter of the comets. Most common parameters detected include comet tail length, % DNA in tail, % DNA in head, and the tail moment (the combination of tail length and % tail DNA). The most recent guidelines suggested % tail DNA as the international standard, covering the widest range of DNA damage, and as reviewed in ref. 32.

3. Automated image analysis: In this cells are automatically selected by the software. However, these are very expensive and based on software learning algorithms, which can be easily biased during their training period of the software. Analysis requires precisely discerning the transition from head to tail of the comet, demanding multifaceted image analysis that can generate invalid results, such as overlapping comets being scored as damaged cells.

4. Micro-patterned agarose arrays of single cells combined with automated image software to identify and analyze comets: Here, the algorithms of the software select comets based on their array registration, eliminating the problem of overlapping comets or debris. Moreover, all cells are trapped in a single focal agarose plane, avoiding manual refocusing for each individual cell. Hence, automated image analysis without any manual user training of the software or intervention during the analysis is currently the best way to quantify a comet experiment [33].

3.6 The Neutral Comet Assay

To conduct the neutral assay four alterations need to be applied to the alkaline Comet assay protocol; other steps will be similar to those followed in Subheadings 3.3–3.5.

1. Use the neutral lysis solution in **step 11** in Subheading 3.3, instead of the alkaline lysis solution.
2. Use neutral electrophoresis buffer for incubation and electrophoresis in **steps 1–3** in Subheading 3.4, instead of the alkaline electrophoresis buffer.
3. Use 1 V/cm during electrophoresis in **step 3** of Subheading 3.4, instead of 0.75 V/cm.
4. Skip all washes performed in **step 5** of Subheading 3.4.

3.7 The Sperm Comet Assay

For conducting the assay in sperm cells, three alterations need to be applied to the alkaline Comet assay protocol; the other steps will remain similar to those followed in Subheadings 3.3–3.5.

1. Prepare 1 and 2 % LMP agarose and maintain at 37 °C in a dry bath.
2. Instead of 1 % LMP agarose in **step 6** of Subheading 3.3, mix the sperm cells with 2 % LMP agarose and add 1 % LMP agarose in **step 9** of Subheading 3.3.
3. In **step 11** of Subheading 3.3, add 10 mM dithiothreitol to the alkaline lysis solution and incubate slides for 1 h at 4 °C. Then add 0.05 g/mL proteinase K to the alkaline lysis solution and incubate slides for 1 h at 4 °C.
4. In **step 3** of Subheading 3.4, electrophorese slides for 20 min.

3.8 The Repair Enzyme Comet Assay

The following modifications need to be incorporated in the alkaline Comet assay protocol. The other steps will be similar to those followed in Subheadings 3.3–3.5.

1. Following **step 12** of Subheading 3.3, incubate cells in repair enzyme buffer for 5 min at room temperature. Repeat twice.
2. Drain excess repair enzyme buffer onto a tissue/Kimwipe©.
3. Pipette 100 μL of enzymes dissolved in repair enzyme buffer onto each gel, add a coverslip, and incubate in a moist chamber at 37 °C for 30 min (FPG) or 40 min (Endo III) (*see* **Notes 19 and 20**).
4. Continue with **step 13** of Subheading 3.3 and further as described for the alkaline Comet assay protocol.

4 Notes

1. A density gradient solution is a solution containing sodium diatrizoate and polysaccharide which retains only lymphocytes while all other leukocytes and erythrocytes pass through it, e.g., Ficoll and Histopaque. We use Lymphoprep© solution by Axis Shield (Oslo, Norway).

2. The addition of DMSO scavenges reactive radicals produced by the iron released from hemoglobin during the lysis step of the cells, e.g., when whole blood is used for preparing the slides. Hence, when using cells or tissue without hemoglobin, DMSO should not be added.

3. Some earlier experimental protocols describe addition of the anionic detergent sodium sarcosinate to the solution. However, this is not necessary for successful lysis and may induce precipitation of the lysis solution at 4 °C [13].

4. Alternative nucleic acid dyes (an appropriate fluorescence filter is needed) include YOYO-1 iodide (absorption: 491 nm, emission: 508 nm) and SYBR Gold (495, 537 nm). These have to be diluted with ddH$_2$O to 1:10,000. Also usage of acridine orange (502, 526 nm) has been reported; however, acridine orange tends to fade after a short time.

5. The described protocol works equally well on other cells besides isolated lymphocytes, e.g., whole blood cells, exfoliated bladder cells, and buccal or nasal mucosal cells, except spermatozoa (*see* Subheading 3.7 for spermatozoa).

6. It is essential that all experimental steps of Subheadings 3.3–3.5 are performed on ice and only with cold (4 °C) solutions, to avoid activation of DNA damage response and subsequent repair. Furthermore, all experiments should be performed only in the dark or in a room with an orange filter to prevent UV light-induced artificial DNA damage.

7. It is also imperative that temperature; voltage; incubation, lysis, and electrophoresis times; electrophoresis buffer volume; and concentration or type of nucleic acid dyes are not changed within matching experiments where more than one electrophoresis run is used.

8. When small volumes of blood are obtained during human biomonitoring studies we use the micro technique for isolating lymphocytes. Briefly, mix 20 μL of whole blood with 1 mL RPMI 1640 in a microcentrifuge tube, and layer 100 μL lymphocyte-separating medium, e.g., Histopaque 1077, below the blood–medium mixture with the help of micropipette. Centrifuge for 3 min at 500×*g*. Take out 100 μL of interface of the media/Histopaque layer carefully with micropipette, mix it with 1 mL of medium, and centrifuge for 3 min at 500×*g*. Centrifuge the pellet again with 1 mL of PBS for 3 min at 500×*g*. Discard the supernatant, resuspend the pellet in PBS, and use for the Comet assay as described.

9. A longer storage of the agarose-pre-coated slides has been shown to increase the possibility of detached agarose layers during lysis or electrophoresis runs.

10. For whole blood as sample: Either mix ~5 μL of whole blood with 75 μL of 0.5 % LMPA OR dilute blood (10–20 μL) with PBS (80–90 μL) and add equal volume (100 μL) of 1 % LMPA. Then layer 80 μL mixture onto an NMA-coated slide and put a coverslip. After 5 min on ice, add a third layer of 0.5 % LMPA, put a coverslip, and keep on ice for 5 min.

11. Dimensions of the agarose gel layers should match the dimension of the coverslip used. Moreover, the gel should not fill the complete slide or contain any air bubbles.

12. An extension of up to 8-h lysis does not influence the outcome of the experiment. Although longer period of lysis greater than 8 h are reported, to our knowledge, this can significantly increase the chances of cell loss. Further, the pH should not be greater than pH 10.5 when increasing the lysis time [24].

13. Rinsing with ddH$_2$O is crucial, as the remaining lysis buffer salt traces can lead to an artificially expressed comet tail during electrophoresis.

14. If too few slides are used in one electrophoresis run, fill gaps with empty slides to allow a homogenous current during electrophoresis.

15. A homogenous and controlled layer of electrophoresis buffer covering all slides is important for reproducible results, as even small differences in electrophoresis buffer volume result in variable strengths of electric fields, subsequently producing inconsistent comet tail lengths and biased results.

16. The incubation times in electrophoresis buffer prior to electrophoresis can vary between 20 and 70 min depending on the cell or the tissue type. An incubation of 30 min for cultured cells and blood samples is generally recommended and is standard in our work group. Shorter incubation periods, however, may reduce background DNA damage levels of control cells, but simultaneously decrease DNA damage detection sensitivity of treated or exposed cells.

17. Alterations of the current of the power supply during electrophoresis result in dramatic increase or decrease of comet tails. The settings described here are international standards and yield the most reproducible results. As a guide, cells not being exposed to DNA-damaging agents (control cells) should have around 5–12 % of their DNA in the tail and the tail lengths of no more than 15–25 μm [5].

18. Generally, ethidium bromide can stain both DNA and RNA. However, under the alkaline conditions of the Comet assay, all the RNA is degraded; hence RNA cannot be stained here.

19. The working concentration of the respective repair enzyme depends on its source. Always use enzyme stocks from the

same batch as well as the same incubation conditions, throughout the entire biomonitoring study.

20. The use of lesion-specific bacterial repair enzymes can increase both sensitivity and selectivity of the Comet assay.

Acknowledgement

The funding from the UK India Education and Research Initiative (UKIERI) standard award to the University of Bradford, UK, and the Institute of Life Sciences, Ahmedabad University, India, is gratefully acknowledged.

References

1. Angerer J, Aylward LL, Hays SM et al (2011) Human biomonitoring assessment values: approaches and data requirements. Int J Hyg Environ Health 214:348–360
2. Bartsch H, Arab K, Nair J (2011) Biomarkers for hazard identification in humans. Environ Health 10(Suppl 1):S11
3. Schoeters GE, Den Hond E, Koppen G et al (2011) Biomonitoring and biomarkers to unravel the risks from prenatal environmental exposures for later health outcomes. Am J Clin Nutr 94:1964S–1969S
4. Anderson D, Zeiger E (1997) Human monitoring. Environ Mol Mutagen 30:95–96
5. Paini A, Scholz G, Marin-Kuan M et al (2011) Quantitative comparison between in vivo DNA adduct formation from exposure to selected DNA-reactive carcinogens, natural background levels of DNA adduct formation and tumour incidence in rodent bioassays. Mutagenesis 26:605–618
6. Woo LL, Egner PA, Belanger CL et al (2011) Aflatoxin B1-DNA adduct formation and mutagenicity in livers of neonatal male and female B6C3F1 mice. Toxicol Sci 122:38–44
7. Sipinen V, Laubenthal J, Baumgartner A et al (2010) In vitro evaluation of baseline and induced DNA damage in human sperm exposed to benzo[a]pyrene or its metabolite benzo[a]pyrene-7,8-diol-9,10-epoxide, using the comet assay. Mutagenesis 25:417–425
8. Neumann HG (2009) Risk assessment of chemical carcinogens and thresholds. Crit Rev Toxicol 39:449–461
9. Dusinska M, Collins AR (2008) The comet assay in human biomonitoring: gene-environment interactions. Mutagenesis 23:191–205
10. Laubenthal J, Zlobinskaya O, Poterlowicz K et al (2012) Cigarette smoke-induced transgenerational alterations in genome stability in cord blood of human F1 offspring. FASEB J. doi:10.1096/fj.11-201194
11. Baumgartner A, Kurzawa-Zegota M, Laubenthal J et al (2012) Comet-assay parameters as rapid biomarkers of exposure to dietary/environmental compounds – an in vitro feasibility study on spermatozoa and lymphocytes. Mutat Res 743:25–35
12. Ostling O, Johanson KJ (1984) Microelectrophoretic study of radiation-induced DNA damages in individual mammalian cells. Biochem Biophys Res Commun 123:291–298
13. Singh NP, McCoy MT, Tice RR et al (1988) A simple technique for quantitation of low levels of DNA damage in individual cells. Exp Cell Res 175:184–191
14. Olive PL, Banath JP, Durand RE (1990) Heterogeneity in radiation-induced DNA damage and repair in tumor and normal cells measured using the "comet" assay. Radiat Res 122:86–94
15. Tice RR, Agurell E, Anderson D et al (2000) Single cell gel/comet assay: guidelines for in vitro and in vivo genetic toxicology testing. Environ Mol Mutagen 35:206–221
16. Albertini RJ, Anderson D, Douglas GR et al (2000) IPCS guidelines for the monitoring of genotoxic effects of carcinogens in humans. International Programme on Chemical Safety. Mutat Res 463:111–172
17. Fairbairn DW, Olive PL, O'Neill KL (1995) The comet assay: a comprehensive review. Mutat Res 339:37–59
18. Klaude M, Eriksson S, Nygren J et al (1996) The comet assay: mechanisms and technical considerations. Mutat Res 363:89–96
19. Collins AR (2004) The comet assay for DNA damage and repair: principles, applications, and limitations. Mol Biotechnol 26:249–261

20. Anderson D, Schmid TE, Baumgartner A et al (2003) Oestrogenic compounds and oxidative stress (in human sperm and lymphocytes in the Comet assay). Mutat Res 544:173–178
21. Cemeli E, Baumgartner A, Anderson D (2009) Antioxidants and the Comet assay. Mutat Res 681(1):51–67. doi:0.1016/j.mrrev.2008.05.002
22. Collins AR, Azqueta A (2012) DNA repair as a biomarker in human biomonitoring studies; further applications of the comet assay. Mutat Res 736(1–2):122–129. doi:10.1016 /j.mrfmmm.2011.03.005
23. Hartmann A, Speit G (1997) The contribution of cytotoxicity to DNA-effects in the single cell gel test (comet assay). Toxicol Lett 90:183–188
24. Collins AR, Oscoz AA, Brunborg G et al (2008) The comet assay: topical issues. Mutagenesis 23:143–151
25. Valverde M, Rojas E (2009) Environmental and occupational biomonitoring using the Comet assay. Mutat Res 681:93–109
26. American Society of Reproductive Medicine (2008) Smoking and infertility. Fertil Steril 90:S254–S259
27. Chuang CH, Hu ML (2004) Use of whole blood directly for single-cell gel electrophoresis (comet) assay in vivo and white blood cells for in vitro assay. Mutat Res 564:75–82
28. Anderson D, Yu TW, Dobrzynska MM et al (1997) Effects in the Comet assay of storage conditions on human blood. Teratog Carcinog Mutagen 17:115–125
29. Anderson D (2005) Male-mediated developmental toxicity. Toxicol Appl Pharmacol 207:506–513
30. Collins AR (2009) Investigating oxidative DNA damage and its repair using the comet assay. Mutat Res 681:24–32
31. Azqueta A, Shaposhnikov S, Collins AR (2009) DNA oxidation: investigating its key role in environmental mutagenesis with the comet assay. Mutat Res 674:101–108
32. Kumaravel TS, Vilhar B, Faux SP et al (2009) Comet assay measurements: a perspective. Cell Biol Toxicol 25:53–64
33. Wood DK, Weingeist DM, Bhatia SN et al (2010) Single cell trapping and DNA damage analysis using microwell arrays. Proc Natl Acad Sci U S A 107:10008–10013

Chapter 19

The Comet Assay in Marine Animals

Giada Frenzilli and Brett P. Lyons

Abstract

Comet assay is a quick and versatile technique for assessing DNA damage in individual cells. It allows the detection of DNA single- and double-strand breaks, as well as the presence of alkali-labile sites and cross-links. Here we describe the protocols for the single-cell gel electrophoresis (Comet) assay in its alkaline (pH > 13), mild alkaline (pH = 12.1), and neutral (pH = 8) versions, when applied in marine animals.

Key words Single-cell gel electrophoresis, DNA strand breaks, Comet assay, Marine animals, Hemocytes, Gills

1 Introduction

The marine environment is known to receive a variety of natural and anthropogenic chemical compounds, some of which are known to be highly persistent and possess genotoxic properties [1, 2]. In the 1970s the genotoxic threat posed to the aquatic environment by these contaminants was first perceived and the risk they posed was monitored using techniques such as the *Salmonella* bioassay [3], on sentinel species, including mussels [4] and fish [5, 6]. Since these initial studies, a wide range of techniques, mainly adapted from mammalian toxicology, have been applied and used to evaluate DNA alterations in aquatic animals [7–9]. A combination of biochemical, molecular, and cytological techniques are now available, which can ultimately predict the genotoxic impact of exposure to environmental contaminants. In general, to be suitable for use in the marine environment, the methods have to be sensitive to a range of contaminant concentrations, have to be applicable to a wide range of species/tissue types, and should have the advantage of detecting and quantifying exposure to genotoxins without a detailed knowledge of the contaminants present. The Comet assay meets these requirements and has been used widely in both laboratory- and field-based studies with a range of marine species [9, 10]. However, unlike mammalian genotoxicology, where the focus is limited to a small number of model species, efforts in the aquatic field have generally lacked coordination and have used an extensive range of sentinel species and target tissue [9–12]. While guidelines relating to the use of the Comet assay have been published for mammalian genotoxicology [13, 14], no standard protocols currently exist for environmental studies. Therefore, in this chapter we focus on detailing a standardized protocol for commonly used marine species. Many different species and cell types have been used previously; however gill cells and hemocytes of bivalve molluscs and the nucleated erythrocytes of teleost fish appear frequently in the scientific literature and will be the focus of the protocols detailed here. While it is expected that these methods will work equally well with other marine species and tissue types, it is recommended that any protocol selected should be validated before undertaking large-scale experimental or biomonitoring studies.

1.1 Basic Protocol Overview

1 % normal-melting agarose (NMA) is spread on glass microscope slides and left to dry. Once set, an 85 µL layer of 0.5 % low-melting agarose (LMA) including 10 µL of cell suspension (approx. 3×10^5) is added to the slide. Finally, the slide is covered with a third layer of 85 µL of LMA. Slides are immersed in ice-cold freshly prepared lysing solution (2.5 M NaCl, 100 mM Na_2EDTA, 10 mM Tris–HCl, 1 % Triton X-100, and 10 % DMSO, pH 10) to lyse the cells and allow DNA unfolding to occur. After at least 1 h at 4 °C in the dark, slides are placed on a horizontal electrophoresis unit. The unit is filled with fresh buffer (1 mM Na_2EDTA, 300 mM NaOH, pH > 13) to cover the slides. The slides are allowed to sit in the high-pH buffer for 10–20 min (depending on cell type and different protocols) to allow DNA unwinding and expression of alkali-labile sites. Electrophoresis is conducted for 5–20 min (tissue/species dependent and should be defined prior to undertaking work for the first time) at 25 V (300 mA). Slides are then gently washed to remove alkali in a neutralization buffer (0.4 M Tris–HCl, pH = 7.5) and stained with 100 µL of ethidium bromide or other dyes (SYBR Green, DAPI, acridine orange, propidium iodide, Hoechst 33258, YOYO-1). Ideally, all steps described above should be conducted under yellow light to prevent additional DNA damage. For each experimental point, 300 cells (50 from each of the two slides, three replicates) are analyzed under a fluorescence microscope (200×) using a software package (or manual scoring). A modified version of the technique includes the use of enzymes able to cut oxidized bases in order to detect the loss of DNA integrity specifically due to oxidative stress [15, 16]. When the amount of DNA damage overwhelms DNA repair capabilities, the process of programmed cell death or apoptosis can occur. A modified version of Comet assay, described by Singh [17], named diffusion assay, is a visual method for apoptotic cells

in which cells are embedded in agarose and lysed and then the DNA is precipitated with ethanol, instead of undergoing electrophoresis. Cells treated with an agent known to induce apoptosis appear with a halo of granular DNA and a hazy outer boundary, related to the presence of diffusing small fragments [16, 18].

2 Materials

1. Artificial seawater (ASW): 0.114 g NaF, 3.4 g $SrCl_2 \cdot 6H_2O$, 1.2 g H_3BO_3, 4.02 g KBr, 27.96 g KCl, 43.98 g $CaCl_2 \cdot 2 H_2O$, weigh and add in the order indicated, while stirring, to 500 mL of deionized water. In another beaker add 159.6 g Na_2SO_4, 199.8 g $MgCl_2 \cdot 5H_2O$, 7.98 g $NaHCO_3$. Unify the content of the two beakers into 60 L of deionized water in a tank, and then add 1,659 g NaCl. Mix until NaCl dissolves. ASW is prepared to be used for in vitro and in vivo studies with mussels (see **Note 1**).

2. Digestion solution: 1.5 mg/mL dispase in HBSS.

3. 25,000 IU/5 mL heparin.

4. 10 N NaOH: Add 200 g of NaOH to 300 mL of dH_2O. Stir and q.s. to 500 mL. Store at room temperature for a maximum of 2 weeks.

5. 200 mM EDTA: Add 14.9 g of Na_2EDTA to 150 mL of dH_2O. Once in solution, maintain pH to 10 and make up the volume to 200 mL. Store at room temperature for a maximum of 2 weeks.

6. Lysing solution: 2.5 M NaCl, 100 mM EDTA disodium salt, 10 mM Trizma base. Add 146.1 g of NaCl, 37.2 g of Na_2EDTA, and 1.2 g of Trizma base to 700 mL dH_2O (see **Note 2**). Begin stirring mixture. Add 8 g of NaOH and allow the mixture to dissolve. Adjust pH to 10 and bring the volume to 900 mL with dH_2O. Store at room temperature for a maximum of 6 months. For final lysing solution, add fresh 1 % Triton X-100 and 10 % DMSO, and then refrigerate for at least 30 min prior to slide addition.

7. Neutralization solution: 0.4 M Tris–HCl. Add 48.5 g Trizma base to 800 mL of dH_2O, adjust pH to 7.5 with concentrated HCl, bring the volume to 1,000 mL with dH_2O, and then store at room temperature for a maximum of 6 months.

8. Low-melting-point agarose (LMA): 0.5 % LMA. Add 0.5 g of LMA in 100 mL of PBS in a heat-resistant container. Microwave until boiling, and then mix gently until the agarose dissolves. Aliquot 5 mL samples into suitable containers and refrigerate until needed. When required briefly melt agarose in microwave or by another appropriate method. Place LMA vial in a 37 °C

water bath to cool and stabilize the temperature just prior to use (*see* **Note 3**).

9. Normal-melting-point agarose (NMA): 1 % NMA: Prepare the gel by adding 1 g of NMA in 100 mL of PBS in a heat-resistant container. Microwave until boiling and mix until the agarose dissolves. While the agarose is hot, pour into a heat-resistant beaker just prior to use.

10. Fpg-Enzyme buffer: 0.5 mM Na_2EDTA, 0.1 M KCl, 0.2 mg/mL bovine serum albumin (BSA), 40 mM *N*-[2-hydroxyethyl]piperazine-*N*-[2-ethanesulfonic acid], pH 8.

11. Repair endonuclease enzyme: 1 % formamidopyrimidine glycosylase (fpg): dilute 2 μL in 200 μL of fpg-enzyme buffer for a working solution.

12. Electrophoresis buffer: For different pH prepare as below.

    (a) 300 mM NaOH/1 mM EDTA, pH > 13. Prepare 1× solution by adding 30 mL of 10 N NaOH and 5 mL of 200 mM EDTA stock solutions and bring the volume to 1,000 mL. Mix well (*see* **Note 4**).

    (b) 300 mM NaOH/1 mM EDTA, pH 12.1. Use the buffer given above, but maintain the pH to 12.1 (*see* **Note 5**).

    (c) 10× TBE buffer, pH 8: 1 M Tris–HCl, 0.9 M boric acid, 0.01 M EDTA, prepared in water. For use, dilute 10× TBE stock to 1× in water.

3 Methods

All steps described below are conducted under yellow light to prevent additional DNA damage.

3.1 General Recommendations and Experimental Design

1. Before laboratory studies commence, it is recommended that pilot experiments are undertaken to characterize the background levels of DNA damage in the species and tissue of choice.

2. If animals used in laboratory studies are collected from the field, a period of acclimation is recommended to ensure that transportation stress or other factors related to animal husbandry do not impact the study design. Since this will be study specific it should be investigated thoroughly before any experimental work is undertaken.

3. For biomonitoring in the field, confounding factors such as sampling location should be investigated, as it is known that for some species, position on the shore line, etc. can impact the level of DNA damage detected by the Comet assay [19].

4. When field-based biomonitoring is undertaken it is often common for animals to be transplanted and caged within the study location. The use of caged sentinel specimens (mainly mussels

and fish) also avoids the possibility that sampling programs spread over relatively wide geographical areas might be influenced by factors such as differences in genetic background, developmental stage, or reproductive status. For such studies it is advised that exposure durations lasting less than 10 days should be avoided, as previous work has indicated that DNA damage may fluctuate initially due to transportation stress, environmental adaptation, and contaminant update (for review *see* ref. 9.

3.2 Obtaining Cell Types for the Assay

1. Mussel hemocytes: After collection of mussels, hemolymph is withdrawn via the posterior adductor muscle (*see* **Note 6**) using a 0.8×40 mm (25 gauge) needle attached into a 1 or a 2 mL capacity syringe, containing 0.1 mL of ASW. Dilute 500 μL of the hemolymph 1:1 with 33% ASW (*see* **Note 7**) and centrifuge at 90×*g* for 10 min in a microcentrifuge. The supernatant is then discarded and the cells are mixed with 75 μL of 0.5 % LMA.

2. Mussel gill cells: After dissection, gills (*see* **Note 8**) are put in 5 mL of 20 % HBSS for 30 min, and then transferred to a 1.5 mg/mL dispase/HBSS solution at 37 °C for 20 min. After digestion, the enzyme is inactivated by diluting with 4 mL of 20% HBSS solution. The resulting digestion product is filtered through a 100 μm nylon mesh filter. The 8–10 mL of cell suspension obtained is centrifuged at 350×*g* for 5 min. For the Comet assay, the pellet is again resuspended in 100 μL of HBSS, microcentrifuged at 1,000 rpm for 10 min, and the resulting pellet mixed with 75 μL of 0.5 % LMA. Cell viability is evaluated by trypan blue exclusion method.

3. Fish erythrocytes: Fish are blooded from the caudal vein using a heparinized syringe with 200 μL of 25,000 IU/5 mL heparin. 10 μL of erythrocytes are used for preparing slides. Care must be taken to follow any local rules regarding conducting experimental procedures with vertebrates.

3.3 Slide Preparation

1. One day before conducting the assay, dip conventional end-frosted slides up to one-half the frosted area in melted 1 % NMA and gently remove. Wipe the underside of slide to remove agarose and lay the slide in a tray on a flat surface to dry (*see* **Note 9**). Store the slides at room temperature until needed; avoid high-humidity conditions.

2. To the pre-coated slide, add 75 μL of 0.5 % LMA (37 °C) mixed with around 10,000 cells in 5–10 μL of cell suspension from hemocytes/gill cells/fish erythrocytes. Do not use more than 10 μL of cell suspension.

3. Put a coverslip and place the slide on a slide tray resting on ice packs until the agarose layer hardens (3–5 min).

4. Gently slide off the coverslip and add a third layer of 85 μL of 0.5 % LMA to the slide. Replace the coverslip and return to the slide tray until the agarose layer hardens (3–5 min).

5. Remove the coverslip and slowly lower the slide into cold freshly made lysing solution. Protect from light and refrigerate for a minimum of 1 h to a maximum of 4 weeks.

3.4 Slide Electrophoresis

The procedure described is for electrophoresis under alkaline (pH > 13) conditions. Electrophoresis can be also conducted at pH 12.1 or under neutral conditions. Due to the high amount of alkali-labile sites in highly condensed chromatin of fish erythrocytes, the electrophoresis is carried out at pH 12.1, to detect only single-strand breaks. Moreover, slides under non-electrophoretic conditions can be used to detect apoptotic/necrotic cells (*see* Subheading 3.7) or can be removed from lysis buffer, treated with repair enzymes specific for different classes of DNA damage (*see* Subheading 3.8).

1. After at least 1 h at 4 °C, gently remove slides from lysing solution. Rinse the slides carefully in 0.4 M Tris–HCl buffer (e.g., three times for 5 min each) to remove detergents and salts. Place slides side by side on the horizontal gel box near one end, sliding them as close together as possible.

2. Fill the buffer reservoirs with freshly made pH > 13 electrophoresis buffer until the liquid level completely covers the slides (*see* **Note 10**).

3. Let slides sit in the alkaline buffer for 10 min to allow for unwinding of the DNA and the expression of alkali-labile damage (*see* **Note 11**).

4. Turn on power supply to 25 V (0.74 V/cm) and adjust the current to 300 mA by raising or lowering the buffer level. Depending on the purpose of the study and on the extent of migration in control samples, conduct electrophoresis for 5–10 min (*see* **Note 12**).

5. Turn off the power. Gently lift the slides from the buffer and place on a tray to drain. Coat the slides with neutralization buffer, and let sit for at least 5 min. Drain the slides and repeat two more times.

6. The slides may be stained with 100 μL of ethidium bromide and scored immediately or dried as in **step 7**.

7. Drain the slides, expose them to cold 100 % methanol for 3–5 min, and dry. Store in a dry area.

3.5 Neutral Gel Electrophoresis

1. Prepare and lyse the slides as usual (*see* **steps 1–5** of Subheading 3.3).

2. Remove the slide from lysing solution and rinse three times in 1× TBE buffer, for 5 min each.

3. Place the slides in an electrophoresis gel box and fill the reservoirs with running 1× TBE buffer to cover the slides.
4. Conduct electrophoresis at 25 V and 40 mA for 10 min.
5. Treat the slides with 0.4 M, Tris–HCl pH 7.4, for 5 min.
6. Stain the slides as usual (see **step 6** of Subheading 3.4).

3.6 Slide Analysis

1. For visualization of DNA damage, observations are made of fluorochrome-stained DNA (2 μL/mL ethidium bromide) using a 20–25× objective on a fluorescent microscope (see **Note 13**).
2. Quantitation of Comet assay data can be performed by an image analyzer or by manual scoring [20]. Visual scoring is based on five recognizable classes of comet, from class 0 (undamaged, no discernible tail) to class 4 (almost all DNA in tail, insignificant head). Each comet is given a value according to the class it is put into so that an overall score can be derived from each gel.
3. Statistical analysis for in vitro data is based on multiple cultures and for in vivo data on a minimum of 4 animals (ideally 5–8 animals) per group. Multifactor Analysis of Variance (MANOVA) is used to analyze data from marine animals, taking into account culture, experiment, species, animal, experiment, gender, animal length, and weight as independent factors. Statistical issues in the use of Comet assay are widely reported by Lovell and Omori [21]. It is recommended that before any experimentation is undertaken statistical experts be consulted to ensure that study design meets the statistical needs of the work.

3.7 Diffusion Assay

The DNA diffusion assay is a modified version of the Comet assay detecting low-molecular-weight DNA which may be lost from agarose matrix during electrophoresis [17]. Slides are prepared as described in Subheading 3.3, but electrophoresis is omitted. Apoptotic cell nuclei have a hazy or an undefined outline without any clear boundary due to nucleosomal sized DNA diffusing into the agarose and a diameter three times over the mean nuclear diameter. Necrotic cell nuclei have a clear, defined outer boundary of the DNA and a relatively homogenous appearance.

1. Prepare slides until the lysis step as described in **steps 1–5** in Subheading 3.3.
2. Place the slides in standard lysis solution for 1–2 h. Remove the slides from lysis solution, briefly rinse free of detergent, and fix using 100 % alcohol as in **step 7** of Subheading 3.4 and allow to dry (see **Note 14**).
3. For scoring, add 100 μL of 2 μL/mL ethidium bromide to each slide, put a coverslip, and allow the dye to spread well for 2 min before scoring (see **Note 15**).

4. Score at least 100 cells per sample into various categories of damage [16, 17].

5. Data from diffusion assay are useful for dose selection and in interpreting the comet data.

3.8 Oxidative DNA Damage

The evaluation of oxidative DNA damage can be obtained by the use of a modified version of the Comet assay [15], which includes digestion with a lesion-specific repair endonuclease (e.g., fpg), which is able to recognize altered DNA bases (e.g., purines) and induces strand breaks in the sugar-phosphate bone at sites of oxidative damage.

1. The enzyme is diluted (2 μL of enzyme in 200 μL of buffer containing 10 % glycerol and stored at −80 °C) in enzyme buffer (0.5 mM Na$_2$EDTA, 0.1 M KCl, 0.2 mg/mL bovine serum albumin, 40 mM N-[2-hydroxyethyl] piperazine-N-[2-ethanesulfonic acid], pH 8).

2. Carry out all **steps 1–5** of Subheading 3.3.

3. After lysis, treat the slides with a total of 50 μL of enzyme solution (or buffer alone, as control) and covered with 22 × 22 mm coverslips.

4. Keep the slides in a moist box, in order to prevent desiccation, incubate at 37 °C for 30 min, and then conduct electrophoresis at pH 12.1.

5. After electrophoresis, steps for neutralization, staining, and image analyses are carried out as previously described in **steps 5** and **6** of Subheading 3.4 and **steps 1–2** of Subheading 3.6, respectively.

4 Notes

1. As soon as ASW is ready it will have pH ≥ 7.5. Before using it, wait until pH is around 7.9. Generally, ASW should be prepared a few hours before use, maintained in aquaria at <20 °C, aerated, and covered.

2. Sodium lauryl sarcosinate is also included by some authors as a component of the lysing solution but it has not turned to be essential in our experience.

3. The recommended concentration of LMA is 0.5 %, while higher concentration of agarose can reduce the extent of DNA migration.

4. The total volume depends on the gel box capacity. Prior to use, measure the pH of the buffer to ensure >13. Electrophoresis buffer needs to be made fresh just prior to use.

5. Use the same molarity of the pH > 13 buffer solution, obtaining pH 12.1 only by adding HCl.

6. Gently prizing open the shell (using a small solid scalpel) will expose the adductor muscle (the compact, slightly translucent or white, collection of filaments attached to the inside of the upper and lower valves) which is located at the posterior end of the shell, lying adjacent to the hinge. The blade width of a solid scalpel should be sufficient to hold the valves apart in order to insert the hypodermic needle. Then be sure to eliminate all the seawater inside the mussel.

7. The objective is to achieve 1 mL of hemolymph; however this may require repeated bleeding in the case of small or difficult mussels in which case extra care should be taken during hemolymph extraction. Smaller specimens may prove difficult to re-bleed and it is suggested that adequately sized animals are chosen to suit the experimental requirements (>3 cm in length).

8. 1–1.5 cm^2 of gill tissue from each valve is enough for a single animal.

9. While preparing the first layer of 1 % NMA by dipping conventional slides into the melted NMA, be sure that the agarose is well melted. As soon as it sets a bit, boil it again in the microwave and go on dipping the slides. The first layer of agarose is a critical step in the Comet assay protocol, in order to avoid gels to slip away from the slide during the electrophoresis step. Once prepared, slides may be air dried or warmed for quicker drying. Be sure to store them under dry conditions. Prepare the slides prior to use. Just before dipping, slides can be labelled with a solvent-resistant marker or with diamond pen.

10. Avoid bubbles over the agarose while placing slides, as well as in the electrophoresis chamber.

11. The longer the exposure to alkali, the greater the expression of alkali-labile damage. We suggest 10-min exposure to electrophoresis buffer for both invertebrates and fish.

12. The goal is to obtain migration among the control cells without it being excessive (e.g., 10–20 % tail DNA) in order to be able to detect both increases and decreases of DNA migration, potentially coming from the action of cross-linking agents. This is another critical step, as controls displaying no DNA migration might interfere with cross-linking detection, while the highly migrated controls might saturate the system interfering with the detection of dose–responses. The percentage of LMA used and electrophoresis duration are the two main parameters to be kept under control while setting up the protocol, to obtain the best controls. The optimal electrophoresis duration differs for different cell types (in our experience 5 min for mussel cells and 10 min for fish cells and should be optimized for specific species before use).

13. Depending on the size of the gels being scored, other objectives can be used (i.e., 16×, 40×).
14. Longer times result in the loss of low-molecular-weight DNA from the agarose gel.
15. The process of DNA degradation associated with apoptosis or necrosis is a continuous one. This methodology detects cells at or near the end of that process. Such cells would normally be lost during the usual electrophoretic conditions used, unless much higher concentrations of agarose are used, along with shorter electrophoretic conditions.

References

1. De Flora S, Bagnasco M, Zanacchi P (1991) Genotoxic, carcinogenic and teratogenic hazards in the marine environment, with special reference to the Mediterranean Sea. Mutat Res 258:285–320
2. Chen GS, White PA (2004) The mutagenic hazards of aquatic sediments: a review. Mutat Res 597(2–3):151–255
3. Ames BN, McCann J, Yamasaki E (1975) Methods for detecting carcinogens and mutagens with the Salmonella/mammalian-microsome mutagenicity test. Mutat Res 31:347–364
4. Parry JM, Tweats DJ, Al-Mossawi MAJ (1976) Monitoring the marine environment for mutagens. Nature 264:538–540
5. Prein AE, Thie GM, Alink GM et al (1978) Cytogenetic changes in fish exposed to water of the river Rhine. Sci Total Environ 9:287–291
6. Alink GM, Frederix-Wolters EMH, van der Gaag MA et al (1980) Induction of sister-chromatid exchanges in fish exposed to Rhine water. Mutat Res 78:369–374
7. Harvey JS, Lyons BP, Waldock M (1997) The application of the 32P–postlabelling assay to aquatic biomonitoring. Mutat Res 378:77–88
8. Jha AN (2004) Genotoxicological studies in aquatic organisms: an overview. Mutat Res 552:1–17
9. Frenzilli G, Nigro M, Lyons BP (2009) The Comet assay for the evaluation of genotoxic impact in aquatic environments. Mutat Res 681:80–92
10. Jha AN (2008) Ecotoxicological applications and significance of the comet assay. Mutagenesis 23(3):207–221
11. Mitchelmore CL, Chipman JK (1998) DNA strand breakage in aquatic organisms and the potential value of the comet assay in environmental monitoring. Mutat Res 399:135–147
12. Lee RF, Steinert S (2003) Use of the single cell gel electrophoresis/comet assay for detecting DNA damage in aquatic (marine and freshwater) animals. Mutat Res 544:43–64
13. Tice RR, Agurell E, Anderson D et al (2000) Single cell gel/comet assay: guidelines for in vitro and in vivo genetic toxicology testing. Environ Mol Mutagen 35:206–221
14. Burlinson B, Tice RR, Speit G et al (2007) Fourth International Workgroup on Genotoxicity testing: results of the in vivo Comet assay workgroup. Mutat Res 627:31–35
15. Collins AR (1999) Oxidative DNA damage, antioxidants, and cancer. Bioessays 21:238–246
16. Frenzilli G, Scarcelli V, Del Barga I et al (2004) DNA damage in eelpout (*Zoarces viviparus*) from Göteborg harbour. Mutat Res 552:187–195
17. Singh NP (2000) A simple method for accurate estimation of apoptotic cells. Exp Cell Res 256:328–337
18. Osman AG, Abuel-Fadl KY, Kloas W (2012) In situ evaluation of the genotoxic potential of the river Nile: II. Detection of DNA strand-breakage and apoptosis in O*reochromis niloticus niloticus* (Linnaeus, 1758) and *Clarias gariepinus* (Burchell, 1822). Mutat Res 747:14–21
19. Halldorsson HP, Ericson G, Svavarsson J (2004) DNA strand breakage in mussels (*Mytilus edulis* L.) deployed in intertidal and subtidal zone in Reykjavik harbour. Mar Environ Res 58:763–767
20. Azqueta A, Meier S, Priestley C et al (2011) The influence of scoring method on variability in results obtained with the comet assay. Mutagenesis 26:393–399
21. Lovell DP, Omori T (2008) Statistical issues in the use of the comet assay. Mutagenesis 23:171–182

Chapter 20

Unscheduled DNA Synthesis (UDS) Test with Mammalian Liver Cells In Vivo

Fabrice Nesslany

Abstract

The unscheduled DNA synthesis test is a short-term genotoxicity assay that allows identification of substances that induce DNA repair in liver cells of treated animals. This endpoint can be assessed by determining the uptake of radiolabelled nucleotide after excision-repair in cells that are not undergoing scheduled (S-phase) DNA synthesis. Here we describe the protocol for performing this well-known primary DNA damage test following the OECD guideline No. 486, EC, and published international recommendations.

Key words Unscheduled DNA synthesis, UDS, In vivo, Primary DNA damage, Liver cells, DNA repair

1 Introduction

Many chemical carcinogens are thought to act by interaction with DNA (genotoxic carcinogens often opposed to "epigenetic" carcinogens). This interaction may result directly or indirectly from the metabolism of the compound in vivo by, e.g., liver mono-oxygenases which produce reactive electrophilic species. Such metabolites will interact with DNA nucleophilic sites within the cell and produce DNA adducts. On the other hand, cells have developed DNA repair systems in order to maintain the integrity of the genome. Therefore, the examination of the cell for DNA repair can indirectly provide a means of studying carcinogen-induced DNA adducts. The in vivo unscheduled DNA synthesis (UDS) test in mammalian liver cells indicates the DNA repair synthesis rate after DNA damage induced in the main organ of metabolism of animals exposed to the test compound.

If a chemical is rapidly eliminated from the body, it would be expected that the peak of any induction of UDS would occur over a relatively short time period (for instance with dimethylhydrazine). Since some test agents may give negative responses at 12–14 h in one experiment, a second experiment is conducted using a 2–4-h sampling time. Animals are treated with the test compound. At time intervals thereafter (12–14 and 2–4 h) in two independent assays, the animals are sacrificed, their liver is dissociated, and the resulting hepatocytes are collected. The cells are exposed to [3H] thymidine which is incorporated into the DNA if UDS is occurring. Normal S-phase DNA synthesis is rare in hepatocytes and can readily be distinguished from UDS autoradiographically. Incorporation is followed by autoradiography of the hepatocytes and grain counting. The technique described here was developed by Mirsalis and Butterworth [1], modified by Ashby et al. [2] and is detailed by Kennelly et al. [3]. The advantage of this model is that it is an in vivo assay which represents the pharmacokinetics and pharmacodynamics of the test item more accurately than in vitro systems and is conducted according to the international guidelines OECD guideline No. 486 [4], and EC [5]. It has been used for many kinds of substances for research purpose, hazard identification, and risk assessment [6–8].

2 Materials

All details concerning volumes or specific material are calculated on the basis of a standard test (*see* **Note 1**).

2.1 Perfusion Components

1. HEPES buffer: 10 mM HEPES, 125 mM NaCl, 3 mM KCl, 1 mM Na_2HPO_4. Weigh 17.872 g of HEPES, 54.787 g of NaCl, 1.678 g of KCl, 1.065 g of Na_2HPO_4 and transfer each into a 10 L glass beaker. Mix until complete dissolution (≈1 h). Adjust pH to 7.65 with HCl. Add water to a final volume of 7,500 mL. Filter on 0.22 μm sterilizing membrane. Store at 4 °C for a maximum of 1 month.

2. HEPES/Ca buffer: 4 mM $CaCl_2$. Weigh 1.176 g of $CaCl_2 \cdot 2H_2O$, add HEPES buffer, pH 7.65 to a final volume of 2,000 mL. Mix for about 3–4 h (*see* **Note 2**). Filter on 0.22 μm sterilizing membrane. Store at 4 °C for a maximum of 3 days.

3. Collagenase Ia buffer: Weigh 500 mg of Collagenase Ia and add HEPES/Ca buffer to a final volume of 2,000 mL. Mix for about 3–4 h (*see* **Note 3**). Filter on 0.22 μm sterilizing membrane and use within 24 h.

2.2 Hepatocyte Culture Components

1. Williams' E Incomplete (WEI) medium: Williams' E medium, 0.03 % L-glutamine, 200 IU/mL penicillin/50 IU/mL streptomycin, 2.4 μg/mL amphotericin B in final concentrations. Store the solution at 4 °C for a maximum of 15 days. Prepare 3,000 mL for one assay.

2. Williams' E Complete (WEC) medium: WEI medium, 10 % fetal calf serum (FCS). To 450 mL WEI medium, add 50 mL of heat-inactivated FCS. Store the medium at 4 °C for a maximum of 10 days. Prepare 3,000 mL for one assay.

3. Collagen solution: 1 mg/mL collagen type VII. Add 40 µl of 5 N chlorhydric acid directly into the vial containing 10 mg collagen type VII, and then add 10 mL of distilled water. Heat it in a water bath at 37 °C ± 1 °C for 1 h. Mix and filter on 0.22 µm sterilizing membrane. Store the solution at 4 °C for a maximum of 6 months (*see* **Note 4** for the coating of coverslips).

4. Trypan blue dye for cytotoxicity: 0.4 % Trypan blue. Weigh 0.4 g of Trypan blue and add 100 mL of phosphate buffer, pH 7.2. Mix and filter on cellulose paper and then on 0.45 µm sterilizing membrane. Store at 4 °C in a vial wrapped with aluminum foil for a maximum of 6 months.

2.3 Radiolabelling of Cultures

1. ^{3}HThymidine solution: 10 µCi/mL ^{3}HThymidine. Just before use, dilute 1 mCi ^{3}HThymidine (e.g., 40–60 Ci/mmol, i.e., around 1 mCi/mL) in 99 mL of WEI medium (*see* **Note 5**).

2. Thymidine solution: 25 mM thymidine. Weigh 120 mg of thymidine and add 19.8 mL of WEI medium to it. Mix and then filter on 0.22 µm membrane and store at 4 °C in a vial wrapped with aluminum foil for a maximum of 10 days. Just before use, dilute 1/100 and in WEI medium, for a final concentration of 250 µM (*see* **Note 6** for use).

2.4 Development and Staining Components

1. D19 developing solution: Put about 3,000 mL distilled water (pre-warmed to 45 °C) to a 10 L glass beaker (*see* **Note 7**). Gently add the whole content of one bag of D19 (Kodak 146 4593). Mix (using magnetic stirring), protect from light until complete dissolution, and then add 800 mL of distilled water. Mix again and check the pH to be between 9 and 12 (an adjustment of the pH is never necessary; if the pH is not within this range, prepare fresh). Filter on cellulose paper and store in glass bottles wrapped with aluminum foil at room temperature, for a maximum of 2 months (*see* **Note 8**).

2. Fixating solution: Put about 3,000 mL of distilled water to a 10 L glass beaker. Gently add the whole content of the sodium fixer sachet. Mix (using magnetic stirring), protect from light until complete dissolution, and then add 800 mL of distilled water. Mix again and check the pH to be between 3 and 6 (an adjustment of the pH is never necessary; if the pH is not within this range, prepare fresh). Filter on cellulose paper and store in glass bottles wrapped with aluminum foil at room temperature, for a maximum of 2 months (*see* **Note 8**).

3. Kodak NTB2 autoradiography emulsion can be used as supplied or may be half-diluted in water as described hereafter: In the darkroom, turn on the water bath at 45 °C. Melt the whole content (118 mL) of the initial Kodak NTB-2 emulsion (Kodak

889 566) for at least 30 min. Pour the molten emulsion in a beaker and add 118 mL of distilled water in order to obtain 1:1 dilution. Mix with a glass rod (*see* **Note 9**). Aliquot 15 mL in amber flasks and store at 4 °C, for a maximum of 6 months (*see* **Note 10**).

4. Carnoy fixing solution: Mix one volume of pure acetic acid with three volumes of absolute ethanol to a 1 L graduated glass cylinder. Transfer to a glass flask and store at 4 °C, for a maximum of 24 h (*see* **Note 11**).

5. The staining components:
   (a) Harris' hematoxylin.
   (b) 95 % Ethanol, absolute ethanol, and xylene.

3 Methods

3.1 Animals

1. Young healthy adult rats are commonly used, although any appropriate mammalian species may be used (after justification). The period of acclimatization is at least 5 days (*see* **Note 12**). The animals receive a clinical examination in order to retain only those which are healthy.

2. If there are no substantial differences in toxicity between sexes, then testing in a single sex, preferably males, is sufficient. The number of animals should be at least three analyzable animals per group to take account of natural biological variation in test response.

3.2 Animal Treatment

1. Formulation of the test substance: Solid substances are dissolved or suspended in appropriate solvents or vehicles and then diluted with the same solvent or vehicle if necessary. Liquid substances are dosed directly or diluted prior to dosing. True solutions are required for intravenous administration.

2. Dose levels: From the results of the preliminary toxicity test two dose levels are chosen for each genotoxicity assay (*see* **Note 13**). An additional and provisional low-dose-level group is also incorporated into the experimental design which would be assessed for UDS in the event of excessively low cell yields or mortality occurring at the high dose as a result of treatment-related effect (*see* Table 1).

3. Dosing of animals: Healthy animals are randomly assigned to the control and treatment groups. Three animals are used per group (negative control, positive control, and treatment groups) and clearly identified (e.g., by numbered earrings). The test item is usually administered in a single treatment by gavage using a stomach tube or a suitable intubation cannula with the standard dose volume of 10 mL/kg (*see* **Note 14**).

Table 1
Standard repartition of treatment groups

Groups	Treatment	Dose (mg/kg)	Number of animals used[a] 2–4 h	12–16 h
1	Vehicle	0	3 (4)	3 (4)
2	Test item	Additional low dose[b]	3 (4)	3 (4)
3		Mid dose	3 (4)	3 (4)
4		High dose	3 (4)	3 (4)
5	Positive reference compounds[c]		3 (4)	–
			–	3 (4)

[a]Number of animals used in the test with early (2–4 h) or normal (12–16 h) expression time: the numbers indicate the number of animals used for hepatocyte culture and, in brackets, the number of animals treated for the test
[b]Group 2: Additional and provisional low-dose-level group which would only be assessed for unscheduled DNA synthesis in the event of unacceptably low cell yields or mortalities occurring in group 4
[c]See **Note 16**

Other routes of exposure may be acceptable when justified (*see* **Note 15**).

Concurrent positive and negative controls should be included in each independently performed part of the experiment. Negative controls are animals treated with vehicle in the same condition as treated animals (same route and vehicle). Positive control animals are always treated by oral route (*see* **Note 16**).

4. Expression/sampling times: Two expression times are performed for the UDS test. The liver cells from treated and control animals are prepared 2–4 h after dosing (earlier expression time) and 12–16 h after dosing (initial/normal expression time).

3.3 Isolation of Hepatocytes

The early mechanical and chemical (e.g., EDTA or citrate) methods for liver cell preparation were relatively successful in converting liver tissue to a suspension of isolated cells but unfortunately nearly all such cells are damaged [9]. The introduction of collagenase as a liver-dispersing enzyme by using physiological liver perfusion greatly facilitated the preparation of intact parenchymal cells in high yield. Almost every worker has incorporated his own modifications. The simple operative routine procedure described step by step hereafter was repeatedly and successfully applied in our lab.

1. Preparation of work plan (*see* Fig. 1): In the laminar flow cabinet, place HEPES, collagenase Ia buffers, and WEC medium in a thermostatically controlled water bath at 37 °C ± 1 °C about 30 min before the perfusion. Under the laminar flow, dip one pipe of the 3-way stopcock (*see* **Note 17**) in HEPES buffer and another in collagenase buffer. The last tube fitted with a Luer

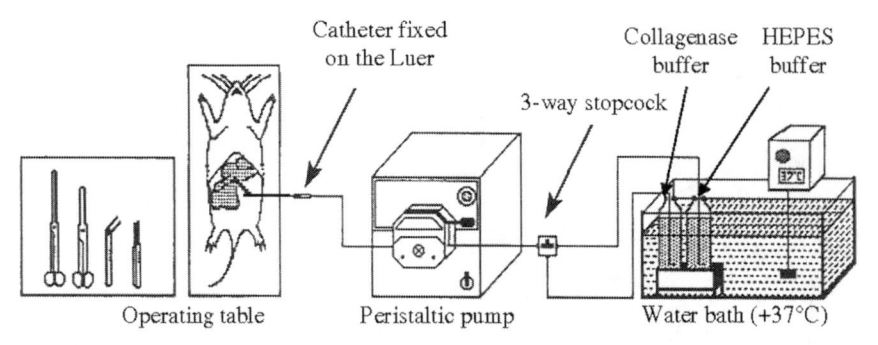

Fig. 1 Preparation of work plan

passes through the peristaltic pump (*see* **Note 18**) and will receive the catheter. Initiate the respective buffer tubes using the peristaltic pump (starting with the collagenase Ia buffer), in order to avoid any passage of air bubbles in the liver during the perfusion step.

2. Prepare a 5 L beaker, dissection instruments, catheters (*see* **Note 19**), and 150 μm sterile nylon filter-funnel placed on 50 mL sterile tubes (one per animal).

3. Deeply anesthetize each animal (i.p. injection of 60 mg/kg b.w. pentobarbital) and shave from abdomen to the chest area (*see* **Note 20**). Place it on the operating table under the laminar flow hood. Brush the abdomen with 70 % ethanol, dissect open a V-shaped transverse incision, and displace the intestines to the left side of the abdominal cavity.

4. Perfusion: Hepatocytes are prepared according to the method developed by Seglen [9] and Williams et al. [10]. For each animal, cannulate the portal vein using a sterile catheter [0.9–1.3 mm (internal/external diameters), 38 mm (length of the needle)] and clamp it. As soon as the perfusion starts, cut the upper vena cava in order to decrease the pressure and to facilitate the washing and the flow of the perfusate. Then connect the cannula to the peristaltic pump and first perfuse the liver with HEPES buffer at 40 mL/min for 5 min. Then turn the 3-way stopcock to perfuse the liver with collagenase Ia buffer at 20 mL/min for 5 min (at the end of perfusion, the liver becomes spongy). Transfer the perfused liver to a 150 μm filter-funnel placed on 50 mL disposable sterile tube (*see* **Note 21**). Carefully cut the Glisson's capsule (3–5 incisions) and add 40 mL of WEC directly on the freshly perfused liver pieces to liberate the isolated hepatocytes into the 50 mL tube. Gently centrifuge the cell suspension at approximately $40 \times g$ for 1 min. Resuspend the resulting pellet in 40 mL of WEC medium. Repeat the centrifugation and resuspension procedures twice.

5. Viability test: After resuspending each pellet with 40 mL fresh WEC, mix 500 μL of each cell suspension with 100 μL of a 0.4 % Trypan blue solution (*see* **Note 22**) in 5 mL tubes. Enumerate the cells using a Malassez hemocytometer: viable cells are refractive, and dead cells appear blue. The percentage of relative survival corresponds to the [percentage of survival cells in treated animal/mean percentage of survival in negative control group] × 100. The percentage viability of the final cell suspensions from negative control group should be higher than 50 % [1, 11].

6. Cell seeding: Dilute the suspensions of hepatocytes in WEC medium to provide approximately 1.5×10^5 viable cells/mL. For each cell suspension, transfer 3 mL (i.e., 4.5×10^5 viable cells/well) to each well of two 6-well multiplates containing collagenase-pre-coated 25 mm round plastic coverslips (for their preparation *see* **Note 4**). Incubate the cultures at 37 °C in an atmosphere of 5 % of CO_2 for approximately 90 min to allow attachment of cells on the coverslips. A shorter time may induce an important loss of cells when supernatant is removed in the next step.

3.4 Culture of Hepatocytes and Radiolabelling of Cultures

1. After the attachment period, remove the supernatant medium, gently wash the cells with approximately 2 mL of WEI medium, and then replace with 1.5 mL of WEI medium containing 10 μCi/mL 3^Hthymidine. Incubate cultures for 3–8 h at 37 °C in an atmosphere of 5 % CO_2.

2. At the end of the labelling period, remove the 3^Hthymidine solution, rinse three times with 3 mL of WEI medium, and finally add 3 mL of 250 μM unlabelled thymidine solution to diminish unincorporated radioactivity (named as "cold-chase," *see* **Note 23**). Incubate the cells overnight at 37 °C in an atmosphere of 5 % CO_2.

3. Remove the thymidine solution, then fix the cells three times using Carnoy fixative (ethanol–acetic acid) for 10 min each, and finally wash five times with distilled water to remove traces of the acid. After that, remove the coverslips from the wells using forceps and place the cell side up on a 90 mm Petri dish to dry in a dust-free location at room temperature. (The use of Petri dish is very convenient because you can use one dish per animal and it could be covered. Petri dish should be annotated with the No. of animal and the corresponding dose in order to avoid any confusion.) When dry, mount the coverslips cell side up on microscope slides with Eukitt. Give a unique identifying number to each slide. Leave the slides to air-dry overnight.

3.5 Autoradiography

All steps involving the emulsion should be done in total darkness in a darkroom (*see* **Note 24**).

1. Turn on the water bath to 45 °C in the darkroom, containing a tray (in which the temperature should be 44 °C ± 1 °C). Melt the half-diluted Kodak NTB-2 emulsion for at least 30 min. Pour slowly the content of the vial of melted emulsion in a dipping-chamber (Pelanne Instruments, E29) placed in the tray kept in the water bath. Wait for about 30 min again, in order to avoid air bubbles (*see* **Note 25**).

2. Once the emulsion is ready to use, dip slides one by one for 1–2 s and then wipe their underside with paper towels before placing them vertically (use the back covers of the lighttight slide boxes, i.e., support-kit). The slides are left for 3–12 h to let the emulsion dry (*see* **Note 26**).

3. Pack the slides in the lighttight slide boxes that contain a false bottom packed with desiccant (Bluegel, i.e., silica gel with moisture indicator).

4. Wrap the boxes in aluminum foil and store them at 4 °C for 10–14 days (*see* **Note 27**).

3.6 Development and Staining

1. Allow the slide boxes to thaw at room temperature for at least 1 h.

2. Fill three trays with the developer D19 solution, distilled water, and sodium fixer and bring the temperature of each solution to 15 °C ± 1 °C (*see* **Note 28**). Once the temperature is maintained, the slides are ready to be developed.

3. Dip the slide rack in the developer D19 solution for 3 min ± 15 s with constant slow stirring.

4. Dip the slide rack in distilled water for 30 s ± 15 s with constant slow stirring.

5. Dip the slide rack in sodium fixer solution for 5 min ± 15 s with gentle agitation every minute.

   The slides are now fixed; it is no longer necessary to work in the dark.

6. Rinse the slides with tap water, in a very thin stream, for 25 min.

7. Let the slides dry at room temperature for 3–4 h.

   Prepare a new set of reagents for the next rack(s) to develop.

8. Dip the slide rack in Harris' hematoxylin for 2 min and 15 s to 2 min and 30 s.

9. Rinse the slides with tap water, in a very thin stream for 15–20 s (*see* **Note 29**).

10. Dehydrate and clear the slides by dipping the slide racks following this schedule:

Ethanol 95 %	A	15 ± 5 s
Ethanol 95 %	B	15 ± 5 s
Ethanol 95 %	C	15 ± 5 s
Absolute ethanol	A	15 ± 5 s
Absolute ethanol	B	15 ± 5 s
Xylene	A	15 ± 5 s
Xylene	B	15 ± 5 s

11. Allow the slides to dry for few minutes and immediately mount a 25 mm square coverslip over the round coverslip using a thin layer of Eukitt.
12. Keep the slides flat overnight to dry after which they are ready for grain counting.

3.7 Analysis

The amount of 3H-TdR incorporation in the nuclei and the cytoplasm of morphologically normal cells, as evidenced by the deposition of silver grains, should be determined by suitable methods (*see* **Note 30**). Grain counting can be done manually but it is a tedious and somehow not a fully objective process. An automated counting system is highly recommended.

1. The slide preparations should contain sufficient cells of normal morphology, to permit a meaningful assessment of UDS. First, examine preparations microscopically for signs of overt cytotoxicity, e.g., pyknosis, and reduced levels of radiolabelling (*see* **Note 31**).
2. The image analysis system is coupled to an optical microscope and a video camera.
3. The following criteria are used to determine if a cell should or should not be counted:
   (a) Cells with abnormal morphology (e.g., pyknotic nuclei) should not be counted.
   (b) Isolated nuclei with no surrounding cytoplasm should not be counted.
   (c) Cells with unusual staining artifacts should not be counted.
   (d) Heavily labelled cells in S-phase should not be counted for UDS assessment but are enumerated (*see* Fig. 2 and **Note 32**).
4. Code the slides before grain counting. Each slide is scanned and for each cell, the following parameters are measured: Nuclear grain count (NC) and the nucleus-equivalent area over the cytoplasm (cytoplasmic grain counts, CC).

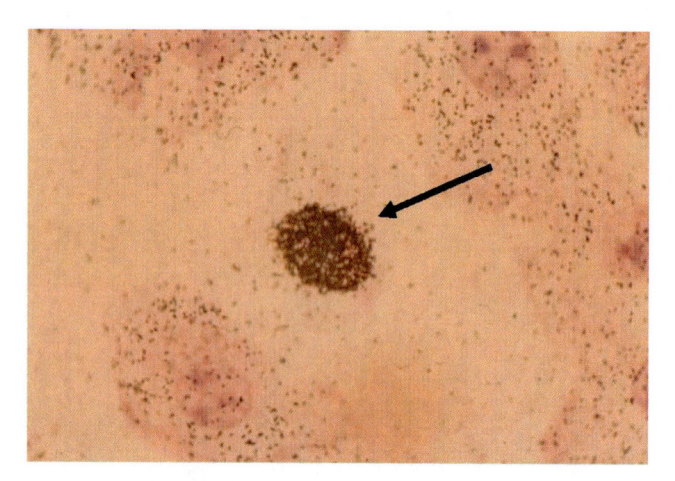

Fig. 2 Example of a cell in S-phase (indicated by the *arrow*)

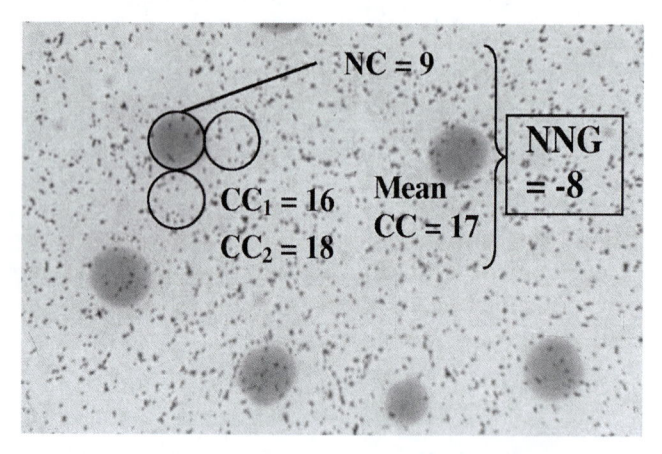

Fig. 3 Example of captured picture and counting. The net nuclear grains (NNG) correspond to the subtraction of the mean CC from the NC

5. At least 100 cells are scored from each animal from at least two slides for each treatment.
6. The net nuclear grains (NNG), corresponding to the subtraction of the mean CC from the NC, can be calculated: NNG = NC − CC (*see* Fig. 3).
7. Cells with NNG ≥ 5 are considered to be in repair.
8. For each cell, the following calculations are performed:
   (a) Population average NNG ± SD (cell to cell).
   (b) Percentage of cells considered to be in repair.
   (c) Population average of NNG ± SD for cells in repair (cell to cell).
9. For each animal, the following calculations are performed:
   (a) Population average NNG ± SD (slide to slide).
   (b) Percentage of cells considered to be in repair (slide to slide).

(c) Population average of NNG ± SD for cells in repair (slide to slide).

10. For each group, the following calculations are performed:

    (a) Population average NNG ± SD (animal to animal).

    (b) Percentage of cells considered to be in repair (animal to animal).

    (c) Population average of NNG ± SD for cells in repair (animal to animal).

3.8 Criteria for Genotoxic Activity

Following Brambilla et al. [12] and Madle et al. [13], a test item is considered positive in this assay, if at any dose and any time:

1. It yields a mean NNG value greater than 0 and at least 10 % of the cells present a mean NNG value ≥ 5.

    OR

2. An increase is observed for both the mean NNG value and the percentage of cells in repair.

4 Notes

1. The volumes of media, buffers, and solutions are calculated for the perfusion of 15 animals corresponding to the number usually used per assay (five groups of three animals each).

2. At least the day before the main assays, dissolve $CaCl_2 \cdot 2H_2O$ in the required volume of HEPES buffer in 2 L glass Erlenmeyer flask. We find that it is best to heat the solution in a water bath at 37 °C ± 1 °C for 3–4 h, stirring occasionally, until precipitation.

3. We find that it is best to heat the collagenase buffer in a water bath at 37 °C ± 1 °C for 3–4 h, before its use.

4. At least one day before each main assay, put one 25 mm round plastic coverslip in each well of 6-well multiplates. Plastic coverslips are easier to use, resistant, available, and cheaper. Deposit 20 μL of the 1 mg/mL collagen solution and spread on the whole surface, using a scrapper. Let dry at room temperature overnight. Just before use, rinse each coverslip with 5 mL WEI. For each assay, thirty 6-well multiplates are required.

5. 1.5 mL of WEI medium with 10 μCi/mL 3^Hthymidine per well, i.e., 270 mL per assay is required. Refer to international and national laws relative to the handling and disposal of radioactive elements.

6. 12 mL of thymidine solution per well, i.e., 2,200 mL per assay is required.

7. Use only glass containers, stainless steel, or plastic. Steel should be avoided as emulsion contains silver bromide that may react with other steels.

8. Use as such, at 15 °C ± 0.5 °C.
9. Avoid the contact of the emulsion with metals.
10. One aliquot of 15 mL is required for 30–40 slides.
11. About 800 mL of Carnoy's fixing agent is required per assay at the end of treatment and after rinsing; e.g., 1.5 mL/well/step of fixation, i.e., 4.5 mL/well is required.
12. Housing and feeding conditions: The temperature in the experimental animal room should be 22 °C (±3 °C). Although the relative humidity should be at least 30 % and preferably not exceed 70 % other than during room cleaning, the aim should be 50–60 %. Lighting should be artificial, the sequence being 12-h light and 12-h dark. For feeding, conventional laboratory diets may be used with an unlimited supply of drinking water.
13. In an effort to reduce the number of laboratory animals required, the toxicity test is performed according to the improved experimental design recommended by Fielder et al. [14] using a large series of doses chosen in accordance with available toxicological data. OECD guideline 486 [1] recommends that the highest tested dose is the maximum tolerated dose (MTD) which is described as the highest dose which causes no mortality, but which may give rise to the appearance of weak signs of toxicity. In a preliminary toxicity assay using the same species, strain, sex, and treatment regimen to be used in the main study, four animals per dose are used. Following dosing, the animals are observed regularly for a period of at least 48 h and any clinical signs and/or mortalities are recorded. The highest dose used for the UDS test is the MTD. For non-toxic products, the maximum dose of 2 g/kg or the maximum, which can be administered practically to the animals, whichever is lower is chosen as the test limit dose. The lower doses should be 50 to 25 % of the highest dose.
14. The maximum volume of liquid that can be administered by gavage or injection at one time depends on the size of the test animal. The volume should not exceed 2 mL/100 g body weight. The use of volumes higher than these must be justified. Except for irritating or corrosive substances which will normally reveal exacerbated effects with higher concentrations, variability in test volume should be minimized by adjusting the concentration to ensure a constant volume at all dose levels.
15. The intraperitoneal route is not recommended due to the direct contact of the liver with the test item.
16. Positive controls should be substances known to produce UDS when administered at exposure levels expected to give a detectable increase over background. Positive controls needing metabolic activation should be used at doses eliciting a moderate response [13]. The doses may be chosen so that the effects are

clear but do not immediately reveal the identity of the coded slides to the reader.

In our lab, both the positive control substances are used by the oral route:

(a) Early sampling time (2–4 h): 1,2-Dimethylhydrazine (DMH, CAS No. 540-73-8) >99 % pure, 1 mg/mL DMH, dissolve in 0.9 % NaCl (prepared in purified water) and administer at a volume of 10 mL/kg giving a final dose of 10 mg/kg.

(b) Late sampling time (12–16 h): 2-acetamidofluorene (2-AAF, CAS No. 53-96-3) >95 % pure. Suspend at 2.5 mg/mL 2-AAF in CMC (at 0.5 % in purified water) and administer at a volume of 10 mL/kg giving a final dose of 25 mg/kg.

Other appropriate positive control substances may be used and, following the OECD guideline, it is acceptable that the positive control may be administered by a route different from the test substance.

Positive controls being genotoxic carcinogens, individual and general protective equipment should be used to avoid any exposure (from weighing to elimination).

17. The 3-way stopcocks are prepared with 3 silicone pipes (1.7 mm internal diameter, including one with a Luer, a small tip that allows connecting the pipe easily with the catheter) and sterilized for 20 min at 120 °C in an autoclave. Note that Luer can be obtained commercially or handmade as described hereafter:

    (a) Cut the extremity of a 1-mL syringe (*e.g.*, CODAN) on *ca.* 1.5–2 cm (*ensure beforehand that the syringe is adapted to the catheter*).

    (b) Insert the pipe into this piece.

18. The peristaltic pumps should be periodically calibrated to ensure the correct volumes.

19. Trays and boards for dissection and dissection sterile equipment (including pairs of scissors, forceps, and bulldog clips).

20. For ethical reasons and to avoid any stress, cages are brought into the room where perfusions are made, while euthanasia is realized in the handling room.

21. 150 μm nylon filters are fixed on glass funnels and then sterilized for 20 min at 120 °C in an autoclave.

22. The trypan blue technique allows distinguishing viable (refringent) and dead (blue) cells and the percentage of viability can be determined as recommended in OECD guideline [1].

23. Several parameters affect the sensitivity of the in vivo UDS test. Incubation in unlabeled thymidine (cold chase) significantly reduces NNG background [15].

24. If absolutely necessary, a Kodak No. 1 red safelight filter may be used sparingly.

25. Dip a test slide, turn on the safelight, and hold the slide up to it to make sure that there is enough emulsion to cover the cells and to check the absence of bubbles in the emulsion, in which case the emulsion is uniform and smooth, while the presence of bubbles results in the formation of holes in the emulsion (then wait for a few minutes more). The use of a reproducible slide-dipping procedure is highly recommended in order to give emulsion layers a uniform thickness. A dipping chamber is a very useful device perfectly adapted to the size of a microscopic slide. This allows to spend emulsion (with around 15 mL, it is possible to dip more than 50 slides).

26. Six of the twelve slides from each animal are actually coated in NTB-2 liquid emulsion (six remaining slides are kept and autoradiographed if required).

27. Shorter times yield lower backgrounds [16].

28. It is important that there is no major difference in temperature between the different solutions (<0.5 °C).

29. At this point of staining, it is necessary to examine the quality of the slides: Cells should have faint blue nuclei and rather pink cytoplasms. Over-stained cells make automatic grain counting difficult. Otherwise, it is recommended to stain the first slide of each control and treated groups because the presence of a large number of grains (in case of a positive response) modifies the incorporation of dye; in such case, time of staining must be increased (about 3 min) and discoloration time decreased (15 s). This technique also applies to positive controls.

30. The principle of the autoradiography is to cover the cell preparations exposed to a labelled nucleotide with a thin layer of photographic emulsion which is sensitive to the emission of radioactive particles. When a particle is emitted, the chemical reaction deposits metallic silver that will be viewed as a black grain after development. The UDS may be measured by either quantitative autoradiography or scintillation. The quantification of thymidine incorporation by scintillation counting can be criticized because of the inability of this method to distinguish between nuclei undergoing replicative or repair DNA synthesis [17]. Therefore, in order to avoid confusion with scheduled DNA synthesis, the UDS is measured by quantitative autoradiographic grain counting as net grains/nucleus.

31. In case of high cytotoxicity, the third dose should be assessed for genotoxicity.

32. The percentage of cells in S-phase is also determined. The increase in the proportion of S-phase cells may correspond to a proliferative effect of the test item.

References

1. Mirsalis J, Butterworth B (1980) Detection of unscheduled DNA synthesis in hepatocytes isolated from rats treated with genotoxic agents: an in vivo/in vitro assay for potential carcinogens and mutagens. Carcinogenesis 1: 621–625
2. Ashby J, Lefevre PA, Burlinson B et al (1985) An assessment of the *in vivo* rat hepatocyte DNA repair assay. Mutat Res 156:1–18
3. Kennelly JC, Waters R, Ashby J et al (1993) In vivo rat liver UDS assay. Supplementary. In: Kirkland DJ, Fox M (eds.) Mutagenicity tests UKEMS recommended procedures. Cambridge University Press, pp 52–77
4. OECD (1997) Guideline for The Testing of Chemicals – Unscheduled DNA synthesis (UDS) test with mammalian liver cells in vivo – Guideline no. 486,.
5. EC – (2008) Commission Regulation No. 440/2008 – B.39. Unscheduled DNA synthesis (UDS) test with mammalian liver cells in vivo – Official Journal No. L 142
6. Nesslany F, Parent-Massin D, Marzin D (2010) Risk assessment of consumption of methylchavicol and tarragon: the genotoxic potential in vivo and in vitro. Mutat Res 696(1):1–9
7. Gaudin J, Le Hegarat L, Nesslany F et al (2009) In vivo genotoxic potential of microcystin-LR: a cyanobacterial toxin, investigated both by the unscheduled DNA synthesis (UDS) and the comet assays after intravenous administration. Environ Toxicol 24(2):200–209
8. Garry S, Nesslany F, el Aliouat M et al (2003) Potent genotoxic activity of benzo[a]pyrene coated onto hematite measured by unscheduled DNA synthesis *in vivo* in the rat. Mutagenesis 18(5):449–455
9. Seglen PO (1976) Preparation of isolated rat liver cells. Methods Cell Biol 13:29–83
10. Williams GM (1977) Detection of chemical carcinogens by unscheduled DNA synthesis in rat liver primary cell cultures. Cancer Res 37:1845–1851
11. Fautz R, Hussain B, Efstathiou E et al (1993) Assessment of the relation between the initial viability and the attachment of freshly isolated rat hepatocytes used for the in vivo/in vitro dna repair assay (UDS). Mutat Res 291:21–27
12. Brambilla G, Martelli A (1992) Grain counting in the in vitro hepatocyte DNA–repair assay. Mutat Res 272(1):9–15
13. Madle S, Dean SW, Andrae U et al (1994) Recommendations for the performance of UDS tests in vitro and in vivo. Mutat Res 312:263–285
14. Fielder RJ, Allen JA, Boobis AR et al (1993) Repor*t of the* BTS/UKEMS Working Group on dose setting in *in vivo* mutagenicity assays. Hum Exp Toxicol 12:189–198
15. Hamilton CM, Mirsalis JC (1987) Factors that affect the sensitivity of the in vivo–in vitro hepatocyte DNA repair assay in the male rat. Mutat Res 189(3):341–347
16. Butterworth BE, Ashby J, Bermudez E et al (1987) A protocol and guide for the in vivo rat hepatocyte DNA–repair assay. Mutat Res 189(2):123–133
17. No authors listed (1990) Omeprazole and genotoxicity. The Lancet. 335(8686):386

Chapter 21

^{32}P-Postlabeling Analysis of DNA Adducts

Heinz H. Schmeiser, Marie Stiborova, and Volker M. Arlt

Abstract

^{32}P-Postlabeling analysis is an ultrasensitive method for the detection and quantitation of DNA adducts and covalent modifications of the DNA. It consists of four main steps: (1) enzymatic digestion of DNA to 3′-monophosphate nucleosides; (2) enrichment of the adducts; (3) 5′OH-labeling of the adducts by T4 polynucleotide kinase-catalyzed transfer of ^{32}P-orthophosphate from [γ-^{32}P]ATP; and (4) chromatographic or electrophoretic separation of labeled adducts and detection and quantification by means of their radioactive decay. The assay requires only micrograms of DNA and is capable of detecting adduct levels as low as one adduct in 10^9–10^{10} normal nucleotides. It is applicable to a wide range of investigations in human, animal, and in vitro studies including monitoring exposure to environmental or occupational carcinogens, determining whether a chemical or a complex mixture has genotoxic properties, elucidation of the toxicological pathways of carcinogens, and monitoring DNA repair.

Key words ^{32}P-postlabeling, DNA adducts, Human biomonitoring, Genotoxicity testing, DNA damage and repair, Environmental and occupational exposures, Thin-layer chromatography

1 Introduction

DNA damage is the first critical step in the carcinogenic process caused by chemicals. In many cases this damage is in the form of chemically stable modified nucleotides, in which the carcinogen or its electrophilic intermediates formed by metabolism bind to nucleophilic centers in DNA termed "DNA adducts" [1, 2]. Because adduct analyses of DNA from cells or tissues is a means to determine the extent of exposure to reactive chemicals at the target site, DNA adducts are used as a biomarker of exposure to carcinogens [3]. DNA adducts are pre-mutagenic lesions, which if not repaired can lead to errors in DNA replication that subsequently can result in tumorigenesis [4].

Thus, measuring DNA adducts is useful for a wide variety of studies such as investigating the genotoxic potential of new compounds, monitoring human exposure to environmental carcinogens for molecular epidemiology studies, studies on elucidating the etiology of human cancer, mechanistic investigations into tumor initiation, and monitoring of DNA repair [5, 6]. There are several methods by which DNA adducts can be detected and identified (reviewed in refs. 7, 8). One of the most sensitive of these which has received wide application is the ^{32}P-postlabeling assay.

The principle underlying the ^{32}P-postlabeling assay is that the radiolabel is introduced into the adduct, after it is formed, by enzymatic transfer of a ^{32}P-containing phosphate group [9]. This results in a much greater sensitivity of detection not achievable with longer lived isotopes and more importantly offers the possibility to examine DNA from any source retrospectively for prior exposure to carcinogens. For most applications, the principal stages of the ^{32}P-postlabeling assay are enzymatic digestion of the DNA to nucleoside 3'-monophosphates, enrichment of the adducts to enhance the sensitivity of the assay, 5'OH-labeling of the deoxynucleotide adducts with ^{32}P-orthophosphate, catalyzed by T4 polynucleotide kinase (T4-PNK), and chromatographic and/or electrophoretic separation of the labeled digest to generate a profile of adducts that is detected and quantitated by means of their radioactive decay.

The method originally developed for simple alkyl-modified DNA adducts [10] in the early 1980s was then improved for the detection of bulky aromatic and/or hydrophobic adducts with high sensitivity (Fig. 1; [11]). One of the two principal enhancement procedures that followed was digestion of the nucleotides with nuclease P_1 prior to labeling, resulting in normal nucleotides, but not many adducts, being converted to nucleosides and thus not substrates for T4-PNK [12]. The other method was extraction of the aromatic/hydrophobic adducts into butanol to separate them from the un-adducted normal nucleotides [13]. For these methods, the labeled adducts are resolved and detected as nucleoside 3',5'-bisphosphates. However, alternative digestion strategies, carried out both before and after ^{32}P-postlabeling, can lead to the production of labeled nucleoside 5'-monophosphates [14]. For some applications, the possession of only one charged phosphate group can result in better resolution of carcinogen–DNA adducts and, additionally, improve the confidence in the assignment of structures to unknown species on the basis of co-chromatography with synthetic standards. Resolution of DNA adducts has been done most commonly by multidirectional thin-layer chromatography (TLC), using polyethyleneimine (PEI)-cellulose plates [11]. Although the TLC procedure is time consuming it enables many samples to be analyzed in parallel and provides the most sensitive method for detecting adducts. Alternatively, high-performance liquid chromatography (HPLC) offers a technique of higher resolution that is used frequently [15–17]. For small DNA lesions, such as those caused by oxidative damage, polyacrylamide gel electrophoresis (PAGE) of DNA digests has also proved useful for resolving the ^{32}P-postlabeled species [18].

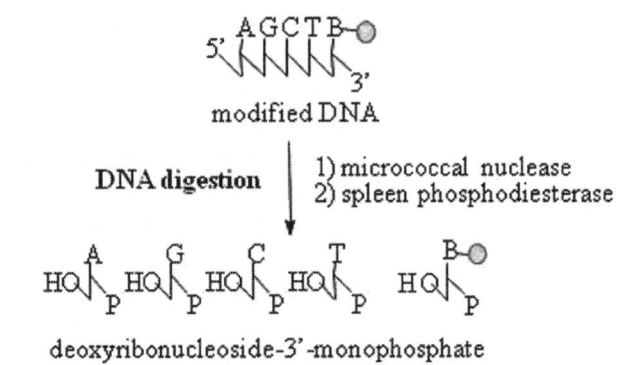

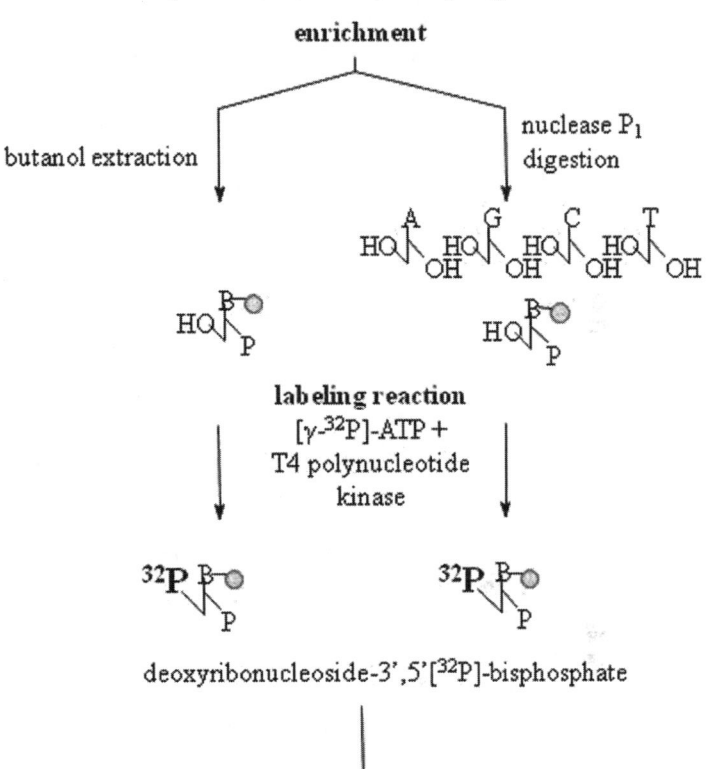

Fig. 1 Outline of the ^{32}P-postlabeling method using nuclease P$_1$ digestion and butanol extraction enrichment procedures. Adducted DNA is digested with micrococcal nuclease and spleen phosphodiesterase to nucleoside 3'-monophosphates. Using nuclease P$_1$ digestion, normal nucleoside 3'-monophosphates are further digested to nucleosides that are not substrates for T4 polynucleotide kinase, whilst most adducted nucleoside 3'-monophosphates are resistant or partially resistant to nuclease P$_1$ treatment. During butanol enrichment, adducted nucleoside 3'-monophosphates are physically partitioned from the normal nucleoside 3'-monophosphates. The removal of the normal nucleosides from the labeling reaction allows the use of high-specific-activity [γ-^{32}P]ATP and greatly increases the sensitivity of the assay. DNA adducts are separated by multidirectional thin-layer chromatography and detected by autoradiography

There are several advantages of ^{32}P-postlabeling over other methods. It does not require the use of radiolabeled test compounds, making it useful in experiments where multiple dosing is required, and is applicable to a wide range of chemicals and types of DNA lesion. Prior structural characterization of adducts is not required, although some assumptions about their likely chromatographic properties may be necessary. It requires only microgram quantities of DNA and is capable of detecting some types of DNA adducts at levels as low as 1 adduct in 10^{10} normal nucleotides. It can be applied to the assessment of the genotoxic potential of complex mixtures of chemicals, such as environmental airborne combustion products, particulates, and industrial or environmental pollutants.

The method is not without its limitations, however [19]. DNA lesions that are not chemically stable as mononucleotides will not be detected reliably. The method does not provide structural information of adducts and identification of adducts often relies on demonstrating their co-chromatography with characterized synthetic standards. Only synthetic standards of adducts (adduct-nucleoside 3'-monophosphates) provide the means for determining the efficiency of labeling and detection, whereas in the absence of standards adduct levels may be underestimated. High labeling efficiencies have been achieved with some adducts like benzo[a]pyrene-derived DNA adducts, whereas some adducts are labeled hardly at all [20, 21]. The method has also been found to detect endogenously derived DNA adducts that may, in some instances, mask the formation of adducts formed by the compound under investigation [19].

In applying ^{32}P-postlabeling to investigate the genotoxic potential of new compounds or a mixture (DNA binding activity), the investigator is faced with a number of decisions concerning the choice of enhancement procedure to be used and the chromatography conditions to be applied. The protocol given here should be regarded as providing guidance for an initial investigation, is broadly applicable to a range of adducts (best suited for bulky and aromatic or hydrophilic adducts), and is based on validated studies of known carcinogen–DNA adducts. Where a new or unknown type of DNA adducts is being investigated, it cannot be asserted that these conditions will be optimal for its detection and quantification.

2 Materials

2.1 DNA Digestion

1. Use double-distilled water or equivalent throughout.
2. Micrococcal nuclease (MN) enzyme: 300 mU/μL MN prepared in water (see **Note 1**).

3. Phosphodiesterase (SPD) enzyme (from calf spleen, Type II): 25 mU/µL SPD prepared in water (*see* **Note 2**).

4. MN/SPD mix: Dilute MN and SPD stock solutions in water to give a final concentration of 150 mU/µL MN and 12.5 mU/µL SPD (*see* **Note 3**).

5. Digestion buffer: 250 mM sodium succinate, pH 6.0, 100 mM $CaCl_2$. Prepare stock solution of 500 mM sodium succinate, adjust to pH 6.0 by NaOH. Prepare stock solution of 200 mM $CaCl_2$; mix equal volumes of stock solutions.

6. All solutions can be stored at –20 °C in small aliquots. Enzymes may be kept for at least 6 months without loss of activity, but repeated freeze–thaw cycles should be avoided.

2.2 Nuclease P_1 Digestion

1. 0.8 M Sodium acetate buffer, pH 5.0. Prepare 0.8 M sodium acetate solution, and adjust pH to 5.0 by glacial acetic acid.
2. 2.0 mM $ZnCl_2$.
3. Nuclease P_1: 4 mg/mL in water (*see* **Note 4**).
4. 0.43 M Tris base.
5. All solutions can be stored at –20 °C in small aliquots.

2.3 Butanol Extraction

1. Buffer A: 11.6 mM ammonium formate, pH 3.5.
2. Buffer B: 10 mM tetrabutylammonium chloride.
3. 1-Butanol (redistilled, water saturated).
4. 1 mL syringe with blunt-ended needle.
5. 250 mM Tris–HCl, pH 9.5.
6. All solutions should be stored at 4 °C.

2.4 DNA Postlabeling

1. T4-PNK with or without 3′-phosphatase activity.
2. Kinase buffer: 400 mM bicine, pH 9.5, 200 mM $MgCl_2$, 200 mM dithiothreitol, 10 mM spermidine. Prepare stock solutions in water of 1.6 M bicine (adjust pH to 9.5 by NaOH), 800 mM $MgCl_2$, 800 mM dithiothreitol, and 40 mM spermidine; mix equal volumes of all four stock solutions.
3. >3,000 Ci/mmol [γ-^{32}P] ATP.
4. 90 µM ATP (cold).
5. All solutions must be stored at –20 °C in small aliquots.

2.5 Thin-Layer Chromatography

1. 20 × 20 cm PEI-impregnated cellulose TLC sheet (*see* **Note 5**).
2. No.1 Whatman filter sheets.
3. D1: 1 M sodium phosphate, pH 6.5. Prepare 1 M solutions of sodium dihydrogen phosphate and disodium hydrogen phosphate; adjust pH by mixing (*see* **Note 6**).

4. D3: 3.5 M lithium formate, 8.5 M urea, pH 3.5. Prepare 3.5 M formic acid, 8.5 M urea solution; adjust pH by adding solid lithium hydroxide (*see* **Notes 7** and **8**).

5. D4: 0.8 M lithium chloride, 0.5 M Tris–HCl, 8.5 M urea; adjust pH to 8.0 using HCl (*see* **Note 8**).

6. D5: 1.7 M sodium phosphate, pH 6.0. Prepare 1.7 M solutions of sodium dihydrogen phosphate and disodium hydrogen phosphate and adjust pH by mixing the two solutions.

7. Efficiency/normals solvent: 250 mM ammonium sulfate, 40 mM sodium phosphate; adjust pH to 6.5 using NaOH.

2.6 Detection and Quantification

1. 10 mM Tris–HCl, pH 9.5 for dilution of normal nucleotides.
2. Autoradiography film or an electronic imaging device (e.g., Canberra Packard InstantImager, Downers Grove, IL, USA).

2.7 HPLC Co-chromatography

1. 4 M pyridinium formate, pH 4.5. Prepare 4 M formic acid solution and adjust pH using pyridine.
2. Methanol.
3. HPLC solvent A: 0.5 M sodium phosphate, pH 3.5/methanol (70:30, v/v).
4. HPLC solvent B: 0.5 M sodium phosphate, pH 3.5/methanol (45:55, v/v).
5. Phenyl-modified reversed-phase column (e.g., 250 × 4.6 mm, particle size 5 μm, Zorbax Phenyl).
6. HPLC system with in-line radioactivity detector.

3 Methods

3.1 DNA Digestion

1. Take 6.25 μg DNA solution in a 1.5 mL tube, evaporate to dryness in a Speedvac evaporator, and redissolve in 3.25 μL water.
2. Add 2.5 μL MN/SPD mix and 0.5 μL of digestion buffer per sample. Vortex to mix and then centrifuge down, using a microcentrifuge.
3. Incubate at 37 °C for 3 h.
4. Remove 1.25 μL of digest for labeling of normal nucleotides (*see* Subheading 3.6). The rest (5 μL) of the DNA digest is enriched for adduct either by nuclease P1 digestion (*see* Subheading 3.2) or by butanol extraction method (*see* Subheading 3.3).
5. Dilute normals digest from **step 4** above with 248.75 μL water and mix on a vortex.
6. Take 10 μL of the diluted normal solution from above, add 190 μL of water, and mix on a vortex and proceed to Subheading 3.6.

3.2 Enrichment of Adducts by Nuclease P₁ Digestion

1. To the above DNA digest (5 μL, from **step 4** of Subheading 3.1), add 0.35 μL of sodium acetate buffer, 0.35 μL of $ZnCl_2$, 0.225 μL of water, and 0.625 μL of nuclease P_1/sample. Incubate at 37 °C for 1 h.
2. Stop the reaction by addition of 1.5 μL of 0.43 M Tris base.

3.3 Enrichment of Adducts by Butanol Extraction

1. Add 215 μL of buffer A and 25 μL of buffer B to the DNA digest (5 μL from **step 4** in Subheading 3.1 above).
2. Immediately add 250 μL of butanol and vortex for 60 s at high speed.
3. Microcentrifuge at $8,000 \times g$ for 90 s. Aspirate the upper butanol layer and keep.
4. Repeat **steps 2** and **3**.
5. To pooled butanol extracts, add 400 μL of butanol. Vortex for 60 s. Microcentrifuge as in **step 3** above.
6. Remove and discard water through the butanol layer using a syringe, being careful not to remove any of the butanol.
7. Repeat **step 5** twice, discarding the water each time.
8. Add 1.5 μL of 250 mM Tris–HCl to washed butanol. Vortex briefly.
9. Speedvac to dryness. Redissolve in 50 μL of butanol by vortexing and speedvac to dryness again.
10. Redissolve in 8 μL of water.

3.4 Labeling of Adducts (See Note 9)

1. Premix stock labeling cocktail (no. of samples + no. of normal samples) for each sample from 0.5 μL kinase buffer, 0.25 μL cold ATP, 10 U T4-PNK, and 50 μCi [γ-³²P] ATP/sample (*see* **Note 10**), and add water to make up the volume to 2 μL.
2. Add 2 μL of labeling cocktail to each solution remaining from **step 2** of Subheading 3.2 or **step 10** of Subheading 3.3 and incubate at 37 °C for 30 min.
3. Staple a 10×12 cm Whatman 3 paper wick to the top edge of a 10×20 cm PEI-cellulose TLC sheet as shown in Fig. 2. Spot the whole of each sample onto the origin of this sheet. Keep tube for efficiency test (*see* Subheading 3.5).
4. Run in D1 overnight with wick hanging outside the tank (*see* **Note 11**).
5. Cut plates down to 10×10 cm as shown in Fig. 2.
6. Wash plates twice in water and dry plates with cool air.
7. Run in D3 and D4 in directions as shown in Fig. 2. Before each run dip lower edge of plate in water to give an even solvent front. D2 is usually omitted.

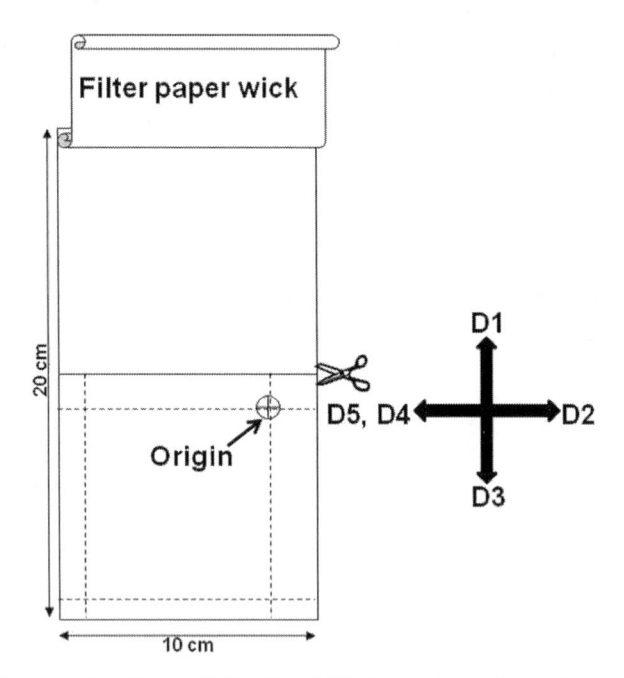

Fig. 2 Diagram showing multidirectional thin-layer chromatography procedures for the resolution of ^{32}P-labeled adducts on polyethyleneimine-cellulose. The whole 10 µL of each labeled sample is spotted onto the origin. During D1 chromatography DNA adducts are retained close to the origin, whereas [γ-^{32}P]ATP and labeled normal nucleotides migrate onto the wick. After cutting the plate DNA adducts are resolved in D3 and D4 directions

8. If necessary run in D5 overnight to reduce background radioactivity.
9. Wash plates twice in water and cool air-dry between solvents.

3.5 Test for Efficiency of Enrichment Techniques

1. Wash the bottom of the tube (**step 3** from Subheading 3.4) with 50 µL water.
2. Vortex and microcentrifuge.
3. Spot 5 µL near lower edge of a 10×20 cm PEI-cellulose TLC sheet.
4. Run in efficiency/normals solvent (250 mM ammonium sulfate, 40 mM sodium phosphate, pH 6.5) to the top edge (*see* **Note 12**).

3.6 Labeling of Normal Nucleotides

Adduct levels are calculated as relative adduct labeling (RAL) values, which represent the ratio of count rates of adducted nucleotides (adducts) over count rates of total (adducted plus normal) nucleotides in the DNA sample analyzed. To determine the amount of total nucleotides in the DNA sample an aliquot (1.25 µL) of the DNA digest (in **step 4** of Subheading 3.1) is removed, diluted, and

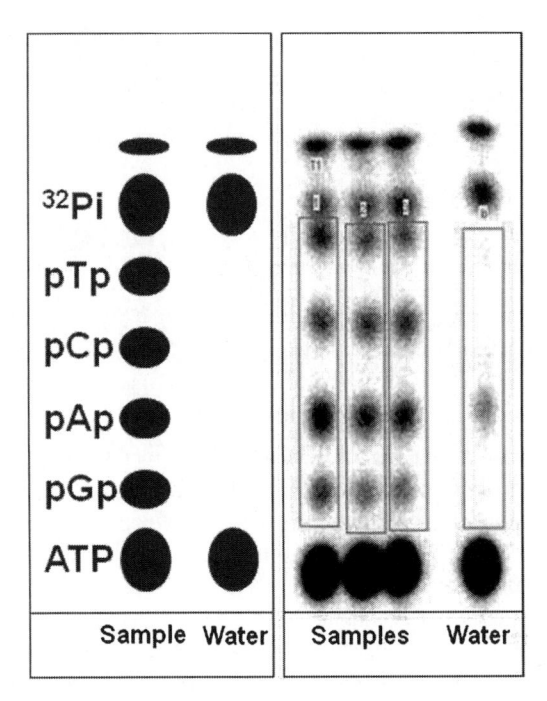

Fig. 3 1-Directional chromatography of ^{32}P-labeled normal nucleotides on polyethyleneimine-cellulose. An aliquot of the DNA digest removed prior to enrichment is diluted and labeled. The presence of [γ-^{32}P]ATP in the aliquot indicates that sufficient ATP was available in the labeling reaction. The level of radioactivity is determined in the four labeled normal bisphosphates in the aliquot and used when calculating DNA adduct levels (i.e., relative adduct labeling)

analyzed separately; this is called labeling of normal nucleotides. Following this protocol the dilution factor for normals is 10^6.

1. Take 5 μL of diluted normals solution from **step 6** of Subheading 3.1 and add 2.5 μL of 10 mM Tris–HCl pH 9.5.
2. Add 1.25 μL of labeling cocktail from **step 1** of Subheading 3.4 and incubate at 37 °C for 30 min.
3. Take out 5.24 μL of the labeled mixture from above, add 750 μL of 10 mM Tris pH 9, vortex, and spin.
4. Spot 5 μL of the mixture 2 cm from the lower edge of a 10×20 cm PEI-cellulose TLC sheet. Run in normals solvent from Subheading 2.5 to top edge.
5. Visualize and count the bisphosphate normals spots as shown in Fig. 3 (*see* **Note 13**). This will give you the sum of counts per minute (cpm) of all four normals spots and a dilution factor of 10^6 compared to the adduct sample.

3.7 Imaging and Quantification

1. Adducts can be visualized by placing plates in cassettes with autoradiography film and keeping at −80 °C for several hours, up to 4 days. Adduct spots can then be cut from the plate and

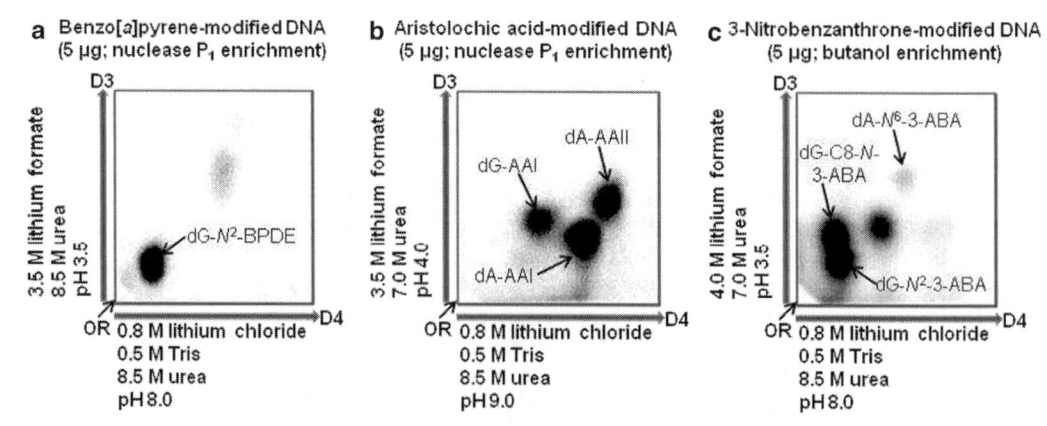

Fig. 4 DNA adduct analysis of selected environmental carcinogens using thin-layer chromatography ^{32}P-postlabeling: (a) benzo[*a*]pyrene [23]; (b) aristolochic acid [24]; and (c) 3-nitrobenzanthrone [25]. The origin (OR) on the *bottom left-hand corner* was cut off before autoradiography. (**a**) Autoradiogram of benzo[*a*]pyrene-modified DNA using the nuclease P₁ enrichment version of the assay: 10-(deoxyguanosin-N^2-yl)-7,8,9-trihydroxy-7,8,9,10-tetrahydrobenzo[*a*]pyrene (dG-N^2-BPDE). (**b**) Autoradiogram of aristolochic acid-modified DNA using the nuclease P₁ enrichment version of the assay: 7-(deoxyadenosin-N^6-yl)aristolactam I (dA-AAI), 7-(deoxyguanosin-N^2-yl)aristolactam I (dG-AAI), and 7-(deoxyadenosin-N^6-yl)aristolactam II (dA-AAII). (**c**) Autoradiogram of 3-nitrobenzanthrone-modified DNA using the butanol extraction enrichment version of the assay: 2-(2′-deoxyguanosin-N^2-yl)-3-aminobenzanthrone (dG-N^2-3-ABA), *N*-(2′-deoxyguanosin-8-yl)-3-aminobenzanthrone (dG-C8-*N*-3-ABA), and 2-(2′-deoxyadenosin-N^6-yl)-3-aminobenzanthrone (dA-N^6-3-ABA)

quantified in a scintillation counter. Alternatively an InstantImager can be used which will give the results in a few minutes (*see* Fig. 4).

2. Adduct levels (RAL) are calculated according to the following formula:

$$\text{RAL} = \frac{\text{cpm in adduct(s)}}{\{\text{cpm in total (normal) nucleosides} \times \text{dilution factor} \\ [\text{i.e.}, 10^6] + \text{cpm in adduct(s)}\}}$$

This will give the adduct level as adducts per 10^6 normal nucleotides.

3. Results can be expressed in this way or as fmol/μg DNA. To arrive at this latter figure multiply the number of adducts per 10^8 normal nucleotides by 0.03, as 33 adducts per 10^8 nucleotides are equivalent to 1 fmol/μg DNA.

3.8 Extraction of Adducts for HPLC Co-chromatography

1. Cut the adduct spot out of the PEI-cellulose TLC sheet and put in a scintillation vial (*see* **Note 14**).

2. Add 500 μL of pyridinium formate and shake gently overnight (*see* **Note 15**).

3. Microcentrifuge extracts at $8,000 \times g$ for 90 s to remove small particles.

4. Speedvac to dryness. Redissolve in 100 µL of water and methanol mix (1:1, v/v).

3.9 HPLC Co-chromatography

1. Aliquots (e.g., 50 µL) of the above extract are analyzed on a phenyl-modified reversed-phase column with a linear gradient of methanol (from 30 to 55 % in 45 min) in 0.5 M sodium phosphate, pH 3.5 (*see* **Note 16**). Radioactivity eluting from the column was measured by monitoring Cherenkov radiation through a radioactivity detector.

4 Notes

1. It is necessary to dialyze the MN solution to remove residual oligonucleotides. Use a 10K Slide-A-Lyzer from Pierce (66425) suspended in 5 L distilled water at 4 °C for 24 h. Change water once.

2. Some SPD preparations must also be dialyzed to remove the ammonium salts, which may inhibit the labeling. Use a 10K Slide-A-Lyzer (Pierce 66425) suspended in 5 L distilled water at 4 °C for 24 h. Change water once.

3. Note that bovine phosphodiesterase has not proved to be active. It may be necessary to vary the amount used depending on the type of adducts to be detected.

4. Nuclease P_1 solutions (in water) should be stored at −20 °C.

5. Plates should be pre-run with distilled water and dried to remove a yellow contaminant, which may lead to an increased background.

6. 1 M sodium phosphate is usually sufficient for many bulky adducts, but smaller or more polar adducts may require higher concentrations (1.7–2.3 M sodium phosphate) to avoid streaking.

7. Use lithium hydroxide to adjust pH to 3.5.

8. D3 and D4 are suitable for many lipophilic bulky adducts, but considerable variation is possible both in concentration and content (*see* Fig. 4).

9. [γ-^{32}P]ATP is a high-energy β-particle emitter and due regard should be given to handling the material. Exposure to ^{32}P should be avoided, by working in a confined laboratory area, with protective clothing, shielding, Geiger counters, and body dosimeter. We normally use 1 cm thick Perspex or glass shielding between the operator and the source material throughout. We routinely wear two pairs of medium-weight rubber gloves

and handle the tubes with 30 cm forceps. An appropriate Geiger counter should be on during the whole procedure and working areas monitored before and after work. All apparatus should be checked for contamination and cleaned where appropriate by immersion in a suitable decontamination fluid (RBS 35 or Decon). Waste must be discarded according to appropriate local safety procedures.

10. It may be possible to use lesser amounts of [γ-^{32}P]ATP/sample (e.g., 10–25 μCi) [22], but the minimum amount that can be used may vary depending on the type of adducts to be detected, and their levels.

11. The top of the tank and the lid should be wrapped around with cling film to avoid contamination with radioactivity.

12. Poor efficiency of either enrichment procedure will be demonstrated by the appearance of the four normal nucleotide spots (*see* Fig. 3). If there is no indication of excess ATP then the sample should be discarded.

13. When quantifying the normals, it is necessary to run a blank using water instead of sample and to subtract the value obtained as background.

14. The origin after D1 can also be cut out of the PEI-cellulose TLC sheet.

15. The extraction can be monitored by measuring Cherenkov radiation in a scintillation counter before and after the extraction procedure.

16. Depending on the adduct type other HPLC conditions may be more suitable.

References

1. Phillips DH (2002) The formation of DNA adducts. In: Allison MR (ed) The cancer handbook. Macmillan, London, pp 293–306
2. Phillips DH, Arlt VM (2009) Genotoxicity: damage to DNA and its consequences. Exs 99:87–110
3. Phillips DH (2005) DNA adducts as markers of exposure and risk. Mutat Res 577:284–292
4. Luch A (2005) Nature and nurture—lessons from chemical carcinogenesis. Nat Rev Cancer 5:113–125
5. Arlt VM, Frei E, Schmeiser HH (2007) ECNIS-sponsored workshop on biomarkers of exposure and cancer risk: DNA damage and DNA adduct detection and 6th GUM-32P-postlabelling workshop, German Cancer Research Center, Heidelberg, Germany, 29–30 September 2006. Mutagenesis 22:83–88
6. Arlt VM, Schwerdtle T (2011) UKEMS/Dutch EMS-sponsored workshop on biomarkers of exposure and oxidative DNA damage & 7th GUM-32P-postlabelling workshop, University of Munster, Munster, Germany, 28–29 March 2011. Mutagenesis 26:679–685
7. Poirier MC, Santella RM, Weston A (2000) Carcinogen macromolecular adducts and their measurement. Carcinogenesis 21:353–359
8. Brown K (2012) Methods for the detection of DNA adducts. In: Parry JM, Parry EM (eds) Genetic toxicology: principles and methods. Heidelberg, Humana, pp 207–230
9. Phillips DH, Arlt VM (2007) The 32P-postlabeling assay for DNA adducts. Nat Protoc 2:2772–2781
10. Randerath K, Reddy MV, Gupta RC (1981) 32P-labeling test for DNA damage. Proc Natl Acad Sci USA 78:6126–6129

11. Gupta RC, Reddy MV, Randerath K (1982) 32P-postlabeling analysis of non-radioactive aromatic carcinogen-DNA adducts. Carcinogenesis 3:1081–1092
12. Reddy MV, Randerath K (1986) Nuclease P1-mediated enhancement of sensitivity of 32P-postlabeling test for structurally diverse DNA adducts. Carcinogenesis 7:1543–1551
13. Gupta RC (1985) Enhanced sensitivity of 32P-postlabeling analysis of aromatic carcinogen:DNA adducts. Cancer Res 45: 5656–5662
14. Randerath K, Randerath E, Danna TF et al (1989) A new sensitive 32P-postlabeling assay based on the specific enzymatic conversion of bulky DNA lesions to radiolabeled dinucleotides and nucleoside 5′-monophosphates. Carcinogenesis 10:1231–1239
15. Pfau W, Lecoq S, Hughes NC et al (1993) Separation of 32P-labelled 3′,5′-bisphosphate adducts by HPLC. In: Phillips DH, Castegnaro M, Bartsch H (eds) Postlabelling methods for detection of DNA adducts. IARC, Lyon, pp 233–242
16. Phillips DH, Hewer A, Horton MN et al (1999) N-Demethylation accompanies α-hydroxylation in the metabolic activation of tamoxifen in rat liver cells. Carcinogenesis 20:2003–2009
17. Nagy E, Cornelius MG, Moller L (2009) Accelerated (32)P-HPLC for bulky DNA adducts. Mutagenesis 24:183–189
18. Jones GD, Dickinson L, Lunec J et al (1999) SVPD—post-labeling detection of oxidative damage negates the problem of adventitious oxidative effects during 32P-labeling. Carcinogenesis 20:503–507
19. Phillips DH, Farmer PB, Beland FA et al (2000) Methods of DNA adduct determination and their application to testing compounds for genotoxicity. Environ Mol Mutagen 35:222–233
20. Beland FA, Doerge DR, Churchwell MI et al (1999) Synthesis, characterization, and quantitation of a 4-aminobiphenyl-DNA adduct standard. Chem Res Toxicol 12:68–77
21. Haack T, Boche G, Kliem C et al (2004) Synthesis, characterization, and 32P-postlabeling of N-(deoxyguanosin)-4-aminobiphenyl 3′-phosphate adducts. Chem Res Toxicol 17:776–784
22. Munnia A, Saletta F, Allione A et al (2007) 32P-Post-labelling method improvements for aromatic compound-related molecular epidemiology studies. Mutagenesis 22:381–385
23. Arlt VM, Stiborova M, Henderson CJ et al (2008) Metabolic activation of benzo[a]pyrene in vitro by hepatic cytochrome P450 contrasts with detoxification in vivo: experiments with hepatic cytochrome P450 reductase null mice. Carcinogenesis 29: 656–665
24. Arlt VM, Zuo J, Trenz K et al (2011) Gene expression changes induced by the human carcinogen aristolochic acid I in renal and hepatic tissue of mice. Int J Cancer 128:21–32
25. Arlt VM (2005) 3-Nitrobenzanthrone, a potential human cancer hazard in diesel exhaust and urban air pollution: a review of the evidence. Mutagenesis 20:399–410

Part IV

Assays in Plants and Alternate Animal Models

Chapter 22

Micronucleus Assay with Tetrad Cells of *Tradescantia*

Miroslav Mišík, Clemens Pichler, Bernhard Rainer, Armen Nersesyan, and Siegfried Knasmueller

Abstract

The *Tradescantia* micronucleus assay is being used since almost 50 years for the detection of genotoxins (including carcinogens) in the environment. A large database on the effects of individual compounds and of complex environmental mixtures (soil, air and water) has accumulated. In contrast to other mutagenicity test systems, the effects of low concentrations of heavy metals, radionuclides, certain herbicides and pesticides, and gaseous mutagens can be detected and it is also possible to use the test for in situ biomonitoring studies. The test system has been validated, and standardized protocols have been developed for laboratory experiments and for field studies, which are described in this chapter.

Key words *Tradescantia*, Tetrads, Genotoxicity studies, Micronucleus

1 Introduction

Already in the 1960s and 1970s, a variety of protocols for mutagenicity assays with higher plants were available, which were based on the detection of mutations in the offspring, somatic mutations, or on the evaluation of chromosomal aberrations [1]. Most of the protocols were time consuming and labor intensive; therefore, only few models survived and are used at present, in particular for environmental monitoring.

The most frequently employed approaches are micronucleus (MN) assays with mitotic root tip cells of *Vicia faba* and *Allium cepa* and experiments with meiotic tetrad cells of *Tradescantia* (*Commelinaceae*), which grows in tropical and subtropical areas. A description of the methods used in root assays and the current databases can be found in the publications of Leme and Marin-Morales [2], Foltete et al. [3], and White and Claxton [4]. The micronucleus assay was first described in *Tradescantia* in the 1950s by Steffensen [5]. The most frequently used clone for genotoxicity experiments is #4430 (Fig. 1), a hybrid between *T. subacaulis* and *T. hirsutifolia*, which was developed in early 1960s in the Brookhaven

Fig. 1 Inflorescences of *Tradescantia* #4430 suitable (**a**) and not suitable (**b**) for treatment and subsequent evaluation of MN

laboratories [6]. Another clone, *T. paludosa* #03, was frequently employed in radiation experiments. In the 1990s, Brazilian groups started to use *T. pallida*, a wild plant, which grows in several South American countries [7, 8].

Hundreds of studies have shown that plant bioassays, in particular the *Tradescantia* micronucleus (Trad MN) test, are valuable tools for the detection of genotoxic carcinogens in the environment. This assay is highly sensitive towards compounds which cause negative or moderate effects in other widely used systems. For example, heavy metals cannot be detected in bacterial mutagenicity assays, which are also not sensitive towards ionizing radiation. Furthermore, it is important to note that, due to the insensitivity of most mutagenicity assays, concentration procedures are required which may lead to loss of active compounds [9, 10]; such procedures can be avoided in Trad MN experiments. Another point, which argues for the use of Trad MN assays, concerns the fact that more than 120 individual compounds and about 100 complex environmental mixtures have been tested [11–13], and therefore, the sensitivity of the assay to different environmental toxicants and to environmental pollution is well known. On the basis of the currently available data, it can be concluded that the Trad MN bioassay is complementarily to other current methods for the detection of environmental genotoxins. Therefore the assay should be included in test batteries that are used in studies concerning environmental pollution with radioactive materials, heavy metals, and air pollution. The test system is based on the detection of micronuclei which refer to clastogenic (chromosome breaking) and aneugenic effects in meiotic pollen tetrad cells. The MN frequencies are scored in buds, which contain early tetrads and are highly synchronized. Plants can be exposed to test substances as stems either in aqueous solutions or in intact form with the roots in soils.

The results which were obtained with *Tradescantia* are summarized in two comprehensive reviews: the article of Rodrigues et al. [14] covers papers which appeared before 1996, while the update by Mišík et al. [15] concerns later publications.

2 Materials

1. Indicator plants: *Tradescantia* clones #4430, #02, and *T. paludosa* #03 should be used for genotoxicity experiments (*see* **Note 1**).

2. Preparation of plant material: Glacial acetic acid, 70 % ethanol, and pure ethanol.

3. Dilution of the test compounds: 1 % dimethyl sulfoxide (DMSO, an organic solvent) or water can be used in stem absorption experiments.

4. Positive controls: 1.0 mM ethyl methanesulfonate (EMS), 0.5 mM arsenic oxide (As_2O_3) or 0.2 mM maleic hydrazide (MH) can be used. These are direct acting mutagens.

5. Staining agent: 1–2 % acetocarmine (Carmine). Heat 50 mL of 45 % acetic acid solution (45 mL glacial acetic acid and 55 mL of distilled water) to boiling. Add 500 mg of carmine and continue heating for 15–20 min while stirring. Cool the resulting solution and filter to remove any precipitate. The solution can be stored in a dark place for several years at room temperature.

6. Pots: The plants should be cultivated in 15 cm diameter plastic pots in pesticide- and herbicide-free soil, or in the case of hydroponic culture, in smaller pots (13/12 cm diameter).

7. Soil/nutrient: 2 parts sand, 4 parts soil, and 1 part peat moss constitute the soil for greenhouse propagation. Two applications of any liquid fertilizer per month are needed (*see* **Note 2**). Standard NPK (nitrogen, phosphor, potassium) liquid fertilizer can be used for hydroponic cultivation.

3 Methods

3.1 Cultivation of Plants

1. Light: If the plants are cultivated in a growth chamber with artificial light, the intensity of the fluorescent light should be about 378 $E/m^2/s$ (1,800 ft candles) and the incandescent light should be around 38 $E/m^2/s$ (180 ft candles) at the tip of the plants. Use of an 18/6-h (light/dark) period is essential for flowering.

2. Temperature: The daytime temperature should be around 21–25 °C, and the nighttime temperature around 16 °C. An increase over 30 °C may have an impact on the background MN frequencies [16].

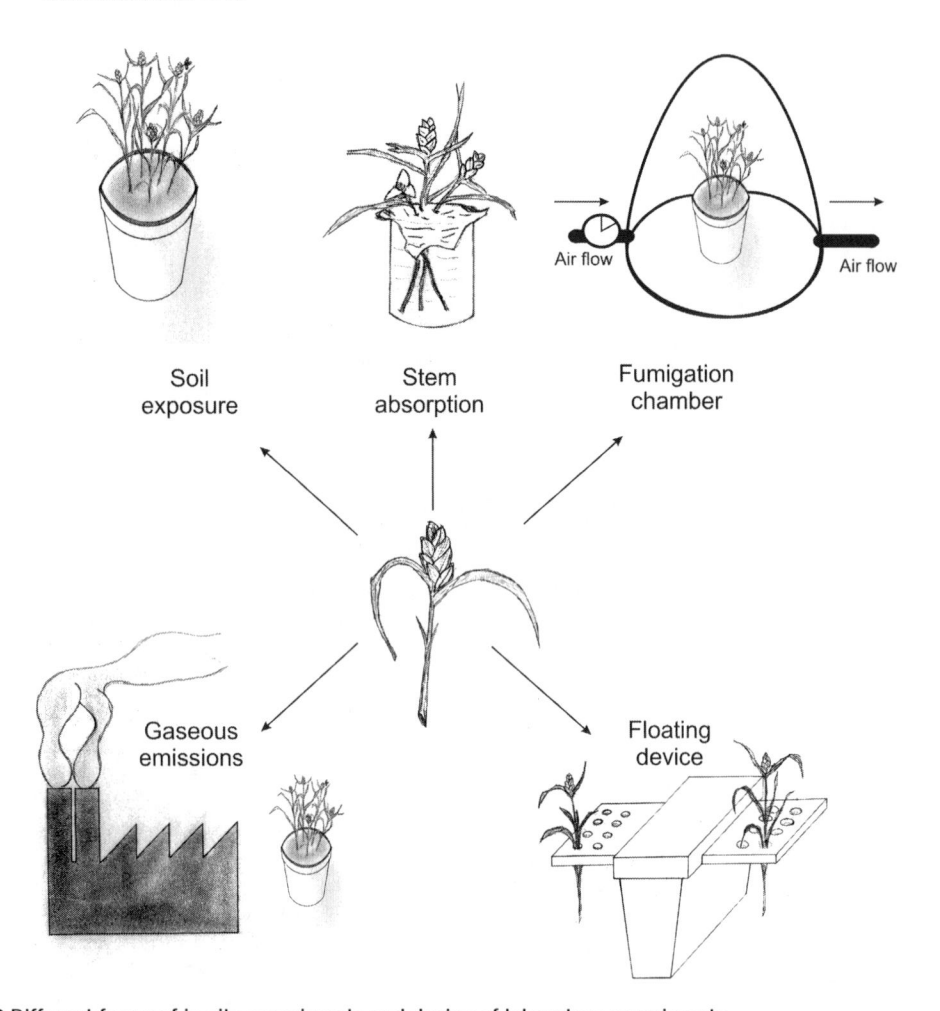

Fig. 2 Different forms of in situ experiments and design of laboratory experiments

3. Propagation of the plants: After 8–12 weeks of cultivation when the pots are full, the plants should be divided and plant rhizomes with leaves are transferred to new pots (*see* **Note 3**).

3.2 Exposure to Chemicals and Complex Mixtures

At least 15 cuttings with immature inflorescences (not with blooming flowers; refer to Fig. 1) should be treated per experimental group. Various types of exposures to the plant are depicted in Fig. 2.

1. Treatments of the plants in the laboratory: In stem absorption experiments, expose the plant cuttings to liquid solution in 250 mL beakers which are covered with aluminum foil.

2. Testing of pure solid or liquid chemicals: The "classical" approach is stem absorption experiment, in which the test compounds are diluted with tap water or an organic solvent (1 % DMSO, maximum final concentration).

3. Testing of soils contaminated with heavy metals: Expose the roots of intact plants in the soils.

4. Testing of gases: Expose the plant cuttings in water to the gaseous agent in fumigation chambers.

5. Testing of contaminated air: It is recommended to expose the plants at different distances from the source of the emissions, in the direction of the wind.

6. Treatment of the plants in situ: Expose the plants as cuttings or as intact plants growing in pots.

7. Testing of surface waters: Use floating devices (perforated swimming plastic plates) with plants (Fig. 2).

3.3 Exposure Time and Recovery Time

1. Exposure time depends on the type of experiment performed.

2. For pure substances (e.g., aqueous solution of herbicides, insecticides, industrial chemicals) in the laboratory, high concentrations can be included and it is suggested to use an exposure time between 3 and 12 h followed by a 24-h recovery period (*see* **Note 4**).

3. For complex mixtures (e.g., polluted water, air, and soil) a continuous exposure up to 30 h is recommended with no recovery periods, since the complex mixtures contain often only low concentrations of mutagens (*see* **Note 5**).

3.4 Fixation

1. Prepare the fixative liquid *ex tempore* (*see* **Note 6**).

2. Fix the inflorescences in 3:1 ratio (v/v) of pure ethanol and acetic acid for 24 h.

3. Subsequently store the fixed tissue in 70 % ethanol (we recommend the use of scintillation tubes; all inflorescences of a single treatment dose can be stored in the same tube in an appropriate volume of fixative (the material should be submerged completely in the liquid).

3.5 Slide Preparation

The different steps are shown in Fig. 3.

1. Only the early tetrad stage is suitable for scoring of MN.

2. First examine one bud of an inflorescence with one to two drops of 1–2 % acetocarmine stain for 5–8 min to see if an early tetrad stage was found.

3. If so, remove the debris using a botanical needle and subsequently put a 24 × 24 mm coverslip over the slide.

4. Slides should be evaluated under a light microscope with 400-fold magnification.

5. Heat slides over a flame for 2–4 s and gently press the coverslips after this treatment so that older tetrads burst and only early ones remain intact.

3.6 Scoring

1. For each experiment, examine slides from at least five buds in the early tetrad stage. From each slide evaluate at least 300 tetrads.

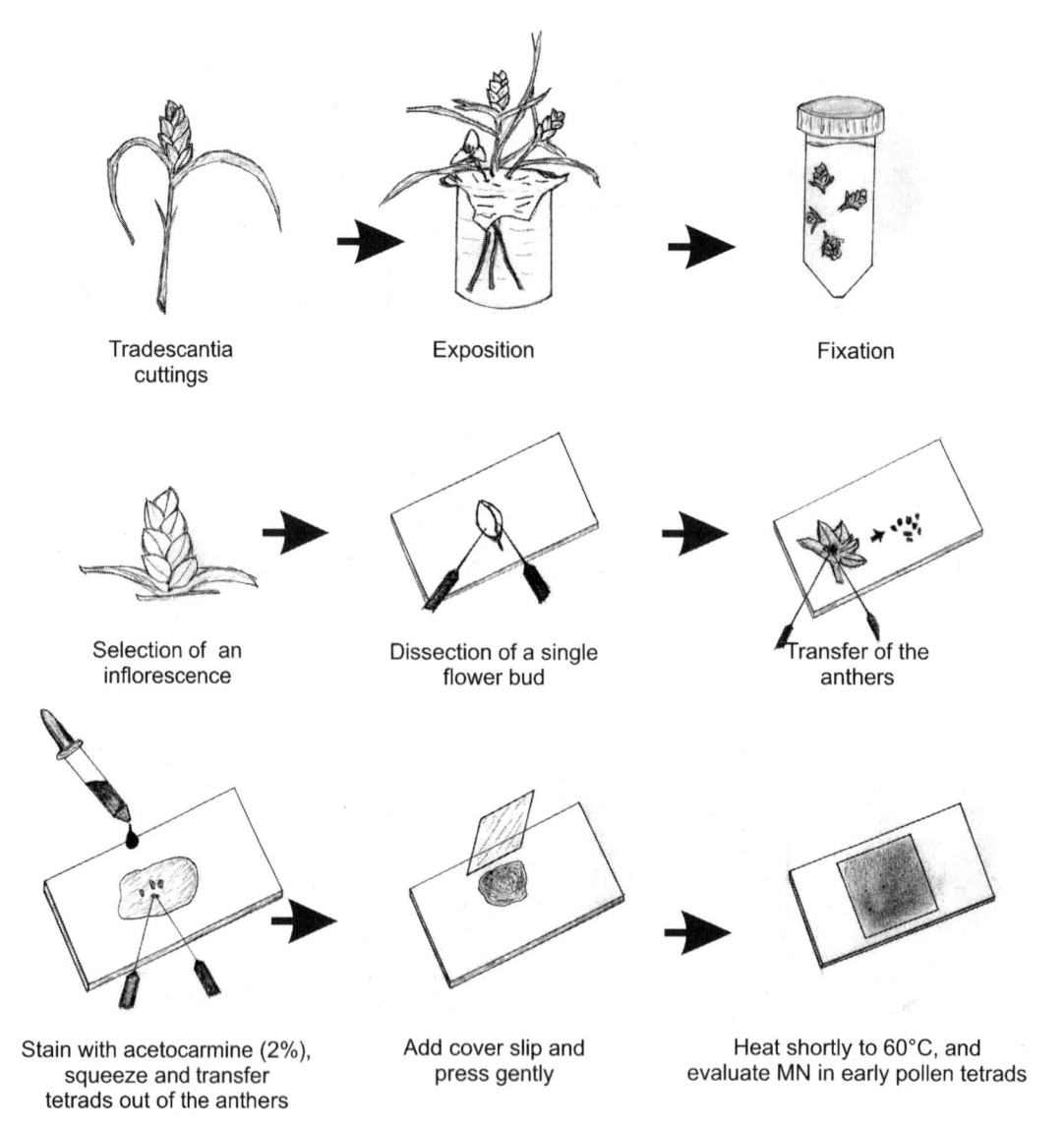

Fig. 3 Different steps of the slide preparation

2. Enter the results of the scoring in scoring sheets (record the overall number of tetrads evaluated, tetrads with one MN, two MN, and three MN).

3. Furthermore, record other anomalies (*see* Fig. 4, *see* **Note 7**).

4. To avoid misinterpretations due to division delays, synchronicity of the cells should also be taken into consideration. Under normal conditions, tetrads within one bud are highly synchronized, but asynchrony (presence of diads, tetrads, and "triads") may be caused by chemical compounds and complex mixtures.

5. Scorers should also check the number of MN in diads, which may be indicative of retarded cell division (*see* **Note 7**).

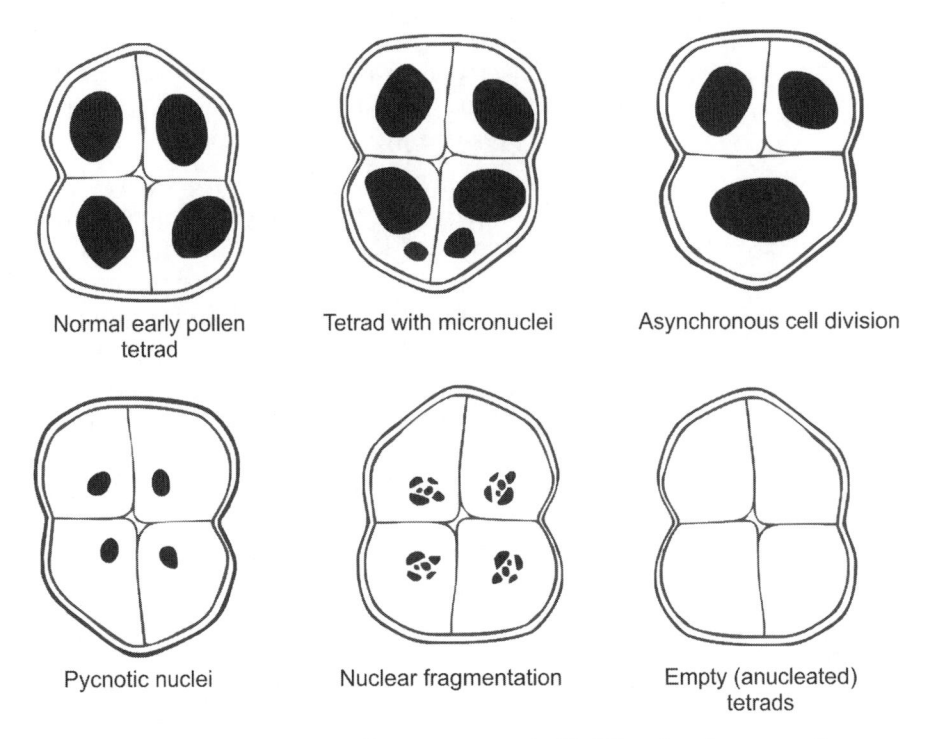

Fig. 4 Schematic representation of cells in different stages and with different nuclear anomalies

| 3.7 Statistical Evaluation | Use one-way ANOVA with Dunnett's post test for statistics in Trad MN assay, alternatively other statistical methods can be applied. |

4 Notes

1. We do not recommend the use of *T. pallida* for several reasons: (a) it has never been validated with a sufficiently large number of model mutagens, (b) it is not known if it is as sensitive as the sterile clone #4430, and (c) as the plants are usually collected in the field it is unclear if specimens from different sites differ in their sensitivity towards genotoxins as a consequence of adaptive responses. Such effects can be excluded when genetically stable sterile hybrids are used.

2. If pesticides or herbicides are used for hydroponic culture, we strongly recommend change of the nutrient solutions after their application. Parasite infections with *Parthenolecanium* species and aphids should be treated with paraffin-based sprays and emulsions. We recommend avoiding synthetic pesticides, which may increase the background MN frequencies.

3. In order to conduct continuous experiments with one test compound per week, with three to four doses and positive and negative controls, 75–150 inflorescences are required which are usually obtained with 120–140 pots.

4. The exposure times and/or the recovery periods have to be sufficiently long to ensure that the cells undergo division which is a prerequisite for MN formation. Some types of exposure may cause inhibition of the cell division; in this case the majority of MN is found in the diad stage and a repetition of the experiment with extended exposure time is mandatory.

5. Negative and positive controls: Tap water can be used to set up the controls but the water has to be tested before the main experiments, to ensure that it does not increase the mutation frequencies. As an alternative, Hoagland's nutrient solution [17] can be used. It is essential to test from each test compound at least three to four concentrations and to include negative (solvent) and positive controls. The background MN frequencies in #4430 vary over a broad range (i.e., between 0.6 and 5.0 MN/100 tetrads). For laboratory experiments, suitable positive controls are directly acting mutagens such as EMS, heavy metals compounds (e.g., As_2O_3), MH, and X-rays or γ-radiation [14, 15]. Results with positive controls are as follows according to our knowledge:

0.2 mM MH (30-h exposure): 5–12.0 MN/100 tetrads

2.5 mM As_2O_3 (6-h exposure, 24-h recovery): 5–10.0 MN/100 tetrads

In environmental monitoring studies, dose–effect relationship can be monitored by collection of samples with different pollution levels or in the case of water or air experiments by exposing plants at different distances from the source of the pollution. The reproducibility of the results should be confirmed in follow-up experiments.

6. Fixed samples can be stored at room temperature up to several years before evaluation.

7. For the scoring of MN it is necessary to identify buds, which contain early-tetrad-stage cells in the inflorescence by use of botanical needles. The procedure is shown in Fig. 3. When a bud is too old, no tetrads are found and younger, smaller buds have to be opened. The pollen cells, which are contained in the buds, represent different meiotic stages. Buds with early stages are in general, smaller and located in lower (proximal) parts of the inflorescence. Figure 5 depicts the different stages of microsporogenesis in *Tradescantia*. MN are formed when pollen mother cells are exposed to DNA-damaging compounds and undergo the first and second meiotic division. In order to form MN, the cells have to undergo one to two cell divisions.

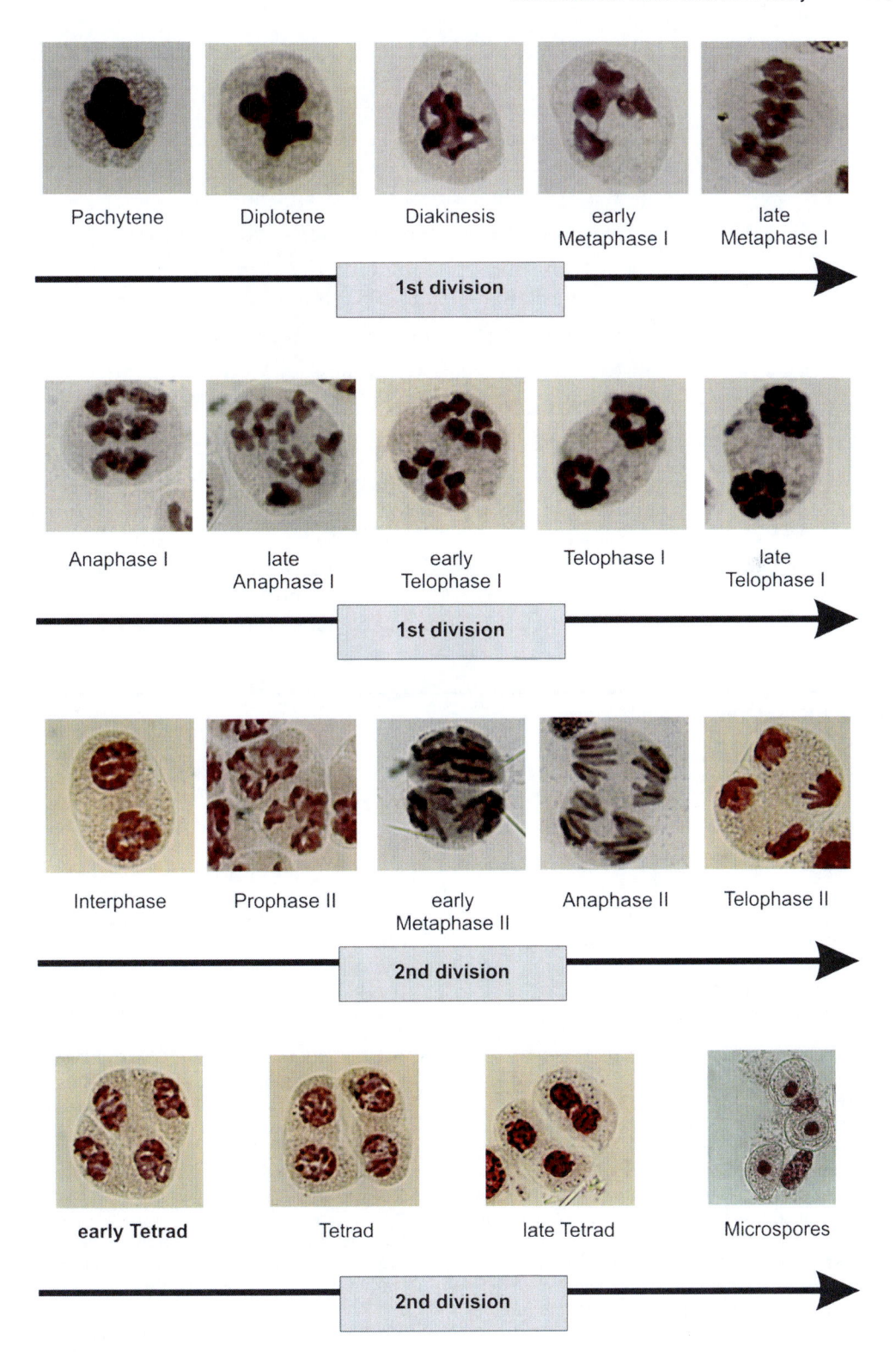

Fig. 5 Different stages of microsporogenesis in *Tradescantia*

Therefore, the precise timing of the experiments is essential to ensure optimal sensitivity. The endpoints of the assay are MN, which are defined as extranuclear bodies containing DNA and a nuclear membrane. The criteria for the identification of MN are the same as described by Tolbert et al. [18] for the evaluation of mammalian cells, as given below:

(a) Rounded smooth perimeter suggestive of a membrane.

(b) Less than one-third of the diameter of the associated nucleus, but large enough to discern shape and color.

(c) Similar staining intensity as the main nucleus.

(d) Texture similar to nucleus.

(e) Same focal plane as nucleus.

(f) Absence of overlap with, or bridge to, nucleus.

Quite often, other types of nuclear aberrations can also be found in the experiments as a consequence of toxic exposure. Sometimes three cells instead of a tetrad are seen, which are indicative of meiotic disturbances; atypical nuclear aberrations are pyknotic cells with condensed (dead) nuclei, and sometimes anuclear cells as well as nuclear fragmentations are found (*see* Fig. 4).

Acknowledgement

We thank P. Breit and E. Gottmann for their help in the preparation of the photographic images.

References

1. Majer BJ, Grummt T, Uhl M et al (2005) Use of plant bioassays for the detection of genotoxins in the aquatic environment. Acta Hydrochim Hydrobiol 33:45–55
2. Leme DM, Marin-Morales MA (2009) Allium cepa test in environmental monitoring: a review on its application. Mutat Res 682:71–81
3. Foltete AS, Dhyevre A, Ferard JF et al (2011) Improvement of Vicia-micronucleus test for assessment of soil quality: a proposal for international standardization. Chemosphere 85:1624–1629
4. White PA, Claxton LD (2004) Mutagens in contaminated soil: a review. Mutat Res 567:227–345
5. Steffensen D (1953) Induction of chromosome breakage at meiosis by a magnesium deficiency in *Tradescantia*. Proc Natl Acad Sci USA 39:613–620
6. Ma TH, Cabrera GL, Chen R et al (1994) *Tradescantia* micronucleus bioassay. Mutat Res 310:221–230
7. Suyama F, Guimaraes ET, Lobo DJ et al (2002) Pollen mother cells of Tradescantia clone 4430 and Tradescantia pallida var. purpurea are equally sensitive to the clastogenic effects of X-rays. Braz J Med Biol Res 35:127–129
8. Batalha JRF, Guimaraes ET, Lobo DJA et al (1999) Exploring the clastogenic effects of air pollutants in Sao Paulo (Brazil) using the *Tradescantia* micronuclei assay. Mutat Res 426:229–232
9. Ohe T, Watanabe T, Wakabayashi K (2004) Mutagens in surface waters: a review. Mutat Res 567:109–149
10. Stahl RG Jr (1991) The genetic toxicology of organic compounds in natural waters and wastewaters. Ecotoxicol Environ Saf 22:94–125
11. Ma TH, Harris MM, Anderson VA et al (1984) Tradescantia-Micronucleus (Trad-MCN) tests on 140 health-related agents. Mutat Res 138:157–167

12. Ma TH (2001) Tradescantia-micronucleus bioassay for detection of carcinogens. Folia Histochem Cytobiol 39(Suppl 2):54–55
13. Ma TH, Cabrera GL, Owens E (2005) Genotoxic agents detected by plant bioassays. Rev Environ Health 20:1–13
14. Rodrigues GS, Ma TH, Pimentel D et al (1997) *Tradescantia* bioassays as monitoring systems for environmental mutagenesis: a review. CRC Crit Rev Plant Sci 16:325–359
15. Mišík M, Ma TH, Nersesyan A et al (2011) Micronucleus assays with *Tradescantia* pollen tetrads: an update. Mutagenesis 26:215–221
16. Klumpp A, Ansel W, Fomin A et al (2004) Influence of climatic conditions on the mutations in pollen mother cells of Tradescantia clone 4430 and implications for the Trad-MCN bioassay protocol. Hereditas 141:142–148
17. Hoagland DR, Arnon DI (1950) The water-culture method for growing plants without soil. Univ Calif Agric Exp Stn Circular 347:1–39
18. Tolbert PE, Shy CM, Allen JW (1991) Micronuclei and other nuclear anomalies in buccal smears: a field test in snuff users. Am J Epidemiol 134:840–850

Chapter 23

The Wing-Spot and the Comet Tests as Useful Assays Detecting Genotoxicity in Drosophila

Ricard Marcos and Erico R. Carmona

Abstract

In spite of its pioneer use in detecting mutational processes, *Drosophila* has yet an important role in studies aiming to detect and quantify the induction of DNA damage. Here we describe two assays, one detecting primary damage (the Comet assay) and the other detecting somatic mutation and recombination effects (wing-spot test). It is important to emphasize that somatic recombination is a key event in cancer and no assays exist to detect and quantify somatic recombination processes, other than the spot tests developed in *Drosophila*.

Key words *Drosophila*, Genotoxicity, Comet assay, Wing-spot test, DNA damage

1 Introduction

An important aspect in toxicity and genotoxicity studies is the selection of the assay system to be used. In this context, in vivo studies offer much more advantage, since in vitro approaches do not completely mimic what happens in the whole organism. To avoid the use of mammals, *Drosophila* appears as a good alternative organism to explore the health risk factors because over 60 % of human disease genes have fly homologs, indicating that the fly response to physiological insults is comparable to humans [1, 2]. This would reinforce the usefulness of the *Drosophila* model as a first-tier in vivo test for genotoxicity studies.

In genotoxicity studies it must be remembered that *Drosophila* was the first organism used to detect the DNA-damaging effect of ionizing radiation [3] and chemicals [4]. These pioneering studies used assays detecting DNA damage in germ cells; but, since these assays are long and tedious they have evolved to the use of somatic cells as a target. In addition to the technical advantages of using somatic cells, as scoring the effects on the exposed individuals without waiting for the two subsequent generations, somatic mutations are directly linked with cancer processes, which have a relevant role on human health.

Thus, at present, somatic assays are the tools usually used in Drosophila for studying DNA damage and detecting agents or exposures suspected to be genotoxic. As in any in vivo test, in *Drosophila* too, absorption, distribution, and metabolism processes take place; it is thus important to emphasize that the metabolic activation system of *Drosophila* is similar to that of mammals [5]. Here we present the methodological aspects of two of the most used genotoxicity assays used at present: a somatic mutation and recombination test (SMART) that uses markers in the wing hair (*the wing-spot test*), and an assay to detect primary DNA damage, mainly strand breaks, in larval hemocyte cells (*the Comet test*).

The wing-spot test was designed to detect the induction of genotoxic damage in a rapid and inexpensive way. It must be recalled that during the embryonic stages of larva, the imaginal disc cells proliferate mitotically, and many genetic events such as point mutation, deletion, somatic recombination, and nondisjunction can thus be determined on the wings of adult fly [6]. If a genetic alteration occurs in one cell of the imaginal discs during the mitotic proliferation, it will form a clone of mutant cells expressing the phenotype regulated by specific genetic markers.

The Comet assay, also known as the single-cell gel (SCG) electrophoresis assay, is a method that allows for the qualitative and quantitative detection of DNA damage in individual eukaryotic cells [7]. This assay has been shown to be highly sensitive, allowing for the detection of low levels of several kinds of DNA damage, such as double- and single-strand DNA breaks, alkali-labile sites, and incomplete repair sites [8]. The alkaline version of the Comet assay has been recently adapted and applied to *Drosophila melanogaster* for the in vivo genotoxicity testing of chemical compounds and environmental pollutants [9, 10]. These studies, employing brain ganglia and mid gut cells from third instar larvae as target cells, demonstrated that *Drosophila* is a suitable model for in vivo genotoxicity evaluation with the Comet assay. Since in mammals, and especially in humans, in vivo studies commonly use lymphocytes from peripheral blood as the main target cells for measuring the induction of DNA damage, we have proposed to use hemocytes of the hemolymph [11]. Hemocytes have the same role as lymphocytes in the blood of mammals and offer several advantages as target cells for in vivo genotoxicity testing, since these cells are highly sensitive to genotoxic agents, and the sampling and processing methodology is quite simple [12].

2 Materials

2.1 Drosophila Strains

Two *D. melanogaster* strains are used for the wing-spot test: the multiple wing hair with genetic constitution *y; mwh j*, and the *flare-3* strain with genetic constitution *flr³/Ln (3LR) TM3, Bd^s*. Both strains carry mutations that visibly affect the phenotype of the trichomes (hair) of the wing cells [13]. The *mwh* mutation is recessive and viable in homozygosis, and it is located on chromosome 3. Its phenotypic expression is characterized by the appearance of three or more hairs in each cell, instead of one per cell, which is the normal phenotype. The *flr³* mutation is also located on the same arm of chromosome 3. It is recessive and produces lethality in homozygosis in the germ line, but not in somatic cells. The *flare* phenotype is quite variable, ranging from short, thick, and deformed to amorphous-like globular hair. More details of the other genetic marker and phenotype descriptions of the strains employed are given by Lindsley and Zimm [14].

The wild-type strain of *D. melanogaster* (Oregon R⁺), efficient for all types of DNA repair mechanisms, is used for Comet assay experiments (*see* **Note 1**).

All the strains are cultured in bottles with standard food medium at a temperature of 25 °C ± 1 °C and at a relative humidity of ~60 %.

2.2 Drosophila Culture Media

Two different media are used for the assay, a standard medium to maintain *Drosophila* strains and an instant medium to carry out the treatments. Although there are different culture media, according to different laboratories, we describe here the components and how to prepare a standard medium.

1. Drosophila standard medium: 170 g of corn flour, 120 g of yeast, 2 g of NaCl, and 10 g of agar. Weigh and mix the components, add 1,200 mL of cold water, mix well, and heat to boiling. Remove from heat, cool, and add to the mixture 4 mL of nipagin fungicide, dissolved in 10 % ethanol and 4 mL of propionic acid. Once prepared, the culture medium is distributed in culture glass bottles of 125 mL. Place approximately 25 mL of medium in each bottle, and allow it to stand for a few hours before use. Place a small piece of paper soaked with the insecticide tetradifon into each bottle in order to control mite populations while maintaining the humidity inside the culture bottles.

2. Instant Drosophila medium: Weigh 4.5 g of instant *Drosophila* medium Formula 4-24® (Carolina Biological Supply Co., USA) and hydrate it with 10 mL of distilled water (dH₂O) or the selected test solution (*see* **Note 2**). When the instant medium is well hydrated (in about 5 min), place Drosophila larvae in it to perform the desired treatments. In this medium, the larvae stay feeding until the adult stage.

2.3 Medium for the Wing-Spot Assay

1. Faure's solution: Mix 30 g Arabic gum and 50 g choral hydrate. Then, add 50 mL of dH$_2$O and dissolve well. Heating—without boiling—is recommended to facilitate dilution. Remove from heat and cool for 20 min. Then, add 20 mL of glycerol and mix well.

   Once prepared, transfer the solution into an amber glass bottle with dropper.

2.4 Preparation of Reagents for the Comet Assay

1. Phosphate-buffered saline (PBS), Ca^{2+}, Mg^{2+} free: Dissolve one tablet (5 g) of Gibco® PBS (Life Technologies, USA) in 500 mL of dH$_2$O. Store the solution at room temperature.

2. Positive control: 1 mM ethyl methanesulfonate (EMS).

3. Phenylthiourea: 0.07 % in cold PBS.

4. Sterilizing solution: 5 % sodium hypochlorite.

5. Lysis solution: 2.5 M NaCl, 100 mM EDTA, 10 mM Trizma base. Add 146.1 g of NaCl, 37.2 g of EDTA, and 1.2 g of Trizma base in 700 mL of dH$_2$O. After stirring the mixture add ~8 g of NaOH, and allow the mixture to dissolve (about 30 min). Adjust to pH 10 using HCl or NaOH, and quantity sufficient (q.s.) to 1 L with dH$_2$O.

6. Final lysis solution: 88 % of fresh lysis buffer (prepared above, 4°C), 10 % dH$_2$O, 1 % Triton X-100, 1 % *N*-lauroylsarcosinate (*see* **Note 3**).

7. Alkaline electrophoresis buffer: 300 mM NaOH, 1 mM EDTA. Prepare a stock solution of 10 N NaOH (200 g in 500 mL of dH$_2$O), and 200 mM EDTA (14.89 g in 200 mL of dH$_2$O). Store both solutions at room temperature and prepare a new one every 2 weeks.

   To prepare 1× buffer (working solution), add 30 mL of 10 N NaOH and 5 mL of 200 mM EDTA, q.s. to 1 L with dH$_2$O, and mix well (*see* **Note 4**).

8. Neutralization buffer: 0.4 M Tris. Add 48.5 g of Tris in 800 mL of dH$_2$O, adjust pH to 7.5 with >10 M HCl, and q.s. to 1 L with dH$_2$O. Store the solution at room temperature.

9. Staining solution: 1 µg/mL 4′,6-diamidine-2-phenylindole (DAPI). Prepare a stock of 200 µg/mL DAPI solution (add 10 mg of DAPI in 50 mL of dH$_2$O). Mix 1 mL of stock solution with 19 mL of dH$_2$O to get 10 µg/mL DAPI solution. Finally, for 1× working solution, mix 1 mL of the 10 µg/mL DAPI with 9 mL of dH$_2$O.

10. Normal-melting-point agarose: 1 % solution in dH$_2$O. Dissolve 500 mg of NMA in 50 mL of dH$_2$O.

11. Low-melting-point agarose: 0.75 % solution in PBS. Dissolve 375 mg of LMPA in 50 mL of PBS. Heat LMA in a microwave and avoid boiling.

12. NMA-precoated slide: Heat the NMA until it is diluted and avoid boiling. Immerse slides gently in hot NMA. Then, drain and clean the back of the slide with tissue paper and dry at room temperature, avoiding high-humidity conditions (*see* **Note 5**).

3 Methods

3.1 Obtaining Trans-heterozygous Larvae for Treatment for the Wing-Spot Assay

1. To get enough number of individuals, the number of individuals (culture bottles) in the *mwh* and *flr³* strains must be increased.
2. Obtain a sufficient number of virgin females from the *flr³* strain (~300 individuals) and mate with males of the *mwh* strain (*see* **Notes 6** and **7**).
3. Collect eggs from this cross during 8-h periods in culture glass bottles containing standard medium (*see* **Note 8**).
4. After 3 days, collect the 72 ± 4-h larvae (third instar), washing the culture bottles softly with tap water. Then, filter the larvae through a sieve of 500 and 400 μm to separate the larvae from the standard culture media.
5. Place the resulting 3-day-old larvae in disposable plastic vials containing 4.5 g of lyophilized instant *Drosophila* medium hydrated with 10 mL of the test solution (or water or solvent as controls). The treatment vials are covered with a cotton plug and kept in an incubator at 25 °C and 60 % humidity until emergence of adult individuals (*see* **Notes 9** and **10**).

3.2 Preparation and Mounting of Drosophila Wing Slides for the Wing-Spot Test

1. Emerging adults are collected (*see* **Note 11**).
2. If wings are not prepared immediately after collection, store the emerged adult individuals in plastic vials with 70 % ethanol at cold temperature (4 °C) until their use.
3. Before removing the wings, wash the flies preserved in ethanol with dH$_2$O in a watch glass. The new adults collected are also washed with water to remove possible rests of culture medium.
4. Place a fly on a well slide in which a drop of Faure's solution has been previously deposited. Proceed to the extraction of its wings with the help of fine forceps under a binocular stereomicroscope.
5. Once the two wings are separated from the body of the fly, take them carefully with fine tweezers, and place them in pairs and aligned on a clean slide.
6. Complete a slide with ~48 wings (24 individuals), in an equal proportion between males and females. Allow the preparation to dry for about 24 h in a dust-free place (e.g., inside a Petri dish).

7. Since trans-heterozygous and balanced heterozygous individuals are obtained in the crosses, both sets of individuals must be used (normal and sawed wings).

8. To make the wing preparation permanent, place a few drops of Faure's solution in the center of the slide and then cover with a coverslip of 24×60 mm, making a light pressure on it for an even distribution of the solution. Finally, place a weight (between 250 and 500 g) on the slide preparation to spread the wings and prevent formation of air bubbles. Wait for about 24 h to dry and carefully remove the weights.

3.3 Microscopic Analysis of the Wings for the Wing-Spot Test

1. The wings are scored at 400× magnification for the presence of mutant spots (see **Note 12**).

2. In each series between 40 and 80 wings are scored (20–40 individuals).

3. Record mutant spots according to their position in each area of the wing (i.e., A, B, C, C′, D, D′, and E, see Fig. 1), category, and size.

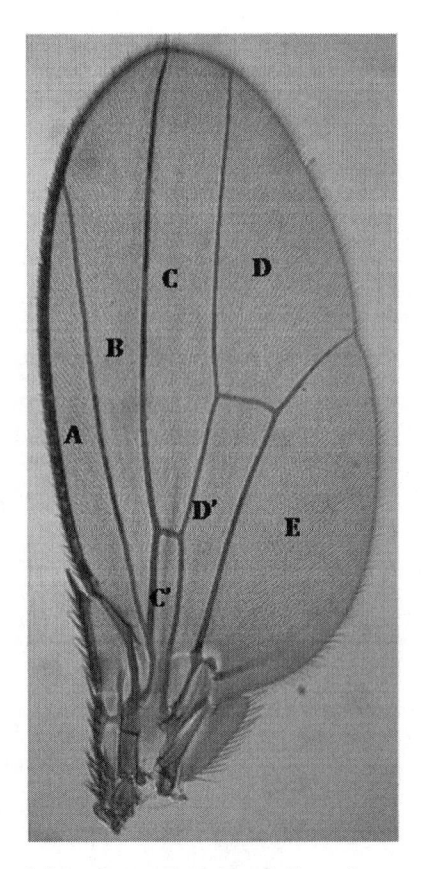

Fig. 1 Dorsal view of *D. melanogaster* wing. **A–E** are the areas analyzed on the surface of the wings

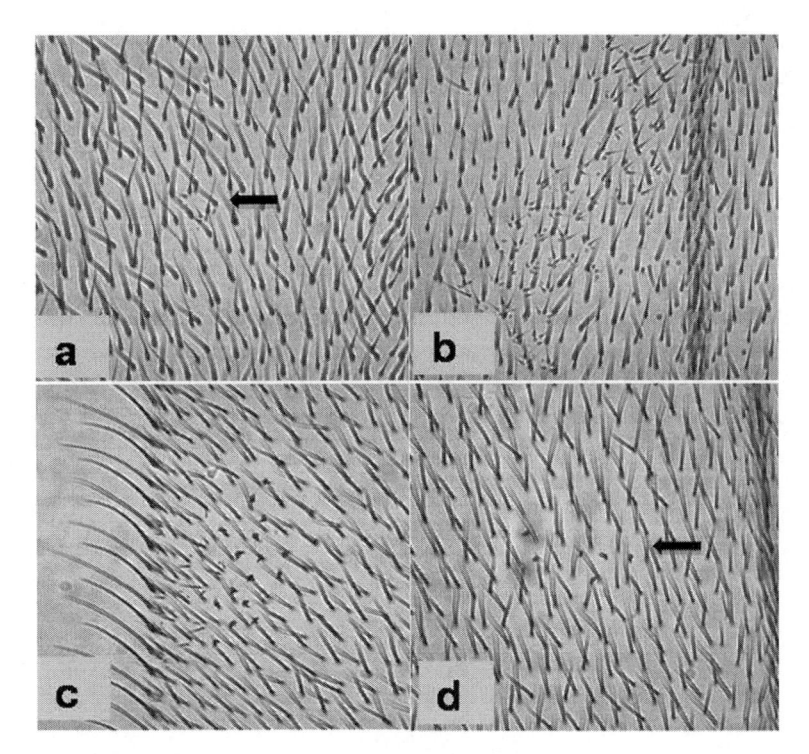

Fig. 2 Trichomes or hair on the wing surface of *D. melanogaster* (×400 magnification). (**a**) Small single *mwh* spot (*arrow* showing two cells affected); (**b**) large single *mwh* spot; (**c**) twin spot; (**d**) single *flr³* spot (*arrow* showing six cells affected). Mutant clones induced by 6 Gy gamma radiation

4. Classify the mutant spots according to the observed categories (*see* Fig. 2):
   (a) Small single *mwh* spots (1–2 cells)
   (b) Large single *mwh* spots (3 or more cells)
   (c) Twin (adjacent *mwh* and *flare*) spots
   (d) Single *flr³* spots

3.4 Data Analysis for the Wing-Spot Test

1. To evaluate differences in the frequency of each type of mutant spot between the treatments and the negative controls, the conditional binomial test of Kastenbaum and Bowman [15] is used, with a significance level of $\alpha = \beta = 0.05$.

2. The multiple-decision procedure described by Frei and Würgler [16] is used to determine the overall response of an agent as positive, weakly positive, negative, or inconclusive.

3. The frequency of clone formation is calculated, without size correction, by dividing the number of *mwh* clones per wings by 24,400, which is the approximate number of cells analyzed per wing [17] (*see* **Note 13**).

3.5 Larvae Treatment for the Comet Assay

1. For treatment, 72 ± 2-h-old larvae (third instar) are placed in plastic vials containing 4.5 g of instant *Drosophila* medium prepared with different test concentration or doses.

2. Negative control larvae receive instant *Drosophila* medium hydrated with distilled water (or solvent), while positive control larvae usually receive 1 mM EMS.

3.6 Hemocyte Collection for the Comet Assay

1. Third-instar larvae of 96 ± 2 h are extracted from the culture medium, washed, sterilized with ~5 % sodium hypochlorite, and dried with filter paper.

2. To collect the hemolymph and circulating hemocytes, the cuticle of each larva is disrupted using two fine forceps, avoiding damage to internal organs. The procedure is performed in a ~40 μL drop of cold PBS containing ~0.07 % of phenylthiourea (PTU) deposited in a well (6 mm ø) of a teflon printed diagnostic slide (Spi Supplies, USA).

3. After collecting the hemolymph from 10 larvae, the drop of PBS/hemocytes is removed from the diagnostic slide with a micropipette and placed within a microcentrifuge tube (1.5 mL). Complete the procedure for up to 40–60 larvae per treatment.

4. The tubes obtained from each treatment are centrifuged at $300 \times g$ for 10 min at 4 °C.

5. Finally, the supernatant is removed and the cell pellet is resuspended in 20 μL of cold PBS containing ~0.07 % PTU.

6. Cell samples from each treatment are carefully resuspended in 75 μL of 0.75 % LMA at 37 °C and mixed by pipetting.

7. Place two drops, each with a volume of 20 μL of the mixture, and place on the slide precoated with 1 % NMA (make one replicate per treatment).

8. Cover each drop with a 22 × 22 mm coverslip. Store at 4 °C for 10 min.

9. Immediately after agarose solidification, the coverslips are removed, and the slides are immersed in cold, freshly prepared final lysis solution for 2 h at 4 °C in a dark chamber.

3.7 Microgel Electrophoresis for the Comet Assay

1. The slides are removed from the lysis solution and transferred into an electrophoresis chamber.

2. Fill the electrophoresis chamber with fresh and cold electrophoresis buffer until the microgel slides are covered.

3. Keep the slides in this alkaline buffer for 25 min at 4 °C to allow DNA unwinding and expression of alkali-labile sites.

4. Turn the power supply to 25 V (0.7 V/cm) and adjust the current to 300 mA by adding or removing alkaline buffer. Electrophoresis is performed for 20 min at 4 °C.

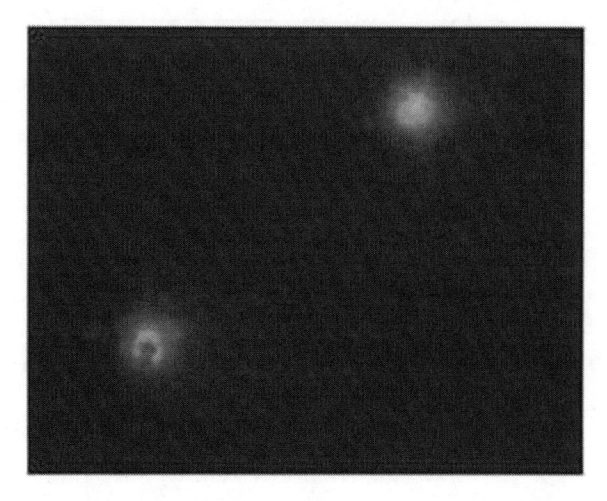

Fig. 3 Comet images of *Drosophila melanogaster* hemocytes (×400 magnification). Genetic damage induced by 1 mM potassium dichromate

5. Finally, the slides are neutralized with 0.4 M Tris (pH 7.5) in two washes of 5 min each.

6. The slides are stained with 20 μL of 1 μg/mL DAPI, per gel. The comet images are analyzed at 400× magnification (Fig. 3) with a Komet 5.5 image analysis system (Kinetic Imaging Ltd., UK) fitted with a fluorescence microscope equipped with a 480–550 wide-band excitation filter and a 590 nm barrier filter (*see* **Notes 14** and **15**).

7. One hundred cells are randomly analyzed (50 cells per two replicates) per treatment. The percentage of DNA in tail (% DNA in tail) is used as a measure of genetic damage. This parameter is widely used and recommended for the analysis of Comet assay results [18]. The % DNA in tail is calculated with the Komet software version 5.5 or 6.0.

3.8 Data Analysis for the Comet Assay

1. A generalized linear model (GLM) is commonly used to analyze differences in the parameter of % DNA in tail (response variable) versus different treatments with the genotoxic agent selected (categorical explanatory factor) using a binomial error structure. The GLM is analogous to traditional ANOVA, but it allows the use of nonparametric and heteroscedastic data, which is the common case of data obtained from Comet assay in *Drosophila* hemocytes.

2. Before analysis with the GLM, the homogeneity of variance and normality assumption of data is tested with the Bartlett and Kolmogorov–Smirnov test, respectively.

3. Finally, for statistical analysis, the statistical package Statgraphic plus version 5.1 is used (Statistical Graphics Corporation, 2001, USA, software package).

4 Notes

1. Any kind of Drosophila strain can be used to carry out this assay, although the background level of DNA damage can depend on the selected strain.

2. This medium ensures a good and uniform distribution of the test compound in the whole volume of medium.

3. Dimethyl sulfoxide (DMSO) is excluded from the lysis solution, because it has been considered unnecessary for *Drosophila* tissues and at low concentrations is cytotoxic for this organism [9, 10].

4. Prepare fresh and cold buffer (4 °C) before each run of electrophoresis and ensure that the pH is >13. The total volume used depends on the capacity of the electrophoresis chamber.

5. Slides must be labelled before storing.

6. To be sure that females are virgin they must be collected no more than 6 h after the pupa hatching.

7. Although both reciprocal crosses can be used, we propose to mate *flr³* virgin females with *mwh* males due to the higher fertility of the *flr³* females.

8. It is better to harvest eggs 2 days after making the crossing, because the egg production is more abundant.

9. Test solutions should be prepared just before use.

10. Check that the temperature does not exceed 25 °C during larvae incubation, which can increase the frequency of spontaneous mutations [19].

11. From the cross, two different types of offspring are obtained: trans-heterozygous individuals (*mwh* +/+ *flr*) and balanced heterozygous (*mwh* +/+ *Bd^S*). These last flies are easily visualized because they have the border of the wing sawed (*Bd^S* Beadle serrate).

12. Since there are two sets of slides (with normal and serrate wings), first score the normal wings. This gives us information whether the treatment has increased the frequency of mutant clones or not. If results are positive, score the serrate set of wings. This will give information about the mechanisms producing such changes. Individuals with the *mwh* +/+ *Bd^S* genetic constitution do not recombine due to the presence of the TM3 inversion. Thus, the difference observed between both sets of flies is due to the role of somatic recombination.

13. Transheterozygote and balanced heterozygous wings data must be analyzed independently. After that, the difference between values indicates the frequency of clones induced by recombination.

14. Other software can be equally used.
15. In the absence of specific software, manual scoring can be used [20] by categorizing the figures according to the level of damage (i.e., from 1 to 5).

Acknowledgements

ER Carmona thanks the support given by FONDECYT (Project No. 11110181) and the Dirección General de Investigación y Postgrado, Universidad Católica de Temuco (Projects No. CD2010 DGIPUCT-01 and 0804 MECESUP UCT).

References

1. Koh K, Evans JM, Hendricks JC et al (2006) A *Drosophila* model for age-associated changes in sleep: wake cycles. Proc Natl Acad Sci USA 103:13843–13847
2. Marsh JL, Thompson LM (2006) *Drosophila* in the study of neurodegenerative disease. Neuron 52:169–178
3. Müller HJ (1927) Artificial transmutation of the gene. Science 143:581–583
4. Auerbach C, Robson JM (1942) Experiments on the action of mustard gas in Drosophila. Production of sterility and mutation. Report to Ministry of Supply, W. 3979
5. Zijlstra JA, Vogel EW, Breimer DD (1987) Pharmacological and toxicological aspects of mutagenicity research in *Drosophila melanogaster*. In: Hodgson E, Bend J, Philpot RM (eds) Reviews in biochemical toxicology, vol 8. Elsevier, North Holland, Amsterdam, pp 121–154
6. Würgler FE, Vogel EW (1986) In vivo mutagenicity testing using somatic cells of Drosophila melanogaster. In: De Serres FJ (ed) Chemical mutagens. Principles and methods for their detection, vol 10. Plenum, New York, pp 1–72
7. Tice RR, Agurell E, Anderson D et al (2000) Single cell gel/Comet assay: guidelines for *in vitro* and *in vivo* genetic toxicology testing. Environ Mol Mutagen 35:206–221
8. Speit G, Hartmann A (2006) The Comet assay: a sensitive genotoxicity test for the detection of DNA damage and repair. Methods Mol Biol 314:275–286
9. Mukhopadhyay I, Chowdhuri DK, Bajpayee M et al (2004) Evaluation of *in vivo* genotoxicity of cypermethrin in *Drosophila melanogaster* using the alkaline Comet assay. Mutagenesis 19:85–90
10. Siddique HR, Chowdhuri DK, Saxena DK et al (2005) Validation of *Drosophila melanogaster* as an *in vivo* model for genotoxicity assessment using modified alkaline Comet assay. Mutagenesis 20:285–290
11. Carmona ER, Guecheva T, Creus A et al (2011) Proposal of an *in vivo* Comet assay using haemocytes of *Drosophila melanogaster*. Environ Mol Mutagen 52:165–169
12. Rigonato J, Mantovani MS, Jordão BQ (2005) Comet assay comparison of different *Corbicula fluminea* (Mollusca) tissues for the detection of genotoxicity. Genet Mol Biol 28:464–468
13. Graf U, Würgler FE, Katz AJ et al (1984) Somatic mutation and recombination test in *Drosophila melanogaster*. Environ Mutagen 6:153–188
14. Lindsley DL, Zimm GG (1992) The genome of *Drosophila melanogaster*. Academic, San Diego, CA
15. Kastenbaum MA, Bowman KO (1970) Tables for determining the statistical significance of mutation frequencies. Mutat Res 9:527–549
16. Frei H, Würgler FE (1988) Statistical methods to decide whether mutagenicity test data from *Drosophila* assays indicate a positive, negative, or inconclusive result. Mutat Res 203:297–308
17. Alonso-Moraga A, Graf U (1989) Genotoxicity testing of antiparasitic nitrofurans in the *Drosophila* wing somatic mutation and recombination test. Mutagenesis 4:105–110
18. Lovell DP, Omori T (2008) Statistical issues in the use of the comet assay. Mutagenesis 23:171–182
19. Rizki M, Kossatz E, Velázquez A et al (2006) Metabolism of arsenic in *Drosophila melanogaster* and the genotoxicity of dimethylarsinic acid in the *Drosophila* wing spot test. Environ Mol Mutagen 47:162–168
20. Pitarque M, Creus A, Marcos R et al (1999) Examination of various biomarkers measuring genotoxic endpoints from Barcelona airport personnel. Mutat Res 440:195–204

Part V

Guidelines

Chapter 24

Genotoxicity Guidelines Recommended by International Conference of Harmonization (ICH)

Gireesh H. Kamath and K.S. Rao

Abstract

Genotoxicity tests are designed to detect the genetic damage by various mechanisms. Several guidelines have provided various tests to be conducted for testing the genotoxicity and each of the regulatory agencies around the world have developed their own requirements for mutagenicity, without realizing that the products developed or registered in one country are also going to be registered and marketed around the world. The ICH guideline of genotoxicity helps to optimize the standard battery for genetic toxicology and to provide guidance on interpretation of results. These suggested standard set of tests does not imply that other genotoxicity tests are inadequate or inappropriate, but they help in improving risk characterization for carcinogenic effects that have their basis in changes in the genetic material.

Key words International Committee of Harmonization, Mutagenicity, Ames, Chromosomal aberration, Micronucleus

1 Introduction

Genotoxicity tests can be defined as in vitro and in vivo tests designed to detect agents/compounds that induce genetic damage by various mechanisms on animals particularly on human beings. Some of these agents can have effects on the deoxyribonucleic acid (DNA) at the gene level (point mutations) or at the chromosomal (clastogenicity) level [1–3]. Such changes are sometimes associated with tumorigenesis. Some genotoxic agents are responsible for initiation of abnormal and uncontrolled growth of cells, the neoplasia. A genotoxic agent may be toxic to nucleic acid or chromosomes, but may not pose specific health hazard. However, it is universally recognized that genetic change is a critical step in the conversion of a normal cell to a transformed cell. Extensive reviews have shown that many compounds that are mutagenic in the bacterial reverse mutation (Ames) test are rodent carcinogens [1, 2]. As a result of this nominal association between mutagenicity and carcinogenesis, newly discovered products (pharmaceuticals, pesticides, food additives, and other chemicals) and products which need reregistration need a thorough investigation of safety, including mutagenicity.

This resulted in looking for mutagenicity tests that are quicker, of lower cost, and highly sensitive. Over the last three to four decades, over two dozen mutagenicity tests have been developed with varying degrees of usefulness and extrapolation to humans. It soon became apparent to scientists and regulatory agencies that there is no single mutagenicity test which can detect all kinds of potential human mutagens with 100 % accuracy or prediction [1–4]. The latter is very true, now than ever, with the knowledge that mutagenesis itself is multifactorial.

Most regulatory agencies, including the US FDA, EPA, and Drug Controller General of India (DCGI—India), require a series of toxicological tests, including mutagenicity testing (3, 5). Several guidelines like ICH, OECD, the US EPA, and the US FDA Red Book have provided various tests to be conducted for testing the genotoxicity. These provide different protocols for the testing of natural and man-made agents for genotoxicity. Each of the regulatory agencies around the world has developed its own requirements for mutagenicity without realizing that the products developed or registered in one country are also going to be registered and marketed around the world. With this realization, there was a need to develop a harmonized guidance for mutagenicity testing, among others.

For most countries, whether or not they had initiated product registration controls earlier, the 1960s and 1970s saw a rapid increase in laws, regulations, and guidelines for reporting and evaluating the data on safety, quality, and efficacy of new medicinal products [3]. The industry, at the time, was becoming more international and seeking new global markets. However, the divergence in technical requirements from country to country was such that the industry found it necessary to duplicate many time-consuming and expensive test procedures, in order to market new products, internationally.

The urgent need to rationalize and harmonize regulation was impelled by concerns over rising costs of health care, escalation of the cost of R&D, and the need to meet the public expectation that there should be a minimum of delay in making safe and efficacious new treatments available to patients in need.

1.1 Initiation of ICH

Harmonization of regulatory requirements was pioneered by the European Community (EC), in the 1980s, as the EC (now the European Union) moved towards the development of a single market for pharmaceuticals. The success achieved in Europe demonstrated that harmonization was feasible. At the same time, there were bilateral discussions between Europe, Japan, and the USA on possibilities for harmonization. It was, however, at the WHO Conference of Drug Regulatory Authorities (ICDRA), in Paris, in

1989, that specific plans for action began to materialize. Soon afterwards, the authorities approached International Federation of Pharmaceutical Manufacturers & Associations (IFPMA) to discuss a joint regulatory-industry initiative on international harmonization, and ICH was conceived [4, 5].

The birth of the ICH took place at a meeting in April 1990, hosted by European Federation of Pharmaceutical Industries and Associations (EFPIA) in Brussels. Representatives of the regulatory agencies and industry associations of Europe, Japan, and the USA met, primarily, to plan an International Conference but the meeting also discussed the wider implications and terms of reference of ICH.

At the first ICH Steering Committee (SC) meeting of ICH, the Terms of Reference were agreed and it was decided that the topics selected for harmonization would be divided into safety, quality, and efficacy to reflect the three criteria which are the basis for approving and authorizing new medicinal products. Since ICH's inception in 1990, the ICH process has gradually evolved. ICH's first decade saw significant progress in the development of Tripartite ICH Guidelines on Safety, Quality, and Efficacy topics. Genotoxicity was established as an ICH topic at the first ICH conference which took place in November1991 in Brussels [4]. Mutagenicity guidelines later were developed as a part of the ICH process which is the subject of this chapter.

The ICH guidelines of genotoxicity help to optimize the standard genetic toxicology battery for prediction of potential human risks and to provide guidance on interpretation of results, with the ultimate goal of improving risk characterization for carcinogenic effects that have their basis in changes in the genetic material [1–3, 5]. The revised guidance describes internationally agreed upon standards for follow-up testing and interpretation of positive results in vitro and in vivo in the standard genetic toxicology battery, including assessment of nonrelevant findings. This guidance is intended to apply only to products being developed as human pharmaceuticals; however conceptually, these guidelines apply equally well for all the products under investigation. This chapter makes an attempt to describe in detail each of the mutagenicity tests that are recommended by the ICH guidance. Testing guidance on genotoxicity and data interpretation for pharmaceuticals intended for human use are described in ICH S2 (R1).

In order to minimize false positives and false negatives, ICH subscribes to the concept of a battery of mutagenicity tests because no single test is capable of detecting all genotoxic mechanisms relevant in tumorigenesis [1, 2].

The general features of a standard test battery are as follows:

1. Assessment of mutagenicity in a bacterial reverse gene mutation test: This test has been shown to detect relevant genetic

changes and the majority of genotoxic rodent and human carcinogens.

2. Genotoxicity should also be evaluated in mammalian cells in vitro and/or in vivo.

ICH has considered the following two options for the standard battery which are considered equally suitable [1, 2, 4]:

1.2 Option 1

1. An in vitro test for gene mutation in bacteria.
2. A cytogenetic test for chromosomal damage (the in vitro metaphase chromosome aberration test or in vitro micronucleus test), or an in vitro mouse lymphoma Tk gene mutation assay.
3. An in vivo test for genotoxicity, generally a test for chromosomal damage using rodent hematopoietic cells, either for micronuclei or for chromosomal aberrations in metaphase cells.

1.3 Option 2

1. An in vitro test for gene mutation in bacteria.
2. An in vivo assessment of genotoxicity with two different tissues, usually an assay for micronuclei using rodent hematopoietic cells.
3. A second in vivo assay. Typically this would be a DNA strand breakage assay in liver, unless otherwise justified.

There is more historical experience with Option 1, partly because it is based on earlier guidance of S2A and S2B. Nevertheless, the reasoning behind considering Options 1 and 2 are equally acceptable: when a positive result occurs in an in vitro mammalian cell assay, clearly negative results in two well-conducted in vivo assays, in appropriate tissues and with demonstrated adequate exposure, are considered sufficient evidence for lack of genotoxic potential in vivo. Thus, a test strategy in which two in vivo assays are conducted is the same strategy that would be used to follow up a positive result in the in vitro assays.

In vivo assessment of mutagenicity can be incorporated as a part of routine acute or sub-chronic exposure toxicity studies [6]. When more than one endpoint is evaluated in vivo, it is preferable that they are incorporated into a single study. Often sufficient information on the likely suitability of the doses for the repeat-dose toxicology study is available before the study begins and can be used to determine whether an acute or an integrated test is suitable.

For compounds that give negative results, the completion of either option of the standard test battery, performed and evaluated in accordance with current recommendations, will usually provide sufficient assurance of the absence of genotoxic activity and no additional tests are warranted [1–3]. Compounds that give positive results in the standard test battery might, depending on their therapeutic use, need to be tested more extensively [5]. There are several in vivo assays that can be used as the second part of the in vivo assessment under Option 2, some of which can be integrated into repeat-dose toxicology studies. The liver is typically the preferred tissue because of exposure and metabolizing capacity, but choice of in vivo tissue and assay should be based on factors such as any knowledge of the potential mechanism, of the metabolism in vivo, or of the exposed tissues thought to be relevant.

Information on numerical changes in chromosome can be derived from the mammalian cell assays in vitro and from the micronucleus assays in vitro or in vivo. Elements of the standard protocols that can indicate such potential are elevations in the mitotic index, polyploidy induction, and micronucleus evaluation. There is also experimental evidence that spindle poisons can be detected in mouse lymphoma assay (MLA). The preferred in vivo cytogenetic test under Option 2 is the micronucleus assay, not a chromosome aberration assay, to include more direct capability for detection of chromosome loss (potential for aneuploidy).

The suggested standard set of tests does not imply that other genotoxicity tests are generally considered inadequate or inappropriate [1, 2]. Additional tests can be used for further investigation of genotoxicity test results obtained in the standard battery. Alternative species, including non-rodents, can also be used if indicated, and if sufficiently validated. Under conditions in which one or more tests in the standard battery cannot be employed for technical reasons, alternative validated tests can serve as substitutes, provided sufficient scientific justification is given.

In the next sections, each of the tests is described in detail of use.

2 Materials

2.1 Common for All In Vitro Protocols

1. Dulbecco's Ca–Mg-free phosphate-buffered saline (PBS, pH 7.4 ± 0.2).
2. S9 homogenate (*see* **Note 2**).
3. Cofactor solution for S9 homogenate: 5 M Glucose-6-phosphate, 33 M potassium chloride, 4 M NADP, 8 M magnesium chloride.
4. S9 Mix: Containing appropriate volume of S9 homogenate in cofactor solution. During use keep this in an ice container.
5. Positive controls.
6. Test tubes.
7. Micropipettes—suitable volume.
8. Micropipette tips—suitable for the micropipettes.
9. Petri plates.
10. Biological safety cabinet (BSC).

2.2 Bacterial Reverse Mutation Test [1, 2, 7–9]

1. Bacterial tester strains (*see* **Note 3**).

2. Vogel-Bonner Medium: 10 g of magnesium sulfate, 100 g of citric acid monohydrate, 500 g of potassium phosphate, 175 g of ammonium sodium phosphate. Using the warm water, solubilize the above salts one after the other, in the order given, and finally make up the volume to 1,000 mL. Sterilize by autoclaving (*see* **Note 1**).

3. 40 % Glucose solution: Dissolve 400 g of glucose/dextrose in 1,000 mL of warm water. Sterilize by autoclaving (*see* **Note 1**).

4. Vogel-Bonner agar (VB agar): Dissolve 15–18 g of agar in 930 mL of water (the amount of water will change according to the additions in the agar). Sterilize by autoclaving (*see* **Note 1**). After sterilization add 20 mL of sterile Vogel-Bonner medium and 50 mL of sterile 40 % glucose solution and pour into plates (*see* **Note 4**).

5. 0.5 % w/v L-Histidine solution: Dissolve 100 mg of L-histidine in 20 mL of water. Sterilize by autoclaving (*see* **Note 1**).

6. 0.1 % Biotin solution: Dissolve 20 mg of biotin in 20 mL of water. Sterilize by autoclaving (*see* **Note 1**).

7. 8 mg/mL Ampicillin solution: Dissolve 32 mg of ampicillin in 4 mL of warm water and sterilize through 0.2 μm membrane filter.

8. 8 mg/mL Tetracycline solution: Dissolve 32 mg of tetracycline in 4 mL of 0.02 N HCl. Sterilize through 0.2 μm membrane filter.

9. 2 mg/mL L-tryptophan: Dissolve 50 mg of L-tryptophan, in 25 mL of water. Sterilize through 0.2 μm membrane filter.

10. Vogel-Bonner agar with biotin (B agar): Add 6 mL of 0.1 % sterile biotin solution to VB agar (from **item 4**) and pour into plates.

11. VB agar with histidine–biotin (HB agar): Add 10 mL of sterile 0.5 % L-histidine and 6 mL of sterile 0.1 % biotin solution to VB agar (from **item 4**) and pour into plates.

12. VB agar with histidine, biotin, tetracycline (HBT agar): Add 0.25 mL of sterile 8 mg/mL tetracycline solution to HB agar (from **item 11**) and pour into plates.

13. VB agar with histidine, biotin, ampicillin (HBA agar): Add 3 mL of sterile 8 mg/mL ampicillin solution to HB agar (from **item 11**) and pour into plates.

14. Vogel-Bonner agar with tryptophan, ampicillin (TA agar): Add 10 mL of sterile 2 mg/mL of tryptophan and 3 mL of sterile 8 mg/mL ampicillin solution to VB agar (from **item 4**) and pour into plates.

15. Vogel-Bonner agar with tryptophan (T agar): Add 10 mL of sterile 2 mg/mL tryptophan to VB agar (from **item 4**) and pour into plates.

16. Oxoid nutrient broth No.2: Dissolve required quantity of the Oxoid nutrient broth No.2 as per the manufacturer's instruction on the label of the container in water. Sterilize by autoclaving (*see* **Note 1**).
17. Top agar: Dissolve 6–7 g of bacto agar, 5 g of NaCl in 1,000 mL of water (the amount of water will change according to the additions to top agar). Sterilize by autoclaving (*see* **Note 1**).
18. Top agar with L-tryptophan solution: Add 2.5 mL of sterile 2 mg/mL L-tryptophan solution to top agar just before use.
19. 0.5 mM Histidine–biotin solution: Weigh required amount of histidine and biotin, as per molecular weight, dissolve in PBS, and maintain pH to 7.4 ± 0.2.
20. Top agar with histidine–biotin solution: Add 100 mL of 0.5 mM histidine–biotin solution to top agar. Sterilize by autoclaving (*see* **Note 1**). Otherwise, sterilize bacto agar, NaCl, and water separately and add 100 mL of sterile 0.5 mM histidine–biotin solution before use.
21. Petri plates.
22. Incubator.
23. Water bath.

2.3 In Vitro Chromosomal Aberration Test [1, 2, 10–13]

1. Test system—Human peripheral blood lymphocytes (HPBL)/CHO cell lines.
2. Basal medium (*see* **Note 5**): Dissolve the required quantity of basal medium in water and sterilize through 0.2 μm membrane filter.
3. Complete medium: Basal medium amended with sterile 10^5 I.U./L penicillin, sterile 100 mg/L streptomycin, sterile 1.5 mg/L amphotericin, and 5 or 10 % fetal bovine serum.
4. Cryoprotective medium: Basal medium:fetal bovine serum:sterile DMSO in the ratio of 4:4:2.
5. 2 % Phytohaemagglutinin (PHA).
6. 10 μg/mL colchicine solution.
7. 0.56 % potassium chloride (*see* **Note 6**).
8. 1:3 ratio of acetic acid:methanol fixative (*see* **Note 7**).
9. Slides (*see* **Note 8**).
10. DPX mountant.
11. 5 % Giemsa stain (prepared in distilled water).
12. Tissue culture flasks.
13. Hemocytometer.
14. CO_2 Incubator.

2.4 In Vitro Cell Gene Mutation Test [1, 2, 14, 15]

1. Test system—Mouse lymphoma L5178Y *Tk* (thymidine kinase).
2. Basal medium—RPMI medium: Dissolve the required quantity of RPMI medium in water and sterilize through 0.2 μm membrane filter.
3. Complete medium: Same as in **item 3** of Subheading 2.3.
4. 6-Thioguanine.
5. Hemocytometer.
6. CO_2 Incubator.
7. Tissue culture flasks.

2.5 In Vitro Micronucleus Test [1, 2, 16, 17]

1. Test system—HPBL/Chinese hamster ovary cell lines.
2. Basal media (*see* **Note 5**): Same as in **item 2** of Subheading 2.3.
3. Complete medium: Same as in **item 3** of Subheading 2.3.
4. 2 % PHA.
5. 0.56 % potassium chloride (*see* **Note 6**).
6. 1:3 ratio of acetic acid:methanol fixative (*see* **Note 7**).
7. Slides (*see* **Note 8**).
8. Cytochalasin B.
9. Hemocytometer.
10. CO_2 Incubator.
11. Tissue culture flasks.
12. DPX mountant.
13. 5 % Giemsa stain prepared in distilled water.

2.6 Common In Vivo Protocols [1, 2, 18–22]

1. Animal room.
2. Animals.
3. Slides (*see* **Note 8**).
4. Cages with cage grills.
5. Cage racks.
6. Water.
7. Feed.
8. Bedding material.
9. Positive controls.
10. Dosing needle.
11. Serum.
12. DPX mountant.
13. 5 % Giemsa stain prepared in distilled water.

In addition to above, for in vivo chromosomal aberration test, the following materials are also required:

14. 0.5 mg/mL Colchicine solution.
15. 0.56 % potassium chloride (*see* **Note 6**).
16. 1:3 ratio of acetic acid:methanol fixative (*see* **Note 7**).

3 Methods

3.1 Bacterial Reverse Mutation Test

1. *Maintenance and genotypic characterization of tester strains*: The tester strains are procured from known repository. Thaw this original culture vial to room temperature. Transfer the contents to Oxoid nutrient broth. Incubate the culture at 37 °C. Grow the culture for up to 16–20 h. Add 9 % v/v DMSO. Distribute the growth in cryovials. Store in deep freezer or in liquid nitrogen (*see* **Note 9**). Shelf life of culture stored in deep freezer would be 2–3 years and that stored in liquid nitrogen would be 9–10 years. This is considered as mother culture stock.

   Thaw a mother culture stock vial in warm water at 37 °C. Place a drop on the appropriate minimal glucose agar plates (*see* **Note 10**). From the drop, streak across the surface of the plates and incubate these plates at 37 °C for 24–72 h. Pick a well-isolated colony and transfer to Oxoid nutrient broth. Incubate the culture at 37 °C. Grow the culture for up to 16–20 h. Add 9 % v/v DMSO at the rate. Distribute the growth in cryovials. Store in deep freezer or in liquid nitrogen (*see* **Note 9**). This is considered as frozen culture stock. Prepare fresh frozen culture stock at least once in 3 years from the mother culture stock.

   The master plates for routine laboratory use are prepared from the frozen culture stock vial. Thaw the frozen culture stock vial in warm water at 37 °C. Place a drop on the appropriate minimal glucose agar plates (*see* **Note 10**). Incubate these plates at 37 °C for 24–72 h. Pick a well-isolated colony and transfer to 0.2–0.4 mL PBS. For each culture, using separate swab, streak on to respective appropriate minimal glucose agar plates (*see* **Note 10**). Incubate these plates at 37 °C for 24–48 h. These are considered as master plates which can be stored at 2–8 °C and used for routine use, for up to a maximum of 2 months (*see* **Note 11**).

   For each batch of master plates test the genotypic characterization, before they are used in the actual experiments, as follows:

   (a) *Histidine/tryptophan requirement*: Streak each strain of *Salmonella typhimurium* on HB plates as well as on B plates. Streak the *Escherichia coli* strains on T plates as well as VB plates. Incubate the plates overnight to 2 days. For the *Salmonella typhimurium* strains, there should be growth on the plates containing histidine and biotin and no growth on plates containing only biotin. For the *Escherichia coli* strains, there should be growth on the plates containing tryptophan and no growth on VB plates.

(b) *rfa Mutation*: Mix 100 μL each of the *Salmonella typhimurium* strains with 2 mL of the molten soft agar separately, spread over the nutrient agar plates, and allow to solidify. Place a sterile filter paper disc dipped in (1 mg/mL) aqueous crystal violet solution at the bacterial lawn and incubate the plates overnight to 2 days. There should be a clear zone of inhibition all around the filter paper disc indicating the presence of rfa mutation.

(c) *uvrB/uvrA Mutation*: Streak each of the tester strains on the nutrient agar plates in parallel streaks. Place a sheet of sterile cardboard/thick paper/aluminum foil over the plate, in such a manner that half the streak in the full line is covered. Expose the plates to a UV lamp for approximately 8–10 s, at approximately 30–40 cm distance. Incubate overnight to 2 days. The strains should show nil to reduced growth in the irradiated area. While in the nonirradiated area in the plate there should be growth.

(d) *R-factor*: Streak R-factor strains (TA 97a, TA 98, TA 100, and TA 102) and non-R-factor strains (TA 1535, TA 1537) in parallel in HBA plate. Similarly, streak R-factor strain, WP2 uvrA (pKM101), and non-R-factor strain, WP2 uvrA, on a TA plate. The R-factor strains become control for the non-R-factor strains and vice versa while comparing results. Incubate the plates overnight to 2 days. Only R-factor strains should grow on the respective plates.

(e) *Spontaneous mutants*: In addition to the characterization for genotypes each strain is tested for its number of spontaneous revertants. Mix 100 μL of each of the tester strains of the *Salmonella typhimurium* with 2 mL of soft agar containing histidine and biotin and tester strains of the *Escherichia coli* with 2 mL of soft agar containing tryptophan and spread over the VB agar plates. Incubate the plates for 48–72 h. Count the number of revertant colonies for each strain manually or using an automated counter.

Compare the results of genotypic characterization against the values provided in Table 1.

2. Procedure

(a) Maximum concentration of 5,000 μg/plate is acceptable for testing. Check the solubility of solid test item using vehicles

Table 1
Requirements and values for genotypic characterization

Tester strains	His./Tryp. requirement	rfa Mutation	uvr Mutation	R-factor	Spontaneous revertants
TA 97a	Histidine	Positive	uvrB	Present	60–180
TA 98	Histidine	Positive	uvrB	Present	3–46
TA 100	Histidine	Positive	uvrB	Present	45–200
TA 102	Histidine	Positive	uvrB	Present	150–320
TA 1535	Histidine	Positive	uvrB	Absent	3–25
TA 1537	Histidine	Positive	uvrB	Absent	4–18
WP2 uvrA	Tryptophan	NA	uvrA	Absent	5–31
WP2 uvrA (pKM101)	Tryptophan	NA	uvrA	Present	30–200

His./Tryp. histidine/tryptophan, *NA* not applicable

suitable for test system. Water, DMSO, dimethylformamide, acetone, methanol, and tetrahydrofuran are some of the vehicles compatible for bacterial test system as well as S9. Liquid test item may be added without dilution or diluted in one of these vehicles. Check the precipitation of test item in the soft agar as well as on the surface of VB agar.

(b) Add a loopful of tester strain from the master plate into Oxoid nutrient broth, and incubate the culture at 37 °C for 16–18 h, to prepare the overnight culture.

(c) Conduct a cytotoxicity test using overnight culture with five to ten concentrations and a concurrent vehicle control. In a tube add 100 µL of test item concentration, 100 µL of overnight culture, and 500 µL of either PBS (for testing in the absence of metabolic activation) or S9 mix at a concentration of 5–30 % S9 v/v in the S9 mix (for testing in the presence of metabolic activation). To this, add 2 mL of soft agar (either amended with histidine–biotin in case of *Salmonella* strains or with tryptophan in case of *Escherichia coli* strains). Mix the contents and pour it onto VB agar plate.

(d) Incubate the plates at 37 °C for 48–72 h.

(e) Observe the bacterial background lawn and count the revertant colonies. Reduction either in the bacterial background lawn or in the number of revertant colonies or combination of the two implies cytotoxic nature of the test item.

(f) The maximum concentration to be tested in the main test is 5,000 µg/plate, and depends on solubility/precipitation and/or on cytotoxicity of the test item (*see* **Note 12**).

Table 2
Examples of positive controls for Ames test

Tester strain	Positive control chemical	
	Presence of S9	Absence of S9
TA 97a	2-Aminoanthracene	ICR 191
TA 98	or Benzo(*a*)pyrene	2-Nitrofluorene
TA 100		Sodium azide
TA 102		Mitomycin C
TA 1535		Sodium azide
TA 1537		9-Aminoacridine
Both strains of *E. coli*		4-Nitroquinoline 1-oxide

(g) In the main test, five concentrations separated by a factor of $\sqrt{10}$, along with concurrent positive and vehicle controls, are plated for each tester strains. These seven groups are tested both in the absence and in the presence of metabolic activation, in triplicate plating.

(h) Appropriate concentrations of different positive controls are used for each tester strains, during experimenting both in the absence and in the presence of metabolic activation.

(i) Some examples of positive controls are given in Table 2.

(j) Incubate the plates at 37 °C for 48–72 h.

(k) Observe the bacterial background lawn and count the revertant colonies.

(l) The demonstration of a concentration-related increase in revertant counts is a dose–response phenomenon. Increase of revertant counts by 2 times (2×) in TA 97, TA 98, TA 100, TA102, and WP2 uvrA pKM101 or 3 times (3×) in TA 1535, TA1537, and WP2 uvrA over the revertant counts of respective vehicle control plates is considered as a positive response.

(m) Clear positive or negative results may not be confirmed. Equivocal or weak positive results might indicate that it would be appropriate to repeat the test, possibly with a modified spacing of concentration levels (*see* **Note 13**).

3.2 In Vitro Chromosomal Aberration Test

1. Maintenance and characterization of cell lines

    (a) The CHO cell line is procured from known repository. Thaw original culture immediately using warm water and transfer the content into T-75 cm² culture flask. Add 10–15 mL of Ham's F12 medium with 10 % FBS. Incubate the flask at 37 °C and 5 % CO_2 under humidified air. After 24–72 h remove the flasks from the incubator, remove the existing medium, trypsinize the cell monolayer, and add required volume of cryoprotective medium. Distribute the growth in cryovials. Store in deep freezer or in liquid nitrogen (*see* **Note 9**). Shelf life of culture stored in deep freezer would be 2–3 years and that stored in liquid nitrogen would be 9–10 years. This is considered as mother culture stock. From the mother culture stock prepare the frozen culture stock and use the frozen culture stock for routine use.

For human lymphocytes, collect the whole blood by venipuncture, using heparinized syringe. Add 4–5 mL of blood to 100 mL of complete media amended with 2 % PHA. Grow the cells at 37 °C and 5 % CO_2 under humidified air. After around 44 h of growth these can be treated.

For each type of cell line and primary cell culture, check for their characteristics as follows:

(b) Microbial contamination check: For each batch of cryopreserved cells, transfer small volume of cell suspension from one of the vials into nutrient broth. Incubate for 48–72 h. Absence of growth in the tube implies that the cell suspension is contamination free (this test is not carried out in case of human lymphocytes).

(c) Mycoplasma contamination check: Test for the absence of mycoplasma using the PCR method or microscopic observation (this test is not carried out in case of human lymphocytes) at least once a year.

(d) Modal chromosome number: Transfer cells to a tissue culture flask. Add required volume of complete medium and incubate the flask. After approximately 18–21 h of incubation, remove the flask, add colchicine final concentration of 0.2 μg/mL of medium, and incubate the flask further for approximately 2–4 h. Remove the cells and suspend them in basic medium. Prepare the cells for chromosomal aberration slides. Count the number of chromosome in at least 100 metaphases. Model chromosome number is tested at least once a year (this test is not carried out in case of human lymphocytes).

(e) Determination of cell viability (cloning efficiency): Transfer 0.5 mL of a cell suspension diluted in complete medium, in a vial. Add one mL of 0.4 % trypan blue stain and mix thoroughly. Allow at room temperature for 2-5 minutes. Count the stained (dead) and unstained (viable) cells using hemocytometer. Calculate the percentage of viable cells at least once a year (this test is not carried out in case of human lymphocytes).

(f) Determination of cell doubling time: Inoculate approximately 10^5 cells per flask into ten T-25 cm^2 tissue culture flasks,

Table 3
Characterization of cell lines

Character tested	Criterion for selection of cell line
Microbial contamination	Absent
Mycoplasma contamination	Absent
Modal chromosome number	More than 95 % metaphases should be with the actual chromosome number
Cell viability	More than 90 %
Cell doubling time	Should be near to actual (CHO cells and human lymphocytes 12–13 h)

containing 5–7 mL of Ham's F12 medium with 10 % FBS and incubate. At approximately 11, 12, 13, 14, and 15 h of incubation count the number of cells in two flasks at each time. Calculate the mean of cell counts from two flasks and time required for doubling. Cell doubling time is determined at least once a year.

(g) Compare the results of cell line characterization as given in Table 3.

2. Procedure

   (a) Maximum concentration of 5,000 μg/mL of medium is acceptable for testing. Check the solubility of solid test item using vehicles suitable for test system. Water, DMSO, acetone, and ethanol are some of the vehicles compatible for cell lines as well as S9. Liquid test item may be added without dilution or diluted in one of these vehicles.

   (b) Add approximately 10^6 cells into the flask, and add required volume of Ham's F12 medium with 10 % FBS (for CHO cell line) or 4–5 mL of whole blood into every 100 mL RPMI with 10 % FBS amended with 2 % PHA (for HPBL). Incubate at 37 °C, 5 % CO_2, under humidified air for 24–25 h for CHO cell line or 44 h for HPBL to get the target cell flask.

   (c) Conduct the cytotoxicity test, by adding test item concentration to the target cell flask in two sets, with five to ten concentrations and a vehicle control. To one set of flasks, add S9 mix at a concentration of 1–10 % v/v S9 in the final test medium (for testing in the presence of metabolic activation). Add Ham's F12 medium with 10 % FBS or RPMI medium with 10 % FBS to all the two sets of flasks. The other set of flasks is treated for testing in the absence of metabolic activation.

   (d) Incubate at 37 °C, 5 % CO_2, under humidified air.

(e) After 3–6 h of treatment, remove the flasks from the incubator. Wash and continue incubation at 37 °C, 5 % CO_2, under humidified air.

(f) In case of CHO cells, 19–21 h after the start of treatment, remove all the flasks from the incubator, to check the cell integrity. Count the cells in each individual concentration.

(g) In case of HPBL, at approximately 18–19 h after the start of the treatment, add colchicine at 0.2 μg/mL of the medium. At approximately 20–21 h after the start of the treatment (equivalent to about 1.5 normal cell cycle length after the beginning of treatment), remove all the flasks from the incubator, check the cell integrity prepare the chromosome slides. Score approximately 2000 cells and calculate mitotic index.

(h) The maximum concentration to be tested is 5,000 μg/mL of medium, depending on the solubility/precipitation and/or on cytotoxicity of the test item (*see* **Note 12**).

(i) In the main test, three concentrations are selected by factor between 2 and $\sqrt{10}$, along with concurrent positive (*see* **Note 14**) and vehicle controls. The target cell flasks are treated in duplicate. These groups are tested both in the presence and in the absence of metabolic activation.

(j) Appropriate concentrations of different positive controls are used for each tester strains, during experiments both in the absence and presence of metabolic activation.

(k) Examples of positive controls are
- Ethyl methane sulfonate in the absence of metabolic activation
- Cyclophosphamide (monohydrate) in the presence of metabolic activation

(l) Incubate at 37 °C, 5 % CO_2, under humidified air.

(m) Terminate the treatment after 3–6 h. Wash and change the media and continue incubation at 37 °C, and 5 % CO_2, under humidified air.

(n) At approximately 18–19 h after the start of the treatment, add colchicine at 0.2 μg/mL of the medium.

(o) At approximately 20–21 h after the start of the treatment (equivalent to about 1.5 normal cell cycle length after the beginning of treatment), remove all the flasks from the incubator, check the cell integrity. The cells are then used to prepare the chromosome slides. Code all the slides before microscopic analysis to avoid possible bias.

(p) At least 200 metaphases are scored per group and tested for the presence of aberrations in chromosomes.

(q) One of the criteria for determining a positive result is concentration-related increase or a reproducible increase in the number of cells with chromosome aberrations. Biological relevance of the results should be considered first. Statistical methods may be used for evaluating the test results. Statistical significance should not be the only determining factor for a positive response. The historical control data is also a deciding factor in concluding the positive response.

(r) Clear positive or negative results may not be confirmed. Equivocal or weak positive results might indicate that it would be appropriate to repeat the test, possibly with a modified spacing of concentration levels (*see* **Note 15**).

3.3 In Vitro Cell Gene Mutation Test

The cell lines are maintained and characterized as per the procedures explained in the in vitro chromosomal aberration test (*see* Subheading 3.2).

1. Maximum concentration of 5,000 μg/mL of medium is acceptable for testing. Check the solubility of solid test item using vehicles suitable for test system. Water, DMSO, acetone, and ethanol are some of the vehicles compatible for cell lines and S9. Liquid test item may be added without dilution or diluted in one of these vehicles.

2. Add approximately 10^6 cells in the flask, add required volume of RPMI medium with 10 % FBS, and incubate at 37 °C, 5 % CO_2, under humidified air for 24–25 h to get the target cell flask.

3. Conduct a cytotoxicity test by adding test item concentration to the target cell flask with five to ten concentrations and a vehicle control. Add 100 μL of test item concentration and 500 μL of either PBS (for testing in the absence of metabolic activation) or S9 mix at a concentration of 1–10 % v/v S9 in the final test medium (for testing in the presence of metabolic activation).

4. Incubate at 37 °C, and 5 % CO_2, under humidified air.

5. After 3–6 h of incubation remove the flasks from the incubator. Remove the medium containing test item concentration. Wash the cell monolayer with PBS two to three times, add fresh RPMI medium with 10 % FBS and continue incubation at 37 °C, 5 % CO_2, under humidified air.

6. About 16–19 h after the start of treatment, remove all the flasks from the incubator, check the cell integrity, trypsinize the cell monolayer, and count the cells in each individual concentration.

7. Either loss of the cell integrity or reduction of cell counts or combination of both implies cytotoxic nature of the test item.

Table 4
Examples of positive controls for in vitro gene mutation test

Metabolic activation	LOCUS	Positive control chemical
Absence	HPRT	Ethylmethanesulfonate
	TK (small and large colonies)	Methylmethanesulfonate
	XPRT	Ethylmethanesulfonate
Presence	HPRT	3-Methylcholanthrene
	TK (small and large colonies)	Benzo(*a*)pyrene
	XPRT	Benzo(*a*)pyrene

8. The maximum concentration to be tested in the main test is 5,000 μg/mL of medium, depending on solubility/precipitation and/or on cytotoxicity of the test item (*see* **Note 12**).

9. In the main test, four concentrations selected by factor between 2 and $\sqrt{10}$, along with concurrent positive (*see* **Note 14**) and vehicle controls, are treated with the target cell flasks in duplicates. These six groups are tested both in the presence and in the absence of metabolic activation.

10. Appropriate concentrations of different positive controls are used for each tester strains, during experimenting with both in the absence and in the presence of metabolic activation.

11. Some examples of positive controls are given in Table 4.

12. Incubate at 37 °C, 5 % CO_2, in humidified air.

13. Wash the cells and incubate again to determine the survival and to allow for expression of the mutant phenotype. Measurement of cytotoxicity is usually initiated after the treatment period, by determining relative cloning efficiency (survival) or relative total growth of the cultures. The minimum time requirement to allow for near-optimal phenotypic expression of newly induced mutants is at least 2 days. Cells are grown in medium with and without selective agent(s) for determination of number of mutants and cloning efficiency, respectively. The measurement of viability (used to calculate mutant frequency) is initiated at the end of the expression time by plating in nonselective medium. If the test item is positive in the L5178Y $TK^{+/-}$ test, colony sizing should be performed on at least one of the test cultures (the highest positive concentration) and on the negative and positive controls. If the test substance is negative in the L5178Y $TK^{+/-}$ test, colony sizing should be performed on the negative and positive controls. In studies using TK6TK$^{+/-}$, colony sizing may also be performed.

14. A concentration-related or a reproducible increase in mutant frequency demonstrates positive result. Biological relevance of

the results should be considered first. Statistical methods may be used as an aid in evaluating the test results. Statistical significance should not be the only determining factor for a positive response.

15. Clear positive or negative results may not be confirmed. Equivocal or weak positive results might indicate that it would be appropriate to repeat the test, possibly with a modified spacing of concentration levels (*see* **Note 15**).

3.4 In Vitro Micronucleus Test

The cell lines are maintained and characterized as per the procedures explained in the in vitro chromosomal aberration test (*see* Subheading 3.2).

1. A maximum concentration of 5,000 μg/mL of medium is acceptable for testing. Check the solubility of solid test item using vehicles suitable for test system. Water, DMSO, acetone, and ethanol are some of the vehicles compatible for cell lines and S9. Liquid test item may be added without dilution or diluted in one of these vehicles.

2. Add approximately 10^6 cells into the flask, and add required volume of Ham's F12 medium with 10 % FBS (for CHO cell line) or 4–5 mL whole blood into every 100 mL of RPMI medium with 10 % FBS amended with 2 % PHA (for HPBL). Incubate at 37 °C, 5 % CO_2, under humidified air for 24–25 h for CHO cell line or 44 h for HPBL, to get the target cell flask.

3. Conduct the cytotoxicity test with four to eight concentrations and a vehicle control, by adding test item concentration to the target cell flask in three sets. Add S9 mix at a concentration of 1–10 % v/v S9 in the final test medium (for testing in presence of metabolic activation) for one set. Add Ham's F12 10 % FBS or RPMI medium with 10 % FBS to the two sets of flasks. The other set of flasks are treated for testing in the absence of metabolic activation.

4. Incubate at 37 °C, 5 % CO_2, under humidified air.

5. After 3–6 h of treatment, remove the flasks from the incubator. Wash, add fresh medium containing cytoB, and continue incubation at 37 °C, 5 % CO_2, under humidified air.

6. About 16–19 h after the start of treatment remove all the flasks from the incubator, check the cell integrity. Process the cells using hypotonic solution followed by fixative. Prepare the slides. Observe at least 200 cells to decide the cytotoxicity by counting the ratio of immature to matured cells.

7. The maximum concentration to be tested in the main test is 5,000 μg/mL of medium, depending on the solubility/precipitation and/or on cytotoxicity of the test item (*see* **Note 12**).

8. In the main test, three concentrations selected by factor between 2 and $\sqrt{10}$, along with concurrent positive (*see* **Note 14**)

and vehicle controls, are added to the target cell flasks in duplicates. These groups are tested both in the absence and in the presence of metabolic activation.

9. Appropriate concentrations of positive controls are used during experimenting in the absence (mitomycin C) and in the presence (cyclophosphamide monohydrate) of metabolic activation.

10. Incubate at 37 °C, 5 % CO_2, under humidified air.

11. Terminate the treatment after 3–6 h. Wash the cells, add fresh medium, and continue incubation at 37 °C, 5 % CO_2, under humidified air.

12. At approximately 21–24 h after the start of the treatment (equivalent to about 1.5–2 normal cell cycle length after the beginning of treatment), remove all the flasks from the incubator, check the cell integrity, and count the number of cells/mL to determine the cytotoxicity.

13. Using the cell preparations prepare slides and stain them. Code all the slides before microscopic analysis to avoid possible bias.

14. At least 2,000 binucleated cells per concentration are scored per group and tested for the presence of micronuclei.

15. Concentration-related increase or a reproducible increase in the number of micronuclei in the cells shows that the test item is mutagenic. Biological relevance of the results should be considered first. Statistical methods may be used for evaluating the test results. Statistical significance should not be the only determining factor for a positive response. The historical control data also is a deciding factor in concluding the positive response.

16. Clear positive or negative results may not be confirmed. Equivocal or weak positive results might indicate that it would be appropriate to repeat the test, possibly with a modified spacing of concentration levels (*see* **Note 15**).

3.5 In Vivo Mammalian Erythrocyte Micronucleus Test

1. Maximum dose of 2,000 mg/kg body weight of animal is acceptable for testing. Check the solubility of solid test item using vehicles suitable for test system. Water, groundnut oil, and carboxy-methyl cellulose (CMC) 1–2 % are some of the vehicles compatible with the animals. Liquid test item may be dosed as such or diluted in one of these vehicles.

2. At least five rats or mice (*see* **Note 16**) of age group 8–12 weeks old are acclimatized for the animal room environment for at least 5 days prior to dosing. Animals of both sexes are dosed using five animals for each sex (*see* **Note 17**).

3. Use sanitized animal rooms. The rooms should be maintained with temperature between 19 and 25 °C and relative humidity

between 30 and 70 %, with dark: artificial light of 12 h each and 15–17 air changes/h. Preferably, avoid individual caging.

4. Provide ad libitum feed and water, for the animals. Use randomly selected apparently healthy animals, with minimal weight variation between the animals of each sex, for dosing (*see* **Note 18**).

5. Dose the animals at the volume not exceeding 2 mL/100 g body weight. Dose the animals through oral, intravenous, or subcutaneous route, consistent with human exposure but modified if appropriate in order to obtain systemic exposure.

6. Availability of suitable data permits conducting study without a range finding test. If performed, it should be in the same laboratory, using the same species, strain, sex, and treatment regimen to be used in the main study.

7. The test item is tested at three dose levels for bone marrow collection at 24 h after dosing and highest additional dose levels for bone marrow collection at 48 h after dosing. These dose levels should cover a range from the maximum to little or no toxicity. Additionally, vehicle and positive controls are dosed for bone marrow collection at 24 h after dosing.

8. In case of limit dose (2,000 mg/kg body weight), the animals are dosed for bone marrow collection at 24 and 48 h after dosing (*see* **Note 19**), along with concurrent vehicle and positive control animals dosed for bone marrow collection at 24 h after dosing. After a laboratory establishes its own historical control data, dose the animals with a positive control (cyclophosphamide) only periodically, and not concurrently with each assay.

9. Consider to keep 3 animals/sex/group at each time point (three to six time points) to test the TK analysis, as satellite group.

10. Dose the animals and house them in the cages provided with ad libitum feed and water. During the acclimatization and treatment observe the animals for clinical signs, once daily. Also, observe them for morbidity and mortality twice daily.

11. Collect bone marrow in bovine serum from both the hind limbs (femurs or tibias) after 24 and 48 h of dosing, from respective animals (*see* **Note 20**).

12. Centrifuge the suspension and prepare the slides. Stain and observe the slides for the presence of micro-nucleated erythrocytes.

13. Code all the slides, before microscopic analysis. For each animal determine the proportion of immature among total (immature + mature) erythrocytes by counting a total of at least 200 erythrocytes for bone marrow. Score at least 2,000 immature erythrocytes per animal for the incidence of micronucleated immature erythrocytes. When analyzing slides, the

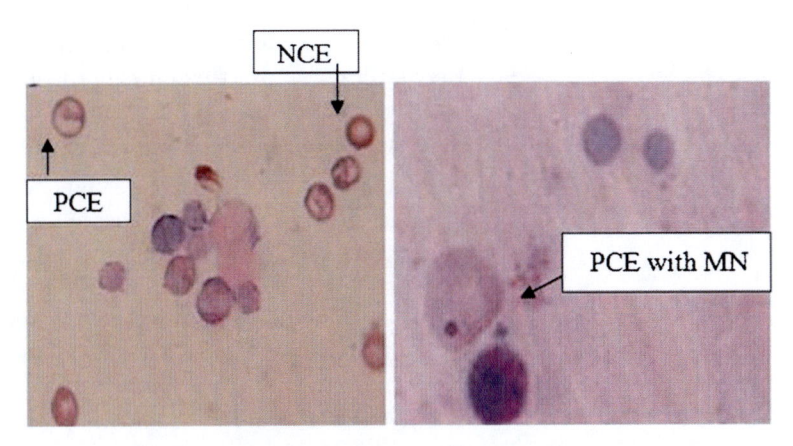

Fig. 1 Polychromatic erythrocytes (PCE) stained with "blue color," normochromatic erythrocytes (NCE) stained with "red/pink color," micronucleus is observed inside the PCEs

proportion of immature erythrocytes among total erythrocytes should not be less than 20 % of the control value.

14. Figure 1 represents the typical Micronucleus present in the PCE.
15. Concentration-related increase or a reproducible increase in the number of micronuclei in the cells shows that the test item is mutagenic. Biological relevance of the results should be considered first. Statistical methods may be used for evaluating the test results. Statistical significance should not be the only determining factor for a positive response. The historical control data is also a deciding factor in concluding the positive response (see **Notes 21** and **22**).

3.6 In Vivo Mammalian Bone Marrow Chromosome Aberration Test

1. Maximum dose of 2,000 mg/kg body weight of animal is acceptable for testing. Check the solubility of solid test item using vehicles suitable for test system. Water, groundnut oil, and CMC 1–2 % are some of the vehicles compatible with the animals. Liquid test item may be dosed as such or diluted in one of these vehicles.
2. At least five rats, mice, or Chinese hamsters (see **Note 16**) of age group 8–12 weeks old are acclimatized for the animal room environment for at least 5 days prior to dosing. Animals of both sexes with five animals for each sex are dosed (see **Note 17**).
3. Use sanitized animal rooms, with temperature between 19 and 25 °C and relative humidity between 30 and 70 %, with dark: artificial light of 12 h each and 15–17 air changes/h. Preferably, avoid individual caging.
4. Provide ad libitum feed and water, for the animals. Use apparently healthy animals, randomly, with minimal weight variation between the animals of each sex, for dosing (see **Note 18**).

5. Dose the animals at the volume not exceeding 2 mL/100 g body weight. Dose the animals through oral, intravenous, or subcutaneous route, consistent with human exposure but modified if appropriate in order to obtain systemic exposure.

6. Availability of suitable data permits conducting study without a range finding test. If performed it should be in the same laboratory, using the same species, strain, sex, and treatment regimen to be used in the main study. Toxic dose of the test item is tested at three dose levels for bone marrow collection at 24 h after dosing. These dose levels should cover a range from the maximum to little or no toxicity. Additionally, vehicle and positive controls are dosed for bone marrow collection at 24 h after dosing.

7. In case of limit dose (2,000 mg/kg body weight), the animals are dosed for bone marrow collection at 24 h after dosing, along with concurrent vehicle and positive control animals (*see* **Note 19**). After a laboratory establishes its own historical control data, dose the animals with a positive control (cyclophosphamide) only periodically, and not concurrently with each assay.

8. Consider to keep 3 animals/sex/group at each time point (three to six time points) to test the TK analysis as satellite group.

9. Dose the animals and house them in the cages provided with ad libitum feed and water. During the acclimatization and treatment observe the animals for clinical signs, once daily. Also, observe them for morbidity and mortality twice daily.

10. Approximately 20–22 h after dosing, the animals are injected intraperitoneally with an appropriate dose of a metaphase arresting agent (e.g., Colcemid® or colchicine at the rate of 4 mg/kg body weight).

11. Collect bone marrow in bovine serum from both the hind limbs (femurs or tibias) after 24 h of dosing.

12. The bone marrow cells in bovine serum are then used to prepare the chromosome slides. Code all the slides, before microscopic analysis.

13. At least 100 metaphases are scored for each animal and tested for the presence of aberrations in chromosomes (Fig. 2). In case of high number of aberrations, scoring may be stopped.

14. One of the criterion for determining a positive result is concentration-related increase or a reproducible increase in the number of cells with chromosome aberrations. Biological relevance of the results should be considered first. Statistical methods may be used for evaluating the test results. Statistical significance should not be the only determining factor for a positive response. The historical control data is also a deciding factor in concluding the positive response (*see* **Note 22**).

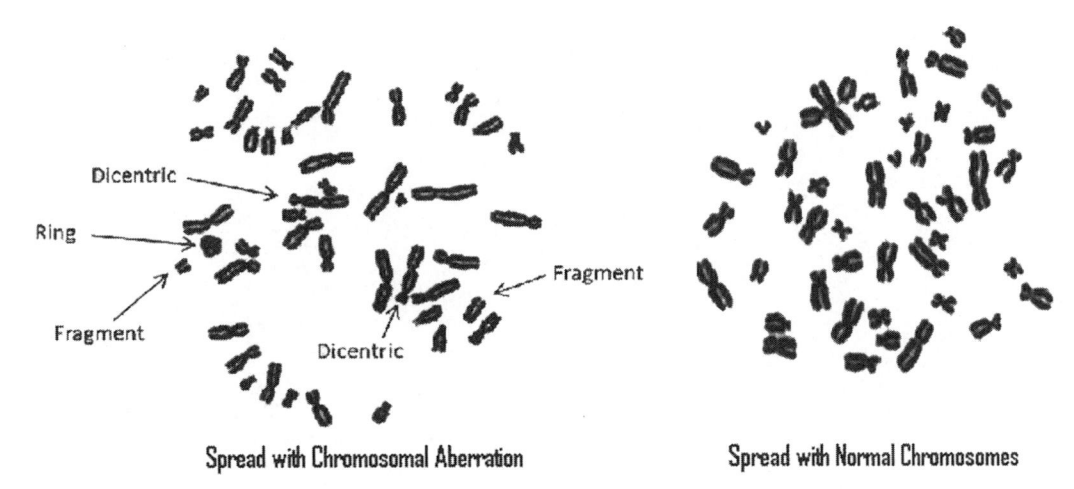

Fig. 2 Chromosome aberrations in mammalian cells along with normal chromosomes

4 Notes

1. To sterilize by autoclaving, load the autoclave with the material and close the door. Keep the material at 121 °C and 15 psi for 15–20 min.

2. S9 homogenate prepared from the livers of rodents treated with enzyme-inducing agents such as Aroclor 1254 or a combination of phenobarbitone and ß naphthoflavone. S9 used should be sterile and should have protein content between 10 and 40 %.

3. Five of the following bacterial tester strains are used for testing:
   (a) *Salmonella typhimurium* TA98
   (b) *Salmonella typhimurium* TA100
   (c) *Salmonella typhimurium* TA1535
   (d) *Salmonella typhimurium* TA1537, TA97, *or* TA97a
   (e) Either *Salmonella typhimurium* TA102 *or Escherichia coli* WP2 *uvrA or Escherichia coli* WP2 *uvrA* (pKM101)

4. While preparing the VB agar sterilize the agar separately and add sterile glucose and sterile Vogel-Bonner medium. Addition of agar, glucose, and Vogel-Bonner medium into water and sterilization may result in charred medium.

5. Grow HPBL in RPMI and Chinese hamster ovary cell line in Ham's F12 medium.

6. Prepare fresh on the day of use and incubate at 37 °C for at least 1 h.

7. Prepare fresh on the day of use and keep at 2–8 °C for at least 1 h, before use. During the use also keep at temperature 2–8 °C.

Table 5
Appropriate minimal agar plates for different tester strains

TA 97a, TA 98, and TA 100	HBA plates
TA 102	HBT plates
TA 1535 and TA 1537	HB plates
WP2 uvrA (pKM 101)	TA plates
WP2 uvrA	T plates

8. Wash the slides with chromic acid, and then rinse in sufficient amount of glass-distilled water to remove the acid. Preferably, use chilled slides.

9. Before storing, cool the tubes gradually at different temperature levels, 2–8 °C, –16 to –20 °C, etc., so that the cultures are protected from cold shock.

10. Appropriate minimal agar plates for different tester strains as follows in Table 5.

11. For tester strain *Salmonella typhimurium* TA102, store the master plate for only up to 15 days.

12. The recommended maximum test concentration for soluble non-cytotoxic substances is 5 mg or 5 mL. For non-cytotoxic substances that are not soluble at this concentration, one or more concentrations tested should be insoluble in the final treatment mixture. Test item that is cytotoxic already below this concentration should be tested up to a cytotoxic concentration and these concentrations preferably cover a range from the maximum to little or no toxicity.

13. In the Ames test, there are some well-characterized examples of artifactual increase in colonies that are not truly revertants. These can occur due to contamination with amino acids (i.e., providing histidine for *Salmonella typhimurium* strains or tryptophan for *Escherichia coli* strains), so the bacterial reversion assay is not suitable for testing a peptide that is likely to degrade. Hence, positive results in the Ames test need extensive follow-up testing to assess whether the in vivo mutagenic and carcinogenic potential would be warranted to assess the potential risk for treatment of patients, unless justified by appropriate risk–benefit analysis. For in vitro negative results further testing should be considered in special cases, such as (the examples given are not exhaustive, but are given as an aid to decision-making) the following: The structure or known metabolism of the test item indicates that standard techniques for in vitro metabolic activation might be inadequate; the structure or known activity of the test item indicates that the use of other test methods/systems might be appropriate.

14. Concurrent positive controls are important, but in vitro mammalian cell tests for genetic toxicity are sufficiently standardized that use of positive controls can generally be confined to a positive control with metabolic activation (when it is done concurrently with the nonactivated test) to demonstrate the activity of the metabolic activation system and the responsiveness of the test system.

15. Any in vitro positive test result in mammalian cell assays should be evaluated based on an assessment of the weight of evidence as indicated below.

    (a) The conditions do not occur in vivo (pH; osmolality; precipitates).
        Note that the 1 mM limit avoids increases in osmolality, and that if the test item alters pH it is advisable to adjust pH to the normal pH of untreated cultures at the time of treatment.

    (b) The effect occurs only at the most toxic concentrations.
        If any of the above conditions apply (this list is not exhaustive, but is given as an aid to decision-making) the weight of evidence indicates a lack of genotoxic potential; the standard battery (Option 1) can be followed. Thus, a single in vivo test is considered sufficient.

16. When peripheral blood is used, use of mice is recommended. Likewise, immature erythrocytes can be used from any other species which has shown an adequate sensitivity to detect clastogens/aneuploidy inducers in bone marrow or peripheral blood, with proper justification.
    Although rats rapidly remove micronucleated erythrocytes from the circulation, it has been established that micronucleus induction by a range of clastogens and aneugens can be detected in rat blood reticulocytes. Rat blood can be used for micronucleus analysis provided methods are used to ensure analysis of the newly formed reticulocytes and the sample size is sufficiently large to provide appropriate statistical sensitivity given the lower micronucleus levels in rat blood than in bone marrow. Whichever method is chosen, each laboratory should determine the appropriate minimum sample size to ensure that scoring error is maintained below the level of animal-to-animal variation.

17. If sex-specific drugs are to be tested, then the assay can be done in the appropriate sex. For the in vivo tests with the acute protocol, use of males alone is considered appropriate. For acute tests, both sexes should be considered only if any existing toxicity, metabolism, or exposure data indicate a toxicologically meaningful sex difference in the species being used. When the genotoxicity test is integrated into a repeat-dose toxicology study in two sexes, samples can be collected from both sexes,

but a single sex can be scored if there is no substantial sex difference evident in toxicity/metabolism. The dose levels for the sex(es) scored should meet the criteria for appropriate dose.

18. At the commencement of the study, the weight variation of animals should be minimal and is not to exceed ±20 % of the mean weight, in case of each of the sex.

19. In case of multiple dose administration, the limit dose may be 1,000 mg/kg body weight, for up to 14-day dosing.

20. When micronucleus analysis is integrated into multi-week studies, collect the sample after a day and after 48 h of the final administration.

21. Many compounds that induce aneuploidy, such as potent spindle poisons, are detectable in in vivo micronucleus assays in bone marrow or blood only within a narrow range of doses approaching toxic doses. This is also true for some clastogens. If toxicological data indicate severe toxicity to the red blood cell lineage (e.g., marked suppression of polychromatic erythrocytes (PCEs) or reticulocytes), doses scored should be spaced not more than about twofold below the top, cytotoxic dose. If suitable doses are not included in a multi-week study, additional data that could contribute to the detection of aneugens and some toxic clastogens could be derived from any one of the following:

    (a) Early blood sampling (at 3–4 days) is advisable when there are marked increases in toxicity with increasing treatment time. For example, when blood or bone marrow is used for micronucleus measurement in a multi-week study (e.g., 28 days), and reticulocytes are scored, marked hematotoxicity can affect the ability to detect micronuclei, i.e., a dose that induces detectable increases in micronuclei after acute treatment might be too toxic to analyze after multiple treatments. The early sample can be used to provide assurance that clastogens and potential aneugens are detected.

    (b) An in vitro mammalian cell micronucleus assay.

    (c) An acute bone marrow micronucleus assay.

22. The in vivo tests have the advantage of taking into account absorption, distribution, and excretion, which are not factors in in vitro tests, but are potentially relevant to human use. In addition metabolism is likely to be more relevant in vivo compared to the systems normally used in vitro. If the in vivo and in vitro results do not agree, then the difference should be considered/explained on a case-by-case basis, e.g., a difference in metabolism and rapid and efficient excretion of a compound in vivo. The in vivo genotoxicity tests also have the potential to give misleading positive results that do not

indicate true genotoxicity. Increase in micronuclei can occur without administration of any genotoxic agent, but due to disturbance in erythropoiesis or DNA adduct data should be interpreted in the light of the known background level of endogenous adducts or indirect, toxicity-related effects could influence the results of the DNA strand break assays (e.g., alkaline elution and Comet assays). Thus it is important to take into account all the toxicological and hematological findings, when evaluating the genotoxicity data. Indirect effects related to toxicological changes could have a safety margin and might not be clinically relevant.

References

1. ICH (2012) Guidance for industry S2 (R1); "Genotoxicity testing and data interpretation for pharmaceuticals intended for human use". U.S. Department of Health and Human Services Food and Drug Administration, Center for Drug Evaluation and Research (CDER) Center for Biologics Evaluation and Research (CBER)
2. ICH (2011) Harmonised tripartite guideline, guidance on genotoxicity testing and data interpretation for pharmaceuticals intended for human use S2 (R1), current step 4 version, dated 9 Nov 2011
3. Jena GB, Kaul CL, Rama RP (2002) Genotoxicity testing, a regulatory requirement for drug discovery and development: impact of ICH guidelines. Indian J Pharmacol 34: 86–99
4. Müller L, Kikuchi Y, Probst G et al (1999) ICH-Harmonised guidances on genotoxicity testing of pharmaceuticals: evolution, reasoning and impact. Mutat Res 436:195–225
5. Eastmond DA, Hatwig A, Anderson D et al (2009) Mutagenicity testing for chemical risk assessment: update of the WHO/IPCS harmonized scheme. Mutagenesis 24:341–349
6. Dertinger SD (2010) Integration of mutation and chromosomal damage endpoints into 28-day repeat dose toxicology studies. Toxicol Sci 115:401–411
7. Ames BN, McCann J, Yamasaki E (1975) Methods for detecting carcinogens and mutagens with *Salmonella*/Mammalian—microsome mutagenicity test. Mutat Res 31: 347–364
8. Maron DM, Ames BN (1983) Revised methods for the *Salmonella* mutagenicity test. Mutat Res 113:173–215
9. Organization for Economic Co-operation and Development (OECD) OECD guidelines for the testing of chemicals; No.471; "Bacterial Reverse Mutation Test"; Adopted 21 July 1997
10. Blakey D, Galloway SM, Kirkland DJ et al (2008) Regulatory aspects of genotoxicity testing: from hazard identification to risk assessment. Mutat Res 657:84–90
11. Hilliard C, Hill R, Amstrong M et al (2007) Chromosome aberrations in Chinese hamster and human cells: a comparison using compounds with various genotoxicity profiles. Mutat Res 616:103–118
12. Organization for Economic Co-operation and Development (OECD) OECD guidelines for the testing of chemicals; No.473; "In vitro mammalian chromosome aberration test"; Adopted 21 July 1997.
13. Swierenga SHH, Heddle JA, Sigal EA et al (1991) Recommended protocols based on a Survey of Current Practice in Genotoxicity Testing Laboratories, IV. chromosome aberration and sister-chromatid exchange in Chinese hamster ovary, V79 Chinese Hamster lung and human lymphocyte cultures. Mutat Res 246:301–322
14. Liber HL, Thilly WG (1982) Mutation assay at the thymidine kinase locus in diploid human lymphoblasts. Mutat Res 94:467–485
15. Organization for Economic Co-operation and Development (OECD) OECD guidelines for the testing of chemicals; No.476; "In vitro mammalian cell gene mutation test"; Adopted 21 July 1997
16. Aardema MJ, Snyder RD, Spicer C et al (2006) SFTG international collaborative study on *in vitro* micronucleus test III using CHO cells. Mutat Res 607:61–87
17. Organization for Economic Co-operation and Development (OECD) OECD guidelines for the testing of chemicals; No.487; "In vitro mammalian cell micronucleus test"; Adopted 22 July 2010

18. Adler ID (1984) Cytogenetic tests in mammals. In: Venitt S, Parry JM (eds) Mutagenicity testing: a practical approach. IRL Press, Oxford, Washington DC, pp 275–306
19. Krishna G, Hayashi M (2000) *In vivo* rodent micronucleus assay: protocol, conduct and data interpretation. Mutat Res 455: 155–166
20. Organization for Economic Co-operation and Development (OECD) OECD guidelines for the testing of chemicals; No.474; "Mammalian erythrocyte micronucleus test"; Adopted 21 July 1997
21. Organization for Economic Co-operation and Development (OECD) OECD guidelines for the testing of chemicals; No.475; "Mammalian bone marrow chromosome aberration test"; Adopted 21 July 1997
22. Schmid W (1975) The micronucleus test. Mutat Res 31:9–15

Index

A

Acridine orange (AO) 127, 131, 180, 188, 277, 285, 286, 359, 364
    coated slide 181, 184
Agar plates
    biotin/histidine/ampicillin plates.....................................8
    biotin/histidine plates..8
    biotin/histidine/tetracycline plates..................................8
    minimal agar plates................................ 5, 7–11, 13, 14, 17, 21, 22, 454
    TB1–kanamycin agar plates........................ 101, 107, 116
    λ-trypticase agar plates 103, 114, 115
Ames test.. 3, 431, 442, 454
Ampicillin... 8, 11, 19, 436
Aneugens...127, 221, 222, 239, 240, 272, 455, 456
Aneuploidy 239, 240, 435, 455, 456
AO. *See* Acridine orange (AO)
Aroclor 1254 ....18, 48, 126, 273, 453

B

Bacterial colony counter 7
Base-pair substitution (BPS) 5
Big Blue® mouse/rat...98–100
Biological dosimetry..168, 209
BPS. *See* Base-pair substitution (BPS)
BrdU. *See* 5-Bromo-2-deoxyuridine (BrdU)
5-Bromo-2-deoxyuridine (BrdU).................... 125, 167, 169, 170, 172, 174, 175
Buds..193, 197, 199–200, 204, 210, 406, 409, 410, 412

C

CBMN assay. *See* Cytokinesis-block micronucleus (CBMN) assay
Centromeric labeling 280–282
CFU. *See* Colony forming units (CFU)
CGH. *See* Comparative genomic hybridization (CGH)
Chinese hamster cells
    chromosomal aberrations........................... 125–126, 451
    lung (CHL, V79).. 125, 126, 327
    micronucleus assay...438
    ovary (CHO) 125, 327, 438, 453

Chromosomal rearrangement.......................................28, 124
Chromosome aberrations
    analysis......................... 124, 128, 143, 147–162, 167, 173
    chromatid-type 124, 139, 141, 149–151, 153
    chromosome-type 124, 135, 140, 150, 151, 157
    classification 133, 135, 148
    human lymphocytes..165–176
    in vitro 123–145, 148, 165, 167, 168, 171, 173, 179, 210, 434, 435, 437, 442–446, 448
    in vivo 148, 165, 167, 168, 173, 179, 435, 439, 451–453
    mouse
        bone marrow cells 148, 149, 151, 153, 155, 165, 451–453
        germ cells 147–162
        spermatocyte cells 148–151, 154, 159–160, 294
        spermatogonial cells...............149, 153–154, 158–159
cII
    assay..97–119
    mutants.. 101–102, 109–110, 117
Clastogens 29, 47, 127, 151, 153, 188, 221, 222, 239, 272, 455, 456
Colcemid 127–129, 169, 170, 266, 452
Colony forming units (CFU) 11, 12
Comet assay
    CHO cells 327, 328, 330, 334
    drosophila 424, 425
    fish erythrocytes...367, 368
    human peripheral lymphocytes...........................168, 327
    in vitro 325–343, 365, 369
    in vivo 325–343, 418
    marine animals 363–372
    mouse multiple organs 332–333
    mussel gill cells 367
    mussel haemocytes..367
    sperm comet assay 354, 358
Comparative genomic hybridization (CGH)
    array CGH 248–253, 255
    chromosomal CGH.............246–248, 250, 251, 261–264
    conventional CGH 247, 257–258
Concanavalin-A (con-A)................................ 81, 82, 84, 86, 89
Crystal violet 8, 13, 19, 440

Cytochalasin-B 127, 239, 240, 242, 243, 270–272, 276, 278, 279, 282, 283
Cytokinesis-block 193, 210, 240, 243, 277
Cytokinesis-block micronucleus (CBMN) assay 191, 192, 239, 271, 276, 278, 282–284, 286, 287
Cytotoxicity
    population doubling time 131, 132, 279
    replicative index 143, 279, 280

D

DDR. *See* DNA damage response (DDR)
Dialysis buffer 107
Digestion buffer 100, 106, 393, 394
DNA adducts
    enrichment
        butanol extraction 391, 393–395, 398
        nuclease P_1 digestion 391, 393–395
    extraction 390, 398–399
    HPLC co-chromatography 394, 398–399
    labeling 390–392, 394–397
DNA damage response (DDR) 124, 311–322, 359
DNA precipitation 261–262, 265
DNA probe 167, 237, 240, 246, 247, 252, 262, 265
DNA staining 44, 211, 322
Drosophila
    comet assay 424, 425
    hemocytes 418, 425
    medium 419, 421, 424
    strains 419, 426
    wing spot test 417–427

E

E. coli. See Escherichia coli
Endogenous mutation 51, 79, 80
Erythrocytes
    mutant 51, 58, 69
    normochromatic 58, 185, 228–230, 451
    wild type 51–54, 58, 69, 80
Escherichia coli
    bacteria 3, 4, 7, 17, 19, 103, 107, 108, 110, 111, 113–115, 442, 454
    G1250 strain (*see also cII* assay)
        liquid culture 107–108
        plating 107, 108
    WL95 (P2) 103, 114, 115
    XL-1Blue MRA 103, 113–114
    XL-1Blue MRA(P2) 103
    YG6020 strain .... 102, 110–111 (*see also gpt* assay)

F

Faure's solution 420–422
FBS. *See* Fetal bovine serum (FBS)

Fetal bovine serum (FBS) 45, 55, 59–63, 73, 75, 81, 105, 166, 169, 170, 174, 175, 183, 193, 194, 196, 204, 214, 216, 220, 222, 223, 225, 226, 231, 296, 314, 328, 332, 337, 339, 341, 437, 442, 444, 446, 448
Flow cytometry
    analysis 52, 56, 211–213, 221, 228–229, 295
    based MN test
        in mouse peripheral blood cells 227
        in primary human lymphocytes 210, 214–216, 221–227
        in TK6 cells 222
    cytograms 305, 306
    data analysis
        γ-H2AX detection 301–305
        interpretation 210, 212
    histograms 303–306
Forward mutation 27–49
    assay 27–30, 32, 33, 35–37, 42–49
Frameshift (FS) mutations 5, 28

G

Genetic toxicology 27, 48, 168, 239, 254, 433
Genomic DNA extraction 100, 105–107
Giemsa stain 127, 131, 148, 153, 154, 156, 159–161, 167–169, 171–175, 181, 183, 188, 194, 196, 270, 272, 277, 285, 437, 438
gpt
    assay 98, 99, 102–103, 107, 110–113, 118
    delta mouse 98–100
    mutants 112, 113
    selection agar 103
    titration agar 103, 111, 112

H

γ-H2AX
    flow cytometric detection 294, 295, 298, 303–306
    in human lymphocytes 319
    in human spermatozoa 319
    immunocytochemical detection 298–301
    immunohistochemical detection 305–309
    laser scanning confocal immunofluorescence microscopy 313
    in mouse bone marrow 301, 308
    in mouse testicular cells 301, 308
High throughput 205, 212, 252–254, 256, 312, 327, 352
Histidine (*his*) operon 5

HPL. *See* Human peripheral lymphocytes (HPL)
HPLC co-chromatography .... 394, 398–399.
   See also DNA adducts
HPRT. *See* Hypoxanthine guanine phosphoribosyl
   transferase (HPRT)
Human biomonitoring 312, 326, 347–361
Human blood
   culture of 170, 173, 175
   sampling of 173, 175
   slides 175
Human cells
   buccal cells 191–206
   lymphocytes 191–206
Human peripheral lymphocytes (HPL)
   culture of 166–169, 173, 327
   micronucleus 168
Hybridisation 238, 240–244, 246–249,
   256, 262–265, 280–281
Hypotonic solution 127–130, 153–155,
   158, 160, 161, 169, 171, 194, 196, 272, 448
Hypoxanthine guanine phosphoribosyl transferase (HPRT)
   cloning efficiency 86, 89
   Hprt assay 84
   Hprt cDNA 81, 83, 90, 93
   Hprt exons 81, 83, 91, 92, 94
   multiplex PCR 81, 83, 91–93
   rat lymphocytes 79–95
   RT-PCR amplification 90

I

ICS. *See* Instrument Calibration Standard (ICS)
Immunocytochemistry 295–296, 299
Immunodetection 298–301, 305–309
Immunohistochemistry 294, 297, 306–307
Immunomagnetic separation 53
Instrument Calibration Standard (ICS) 57, 58,
   60, 62, 64–69, 71–74
In vitro DNA packaging 100–101, 107
Ionomycin 81, 82, 84, 88, 89

K

Kanamycin 101, 102, 107, 108,
   110, 111, 116
Kinetochore staining 269–288

L

Lambda (λ) phage 98, 99
Laser scanning confocal immunofluorescence
   microscopy (LSCIM) 311–322
Leukodepletion 59–60, 72
LSCIM. *See* Laser scanning confocal immunofluorescence
   microscopy (LSCIM)

M

Metabolic activation 6, 32–36, 47–48,
   126–129, 131, 210, 273–274,
   327, 339, 340, 384, 418, 441, 442,
   444–449, 454, 455
Micronuclei
   criteria for scoring 133, 186, 193
   scoring 124, 143, 449
Micronucleus assay
   automated 278, 284–285
   in buccal cells (BMNcyt)
      collection 200, 205
      harvesting 201
      slide preparation 192, 193, 196–197, 201
      staining 192, 196–197, 206
   fluorescent in situ hybridization (FISH) 239
   in vitro 125, 210, 269–288, 435, 456
   in vivo 179–189, 212, 456
   micronucleus cytome assay 191, 203–205
   in peripheral blood cells (CBMNcyt)
      collection 195–196
      culture 266
      harvesting 196, 273
      slide preparation 196–197
      staining 196–197
   in primary human lymphocytes 213–216, 221–227
   in rodents 179, 434
   in TK6 cell line 213, 217–221
   in *Tradescantia* 405–414
   using flow cytometry 213–227
   96-well format 217, 220
Microsome mutagenicity test 3
Mitotic recombination 28
   non disjunction 28
Mononuclear cells 81, 86, 94, 270, 283, 287
Mouse bone marrow
   erythrocytes 179
   micronucleus assay 179–189
   normochromatic erythrocytes (NCE) 185, 186
   peripheral blood 179–189
   polychromatic erythrocyte (PCE) 185, 186
Mouse lymphoma
   assay 27–49
   continuous treatment 35–36
   L5178Y (*Tk*[+/−]-3.7.2C) cells 27–49
   micro-well cloning 42
   mutant frequency 42
   short treatment 32–35
   soft agar cloning 37–40
   thymidine kinase 28, 438
Multicolor laser scanning confocal
   immunofluorescence microscopy 311–322

GENOTOXICITY ASSESSMENT: METHODS AND PROTOCOLS
Index

Mutagenicity testing 3, 4, 14, 21, 27, 248, 432, 433
Muta™ mouse 98–100
Mutant
    cloning 28, 42, 46, 81–83, 91, 94, 112, 418, 423, 426, 447
    efficiency 42, 46, 447
    frequency 15, 42–45, 48, 53, 54, 56, 69–71, 80–82, 85, 89, 94, 98, 109–110, 112, 115, 117, 426, 447
    index 42, 54
    reticulocytes 51, 69
    selection plates 37–42, 109–112
Mutation assay
    bacterial reverse mutation test 27, 431, 436–437, 439–442
    cII assay 98, 99, 107–109
    gpt assay 98
    Pig-a assay 54
    Spi-assay 97–119

N

β-Naphthoflavone 6, 18, 48, 126, 340, 453
Nobel agar 30
Nucleoplasmic bridge (NPB) 193, 197, 200, 205, 210, 286
Nutrient agar plates 7, 9, 11, 13, 440
Nutrient broth 7, 9, 11–13, 437, 439, 441, 443

O

OECD Test guidelines
    No. 471 13
    No. 474 .... 180, 209, 269, 276
    No. 475 148
    No. 483 148
    No. 486 .... 374, 384
    No. 487 ....209, 269, 282

P

pAQ1 plasmid strain 5, 9, 11, 19
PHA. *See* Phytohemagglutinin (PHA)
Phenobarbital 6, 18
Phorbol-12-myristate-13-acetate (PMA) 81, 82, 84, 87, 89
Phosphatidyl inositol glycan (*Pig-a*)
    cDNA nested PCR 90–91, 93
    cDNA RT-PCR amplification 90
    cDNA synthesis 83, 90
    erythrocyte-based assay 51–76
    flow cytometry immunomagnetic separation 53
    gene 51–76, 84
    glycosylphosphatidylinositol (GPI) anchor 51, 80
    lymphocytes 79–94
    mutation assay 51–76, 84
    rat reticulocytes 51
Phytohemagglutinin (PHA) 165–167, 170, 173, 194, 196, 216, 223, 243, 267, 437, 438, 443, 444, 448
pKM101 plasmid 5, 6, 9, 11, 19
Plate incorporation test 21
PMA. *See* Phorbol-12-myristate-13-acetate (PMA)
Positive control 9, 10, 13, 16, 19–20, 23, 33–35, 40, 43, 47, 127, 128, 133, 144, 145, 157, 180, 182, 223, 224, 272, 274–276, 331, 332, 340, 376, 377, 384–386, 407, 412, 420, 424, 435, 438, 442, 445, 447, 449, 450, 452, 455
^{32}P-Postlabeling assay 389–400
Pre-incubation assay/test 16, 21
proAER. *See* Proaerolysin (proAER)
Proaerolysin (proAER)
    resistant mutant 81, 92
    solution 82
Purine analogue 80

R

Reporter gene 79, 80, 97
RET. *See* Reticulocyte (RET)
Reticulocyte (RET) 51, 54, 69, 70, 74, 186–189, 213, 228–230, 455, 456
    wild type 58, 69
Rfa (deep rough) mutation 13

S

S9
    co-factors 8
    homogenate 6, 30, 435, 453
    mix 8, 14, 15, 17–19, 21–23, 30, 34, 126–128, 273, 340, 441, 444, 446, 448
Saccomanno's fixative 195, 200, 206
Salmonella tester strain
    TA97 (or TA1537) 5, 10, 14, 453
    TA98 .... 5, 10, 14, 453
    TA100 .... 5, 10, 14, 453
    TA102 .... 5, 10, 14, 453, 454
    TA1535 .... 5, 10, 14, 453
Schiff's reagent 195, 202, 206
Single cell gel electrophoresis 325, 347
Sodium phosphate buffer 8, 30, 181, 183, 273, 340
Somatic recombination 418, 426

T

Tetracycline 8, 11, 19, 436
6 TG. *See* 6-Thioguanine (6 TG)
Thin-layer chromatography 390, 391, 393–394, 396, 398

6-Thioguanine (6 TG) 80–82, 85, 86, 91, 103, 111, 112, 118, 438
3^{H}Thymidine 375, 379, 383
Thymidine kinase 28, 438
Top agar 8, 12, 14, 15, 21, 22, 101, 103, 108–112, 114–116, 118, 119, 437
Tradescantia
   clones 407
   cultivation 407–408
   exposure 408–409
   fixation 409
   micronucleus assay 405–414
   scoring micronuclei 409–411
   slide preparation 409
   tetrads 405–414
Transgenes 98, 99
Transgenic
   assays 97–99, 118, 119

rodents 80, 100
Tryptophan (*trp*) operon 5
Two-fold rule 15, 48

U

Unscheduled DNA synthesis
   autoradiography 380
   hepatocytes
      culture 374–375, 379
      radio-labelling 375, 379
   in vivo 373–386
   3^{H}Thymidine 375, 379, 383
   in liver cells 373–376
   nuclear grain count 381

V

Vogel-Bonner (VB) minimal salts 7